SCIENCE AND DECISIONS

Advancing Risk Assessment

Committee on Improving Risk Analysis Approaches Used by the U.S. EPA

Board on Environmental Studies and Toxicology

Division on Earth and Life Studies

NATIONAL RESEARCH COUNCIL
OF THE NATIONAL ACADEMIES

THE NATIONAL ACADEMIES PRESS
Washington, D.C.
www.nap.edu

THE NATIONAL ACADEMIES PRESS 500 Fifth Street, NW Washington, DC 20001

NOTICE: The project that is the subject of this report was approved by the Governing Board of the National Research Council, whose members are drawn from the councils of the National Academy of Sciences, the National Academy of Engineering, and the Institute of Medicine. The members of the committee responsible for the report were chosen for their special competences and with regard for appropriate balance.

This project was supported by Contract EP-C-06-056 between the National Academy of Sciences and the U.S. Environmental Protection Agency. Any opinions, findings, conclusions, or recommendations expressed in this publication are those of the authors and do not necessarily reflect the view of the organizations or agencies that provided support for this project.

Library of Congress Cataloging-in-Publication Data

National Research Council (U.S.). Committee on Improving Risk Analysis Approaches Used by the U.S. EPA.
Science and decisions : advancing risk assessment / Committee on Improving Risk Analysis Approaches Used by the U.S. EPA, Board on Environmental Studies and Toxicology, Division on Earth and Life Studies.
p. cm.
Includes bibliographical references.
ISBN-13: 978-0-309-12046-3 (pbk.)
ISBN-10: 0-309-12046-2 (pbk.)
ISBN-13: 978-0-309-12047-0 (pdf)
ISBN-10: 0-309-12047-0 (pdf)
1. Environmental risk assessment—United States. 2. Technology—Risk assessment—United States. 3. Health risk assessment—United States. I. National Research Council (U.S.). Board on Environmental Studies and Toxicology. II. National Research Council (U.S.). Division on Earth and Life Studies. III. Title.
GE150.N37 2009
361.1—dc22

2008055771

Additional copies of this report are available from

The National Academies Press
500 Fifth Street, NW
Box 285
Washington, DC 20055

800-624-6242
202-334-3313 (in the Washington metropolitan area)
http://www.nap.edu

Printed in the United States of America.

First Printing, August 2009

Second Printing, July 2010

THE NATIONAL ACADEMIES

Advisers to the Nation on Science, Engineering, and Medicine

The **National Academy of Sciences** is a private, nonprofit, self-perpetuating society of distinguished scholars engaged in scientific and engineering research, dedicated to the furtherance of science and technology and to their use for the general welfare. Upon the authority of the charter granted to it by the Congress in 1863, the Academy has a mandate that requires it to advise the federal government on scientific and technical matters. Dr. Ralph J. Cicerone is president of the National Academy of Sciences.

The **National Academy of Engineering** was established in 1964, under the charter of the National Academy of Sciences, as a parallel organization of outstanding engineers. It is autonomous in its administration and in the selection of its members, sharing with the National Academy of Sciences the responsibility for advising the federal government. The National Academy of Engineering also sponsors engineering programs aimed at meeting national needs, encourages education and research, and recognizes the superior achievements of engineers. Dr. Charles M. Vest is president of the National Academy of Engineering.

The **Institute of Medicine** was established in 1970 by the National Academy of Sciences to secure the services of eminent members of appropriate professions in the examination of policy matters pertaining to the health of the public. The Institute acts under the responsibility given to the National Academy of Sciences by its congressional charter to be an adviser to the federal government and, upon its own initiative, to identify issues of medical care, research, and education. Dr. Harvey V. Fineberg is president of the Institute of Medicine.

The **National Research Council** was organized by the National Academy of Sciences in 1916 to associate the broad community of science and technology with the Academy's purposes of furthering knowledge and advising the federal government. Functioning in accordance with general policies determined by the Academy, the Council has become the principal operating agency of both the National Academy of Sciences and the National Academy of Engineering in providing services to the government, the public, and the scientific and engineering communities. The Council is administered jointly by both Academies and the Institute of Medicine. Dr. Ralph J. Cicerone and Dr. Charles M. Vest are chair and vice chair, respectively, of the National Research Council.

www.national-academies.org

COMMITTEE ON IMPROVING RISK ANALYSIS APPROACHES USED BY THE U.S. ENVIRONMENTAL PROTECTION AGENCY

Members

THOMAS A. BURKE *(Chair)*, Johns Hopkins Bloomberg School of Public Health, Baltimore, MD
A. JOHN BAILER, Miami University, Oxford, OH
JOHN M. BALBUS, Environmental Defense, Washington, DC
JOSHUA T. COHEN, Tufts Medical Center, Boston, MA
ADAM M. FINKEL, University of Medicine and Dentistry of New Jersey, Piscataway, NJ
GARY GINSBERG, Connecticut Department of Public Health, Hartford, CT
BRUCE K. HOPE, Oregon Department of Environmental Quality, Portland, OR
JONATHAN I. LEVY, Harvard School of Public Health, Boston, MA
THOMAS E. MCKONE, University of California, Berkeley, CA
GREGORY M. PAOLI, Risk Sciences International, Ottawa, ON, Canada
CHARLES POOLE, University of North Carolina School of Public Health, Chapel Hill, NC
JOSEPH V. RODRICKS, ENVIRON International Corporation, Arlington, VA
BAILUS WALKER, JR., Howard University Medical Center, Washington, DC
TERRY F. YOSIE, World Environment Center, Washington, DC
LAUREN ZEISE, California Environmental Protection Agency, Oakland, CA

Staff

EILEEN N. ABT, Project Director
JENNIFER SAUNDERS, Associate Program Officer (through December 2007)
NORMAN GROSSBLATT, Senior Editor
RUTH CROSSGROVE, Senior Editor
MIRSADA KARALIC-LONCAREVIC, Manager, Technical Information Center
RADIAH A. ROSE, Editorial Projects Manager
MORGAN R. MOTTO, Senior Program Assistant (through February 2008)
PANOLA GOLSON, Senior Program Assistant

Sponsor

U.S. ENVIRONMENTAL PROTECTION AGENCY

BOARD ON ENVIRONMENTAL STUDIES AND TOXICOLOGY[1]

[1]This study was planned, overseen, and supported by the Board on Environmental Studies and Toxicology.

OTHER REPORTS OF THE BOARD ON ENVIRONMENTAL STUDIES AND TOXICOLOGY

Phthalates and Cumulative Risk Assessment: The Tasks Ahead (2008)
Estimating Mortality Risk Reduction and Economic Benefits from Controlling Ozone Air Pollution (2008)
Respiratory Diseases Research at NIOSH (2008)
Evaluating Research Efficiency in the U.S. Environmental Protection Agency (2008)
Hydrology, Ecology, and Fishes of the Klamath River Basin (2008)
Applications of Toxicogenomic Technologies to Predictive Toxicology and Risk Assessment (2007)
Models in Environmental Regulatory Decision Making (2007)
Toxicity Testing in the Twenty-first Century: A Vision and a Strategy (2007)
Sediment Dredging at Superfund Megasites: Assessing the Effectiveness (2007)
Environmental Impacts of Wind-Energy Projects (2007)
Scientific Review of the Proposed Risk Assessment Bulletin from the Office of Management and Budget (2007)
Assessing the Human Health Risks of Trichloroethylene: Key Scientific Issues (2006)
New Source Review for Stationary Sources of Air Pollution (2006)
Human Biomonitoring for Environmental Chemicals (2006)
Health Risks from Dioxin and Related Compounds: Evaluation of the EPA Reassessment (2006)
Fluoride in Drinking Water: A Scientific Review of EPA's Standards (2006)
State and Federal Standards for Mobile-Source Emissions (2006)
Superfund and Mining Megasites—Lessons from the Coeur d'Alene River Basin (2005)
Health Implications of Perchlorate Ingestion (2005)
Air Quality Management in the United States (2004)
Endangered and Threatened Species of the Platte River (2004)
Atlantic Salmon in Maine (2004)
Endangered and Threatened Fishes in the Klamath River Basin (2004)
Cumulative Environmental Effects of Alaska North Slope Oil and Gas Development (2003)
Estimating the Public Health Benefits of Proposed Air Pollution Regulations (2002)
Biosolids Applied to Land: Advancing Standards and Practices (2002)
The Airliner Cabin Environment and Health of Passengers and Crew (2002)
Arsenic in Drinking Water: 2001 Update (2001)
Evaluating Vehicle Emissions Inspection and Maintenance Programs (2001)
Compensating for Wetland Losses Under the Clean Water Act (2001)
A Risk-Management Strategy for PCB-Contaminated Sediments (2001)
Acute Exposure Guideline Levels for Selected Airborne Chemicals (seven volumes, 2000-2008)
Toxicological Effects of Methylmercury (2000)
Strengthening Science at the U.S. Environmental Protection Agency (2000)
Scientific Frontiers in Developmental Toxicology and Risk Assessment (2000)
Ecological Indicators for the Nation (2000)
Waste Incineration and Public Health (2000)
Hormonally Active Agents in the Environment (1999)
Research Priorities for Airborne Particulate Matter (four volumes, 1998-2004)

The National Research Council's Committee on Toxicology: The First 50 Years (1997)
Carcinogens and Anticarcinogens in the Human Diet (1996)
Upstream: Salmon and Society in the Pacific Northwest (1996)
Science and the Endangered Species Act (1995)
Wetlands: Characteristics and Boundaries (1995)
Biologic Markers (five volumes, 1989-1995)
Science and Judgment in Risk Assessment (1994)
Pesticides in the Diets of Infants and Children (1993)
Dolphins and the Tuna Industry (1992)
Science and the National Parks (1992)
Human Exposure Assessment for Airborne Pollutants (1991)
Rethinking the Ozone Problem in Urban and Regional Air Pollution (1991)
Decline of the Sea Turtles (1990)

Copies of these reports may be ordered from the National Academies Press
(800) 624-6242 or (202) 334-3313
www.nap.edu

Preface

Risk assessment has become a dominant public-policy tool for informing risk managers and the public about the different policy options for protecting public health and the environment. Risk assessment has been instrumental in fulfilling the missions of the U.S. Environmental Protection Agency (EPA) and other federal and state agencies in evaluating public-health concerns, informing regulatory and technologic decisions, setting priorities for research and funding, and developing approaches for cost-benefit analyses.

However, risk assessment is at a crossroads. Despite advances in the field, it faces a number of substantial challenges, including long delays in completing complex risk assessments, some of which take decades to complete; lack of data, which leads to important uncertainty in risk assessments; and the need for risk assessment of many unevaluated chemicals in the marketplace and emerging agents. To address those challenges, EPA asked the National Academies to develop recommendations for improving the agency's risk-analysis approaches.

In this report, the Committee on Improving Risk Analysis Approaches Used by the U.S. EPA conducts a scientific and technical review of EPA's current risk-analysis concepts and practices and offers recommendations for practical improvements that EPA could make in the near term (2-5 y) and in the longer term (10-20 y). The committee focused on human health risk assessment but considered the implications of its conclusions and recommendations for ecologic risk assessment.

This report has been reviewed in draft form by persons chosen for their diverse perspectives and technical expertise in accordance with procedures approved by the National Research Council's Report Review Committee. The purpose of this independent review is to provide candid and critical comments that will assist the institution in making its published report as sound as possible and to ensure that the report meets institutional standards of objectivity, evidence, and responsiveness to the study charge. The review comments and draft manuscript remain confidential to protect the integrity of the deliberative process. We wish to thank the following for their review of this report: Lawrence W. Barnthouse, LWB Environmental Services, Inc.; Roger G. Bea, University of California, Berkeley; Allison C. Cullen, University of Washington; William H. Farland, Colorado State University; J. Paul

Gilman, Convanta Energy Corporation; Bernard D. Goldstein, University of Pittsburgh; Lynn R. Goldman, Johns Hopkins University; Dale B. Hattis, Clark University; Carol J. Henry, American Chemistry Council (retired); Daniel Krewski, University of Ottawa; Amy D. Kyle, University of California, Berkeley; Ronald L. Melnick, National Institute of Environmental Health Sciences; Gilbert S. Omenn, University of Michigan Medical School; Louise Ryan, Harvard School of Public Health; and Detlof von Winterfeldt, University of Southern California.

Although the reviewers listed above have provided many constructive comments and suggestions, they were not asked to endorse the conclusions or recommendations, nor did they see the final draft of the report before its release. The review of the report was overseen by the review coordinator, William Glaze, Georgetown, TX and the review monitor, John Ahearne, Sigma Xi. Appointed by the National Research Council, they were responsible for making certain that an independent examination of the report was carried out in accordance with institutional procedures and that all review comments were carefully considered. Responsibility for the final content of the report rests entirely with the committee and the institution.

The committee gratefully acknowledges the following for making presentations to the committee: Nicholas Ashford, Massachusetts Institute of Technology; Robert Brenner, Michael Callahan, George Gray, Jim Jones, Tina Levine, Robert Kavlock, Al McGartland, Peter Preuss, Michael Shapiro, Glenn Suter, and Harold Zenick, EPA; Douglas Crawford-Brown, University of North Carolina; Kenny Crump, ENVIRON International Corporation; Robert Donkers, Delegation of the European Commission to the United States; William Farland, Colorado State University; James A. Fava, Five Winds International; Penny Fenner-Crisp, International Life Sciences Institute Research Foundation; Dale Hattis, Clark University; Amy D. Kyle, University of California, Berkeley; Rebecca Parkin, George Washington University; Chris Portier, National Institute of Environmental Health Sciences; Lorenz Rhomberg, Gradient Corporation; Jennifer Sass, Natural Resources Defense Council; Jay Silkworth, General Electric Company; and Thomas Sinks, Centers for Disease Control and Prevention.

The committee is thankful for the useful input of Roger Cooke, Resources for the Future and Dorothy Patton, Environmental Protection Agency (retired) in the early deliberations of this study. The committee is also grateful for the assistance of the National Research Council staff in preparing this report. Staff members who contributed to this effort are Eileen Abt, project director; James Reisa, director of the Board on Environmental Studies and Toxicology; Jennifer Saunders, associate program officer; Norman Grossblatt and Ruth Crossgrove, senior editors; Mirsada Karalic-Loncarevic, manager of the Technical Information Center; Radiah Rose, editorial projects manager; and Morgan Motto and Panola Golson, senior program assistants.

I would especially like to thank the committee members for their efforts throughout the development of this report.

Thomas Burke, *Chair*
Committee on Improving Risk Analysis Approaches
Used by the U.S. EPA

Abbreviations

ARARs	Applicable or Relevant and Appropriate Requirements
ATSDR	Agency for Toxic Substances and Disease Registry
BMD	benchmark dose
CARE	Community Action for a Renewed Environment
CASAC	Clean Air Scientific Advisory Committee
CBPR	community-based participatory research
CERCLA	Comprehensive Environmental Response Compensation and Liability Act
CTE	central tendency exposure
DBP	dibutyl phthalate
DBPs	disinfection byproducts
EPA	U.S. Environmental Protection Agency
EPHT	Environmental Public Health Tracking Program
FIFRA	Federal Insecticide, Fungicide and Rodenticide Act
FQPA	Food Quality Protection Act
GAO	Government Accountability Office
GIS	geographic information systems
HAPs	hazardous air pollutants
HI	hazard index
IARC	International Agency for Research on Cancer
IPCS	International Program on Chemical Safety
IRIS	Integrated Risk Information System
LNT	linear, no-threshold
MACT	maximum achievable control technology
MCL	maximum contaminant level

MCLG	maximum contaminant level goal
$MeCl_2$	methylene chloride
MEI	maximally exposed individual
MOA	mode of action
MOE	margin of exposure
MTD	maximum tolerated dose
NAAQS	National Ambient Air Quality Standards
NCEA	National Center for Environmental Assessment
NEJAC	National Environmental Justice Advisory Council
NER	National Exposure Registry
NHANES	National Health and Nutrition Examination Survey
NOAEL	no-observed-adverse-effect level
NPL	National Priorities List
NRC	National Research Council
NTP	National Toxicology Program
OAR	Office of Air and Radiation
OP	organophosphate
OPPTS	Office of Prevention, Pesticides and Toxic Substances
OSWER	Office of Solid Waste and Emergency Response
OW	Office of Water
PBPK	physiologically based pharmacokinetic
PD	pharmacodynamic
PDF	probability density function
PK	pharmacokinetic
POD	point of departure
PPDG	Pesticide Program Dialogue Group
RAGS	Risk Assessment Guidance for Superfund
Red Book	*Risk Assessment in the Federal Government: Managing the Process*
RfC	reference concentration
RfD	reference dose
RI/FS	remedial investigation and feasibility study
RME	reasonable maximum exposure
ROD	record of decision
RR	relative risk
RRM	relative risk model
SDWA	Safe Drinking Water Act
SEP	socioeconomic position
TCA	1,1,1-trichloroethane
TCE	trichloroethylene
TSCA	Toxic Substances Control Act
UF	uncertainty factor
VOI	value-of-information
WHO	World Health Organization
WOE	weight-of-evidence

Contents

BOXES, FIGURES, AND TABLES

Boxes

Figures

Tables

SCIENCE AND DECISIONS

Advancing Risk Assessment

Summary

Virtually every aspect of life involves risk. How we deal with risk depends largely on how well we understand it. The process of risk assessment has been used to help us understand and address a wide variety of hazards and has been instrumental to the U.S. Environmental Protection Agency (EPA), other federal and state agencies, industry, the academic community, and others in evaluating public-health and environmental concerns. From protecting air and water to ensuring the safety of food, drugs, and consumer products such as toys, risk assessment is an important public-policy tool for informing regulatory and technologic decisions, setting priorities among research needs, and developing approaches for considering the costs and benefits of regulatory policies.

Risk assessment, however, is at a crossroads, and its credibility is being challenged (Silbergeld 1993; Montague 2004; Michaels 2008).[1] Because it provides a primary scientific rationale for informing regulations that will have national and global impact, risk assessment is subject to considerable scientific, political, and public scrutiny. The science of risk assessment is increasingly complex; improved analytic techniques have produced more data that lead to questions about how to address issues of, for example, multiple chemical exposures, multiple risks, and susceptibility in populations. In addition, risk assessment is now being extended to address broader environmental questions, such as life-cycle analysis and issues of costs, benefits, and risk-risk tradeoffs.

The regulatory risk assessment process is bogged down; major risk assessments for some chemicals take more than 10 years. In the case of trichloroethylene, which has been linked to cancer, the assessment has been under development since the 1980s, has undergone multiple independent reviews, and is not expected to be final until 2010. Assessments of formaldehyde and dioxin have had similar timelines. EPA is struggling to keep up with demands for

[1]Silbergeld, E.K. 1993. Risk assessment: The perspective and experience of U.S. environmentalists. Environ. Health Perspect. 101(2):100-104; Montague, P. 2004. Reducing the harms associated with risk assessment. Environ. Impact Assess. Rev. 24:733-748; Michaels, D. 2008. Doubt Is Their Product: How Industry's Assault on Science Threatens Your Health. New York: Oxford University Press.

hazard and dose-response information but is challenged by a lack of resources, including funding and trained staff.

Decision-making based on risk assessment is also bogged down. Uncertainty, an inherent property of scientific data, continues to lead to multiple interpretations and contribute to decision-making gridlock. Stakeholders—including community groups, environmental organizations, industry, and consumers—are often disengaged from the risk-assessment process at a time when risk assessment is increasingly intertwined with societal concerns. Disconnects between the available scientific data and the information needs of decision-makers hinder the use of risk assessment as a decision-making tool.

Emerging scientific advances hold great promise for improving risk assessment. For example, new toxicity-testing methods are being developed that will probably be quicker, less expensive, and more directly relevant to human exposures, as described in the National Research Council's *Toxicity Testing in the 21st Century: A Vision and a Strategy* (2007). However, the realization of the promise is at least a decade away.

To address current challenges, EPA asked the National Research Council to perform an independent study on improving risk-analysis approaches, one of a number of studies by the National Research Council that have examined risk assessment in EPA. Specifically, the committee selected by the National Research Council was charged to identify practical improvements that EPA could make in the near term (2-5 years) and in the longer term (10-20 years). The committee focused primarily on human health risk assessment but also considered the implications of its conclusions and recommendations for ecologic risk assessment. The committee conducted its data gathering for this study between fall 2006 and winter 2008, so materials published after this were not considered in the committee's evaluation.

COMMITTEE'S EVALUATION

The committee focused on two broad elements in its evaluation: (1) improving the *technical analysis* that supports risk assessment (addressed in Chapters 4-7) and (2) improving the *utility* of risk assessment (addressed in Chapters 3 and 8). Improving technical analysis entails the development and use of scientific knowledge and information to promote more accurate characterizations of risk. Improving utility entails making risk assessment more relevant to and useful for risk-management decisions.

Regarding improvement in technical analysis, the committee considered such issues as how to improve uncertainty and variability analysis and dose-response assessment to ensure the best use of scientific data, and it concluded that technical improvements are necessary. The committee concluded that EPA's overall concept of risk assessment, which is generally based on the National Research Council's *Risk Assessment in the Federal Government: Managing the Process* (1983), also known as the Red Book, should be retained. The four steps of risk assessment (hazard identification, dose-response assessment, exposure assessment, and risk characterization) have been adopted by numerous expert committees, regulatory agencies, public-health institutions, and others.

With respect to improving utility, the committee considered such issues as how risk-related problems are identified and formulated before the development of risk assessments and how a broad set of options might be considered to ensure that risk assessments are most relevant to the problems.

CONCLUSIONS AND RECOMMENDATIONS

A number of improvements are needed to streamline EPA's risk-assessment process to ensure that risk assessments make better use of appropriate available science and are more relevant to decision-making. Implementing improvements will require building on EPA's current practices and developing a long-term strategy that includes greater coordination and communication within the agency, training and building a workforce with the requisite expertise, and a commitment by EPA, the executive branch, and Congress to implement the framework for risk-based decision-making recommended in this report and to fund the needed improvements.

The committee recommends an important extension of the Red Book model to meet today's challenges better—that risk assessment should be viewed as a method for evaluating the relative merits of various options for managing risk rather than as an end in itself. Risk assessment should continue to capture and accurately describe what various research findings do and do not tell us about threats to human health and to the environment, but only *after* the risk-management questions that risk assessment should address have been clearly posed, through careful evaluation of the options available to manage the environmental problems at hand, similar to what is done in ecologic risk assessment. That alteration in the current approach to risk assessment has the potential to increase its influence on decisions because it requires greater up-front planning to ensure that it is relevant to the specific problems being addressed and that it will cast light on a wider range of decision options than has traditionally been the case.

A second recommended shift in thinking is seen in the technical recommendations in this report that call for improvements in uncertainty and variability analysis and for a unified approach to dose-response assessment that will result in risk estimates for both cancer and noncancer end points. Just as a risk assessment itself should be more closely tied to the questions to be answered, so should the technical analyses supporting it. For example, descriptions of the uncertainty and variability inherent in all risk assessments may be complex or relatively simple; the level of detail in the descriptions should align with what is needed to inform risk-management decisions. Similarly, the results of a dose-response assessment should be relevant to the problem being addressed, whether it is informing risk-risk tradeoffs or a cost-benefit analysis. Ensuring that the technical analyses supporting a risk assessment are supported by the science and are relevant to the problem being addressed will go a long way toward improving the value, timeliness, and credibility of the assessment.

The committee's most important conclusions and recommendations are summarized below. The committee believes that implementation of its recommendations will do much to enhance the credibility and usefulness of risk assessment.

Design of Risk Assessment

The process of planning risk assessment and ensuring that its level and complexity are consistent with the needs to inform decision-making can be thought of as the "design" of risk assessment. The committee encourages EPA to focus greater attention on design in the formative stages of risk assessment, specifically on planning and scoping and problem formulation, as articulated in EPA guidance for ecologic and cumulative risk assessment (EPA 1998, 2003).[2] Good design involves bringing risk managers, risk assessors, and various stakeholders together early in the process to determine the major factors to be considered,

[2]EPA (U.S. Environmental Protection Agency). 1998. Guidelines for Ecological Risk Assessment. EPA/630/R-95/002F. Risk Assessment Forum, U.S. Environmental Protection Agency, Washington, DC; EPA (U.S. Environmental Protection Agency). 2003. Framework for Cumulative Risk Assessment. EPA/600/P-02/001F. National Center for Environmental Assessment, Risk Assessment Forum, U.S. Environmental Protection Agency, Washington, DC.

the decision-making context, and the timeline and depth needed to ensure that the right questions are being asked in the context of the assessment.

Increased emphasis on planning and scoping and on problem formulation has been shown to lead to risk assessments that are more useful and better accepted by decision-makers (EPA 2002, 2003, 2004);[3] however, incorporation of these stages in risk assessment has been inconsistent, as noted by their absence from various EPA guidance documents (EPA 2005a,b).[4] An important element of planning and scoping is definition of a clear set of options for consideration in decision-making where appropriate. This should be reinforced by the up-front involvement of decision-makers, stakeholders, and risk assessors, who together can evaluate whether the design of the assessment will address the identified problems.

Recommendation: Increased attention to the design of risk assessment in its formative stages is needed. The committee recommends that planning and scoping and problem formulation, as articulated in EPA guidance documents (EPA 1998, 2003),[2] should be formalized and implemented in EPA risk assessments.

Uncertainty and Variability

Addressing uncertainty and variability is critical for the risk-assessment process. Uncertainty stems from lack of knowledge, so it can be characterized and managed but not eliminated. Uncertainty can be reduced by the use of more or better data. Variability is an inherent characteristic of a population, inasmuch as people vary substantially in their exposures and their susceptibility to potentially harmful effects of the exposures. Variability cannot be reduced, but it can be better characterized with improved information.

There have been substantial differences among EPA's approaches to and guidance for addressing uncertainty in exposure and dose-response assessment. EPA does not have a consistent approach to determine the level of sophistication or the extent of uncertainty analysis needed to address a particular problem. The level of detail for characterizing uncertainty is appropriate only to the extent that it is needed to inform specific risk-management decisions appropriately. It is important to address the required extent and nature of uncertainty analysis in the planning and scoping phases of a risk assessment. Inconsistency in the treatment of uncertainty among components of a risk assessment can make the communication of overall uncertainty difficult and sometimes misleading.

Variability in human susceptibility has not received sufficient or consistent attention in many EPA health risk assessments although there are encouraging exceptions, such as those for lead, ozone, and sulfur oxides. For example, although EPA's 2005 *Guidelines for Carcinogen Risk Assessment* acknowledges that susceptibility can depend on one's stage in life,

---

[3]EPA (U.S. Environmental Protection Agency). 2002. A Review of the Reference Dose and Reference Concentration Processes. EPA/630/P-02/002F. Risk Assessment Forum, U.S. Environmental Protection Agency, Washington, DC; EPA (U.S. Environmental Protection Agency). 2003. Framework for Cumulative Risk Assessment. EPA/600/P-02/001F. National Center for Environmental Assessment, Risk Assessment Forum, U.S. Environmental Protection Agency, Washington, DC; EPA (U.S. Environmental Protection Agency). 2004. Risk Assessment Principles and Practices. Staff Paper. EPA/100/B-04/001. Office of the Science Advisor, U.S. Environmental Protection Agency, Washington, DC.

[4]EPA (U.S. Environmental Protection Agency). 2005a. Guidelines for Carcinogen Risk Assessment. EPA/630/P-03/001F. Risk Assessment Forum, U.S. Environmental Protection Agency, Washington, DC; EPA (U.S. Environmental Protection Agency). 2005b. Supplemental Guidance for Assessing Susceptibility for Early-Life Exposures to Carcinogens. EPA/630/R-03/003F. Risk Assessment Forum, U.S. Environmental Protection Agency, Washington, DC.

greater attention to susceptibility in practice is needed, particularly for specific population groups that may have greater susceptibility because of their age, ethnicity, or socioeconomic status. The committee encourages EPA to move toward the long-term goal of quantifying population variability more explicitly in exposure assessment and dose-response relationships. An example of progress that moves toward this goal is EPA's draft risk assessment of trichloroethylene (EPA 2001; NRC 2006),[5] which considers how differences in metabolism, disease, and other factors contribute to human variability in response to exposures.

Recommendation: EPA should encourage risk assessments to characterize and communicate uncertainty and variability in all key computational steps of risk assessment—for example, exposure assessment and dose-response assessment. Uncertainty and variability analysis should be planned and managed to reflect the needs for comparative evaluation of the risk management options. In the short term, EPA should adopt a "tiered" approach for selecting the level of detail to be used in the uncertainty and variability assessments, and this should be made explicit in the planning stage. To facilitate the characterization and interpretation of uncertainty and variability in risk assessments, EPA should develop guidance to determine the appropriate level of detail needed in uncertainty and variability analyses to support decision-making and should provide clear definitions and methods for identifying and addressing different sources of uncertainty and variability.

Selection and Use of Defaults

Uncertainty is inherent in all stages of risk assessment, and EPA typically relies on assumptions when chemical-specific data are not available. The 1983 Red Book recommended the development of guidelines to justify and select from among the available inference options, the assumptions—now called defaults—to be used in agency risk assessments to ensure consistency and avoid manipulations in the risk-assessment process. The committee acknowledges EPA's efforts to examine scientific data related to defaults (EPA 1992, 2004, 2005a),[6] but recognizes that changes are needed to improve the agency's use of them. Much of the scientific controversy and delay in completion of some risk assessments has stemmed from the long debates regarding the adequacy of the data to support a default or an alternative approach. The committee concludes that established defaults need to be maintained for the steps in risk assessment that require inferences and that clear criteria should be available for judging whether, in specific cases, data are adequate for direct use or to support an inference in place of a default. EPA, for the most part, has not yet published clear, general guidance on what level of evidence is needed to justify use of agent-specific data and not resort to a default. There are also a number of defaults (missing or implicit defaults) that are engrained in EPA risk-assessment practice but are absent from its risk-assessment guidelines. For ex-

---

[5]EPA (U.S. Environmental Protection Agency). 2001. Trichloroethylene Health Risk Assessment: Synthesis and Characterization. External Review Draft. EPA/600/P-01/002A. Office of Research and Development, Washington, DC. August 2001 [online]. Available: http://rais.ornl.gov/tox/TCEAUG2001.PDF [accessed Aug. 2, 2008]; NRC (National Research Council). 2006. Assessing the Human Risks of Trichloroethylene. Washington, DC: The National Academies Press.

[6]EPA (U.S. Environmental Protection Agency). 1992. Guidelines for Exposure Assessment. EPA/600/Z-92/001. Risk Assessment Forum, Office of Research and Development, U.S. Environmental Protection Agency, Washington, DC; EPA (U.S. Environmental Protection Agency). 2004. Risk Assessment Principles and Practices. Staff Paper. EPA/100/B-04/001. Office of the Science Advisor, U.S. Environmental Protection Agency, Washington, DC; EPA (U.S. Environmental Protection Agency). 2005a. Guidelines for Carcinogen Risk Assessment. EPA/630/P-03/001F. Risk Assessment Forum, U.S. Environmental Protection Agency, Washington, DC.

ample, chemicals that have not been examined sufficiently in epidemiologic or toxicologic studies are often insufficiently considered in or are even excluded from risk assessments; because no description of their risks is included in the risk characterization, they carry no weight in decision-making. That occurs in Superfund-site and other risk assessments, in which a relatively short list of chemicals on which there are epidemiologic and toxicologic data tends to drive the exposure and risk assessments.

Recommendation: EPA should continue and expand use of the best, most current science to support and revise default assumptions. EPA should work toward the development of explicitly stated defaults to take the place of implicit defaults. EPA should develop clear, general standards for the level of evidence needed to justify the use of alternative assumptions in place of defaults. In addition, EPA should describe specific criteria that need to be addressed for the use of alternatives to each particular default assumption. When EPA elects to depart from a default assumption, it should quantify the implications of using an alternative assumption, including how use of the default and the selected alternative influences the risk estimate for risk management options under consideration. EPA needs to more clearly elucidate a policy on defaults and provide guidance on its implementation and on evaluation of its impact on risk decisions and on efforts to protect the environment and public health.

A Unified Approach to Dose-Response Assessment

A challenge to risk assessment is to evaluate risks in ways that are consistent among chemicals, that account adequately for variability and uncertainty, and that provide information that is timely, efficient, and maximally useful for risk characterization and risk management. Historically, dose-response assessments at EPA have been conducted differently for cancer and noncancer effects, and the methods have been criticized for not providing the most useful results. Consequently, noncancer effects have been underemphasized, especially in benefit-cost analyses. A consistent approach to risk assessment for cancer and noncancer effects is scientifically feasible and needs to be implemented.

For cancer, it has generally been assumed that there is no dose threshold of effect, and dose-response assessments have focused on quantifying risk at low doses and estimating a population risk for a given magnitude of exposure. For noncancer effects, a dose threshold (low-dose nonlinearity) has been assumed, below which effects are not expected to occur or are extremely unlikely in an exposed population; that dose is a reference dose (RfD) or a reference concentration (RfC)—it is thought "likely to be without an appreciable risk of deleterious effects" (EPA 2002).[7]

EPA's treatment of noncancer and low-dose nonlinear cancer end points is a major step by the agency in an overall strategy to harmonize cancer and noncancer approaches to dose-response assessment; however, the committee finds scientific and operational limitations in the current approaches. Noncancer effects do not necessarily have a threshold, or low-dose nonlinearity, and the mode of action of carcinogens varies. Background exposures and underlying disease processes contribute to population background risk and can lead to linearity at the population doses of concern. Because the RfD and RfC do not quantify risk for different magnitudes of exposure but rather provide a bright line between possible harm and safety,

[7]EPA (U.S. Environmental Protection Agency). 2002. A Review of the Reference Dose and Reference Concentration Processes. EPA/630/P-02/002F. Risk Assessment Forum, U.S. Environmental Protection Agency, Washington, DC.

their use in risk-risk and risk-benefit comparisons and in risk-management decision-making is limited. Cancer risk assessments usually do not account for differences among humans in cancer susceptibility other than possible differences in early-life susceptibility.

Scientific and risk-management considerations both support unification of cancer and noncancer dose-response assessment approaches. The committee therefore recommends a consistent, unified approach for dose-response modeling that includes formal, systematic assessment of background disease processes and exposures, possible vulnerable populations, and modes of action that may affect a chemical's dose-response relationship in humans. That approach redefines the RfD or RfC as a risk-specific dose that provides information on the percentage of the population that can be expected to be above or below a defined acceptable risk with a specific degree of confidence. The risk-specific dose will allow risk managers to weigh alternative risk options with respect to that percentage of the population. It will also permit a quantitative estimate of benefits for different risk-management options. For example, a risk manager could consider various population risks associated with exposures resulting from different control strategies for a pollution source and the benefits associated with each strategy. The committee acknowledges the widespread applications and public-health utility of the RfD; the redefined RfD can still be used as the RfD has been to aid risk-management decisions.

Characteristics of the committee's recommended unified dose-response approach include use of a spectrum of data from human, animal, mechanistic, and other relevant studies; a probabilistic characterization of risk; explicit consideration of human heterogeneity (including age, sex, and health status) for both cancer and noncancer end points; characterization (through distributions to the extent possible) of the most important uncertainties for cancer and noncancer end points; evaluation of background exposure and susceptibility; use of probabilistic distributions instead of uncertainty factors when possible; and characterization of sensitive populations.

The new unified approach will require implementation and development as new chemicals are assessed or old chemicals are reassessed, including the development of test cases to demonstrate proof of concept.

Recommendation: The committee recommends that EPA implement a phased-in approach to consider chemicals under a unified dose-response assessment framework that includes a systematic evaluation of background exposures and disease processes, possible vulnerable populations, and modes of action that may affect human dose-response relationships. The RfD and RfC should be redefined to take into account the probability of harm. In developing test cases, the committee recommends a flexible approach in which different conceptual models can be applied in the unified approach.

Cumulative Risk Assessment

EPA is increasingly asked to address broader public-health and environmental-health questions involving multiple exposures, complex mixtures, and vulnerability of exposed populations—issues that stakeholder groups (such as communities affected by environmental exposures) often consider to be inadequately captured by current risk assessments. There is a need for cumulative risk assessments as defined by EPA (EPA 2003)[8]—assessments that

[8]EPA (U.S. Environmental Protection Agency). 2003. Framework for Cumulative Risk Assessment. EPA/600/P-02/001F. National Center for Environmental Assessment, Risk Assessment Forum, U.S. Environmental Protection Agency, Washington, DC.

include combined risks posed by aggregate exposure to multiple agents or stressors; aggregate exposure includes all routes, pathways, and sources of exposure to a given agent or stressor. Chemical, biologic, radiologic, physical, and psychologic stressors are considered in this definition (Callahan and Sexton 2007).[9]

The committee applauds the agency's move toward the broader definition in making risk assessment more informative and relevant to decisions and stakeholders. However, in practice, EPA risk assessments often fall short of what is possible and is supported by agency guidelines in this regard. Although cumulative risk assessment has been used in various contexts, there has been little consideration of nonchemical stressors, vulnerability, and background risk factors. Because of the complexity of considering so many factors simultaneously, there is a need for simplified risk-assessment tools (such as databases, software packages, and other modeling resources) that would allow screening-level risk assessments and could allow communities and stakeholders to conduct assessments and thus increase stakeholder participation. Cumulative human health risk assessment should draw greater insights from ecologic risk assessment and social epidemiology, which have had to grapple with similar issues. A recent National Research Council report on phthalates addresses issues related to the framework within which dose-response assessment can be conducted in the context of simultaneous exposures to multiple stressors.

Recommendation: EPA should draw on other approaches, including those from ecologic risk assessment and social epidemiology, to incorporate interactions between chemical and nonchemical stressors in assessments; increase the role of biomonitoring, epidemiologic, and surveillance data in cumulative risk assessments; and develop guidelines and methods for simpler analytical tools to support cumulative risk assessment and to provide for greater involvement of stakeholders. In the short-term, EPA should develop databases and default approaches to allow for incorporation of key nonchemical stressors in cumulative risk assessments in the absence of population-specific data, considering exposure patterns, contributions to relevant background processes, and interactions with chemical stressors. In the long-term, EPA should invest in research programs related to interactions between chemical and nonchemical stressors, including epidemiologic investigations and physiologically based pharmacokinetic modeling.

Improving the Utility of Risk Assessment

Given the complexities of the current problems and potential decisions faced by EPA, the committee grappled with designing a more coherent, consistent, and transparent process that would provide risk assessments that are relevant to the problems and decisions at hand and that would be sufficiently comprehensive to ensure that the best available options for managing risks were considered. To that end, the committee proposes a framework for risk-based decision-making (see Figure S-1). The framework consists of three phases: I, enhanced problem formulation and scoping, in which the available risk-management options are identified; II, planning and assessment, in which risk-assessment tools are used to determine risks under existing conditions and under potential risk-management options; and III, risk management, in which risk and nonrisk information is integrated to inform choices among options.

The framework has at its core the risk-assessment paradigm (stage 2 of phase II) estab-

[9]Callahan, M.A., and K. Sexton. 2007. If 'cumulative risk assessment' is the answer, what is the question? Environ. Health Perspect. 115(5):799-806.

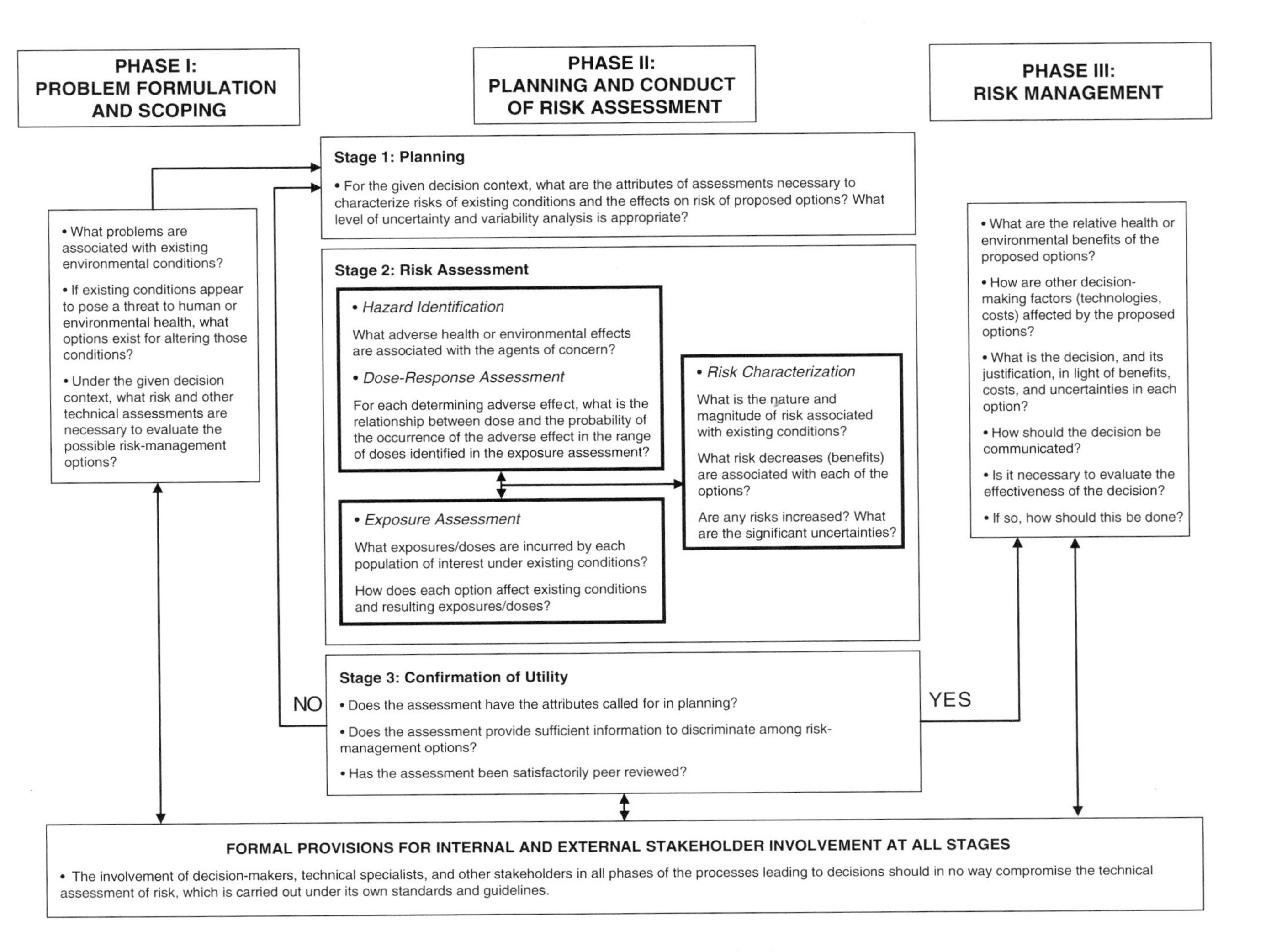

FIGURE S-1 A framework for risk-based decision-making that maximizes the utility of risk assessment.

lished in the Red Book (NRC 1983).[10] However, the framework differs from the Red Book paradigm, primarily in its initial and final steps. The framework begins with a "signal" of potential harm (for example, a positive bioassay or epidemiologic study, a suspicious disease cluster, or findings of industrial contamination). Under the traditional paradigm, the question has been, What are the probability and consequence of an adverse health (or ecologic) effect posed by the signal? In contrast, the recommended framework asks, implicitly, What *options* are there to reduce the *hazards* or *exposures* that have been identified, and how can risk assessment be used to evaluate the merits of the various options? The latter question focuses on the risk-management options (or interventions) designed to provide adequate public-health and environmental protection and to ensure well-supported decision-making. Under this framework, the questions posed arise from early and careful planning of the types of assessments (including risks, costs, and technical feasibility) and the required level of scientific depth that are needed to evaluate the relative merits of the options being considered.[11] Risk management involves choosing among the options after the appropriate assessments have been undertaken and evaluated.

The framework begins with enhanced problem formulation and scoping (phase I), in which risk-management options and the types of technical analyses, including risk assessments, needed to evaluate and discriminate among the options are identified. Phase II consists of three stages: planning, risk assessment, and confirmation of utility. Planning (stage 1) is done to ensure that the level and complexity of risk assessment (including uncertainty and variability analysis) are consistent with the goals of decision-making. After risk assessment (stage 2), stage 3 evaluates whether the assessment was appropriate and whether it allows discrimination among the risk-management options. If the assessment is determined not to be adequate, the framework calls for a return to planning (phase II, stage 1). Otherwise, phase III (risk management) is undertaken: the relative health or environmental benefits of the proposed risk-management options are evaluated for the purpose of reaching a decision.

The framework systematically identifies problems and options that risk assessors should evaluate at the earliest stages of decision-making. It expands the array of impacts assessed beyond individual effects (for example, cancer, respiratory problems, and individual species) to include broader questions of health status and ecosystem protection. It provides a formal process for stakeholder involvement throughout all stages but has time constraints to ensure that decisions are made. It increases understanding of the strengths and limitations of risk assessment by decision-makers at all levels, for example, by making uncertainties and choices more transparent.

The committee is mindful of concerns about political interference in the process, and the framework maintains the conceptual distinction between risk assessment and risk management articulated in the Red Book. It is imperative that risk assessments used to evaluate risk-management options not be inappropriately influenced by the preferences of risk managers.

With a focus on early and careful planning and problem formulation and on the options for managing the problem, implementation of the framework can improve the utility of risk assessment for decision-making. Although some aspects of the framework are achievable in the short term, its full implementation will require a substantial transition period. EPA should phase in the framework with a series of demonstration projects that apply it and

[10] NRC (National Research Council). 1983. Risk Assessment in the Federal Government: Managing the Process. Washington, DC: National Academy Press.

[11] The committee notes that not all decisions require or are amenable to risk assessment and that in most cases one of the options explicitly considered is "no intervention."

that determine the degree to which it meets the needs of the agency risk managers, how risk-management conclusions differ as a result of its application, and the effectiveness of measures to ensure that risk managers and policy-makers do not inappropriately influence the scientific conduct of risk assessments.

Recommendation: To make risk assessments most useful for risk-management decisions, the committee recommends that EPA adopt a *framework for risk-based decision-making* (see Figure S-1) that embeds the Red Book risk-assessment paradigm into a process with initial problem formulation and scoping, upfront identification of risk-management options, and use of risk assessment to discriminate among these options.

Stakeholder Involvement

Many stakeholders believe that the current process for developing and applying risk assessments lacks credibility and transparency. That may be partly because of failure to involve stakeholders adequately as active participants at appropriate points in the risk-assessment and decision-making process rather than as passive recipients of the results. Previous National Research Council and other risk-assessment reports (for example, NRC 1996; PCCRARM 1997)[12] and comments received by the committee (Callahan 2007; Kyle 2007)[13] echo such concerns.

The committee agrees that greater stakeholder involvement is necessary to ensure that the process is transparent and that risk-based decision-making proceeds effectively, efficiently, and credibly. Stakeholder involvement needs to be an integral part of the risk-based decision-making framework, beginning with problem formulation and scoping.

Although EPA has numerous programs and guidance documents related to stakeholder involvement, it is important that it adhere to its own guidance, particularly in the context of cumulative risk assessment, in which communities often have not been adequately involved.

Recommendation: EPA should establish a formal process for stakeholder involvement in the framework for risk-based decision-making with time limits to ensure that decision-making schedules are met and with incentives to allow for balanced participation of stakeholders, including impacted communities and less advantaged stakeholders.

Capacity-Building

Improving risk-assessment practice and implementing the framework for risk-based decision-making will require a long-term plan and commitment to build the requisite capacity of information, skills, training, and other resources necessary to improve public-health and environmental decision-making. The committee's recommendations call for considerable modification of EPA risk-assessment efforts (for example, implementation of the risk-based decision-making framework, emphasis on problem formulation and scoping as a discrete

---

[12]NRC (National Research Council). 1996. Understanding Risk: Informing Decisions in a Democratic Society. Washington, DC: National Academy Press; PCCRARM (Presidential/Congressional Commission on Risk Assessment and Risk Management). 1997. Framework for Environmental Health Risk Management - Final Report, Vol. 1.

[13]Callahan, M.A. 2007. Improving Risk Assessment: A Regional Perspective. Presentation at the Third Meeting of Improving Risk Analysis Approaches Used by EPA, February 26, 2007, Washington, DC; Kyle, A. 2007. Community Needs for Assessment of Environmental Problems. Presentation at the Fourth Meeting of Improving Risk Analysis Approaches Used by EPA, April 17, 2007, Washington, DC.

stage in risk assessment, and greater stakeholder participation) and of technical aspects of risk assessment (for example, unification of cancer and noncancer dose-response assessments, attention to quantitative uncertainty analysis, and development of methods for cumulative risk assessment). The recommendations are tantamount to "change-the-culture" transformations in risk assessment and decision-making in the agency.

EPA's current institutional structure and resources may pose a challenge to implementation of the recommendations, and moving forward with them will require a commitment to leadership, cross-program coordination and communication, and training to ensure the requisite expertise. That will be possible only if leaders are determined to reverse the downward trend in budgeting, staffing, and training and to making high-quality, risk-based decision-making an agencywide goal.

Recommendation: EPA should initiate a senior-level strategic re-examination of its risk-related structures and processes to ensure that it has the institutional capacity to implement the committee's recommendations for improving the conduct and utility of risk assessment for meeting the 21st century environmental challenges. EPA should develop a capacity building plan that includes budget estimates required for implementing the committee's recommendations, including transitioning to and effectively implementing the framework for risk-based decision-making.

CONCLUDING REMARKS

Global impacts are combining with the high financial and political stakes of risk management to place unprecedented pressure on risk assessors in EPA. But risk assessment remains essential to the agency's mission to ensure protection of public health and the environment. Much work is needed to improve the scientific status, utility, and public credibility of risk assessment. The committee's recommendations focus on designing risk assessments to ensure that they make the best possible use of available science, are technically accurate, and address the appropriate risk-management options effectively to inform risk-based decision-making. The committee hopes that the recommendations and the proposed framework for risk-based decision-making will provide a template for the future of risk assessment in EPA and strengthen the scientific basis, credibility, and effectiveness of future risk-management decisions.

1

Introduction

In response to a request from the Environmental Protection Agency (EPA) National Center for Environmental Assessment (NCEA), the National Research Council established the Committee on Improving Risk Analysis Approaches Used by the EPA. The committee was charged with developing recommendations that, if implemented, could assist the agency in developing risk assessments[1] that are both consistent with current and evolving scientific understanding and relevant to the many risk-management missions of the agency. Recommendations were to focus on both short- and long-term objectives.

The importance of risk assessment to the mission of EPA—indeed to the mission of many other federal agencies and to their state counterparts—is attested to by a long series of major efforts by the National Academies and other expert bodies to strengthen the technical content and utility of risk assessment and to ensure its scientific integrity. As EPA has attempted to respond to the recommendations that have resulted from the various efforts, both the science underlying risk assessment and the decision contexts in which risk assessments are used have been increasingly complex. As will be revealed later in this report, the committee perceives that risk assessment is now at a crossroads and its value and relevance are increasingly questioned (Silbergeld 1993; Montague 2004). Nonetheless, the committee believes strongly that risk assessment remains the most appropriate available method for measuring the relative benefits of the many possible interventions available to improve human health and the environment and that its absence or its inappropriate application will result in seriously flawed decisions. The committee believes that implementation of the recommendations set forth in this report will do much to enhance the power and usefulness of risk assessment and will be the appropriate road forward.

[1]EPA's charge to the committee used the phrase *risk analysis*. The latter is sometimes used synonymously with *risk assessment* but sometimes used more broadly. The committee will use *risk assessment* to describe the process leading to a characterization of risk. Risk as defined by NRC (2007a) can be a hazard, a probability, a consequence, or a combination of probability and severity of consequence.

BACKGROUND

Since the 1983 publication of the National Research Council's report *Risk Assessment in the Federal Government: Managing the Process* (the so-called Red Book), EPA has made efforts to advance risk assessment with the generation of risk-assessment guidelines, the establishment of intra-agency and cross-agency science-policy panels, and improvements in peer-review standards for agency risk assessments. The Red Book committee demonstrated how risk assessment could fill the gap between results emerging from the research setting and their use in risk management. A framework for systematically carrying out the process of risk assessment was established, and the Red Book's risk-assessment framework remains in place today. The Red Book also revealed how the development of what were called inference guidelines (see below) was necessary to ensure the scientific integrity of the process by which risk assessments were conducted and of the product of that process.

Various closely related forms of the risk-assessment framework have been widely used by international organizations and other federal agencies, including the Consumer Product Safety Commission, the Nuclear Regulatory Commission, the Food and Drug Administration, the Occupational Safety and Health Administration, the U.S. Department of Agriculture, the Department of Defense, and the Department of Energy. OSTP (50 Fed. Reg. 10371[1985]) adopted the Red Book framework for carcinogen analysis and provided agencies a basis for developing the guidelines recommended by NRC (1983).

Publication of the Red Book was followed by an intensification of risk-assessment activity in EPA. EPA endorsed the Red Book in the publication, *Risk Assessment and Management: Framework for Decision Making* (EPA 1984). The agency established in 1984 what is now called the Risk Assessment Forum and in 1993 added a Science Policy Council (see Appendix C for a timeline of selected risk-assessment activities)—evidence that the Red Book and EPA's efforts to advance risk assessment fell on fertile ground (Goldman 2003). William Ruckelshaus, during his second tour as administrator of EPA (1983-1985), used the Red Book as the basis of a main theme of his tenure: strengthening risk assessment as a tool to inform decision-making. EPA initially focused on human health risk assessment with the *Guidelines for Carcinogen Risk Assessment* (EPA 1986) and the agency's *Unfinished Business: A Comparative Assessment of Environmental Problems* (EPA 1987), which compared the magnitude of environmental risks with EPA's resource allocations to programs that address them. The agency's Science Advisory Board evaluated the latter document in another key report, *Reducing Risk: Setting Priorities and Strategies for Environmental Protection* (EPA SAB 1990), and EPA was involved in a 1992 conference that evaluated the risk-based model for setting national priorities against several alternatives that incorporated information about solutions, environmental justice, and other factors (Finkel and Golding 1994).

In the 1990s, the four-step approach outlined in the Red Book was adapted to ecologic risk assessment to address evaluations in which human health is not the primary focus (EPA 2004). Ecologic risk assessors pioneered new approaches to complex risk problems by delineating the need for "planning and problem formulation" to address technically challenging assessments of ecosystems, chemical mixtures, and cumulative risk. In the planning step, the risk managers—in consultation with risk assessors and other interested parties—frame management goals, management options, and the scope and necessary level of complexity for the risk assessment. Problem formulation is the phase in which the risk managers' charge to the assessors is converted into an actionable plan for performing the assessment (EPA 1998; Suter 2007).

Several National Research Council and other expert panels expanded on the risk-assessment principles presented in the Red Book with the publication of reports that included

Pesticides in the Diets of Infants and Children (NRC 1993), *Science and Judgment in Risk Assessment* (NRC 1994), and *Understanding Risk: Informing Decisions in a Democratic Society* (NRC 1996). In 1997, another expert panel issued its report, *Presidential/Congressional Commission on Risk Assessment and Risk Management* (PCCRARM 1997).

EPA has also recently upgraded its standards for peer review of technical documents with the Science Policy Council's *Peer Review Handbook* (EPA 2000) and guidance (EPA 2002) to conform with the Office of Management and Budget's *Final Information Quality Bulletin for Peer Review* (OMB 2004).

CHALLENGES

As risk assessment has come to be widely used in a fairly consistent framework, EPA practices have continued to draw scrutiny in that competing pressures are pushing the agency to improve the timeliness and quality of its risk assessments. It is now evident that many risk assessments are taking 10-20 years to complete including assessments on chemicals such as dioxin, formaldehyde, and trichloroethylene (GAO 2008). There are a myriad of reasons for delays in the completion of risk assessments including controversy surrounding the science, uncertainties in the data, regulatory requirements, political priorities, and economic factors. In the absence of completed risk assessments, risk management decisions continue to be made by state and federal agencies; however it is not known whether the decisions being made are health protective. To the extent that this practice continues, the value of risk assessment will erode.

For example, trichloroethylene, the most common organic contaminant in groundwater, which has been linked to cancer, does not have a completed EPA toxicity assessment. The EPA assessment has been under development since the 1980s and subjected to multiple independent reviews including EPA's Science Advisory Board. Key issues were evaluated by the National Research Council in 2006. NRC (2006) urged that the toxicity assessment be finalized with currently available data, but the assessment is not anticipated to be finalized until 2010 (GAO 2008). Another example is formaldehyde, which the World Health Organization classified as a known human carcinogen and whose assessment was begun by EPA in 1997 but is not expected to be completed until 2010 (IARC 2006). The lack of an updated toxicity assessment for formaldehyde has impacted EPA's regulatory decisions (GAO 2008).[2]

In recent years, a number of federal agencies have raised concerns about EPA risk assessments of contaminants and are now playing a more formal role in risk policy-making at the federal level. Some of the agencies are also potentially responsible parties facing cleanup responsibilities and are seeking more input as EPA moves toward final reviews. Those other agencies and other public and private stakeholders often assert that they are inadequately involved in EPA processes (for example, Risk Policy Report 2005, 2007).

The Integrated Risk Information System (IRIS) is an important compendium of chemical toxicity values in which new EPA science policies are often implemented for the first time. However, IRIS has been criticized because of limitations, including a lack of funding and delays in updating toxicity values. EPA is now seeking greater science-policy input on its chemical reviews earlier in the process so that critical issues can be identified and adjustments made in response to new scientific and science-policy information.

[2]GAO (2008) acknowledges that because there was no updated EPA cancer risk estimate, EPA's Office of Air and Radiation used an alternative estimate in establishing a National Emissions Standard for Hazardous Air Pollutants covering facilities in the plywood and composite wood industries.

Those types of problems are exacerbated by the fact that the scientific issues underlying risk assessments and the decisions that risk assessments are developed to support are increasingly complex, as a result of a greater quantity and diversity of data stemming from advancements such as in genomics and biomarkers. This report is intended to assist EPA as it attempts to deal with those and other challenges.

TRADITIONAL AND EMERGING VIEWS OF THE ROLES OF RISK ASSESSMENT

A large community of public-health research scientists in many disciplines is involved in the development of knowledge about how agents in the environment—whether chemical, biologic, radiologic, or physical and whether of natural origin or resulting from human activity—can harm human health and about the conditions under which they may do so. As this type of knowledge emerges from research, policy-makers in government and many other institutions concerned with public health begin to focus on whether some type of action is needed to protect public health and, if so, whether some courses of action yield better results than others. Societal support for action is found in the many laws that guide regulatory and public-health agencies. This support is evident in the relationship between the research community concerned with understanding threats to ecosystems and people responsible for protecting them.

It is clear that research findings are rarely directly suitable for decision-making. Results of different studies of the same phenomena often conflict, uncertainties can be large, and the conditions under which health and ecosystem threats are studied (or can be studied) usually do not match the conditions of interest for public-health or ecosystem protection. Research findings need to be interpreted. In matters related to public and ecosystem health, the interpretive process is called risk assessment. Risk assessment has come to be seen as an essential component of regulatory and related types of decision-making, and its scientific underpinnings and its roles in decision-making are the central subjects of this report.

Much scholarly work that has appeared since the publication of the Red Book has been devoted to countering a tendency to view risk assessment, in its practical applications, both as the sole source of information on the problems to be managed and as providing the management choice. To the extent that that tendency exists, we urge that it be resisted. Risk assessment, we propose, should certainly continue to capture and accurately describe what various bodies of research findings do and do not tell us about various threats to human health and to the environment, but it should do so only *after* the questions that risk assessment is supposed to address have been posed, through careful evaluation of the options available to manage the environmental problem at hand, similar to what is done in ecologic risk assessment. In this context, risk assessment is seen as a method for evaluating the relative merits of various options (or interventions) for managing risk.

Risk assessment, in that decision-making context, is an essential tool for understanding what public-health and environmental goals can be achieved or have been achieved by the actions taken. As will be seen later in this report, early emphasis on identifying risk-management options and on seeking, through risk assessment, analyses that are most useful for evaluating the options is somewhat at variance with the risk-assessment–risk-management model first proposed in the Red Book in that the management options are no longer driven by whatever risk-assessment findings happen to emerge. The new model does not alter the technical content of risk assessment from that set out in the Red Book, and, if appropriate precautions are taken, it does not lead to inappropriate intrusions by risk managers into the risk-assessment process (an issue of much concern to the Red Book authors; see Chapter 2). But it has great potential to increase the influence of risk assessment on ultimate decisions because it is asked to cast light on a wider range of decision options than has traditionally been the case. We see this as a necessary and worthwhile extension of the Red Book model, one better suited to today's challenges. Its full scope is elucidated in Chapters 3 and 8, which focus on increasing the utility of risk assessments.

Regulatory decision-makers, including those in EPA, do not routinely approach public-health and environmental problems by arraying a wide range of options for dealing with them and then setting into motion the various technical analyses (risk assessments, control-technology analyses, analyses of resource costs, and so on) that are necessary to achieve the optimal outcome. The various laws administered by EPA and other regulatory agencies appear to constrain, or have traditionally been interpreted as constraining, the options to be considered for risk management. The broader decision context that we propose (discussed in Chapters 3 and 8) recommends the consideration of other tools now being used or under development (such as life-cycle analysis [LCA] and sustainability evaluation) that are directed at environment-related problems of broader scope than those traditionally considered by EPA and related institutions. The integration of the scientific power of risk assessment with the broader reach of LCA, for example, should enlarge the influence of risk assessment and increase its utility for managing the most urgent and far-reaching problems—those having both human and environmental health components.

Whether operating in a broad or more narrowly constrained decision context, risk assessment is essential for the reasons described above. Whatever the decision context, the goal of risk assessment is to describe the probability that adverse health or ecosystem effects of specific types will occur under specified conditions of exposure to an activity or an agent (chemical, biologic, radiologic, or physical), to describe the uncertainty in the probability estimate, and to describe how risk varies among populations. To be most useful in decision-making, risk assessment would consider the risks associated with existing conditions (that is, the probability of harm under the "take no action" alternative) and the risks that would remain if each of various possible actions were taken to alter the conditions. There would also be a need for some commonality in the uncertainty analysis goals and assumptions that are applied to each of the analyses so that the different policy options can be compared. The conduct of risk assessment in the broadest practicable risk-management context brings to light the fullest possible picture of net public-health and environmental benefits. That does not mean that other options cannot surface during the conduct of a risk assessment; in fact, improved stakeholder engagement in the process may make this possible.

Achieving such results requires the use of the framework for the conduct of risk assessment set forth in the 1983 Red Book, which has been adopted by numerous expert committees, regulatory agencies, and public-health institutions and which this committee sees no reason to alter. The framework includes three well-known analytic steps—hazard identification, dose-response assessment, and exposure assessment—and a fourth step, risk characterization, in which results of the first three steps are integrated to yield information on the probability that the adverse effects described in hazard identification will occur under the conditions described in exposure assessment. Uncertainty findings from the first three steps are also integrated into risk characterization. Many other types of review of human-health or ecologic data emerge from regulatory and public-health institutions, but only those which in some way incorporate all four of the above steps can properly be called risk assessments.

Although all risk assessments include the four steps, it is critical to recognize that risk assessments can be undertaken at various levels of technical detail. Given a sufficiently rich database, highly quantitative estimates of risk can be developed, sometimes involving probabilistic modeling and substantial biologic data. In other cases, risk assessments may be semiquantitative. Similarly, descriptions of the uncertainties inherent in all risk assessments may be complex or relatively simple. Because risk assessments can vary in detail and complexity, it is important to know how a risk assessment will be used before it is undertaken so that it can be designed and carried out at the level of technical detail appropriate to the problem at hand. Risk-assessment design is the subject of Chapter 3.

Decisions regarding risks and risk changes expected under various risk-management options are informed by the availability of risk assessments. The goal of achieving accurate, highly quantitative estimates of risk, however, is hampered by limitations in scientific understanding and the availability of relevant data, which can be overcome only by the advance of relevant research. Decisions to protect public health and the environment cannot await "perfection" in scientific knowledge (an unachievable goal in any case); in the absence of the understanding that risk assessments, however imperfect, can bring, it will not be possible to know the public-health or environmental value of whatever decisions are ultimately made. It is therefore important that risk assessments incorporate the best available scientific information in scientifically rigorous ways and that they capture and describe the uncertainties in the information in ways that are useful for decision-makers. Moreover, the goal of timeliness is as important as (sometimes more important than) the goal of a precise risk estimate. The need to seek improvements in EPA's regulatory decision-making by improving the quality and utility of risk assessment is the impetus for the current study.

TECHNICAL IMPEDIMENTS TO RISK ASSESSMENT

It is useful to describe some of the types of obstacles that hamper the risk-assessment process and that limit the utility of its results. It should be kept in mind that risk assessments should not be blamed for a lack of relevant scientific data and knowledge; such a lack reflects inadequate support for research. But inadequacies in the use of whatever data and knowledge are available clearly are a problem for risk assessment. The following questions will receive much attention in this report because they reflect identifiable impediments to risk assessment and its most important use—for informing decision-making.

1. Are the decision contexts in which risk assessments are to be developed well defined in advance? It is important to understand the context in which a risk assessment will be used, so that the appropriate options for addressing a problem can be considered. It seems that current regulatory thinking on this matter may be overconstrained and often fails even to begin to incorporate a full range of decision options, perhaps because of limitations, or perceived limitations, embodied in laws. In any event, the utility of risk assessments may be less than ideal because of a failure to achieve clarity regarding the options for decision-making in advance of identifying the types of risk assessments that will be of value.
2. What is the right level of detail for a risk assessment? Early delineation of problems and options for managing them allows—through the necessary interactions among risk managers, risk assessors and other technical analysts, and other stakeholders—the development of risk assessments whose level of detail and scientific completeness match the decision-making requirements and so can maximize the efficiency of the process.
3. Are the criteria for selecting the "defaults" necessary to complete risk assessments and for departing[3] from them fully specified and set forth in agency guidelines? Because of the need for a variety of inferences in risk assessment and because the rationales for drawing the inferences are not always distinguishable on purely scientific grounds, the choice of

[3]The committee recognizes that the current EPA policy on defaults uses the term "invokes" rather than "departs." EPA's current policy on defaults is presented in Chapter 6.

default options to be used involves an element of policy (see the discussion of the Red Book in Chapter 2). The inferences selected, which are commonly referred to as defaults, can have substantial effects on the results of risk assessments. Their selection and the criteria for judging when, in a specific case, a default can be replaced with an alternative inference based on chemical-specific information are among the most contentious elements of the risk-assessment process and a cause of sometimes great delays in their completion.

4. Are the best available scientific information and defaults used to deal with the problem of variability? Variability in exposures to hazardous agents and in biologic responses to them is a fact of nature. Scientific knowledge of variability is highly limited, and current risk-assessment approaches to the problem rely heavily on uncertainty factors and other assumptions. It is important for the advance of risk assessment to consider the types of scientific knowledge now available and their use for improving the quantitative characterization of variability.

5. What methods should be used to describe and express the uncertainties that accompany all risk assessments? Failure to deal adequately with this matter is a source of much contention and hampers the goals of decision-making. An issue of central concern is the relative utility for decision-makers of the various methods available to express uncertainties.

6. Is information about the hazardous properties of chemicals and other agents given adequate attention in risk assessment? The toxic or carcinogenic properties of substances under assessment are now typically described in qualitative terms (a weight-of-evidence evaluation), and without quantitative expressions of the probability that the adverse effect is relevant to the human population that is the subject of the risk assessment. The possible importance of this limitation in risk assessment has been little discussed.

7. Are current methods for dealing with substances thought to act through threshold mechanisms (for example, the development of toxicity reference doses) yielding the most useful information for decision-making? Current "bright-line" approaches, while valuable in certain public-health decision-making contexts, clearly lack utility in other contexts.

8. Do current methods for integrating and weighing evidence from different sources (for example, from epidemiology and experimental studies) ensure that subjective influences are minimized and transparency maximized?

9. What are the appropriate scientific and policy approaches for dealing with substances on which very little health-effects or exposure information is available so that the risks they pose are not ignored relative to those posed by better-studied substances?

10. What approaches should be pursued for defining the risk assessments necessary to address broad questions of communitywide and cumulative risks (which may involve many exposure sources and pathways)? Given an ability to formulate appropriate risk questions in such broad contexts, how can risk information best serve decisions needed to reduce burdens on public health and the environment?

IMPROVING RISK ANALYSIS

Based on the above questions, improvements in risk analysis can be considered at two broad levels. First, consideration can be given to improvements in the *utility* of risk assessments for decision-making. Second, improvements in the *technical analysis* supporting one or more of the steps of risk assessment can also be feasible, as new scientific knowledge becomes available. The committee understands its charge to encompass both types of improvements.

Improved utility can be achieved in several ways. As has been noted, there are opportunities to improve the processes through which risk-related problems and options for intervention are identified and formulated prior to the development of risk assessments. Similar opportunities arise for improvements in the interactions among risk managers and other stakeholders and risk assessors during the development of assessments. Utility might also be enhanced by improvements in the ways risks are characterized and uncertainties expressed, to ensure they are adequately understood by decision-makers. Can the public health be better served in certain circumstances, for example, by probabilistic expressions of risk and uncertainties, for toxicity information, than they are by "bright line" estimates such as toxicity reference doses and concentrations? Can assessments in which the results of applying different default options are presented, each with a description of its scientific strengths and weaknesses, better serve decision-makers than those that rely primarily upon pre-assigned defaults? These types of questions pertain to improvements that might increase the utility of risk assessments for decision-making.

Improving the technical analysis involved in each of the steps of risk assessment generally refers to the development and use of scientific knowledge and information that, for a number of reasons, might lead to more accurate characterizations of risk. Because there are generally no means empirically to verify the results of most risk assessments, it is difficult to assess whether "accuracy" has been improved. But there nevertheless seems to be a basis for believing that greater understanding of the biological processes underlying the production of toxicity or other types of adverse health effects can, if properly applied, increase confidence in risk-assessment results. Indeed, much of the current research in toxicology is directed at gaining that understanding, and with that understanding can come reduced reliance upon defaults. In addition, the development of databases of empirical observations relevant to specific uncertainty factors can be used to replace single-point uncertainty factors with distributions. Increased confidence in risk assessments might also arise from increased development and use of human data—both epidemiology and in vitro data (NRC 2007b).

It should be noted that, while improvements in the utility of risk analysis are always desirable, the quest for improvements in scientific accuracy may not always be necessary or desirable in the context of specific risk assessments. The latter usually requires investment in significant research, and so will necessarily be limited to substances of significant social or economic importance. Default-based risk assessments will continue to have significant roles because decisions must be efficiently made on large numbers of hazards for which resources will not be available to corroborate the validity of each default, or to explore specific alternatives, and because as experience accrues, many of the defaults are viewed as a culmination of scientific understanding about general phenomena (for which exceptions may apply in particular cases). It is, of course, possible that, as new scientific understanding becomes available, certain alternatives to established defaults may prove to be supportable on a general basis, and this would increase confidence in risk assessments based on them. But default-based risk assessments will remain necessary for many substances and situations.

Much of what follows in the remaining chapters of the report derives from the committee's view of these two broad ways in which improvements in risk analysis might be achieved.

THE NATIONAL RESEARCH COUNCIL COMMITTEE

In response to the study request from EPA, the NRC established the Committee on Improving Risk Analysis Approaches Used by EPA. Committee members were selected for their expertise in biostatistics, dose-response modeling, ecotoxicology, environmental transport and fate modeling, environmental health, environmental regulation, epidemiology, exposure assessment, risk assessment, toxicology, and uncertainty analysis. Members come from universities and other organizations and serve pro bono. Committee members were asked to serve as individual experts, not as representatives of any organization.

The committee was charged with developing scientific and technical recommendations for improving risk analysis approaches used by EPA, including providing practical improvements that EPA could make in the near term (2-5 years) and in the longer term (10-20 years). The committee focused primarily on human health risk assessment, but considered the implications of its findings and recommendations to ecological risk analysis. In reviewing EPA's risk analysis concepts and practices, the committee considered past evaluations and ongoing studies by NRC and others, and risk analyses involving different exposure pathways and environmental media. In its evaluation, the committee was asked to consider a number of topics relating to uncertainty, variability, modeling, and mode of action[4] (see Appendix B for complete statement of task).

To address its task, the committee held five public sessions in which it heard presentations from officials from EPA's Office of Research and Development, its policy, program and regional offices; the Centers for Disease Control and Prevention; representatives from industry and environmental organizations; consultants; and academia.

In addressing its charge, the committee considered carefully the concerns expressed by the presenters regarding the challenges and limitations of risk assessment (Callahan 2007; Kyle 2007). Peter Preuss, the director of NCEA urged the committee to consider three specific questions (Preuss 2006): 1) What improvements can be made to risk assessment in the present? 2) What improvements can be made to risk assessment in the longer term? 3) What alternative risk paradigms should be considered? Although the charge is focused on risk assessment at EPA, it is the committee's hope that the recommendations have influence over risk assessment wherever it is practiced and used.

ORGANIZATION OF THE REPORT

The body of this report is organized into nine chapters. Chapter 2 presents an evolution of risk assessment and its applications since the 1980s. Chapter 3 addresses the design of risk assessment, emphasizing the role of planning and scoping and problem formulation in the process. Chapter 4 considers uncertainty and variability in risk assessment, addressing both EPA's methodologies and needs for improvement. Chapter 5 presents a unified approach for non-cancer and cancer dose-response modeling that explicitly incorporates uncertainty and variability into the process. Chapter 6 addresses an important area of uncertainty, selection and use of defaults. Chapter 7 discusses the need and methods for considering a broader range of factors in risk assessment, that is cumulative risk assessment, including chemical and non-chemical stressors, vulnerability of the exposed population, and the impact of actions on stakeholders, in particular communities. Chapter 8 presents a framework for risk-based decision-making that is intended to improve the utility of risk assessment. Chapter 9 presents the committee's conclusions and recommendations along with a strategy for implementing them.

[4]A description of observable key events or processes from interaction of an agent with a cell or tissue through operational and anatomical changes to the disease state (EPA 2005).

REFERENCES

Callahan, M.A. 2007. Improving Risk Assessment: A Regional Perspective. Presentation at the Third Meeting of Improving Risk Analysis Approaches Used by EPA, February 26, 2007, Washington, DC.

EPA (U.S. Environmental Protection Agency). 1984. Risk Assessment and Risk Management: Framework for Decision Making. EPA 600/9-85-002. Office of the Administrator, U.S. Environmental Protection Agency, Washington, DC. December 1984.

EPA (U.S. Environmental Protection Agency). 1986. Guidelines for Carcinogen Risk Assessment. EPA/630/R-00/004. Risk Assessment Forum, U.S. Environmental Protection Agency, Washington, DC. September 1986 [online]. Available: http://www.epa.gov/ncea/raf/car2sab/guidelines_1986.pdf [accessed Jan. 7, 2007].

EPA (U.S. Environmental Protection Agency). 1987. Unfinished Business: A Comparative Assessment of Environmental Problems Overview Report. EPA 230287025a. Office of Policy, Planning and Evaluation, U.S. Environmental Protection Agency, Washington, DC. February 1987.

EPA (U.S. Environmental Protection Agency). 1998. Guidelines for Ecological Risk Assessment. EPA/630/R-95/002F. Risk Assessment Forum, U.S. Environmental Protection Agency, Washington, DC. April 1998 [online]. Available: http://oaspub.epa.gov/eims/eimscomm.getfile?p_download_id=36512 [accessed Feb. 9, 2007].

EPA (U.S. Environmental Protection Agency). 2000. Science Policy Peer Review Handbook, 2nd Ed. EPA 100-B-00-001. Office of Science Policy, Office of Research and Development, U.S. Environmental Protection Agency, Washington, DC. December 2000 [online]. Available: http://www.epa.gov/osa/spc/pdfs/prhandbk.pdf [accessed Feb. 9, 2007].

EPA (U.S. Environmental Protection Agency). 2002. Guidelines for Ensuring and Maximizing the Quality, Utility and Integrity of Information Disseminated by the Environmental Protection Agency. EPA/260R-02-008. Office of Environmental Information, U.S. Environmental Protection Agency, Washington, DC. October 2002 [online]. Available: http://www.epa.gov/QUALITY/informationguidelines/documents/EPA_InfoQualityGuidelines.pdf [accessed Feb. 9, 2007].

EPA (U.S. Environmental Protection Agency). 2004. Risk Assessment Principles and Practices: Staff Paper. EPA/100/B-04/001. Office of the Science Advisor, U.S. Environmental Protection Agency, Washington, DC. March 2004 [online]. Available: http://www.epa.gov/osa/pdfs/ratf-final.pdf [accessed Feb. 9, 2007].

EPA (U.S. Environmental Protection Agency). 2005. Guidelines for Carcinogen Risk Assessment. EPA/630/P-03/001F. Risk Assessment Forum, U.S. Environmental Protection Agency, Washington, DC. March 2005 [online]. Available: http://cfpub.epa.gov/ncea/cfm/recordisplay.cfm?deid=116283 [accessed Feb. 7, 2007].

EPA SAB (U.S. Environmental Protection Agency Science Advisory Board). 1990. Risk Assessment: Setting Priorities and Strategies for Environmental Protection. EPA/SAB-EC-90-021. U.S. Environmental Protection Agency Science Advisory Board, Washington, DC.

Finkel, A.M. and D. Golding, eds. 1994. Worst Things First? The Debate Over Risk-Based National Environmental Priorities. Washington, DC: Resources for the Future Press.

GAO (U.S. General Accountability Office). 2008. Chemical Assessments: Low Productivity and New Interagency Review Process Limit the Usefulness and Credibility of EPA's Integrated Risk Information System. GAO-08-440. U.S. General Accountability Office, Washington, DC. March 2008 [online]. Available: http://www.gao.gov/new.items/d08440.pdf [accessed June 11, 2008].

Goldman, L.R. 2003. The Red Book: A reassessment of risk assessment. Hum. Ecol. Risk Assess. 9(5):1273-1281.

IARC (International Agency for Research on Cancer). 2006. IARC Monographs on the Evaluation of Carcinogenic Risks to Humans. Volume 88. Formaldehyde, 2-Butoxyehtanol and 1-*tert*-Butoxypropan-2-ol. Lyon: International Agency for Research on Cancer Press.

Kyle, A. 2007. Community Needs for Assessment of Environmental Problems. Presentation at the Fourth Meeting of Improving Risk Analysis Approaches Used by EPA, April 17, 2007, Washington, DC.

Montague, P. 2004. Reducing the harms associated with risk assessment. Environ. Impact. Asses. Rev. 24:733-748.

NRC (National Research Council). 1983. Risk Assessment in the Federal Government: Managing the Process. Washington, DC: National Academy Press.

NRC (National Research Council). 1993. Pesticides in the Diets of Infants and Children. Washington, DC: National Academy Press.

NRC (National Research Council). 1994. Science and Judgment in Risk Assessment. Washington, DC: National Academy Press.

NRC (National Research Council). 1996. Understanding Risk: Informing Decisions in a Democratic Society. Washington, DC: National Academy Press.

NRC (National Research Council). 2006. Assessing the Human Health Risks of Trichloroethylene: Key Scientific Issues. Washington, DC: National Academies Press.

NRC (National Research Council). 2007a. Scientific Review of the Proposed Risk Assessment Bulletin from the Office of Management and Budget. Washington, DC: National Academies Press.

NRC (National Research Council). 2007b. Toxicity Testing in the Twenty-First Century: A Vision and a Strategy. Washington, DC: National Academies Press.

OMB (Office of Management and Budget). 2004. Final Information Quality Bulletin for Peer Review. December 15, 2004 [online]. Available: http://cio.energy.gov/documents/OMB_Final_Info_Quality_Bulletin_for_peer_bulletin(2).pdf [accessed Jan. 4, 2007].

PCCRARM (Presidential/Congressional Commission on Risk Assessment and Risk Management). 1997. Framework for Environmental Health Management-Final Report, Vol. 1 [online]. Available: http://www.riskworld.com/nreports/1997/risk-rpt/pdf/EPAJAN.PDF [accessed Jan. 7, 2008].

Preuss, P. 2006. Human Health Risk Assessment at EPA: Background, Current Practice, Future Directions. Presentation at the First Meeting of Improving Risk Analysis Approaches Used By the U.S. EPA, November 20, 2006, Washington, DC.

Risk Policy Report. 2005. EPA Plan for Expanded Risk Reviews Draws Staff Criticism, DOD Backing. Inside EPA's Risk Policy Report 12(17):1, 8. May 3, 2005.

Risk Policy Report. 2007. GAO Inquiry Prepares to Focus on EPA Delay of New Risk Review Plan. Inside EPA's Risk Policy Report 14(38):1, 6. September 18, 2007.

Silbergeld, E.K. 1993. Risk assessment: The perspective and experience of U.S. environmentalists. Environ. Health Perspect. 101(2):100-104.

Suter, G.W. 2007. Ecological Risk Assessment, 2nd Ed. Boca Raton, FL: CRC Press.

2

Evolution and Use of Risk Assessment in the Environmental Protection Agency: Current Practice and Future Prospects

OVERVIEW

EPA risk-assessment concepts, principles, and practices are products of many diverse factors, and each agency program is based on a "unique mixture of statutes, precedents, and stakeholders" (EPA 2004a, p. 14). With respect to statutes, Congress established the basic plan through a series of environmental laws, most enacted during the 1970s and most authorizing science-based regulatory action to protect public health and the environment. Another factor is EPA's case-by-case experience with implementing these laws and the resulting supplementary principles and practices. Equally important, advisory bodies have drawn on the expertise of scientists and other environmental professionals in universities, private organizations, and other government agencies to recommend corrections and improvements. The net result is that risk assessment in EPA is a continually evolving process that has a stable common core but takes several forms.

This chapter traces the origins and evolution of risk assessment in EPA with an emphasis on *current* processes and procedures as a stepping-off point for the *future* improvements envisioned in later chapters. This chapter first describes the diverse statutory requirements that have led to a broad array of agency programs with correspondingly varied approaches to risk assessment; it then highlights current concepts and practices, outlines EPA's multifaceted institutional arrangements for managing the process, and identifies extramural influences. The record shows that EPA continually updates the process with new scientific information and policies, often in response to new laws or advice from advisory bodies as to general principles or individual assessments. Not all external recommendations necessarily warrant agency action, but it is clear that implementation of some recommendations has been incomplete. The chapter closes with process recommendations for implementing some of the substantive recommendations in the chapters that follow.

STATUTORY PLAN AND REGULATORY STRUCTURE

The environmental laws enacted by Congress shape EPA's regulatory structure, which, in turn, influence EPA risk-assessment practices and perspectives. The statutes give EPA authority to regulate many forms of pollution (for example, pesticides, solid wastes, and industrial chemicals) as they affect different aspects of the environment (for example, air quality, water quality, human health, and plant and animal wildlife). The premise central to EPA risk-assessment practices can be found in enabling legislation for its four major program offices: air and radiation, water, solid waste and emergency response, and prevention, pesticides, and toxic substances. Selected provisions appear below.

- The Clean Water Act calls for standards "adequate to protect public health and the environment from any reasonably anticipated adverse effects" (CWA § 405 (d)(2)(D)).
- The Clean Air Act, when addressing criteria pollutants, directs the agency to develop criteria "reflecting the latest scientific knowledge" and, on the basis of those criteria, to issue "national primary ambient air quality standards to . . . protect public health with an adequate margin of safety" (CAA §§ 108,109).
- The primary purpose of the Toxic Substances Control Act is "to assure [that technologic] innovation and commerce in such chemical substances and mixtures do not present an unreasonable risk of injury to health or the environment" (TSCA § 2 (b)(3)).
- Under the Federal Insecticide, Fungicide, and Rodenticide Act (FIFRA), one criterion for registering (licensing) a pesticide is that "it will perform its intended function without unreasonable adverse effects on human health and the environment" (FIFRA § 3).
- The Superfund National Contingency Plan specifies that "criteria and priorities [for responding to releases of hazardous substances] shall be based upon relative risk or danger to public health or welfare or the environment" (CERCLA § 105 (a)(8)(A)).

The term *risk assessment* does not appear often in the statutes, and it is important to note that these statues were enacted prior to the emergence of risk analysis as an integrative discipline in the late 1970s and early 1980s. Rather, EPA risk-assessment principles and practices stem from statutory provisions calling for information on "adverse effects" (EPA 2004a, p. 14), "relative risk" (p. 82), "unreasonable risk" (p. 14), and "the current scientific knowledge" (p. 104) and for regulatory decisions on protecting human health and the environment. The statutes provide various standards and procedures related to the scientific analyses used to evaluate the risk potential of pollutants subject to the statutes.[1,2]

---

[1]Different emphases and terminology lead to different risk-assessment approaches, sometimes for the same pollutant, in different agency programs. That can confuse and confound observers. For example, Clean Air Act provisions related to four air-pollution topics use different terms for what is essentially the same statutory finding:

- Clean Air Act provisions related to pollutants regulated as national ambient air quality standards are designed to "protect the public health with an adequate margin of safety" (CAA § 109, emphasis added).
- For welfare (environmental) effects, this provision directs the office to "protect the public welfare from any *known or anticipated adverse effects*" (CAA § 109, emphasis added).
- Standards for "hazardous" pollutants from stationary sources (for example, factories) are to "provide an *ample* margin of safety to protect public health or prevent an *adverse environmental* effect" (CAA § 112, emphasis added).
- Regarding mobile sources (for example, cars), the statute calls for ensuring that these vehicles do not "cause or contribute to an *unreasonable risk* to public health, welfare or safety" (CAA § 202 (a)(4), emphasis added).

[2]Some statutes call for technology-based standards that require, for example, specific control techniques or technology-forcing standards that specify emission limits to be achieved within given periods. Such standards are based on costs, engineering feasibility, and related technical considerations. Examples include Clean Air Act Sections 111 (new-source review) and 202 (mobile-source emissions).

The existence of several medium-oriented statutes explains why EPA has multiple risk-assessment programs. This circumstance often draws criticism as "stovepiping" that leads to delay and inconsistency in both risk assessment and regulation. In the early 1990s, Congress considered but did not pass legislation to incorporate common risk-assessment terminology, concepts, and requirements into comprehensive risk-assessment legislation.[3] Instead, recent enactments are notable for precise terms that amplify and clarify legislative objectives in individual statutes by specifying elements that assessments subject to particular statutes must include

- The 1996 Food Quality Protection Act specifies that "in the case of threshold effects . . . an additional ten-fold margin of safety for the pesticide chemical residues shall be applied for infants and children" (FFDCA § 408 (b)(2)(C)).
- 1996 amendments to the Safe Drinking Water Act are similarly explicit about the presentation of risk estimates and uncertainty: "The Administrator shall, in a document made available to the public in support of a regulation promulgated under this section, specify, to the extent practicable
  - Each population addressed by any estimate of public health effects
  - The expected risk or central estimate of risk for the specific populations
  - Each appropriate upper-bound or lower-bound estimate of risk" (SDWA § 300g-1 (b)(3)).

Provisions like those that apply to individual programs (the examples above appear in pesticide and water legislation, respectively) account for some of the variation in risk-assessment practices and results. However, although the new terms apply directly only to the program governed by the statute, other programs have adopted some of the changes.

Despite differences in statutory language, environmental media, and pollutants, several factors common to the major statutes continue to shape EPA's regulatory structure and function and its perspectives on risk assessment:

- The emphasis in each statute on protecting human health and the environment provides the basis of EPA's purported conservative approach to risk assessment. Examples range from generic "adequate margin of safety" language in the Clean Air Act (CAA) amendments of 1971 (§ 109) to the required additional safety factor of 10 for protection for infants and children in the 1996 Food Quality Protection Act (FQPA; FFDCA § 408 (b)(2)(C)). As explained recently, "consistent with its mission, EPA risk assessments tend towards protecting public and environmental health by preferring an approach that does not underestimate risk in the face of uncertainty and variability" (EPA 2004a, p. 11).
- Except as noted above (footnote 2) and later in this chapter (page 51), the statutory provisions related to EPA's main standards for protecting human health and the environment treat scientific analysis as a central element in regulatory decision-making and call for collection and evaluation of scientific information related to the pollutant undergoing regulatory review. Statutes often detail the kinds of information, analyses, and formal documentation required in the rule-making record.

[3]A bipartisan coalition of senators sponsored the Thompson-Levin bill (S981), titled "Regulatory Improvement Bill," which would have codified the Office of Management and Budget (OMB) role in review of agency regulations; some provisions later appeared in the OMB *Bulletin* (70 Fed. Reg. 2664 [2005]). The Moynihan bill (S123) called for comparative risk assessment.

- Although some sections of statutes focus solely on health-effect considerations,[4] many also identify information and analyses from other fields—such as economic analysis, technical feasibility, and societal impacts—for use in making regulatory decisions. "It is generally recognized—by the science community, by the regulatory community, and by the courts—that it is important to consider other factors along with the science when making decisions about risk management" (EPA 2004a, p. 3).

The resulting decisions—whether or not to regulate and, if so, the nature and form of regulation—seek to protect human health and the environment where appropriate, in part on the basis of scientific analysis and in part on the basis of consideration of information on costs, societal values, legal requirements, and other factors. As the proponent of any new regulation, EPA generally[5] has the burden of proving that the proposed regulation meets statutory standards. That is not a requirement for EPA to *prove* "cause and effect" in the customary scientific sense, but rather to demonstrate by way of science-based analysis that the proposed regulation meets statutory criteria related to adverse effects, unreasonable risks, and other statutory thresholds for regulation:

> Although regulatory agencies do not have the technical burden of proving that a particular company's products or activities have caused or will cause a particular person's disease, they do have the practical burden of assembling a record containing sufficient scientific information and analysis to survive a reviewing court's "hard look" review under the "substantial evidence" or "arbitrary and capricious" tests for judicial review of administrative action [McGarity 2004].

The environmental statutes administered by EPA and general administrative law require documentation and review of relevant data and analyses. Some statutory provisions for pesticides facilitate gathering data for risk assessment by enabling the agency to impose data requirements on producers and others (for example, FIFRA § 3); the agency's ability to impose data requirements has proved far more limited under the Toxic Substances Control Act (TSCA; GAO 2005) and other statutes.

As the primary scientific rationale for many EPA regulations, risk assessment is subject to scientific, political, and public controversy. Building on the statutory foundation, the 1983 Red Book introduced principles, terminology, and practices that have become mainstays of the process. That report, which provided for a common framework for reconciling, to some extent, the differing requirements of the statutes, led to changes in the 1980s and 1990s and continues to shape the process today.

THE PIVOTAL ROLE OF THE RED BOOK

The 1983 National Research Council Report

During the 1970s, the scientific assessment practices of EPA and other federal agencies faced with similar responsibilities—the Occupational Safety and Health Administration, the Food and Drug Administration (FDA), and the Consumer Products Safety Commis-

---

[4]Section 109 of the CAA of 1970 is the most often cited example; note, however, that the statute expressly provides for consideration of costs, feasibility, and other factors in state implementation plans (§ 110). Such considerations influence the time allowed for compliance with the standards.

[5]The situation differs for pesticides. The pesticide statute, FIFRA, requires manufacturers to submit data showing a "reasonable certainty of no harm" before pesticides can be registered and marketed and to maintain the registration.

sion—came under close scrutiny as decisions resulting from those practices took on greater social importance. In 1981, Congress (PL-96528) directed that FDA support a National Research Council study of the "merits of an institutional separation of the scientific functions of developing objective risk assessments from the regulatory process of making public and social policy decisions and the feasibility of unifying risk assessment functions." The National Research Council organized the Committee on the Institutional Means for Assessment of Risks to Public Health in October 1981, and the committee's report, the Red Book, was issued on March 1, 1983. In his letter transmitting the report to the commissioner of FDA, the chairman of the National Research Council, Frank Press, stated,

> The Congress made provision for this study to strengthen the reliability and objectivity of scientific assessment that forms the basis for federal regulatory policies applicable to carcinogens and other public health hazards. Federal agencies that perform risk assessments are often hard pressed to clearly and convincingly present the scientific basis for their regulatory decision. In the recent past, for example, decisions on saccharin, nitrites in food, formaldehyde use in home insulation, asbestos, air pollutants and a host of other substances have been called into question.
>
> The report recommends no radical changes in the organizational arrangements for performing risk assessments. Rather, the committee finds that the basic problem in risk assessment is the incompleteness of data, a problem not remedied by changing the organizational arrangement for performance of the assessments. Instead, the committee has suggested a course of action to improve the process within the practical constraints that exist.

As noted in Press's letter, the "course of action" recommended by the committee focused primarily on the *process* through which complex and uncertain, and often contradictory, scientific information derived from laboratory and other types of research could be made useful for regulatory and public-health decision-making. The committee was also sensitive to the concern, expressed in the congressional language, that scientific assessments should be "objective" and free of policy (and political) influences. Because all assessments of scientific data are subject to uncertainties and because scientific knowledge is incomplete, it is possible for different analysts to arrive at different interpretations of the same set of data. If the assessment involves risks to human health from chemical toxicity or other types of hazards, the differences in interpretation can be large. The committee therefore recognized that risk assessments could be easily manipulated to achieve some predetermined risk-management (policy) outcome. Much of the work of the committee was directed at finding ways to minimize that potential problem while avoiding the undesirable step of institutional separation of scientific assessment from decision-making.

The 1983 report was not directed at the technical analyses involved in risk assessment. Rather, it offered a coherent and generally applicable *framework* within which the process of risk assessment could be undertaken. That framework was shown to be necessary to fill the gap between the *research* setting within which general scientific knowledge and diverse types of information on specific threats to human health are developed and the various types of *risk-management* activities undertaken by regulatory and public-health agencies to minimize those threats. The committee's recommendations gave order to the developing field of risk assessment by defining terms and elucidating the four (now well-known) steps of the risk-assessment process. The committee chose the term *risk characterization* to describe the fourth and final step of the risk-assessment process, in which there is an integration and synthesis of the information and analysis contained in the first three steps (see Figure 2-1). The committee stated that the term *characterization* was chosen to convey the idea that both quantitative and qualitative elements of the risk analysis, and of the scientific uncertainties

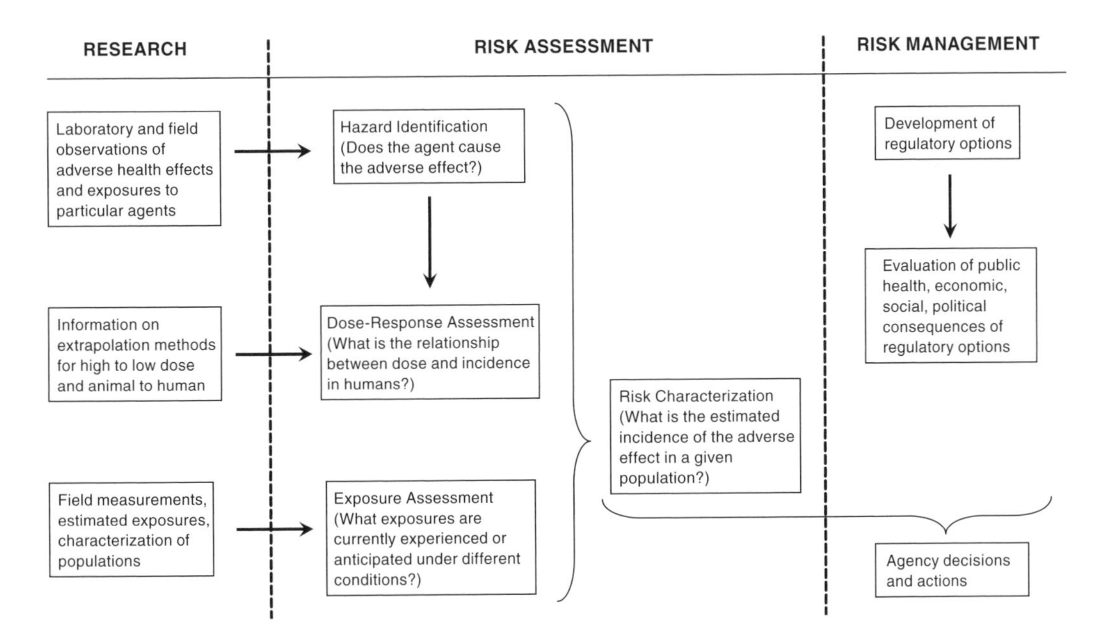

FIGURE 2-1 The National Research Council risk-assessment–risk-management paradigm. Source: NRC 1983.

in it, should be fully captured for the risk manager. Risks associated with chemical toxicity necessarily involve biologic data and uncertainties, many of which are not readily expressed in quantitative terms. Again, it was beyond the charge of the committee to offer specific technical guidance on the modes of scientific analysis appropriate for each of the steps of risk assessment.

The first recommendation of the Red Book is the following (NRC 1983, p. 7):

> We recommend that regulatory agencies take steps to establish and maintain a clear conceptual distinction between assessment of risks and consideration of risk management alternatives; that is, the scientific findings and policy judgments embodied in risk assessments should be explicitly distinguished from the political, economic, and technical considerations that influence the design and choice of regulatory strategies.

Two aspects of that critical recommendation are especially noteworthy. First, the committee emphasized that the distinction between risk assessment and risk management is a *conceptual* one; that is, it concerns the fact that the content and goals of the two activities are distinguishable on a conceptual level. The Red Book nowhere calls for any other type of "separation" of the two activities.

Second, the phrase "policy judgments embodied in risk assessment" (which are said to be different in kind from those involved in risk management) points to one of the most important insights of the committee. In particular, the committee recognized that almost no risk assessment can be completed unless scientific information (data and knowledge) is supplemented with assumptions that have not been documented in relation to the particular risk assessment at hand, although they have probably been supported by substantial evidence or theory for the general case.[6] The clearest examples of such assumptions related to risks posed by chemical toxicity concern the shape of dose-response curves in the region of very low doses and the relevance to humans of various toxicity responses observed in high-dose animal experiments; assumptions regarding these and many other aspects of the data used for risk assessment are necessary to provide risk managers useful risk characterizations based on consistent approaches.

The Red Book committee recognized that for a given analytic component of any of the steps of a risk assessment for which an assumption is necessary, several scientifically plausible assumptions might be available. The committee used the phrase "inference options" to describe the array of possibilities. To bring order and consistency to risk assessments conducted by the federal government and to minimize case-by-case manipulations of risk-assessment outcomes, the committee recommended the development of specific "inference guidelines"; these were to contain "an explicit statement of a predetermined choice among alternative inference options" (NRC 1983, p. 4) (see Box 2-1). Thus, agencies should take steps to describe, in explicit guidelines, the technical approaches used to conduct risk assessments, and these guidelines should include specification of the assumptions (including, in some cases, models) that would be consistently used to draw inferences in all the analytic components of the risk-assessment process where they are needed. Inference options have come to be called default options, and the inferences selected for risk assessments have come to be called defaults. The development and consistent use of technical guidelines for risk assessment, with the specification of all the necessary defaults, were seen by the Red Book committee as necessary to avoid the institutional separation of scientific assessment from policy development and implementation while minimizing inappropriate and sometimes invisible policy influences on the risk-assessment process.

As noted later in this chapter, some critics of the Red Book have raised the concern that the committee's commendable effort to avoid "inappropriate influences" can readily be taken to mean "no influence" from risk managers and other stakeholders.

One additional feature of the Red Book's recommendations bears on the current committee's task. Thus, as part of the statement of Recommendation 6, which concerns the criteria for useful risk-assessment guidelines, can be found the following (NRC 1983, p. 165):

> *Flexibility*
>
> The committee espouses flexible guidelines. Rigid guidelines, which permit no variation, might preclude the consideration of relevant scientific information peculiar to a particular chemical and thus force assessors to use inference options that are not appropriate in a given case. Also, rigid guidelines might mandate the continued use of concepts that become obsolete with new scientific developments. Large segments of the scientific community would undoubtedly object to such guidelines as incompatible with the use of the best scientific judgment for policy decisions.
>
> Flexibility can be introduced by the incorporation of default options. The assessor would be instructed to use a designated (default) option unless specific scientific evidence suggested otherwise. The guidelines would thus permit exceptions to the general case, as long as each exception could be justified scientifically. Such justifications would be reviewed by the sci-

[6] No scientific knowledge is without uncertainty, but it is generally subject to empirical verification; when the empirical evidence is supportive and no contrary evidence can be found, documentation is said to have been established, at least tentatively. The assumptions needed to complete risk assessments are generally well supported for the relevant set of past assessments; however, in any specific case it will often be difficult, if not impossible, to verify empirically that a given assumption also holds for the substance at issue.

BOX 2-1 Agencywide[a] Risk-Assessment Guidelines

1986	Guidelines for Carcinogen Risk Assessment (EPA 1986a)
	Guidelines for Health Assessment of Suspect Developmental Toxicants (51 Fed. Reg. 34028 [1986])
	Guidelines for Mutagenicity Risk Assessment (EPA 1986b)
	Guidelines for Estimating Exposures (51 Fed. Reg. 34042 [1986])
	Guidelines for Health Assessment of Chemical Mixtures (EPA 1986c)
1991	Developmental Toxicity Risk Assessment (revised and updated) (EPA 1991)
1992	Guidelines for Exposure Assessment (EPA 1992a)
1996	Guidelines for Reproductive Toxicity Risk Assessment (EPA 1996a)
1998	Guidelines for Ecological Risk Assessment (EPA 1998a)
	Guidelines for Neurotoxicity Risk Assessment (EPA 1998b)
2000	Supplementary Guidance for Health Risk Assessment of Chemical Mixtures (EPA 2000a)
2005	Guidelines for Carcinogen Risk Assessment and Supplemental Guidance for Assessing Susceptibility from Early-Life Exposure to Carcinogens (EPA 2005a,b)

These guidelines, which are consistent with Red Book recommendations (NRC 1983, p. 7), "structure the interpretation of scientific and technical information relevant to the assessment" and "address all elements of risk assessment, but allow flexibility to consider unique scientific evidence in particular instances."

Each guideline is a multiyear project developed by multioffice teams composed of scientists in EPA laboratories, centers, program offices, and regional offices. Draft guidelines are peer-reviewed in open public meetings and published for comment in the *Federal Register*. In general, each guideline follows the 1983 Red Book paradigm, providing guidance on the use and interpretation of information in each field of analysis, including the role of defaults and assumptions and approaches to uncertainties and risk characterization. Some guidelines are accompanied by supplementary reports on special topics, for example, "Assessing Susceptibility from Early-life Exposure to Carcinogens" (EPA 2005b) and "Guiding Principles for Monte Carlo Analysis" (EPA 1997a).

[a]EPA's guideline library includes many other guidance documents and policies, including those specific to individual programs (see, for example, Tables C-1 and D-1 and references).

entific review panels and by the public under procedures described above. Guidelines could profitably highlight subjects undergoing relatively rapid scientific development (for example, the use of metabolic data for interspecies comparisons) and any other components in which exceptions to particular default options were likely to arise. They should also attempt to present criteria for evaluating whether an exception is justified.

As will be evident throughout this report, it has proved difficult to achieve scientific consensus on judgments regarding the adequacy of scientific evidence to justify, in specific cases, departures from one or more defaults.

One of the objectives of the present committee's work might be seen as determining whether 25 years of scientific research and of scholarly thinking about the conduct of risk assessments provides new insights into whether there might be better ways of approaching the uncertainties that give rise to the need for defaults.

Later National Research Council Studies

NRC (1993a) advocated the integration of ecological risk assessment into the 1983 Red Book framework. The framework for risk assessment and its four-step analytic process were adopted and promoted in the National Research Council's *Science and Judgment in Risk Assessment* (NRC 1994) and *Understanding Risk: Informing Decisions in a Democratic Society* (NRC 1996). Indeed, the framework has been widely adopted in other expert studies of risk assessment (see PCCRARM 1997 and references cited therein) and has been adopted outside the United States (in the European Union and the World Health Organization) (see Figure 2-2). Moreover, as regulatory and public-health institutions have had to bring a greater degree of scientific analysis and consistency to health threats posed by microbial pathogens (Parkin 2007), excessive nutrient intakes (IOM 1997, 1998, 2003; WHO 2006), and other environmental stressors, they have found the Red Book framework both scientifically appropriate and useful.

One additional theme regarding the risk-assessment process is given great attention by the National Research Council in *Understanding Risk* (NRC 1996, p. 6):

> The analytic-deliberative process leading to a risk characterization should include early and explicit attention to *problem formulation*; representation of the spectrum of interested and

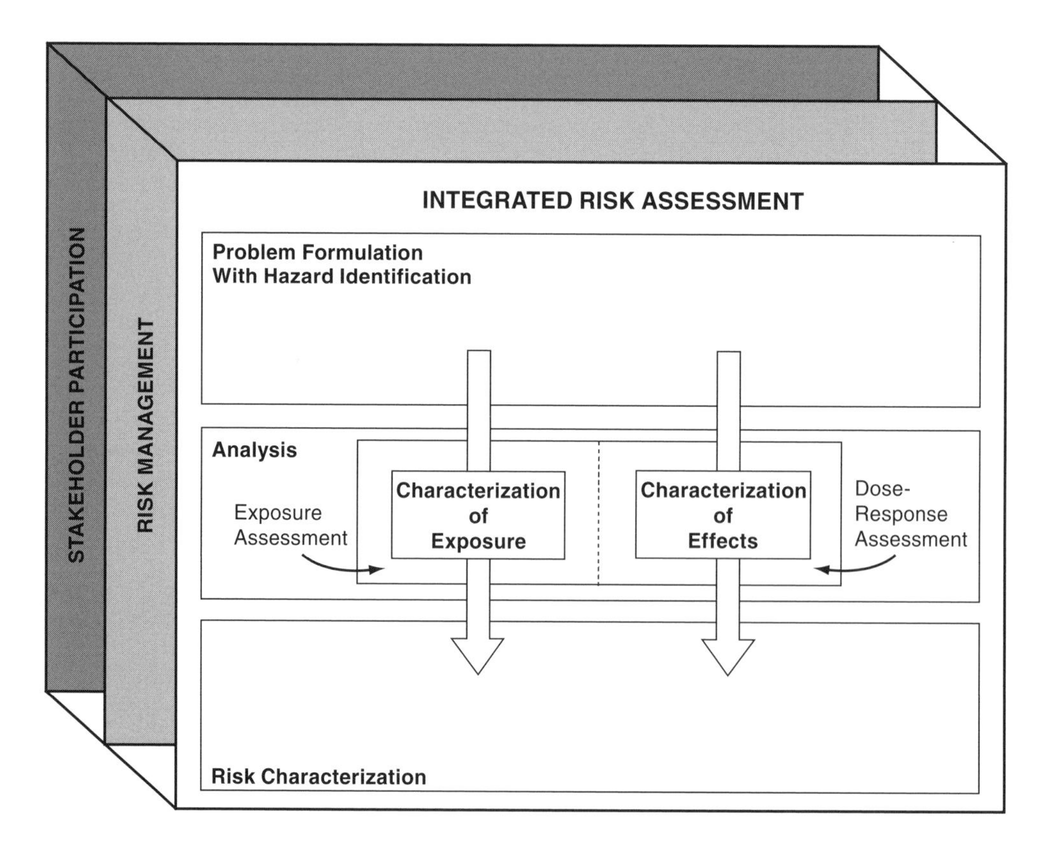

FIGURE 2-2 The World Health Organization's framework for integrated health and ecologic risk assessment. NOTE: Figures 2-1 and 2-2 show different renditions and evolving emphases as to the basic elements of the Red Book paradigm. Source: Suter et al. 2001.

affected parties at this early stage is imperative. The analytic-deliberative process should be *mutual and recursive*. Analysis and deliberation are complementary and must be integrated throughout the process leading to risk characterization: deliberation frames analysis, analysis informs deliberation, and the process benefits from feedback between the two.

That recommendation provides nuance to the Red Book's call for "separation" of assessment and management to facilitate the supreme goal of risk assessment: to provide the scientific basis for public-health and regulatory decisions. As long as "analysis and deliberation" does not involve efforts by risk managers to shape risk-assessment outcomes to match their policy preferences, but rather involves efforts to ensure that assessments (whatever their outcomes) will be adequate for decision-making, interactive processes involving "the spectrum of interested and affected parties" are seen as imperative.

The 1994 National Research Council report *Science and Judgment in Risk Assessment* evaluated EPA's risk-assessment practices as they apply to hazardous air pollutants from sources subject to Section 112 of the CAA amendments of 1990. That report did not alter the principles for risk assessment set forth by the Red Book but rather examined EPA guidelines and practices and then recommended ways in which various technical improvements in the conduct of risk assessments and in the presentation of risk characterizations might be accomplished. Thus, the present committee's efforts resemble in many ways those undertaken by the *Science and Judgment* committee.

The issue of default options was given much consideration (see Box 2-2). Indeed, the 1994 National Research Council committee found EPA's existing technical guidelines for risk assessment to be deficient with respect to their justifications for defaults and with respect to evidentiary standards and scientific criteria to be met for case-specific departures from them.[7] The committee offered a long series of recommendations, each preceded by a discussion of the state of technical understanding, on issues of data needs for risk assessment, uncertainty, variability, aggregation of exposures and risk, and model development.

The 1994 committee's recommendations extended beyond the technical content of risk assessment and included issues of process, institutional arrangements, and even problems of risk communication. Although there was much focus on air-pollutant risks, particularly the technical issues related to exposure assessment, most of that committee's recommendations had broad applicability to risk assessment.

In Appendix D to the present report, the committee has selected representative recommendations contained in the three National Research Council reports cited above and attempted to provide a view of how EPA has responded to many of them. It can be seen that EPA has devoted considerable effort to ensuring that its guidelines conform to many National Research Council recommendations, although the record on accepting and implementing recommendations is uneven and incomplete (see, for example, Boxes 2-4 and 2-5 and Chapter 6).

The present committee has been asked to review current EPA "concepts and practices," taking into account the previous National Research Council studies and studies in which new scientific approaches are being evaluated. The present committee is not specifically charged with modifying the fundamental concepts first elucidated in the Red Book unless the scientific understanding on environmental hazards and the research on the conduct of risk assessment that have developed over the past 25 years demand such a modification. Thus, as

[7]Appendix N to the 1994 report contains two views of the issue of defaults, one of committee member Adam Finkel and one of members Roger McClellan and D. Warner North; their papers represent a range of committee perspectives on the appropriate balance of science and policy considerations in a system for departure from default assumptions.

BOX 2-2 Science Policy and Defaults

Science and Judgment (NRC 1994) describes defaults as the "science policy components of risk assessment" (p. 40) and points out that "if the choice of inference options is not governed by guidelines, the written assessment itself should make explicit the assumptions used to interpret data or support conclusions reached in the absence of data" (p. 15). The report recognizes "choice" as an aspect of science policy (p. 27):

> The [1983 Red Book] committee pointed out that selection of a particular approach under such circumstances involves what it called a science-policy choice. Science-policy choices are distinct from the policy choices associated with ultimate decision-making. . . . The science-policy choices that regulatory agencies make in carrying out risk assessments have considerable influence on the results.

Those principles are the basis of EPA's call for "transparency," "full disclosure," and "scientific conclusions identified separately from default assumptions and policy calls" in the *Risk Characterization Handbook* (EPA 2000b). EPA's recent *Staff Paper* (EPA 2004a, p. 12) embraces and expands on the principles: "Science policy positions and choices are by necessity utilized during the risk assessment process."

The Superfund program's supplemental guidance document *Standard Default Exposure Factors* was developed in response to requests to make Superfund assessments more transparent and their assumptions more consistent. The guidance states that defaults are used when "there is a lack of site-specific data or consensus on which parameters to choose, given a range of possibilities" (EPA 2004a, p. 105).

as the committee undertook its technical evaluations, it remained sensitive to the question of whether the Red Book's framework for risk assessment and its conceptual underpinnings are adequate to meet the challenges of understanding and managing the array of environmental threats to health and the environment that we are expected to face in the foreseeable future. These considerations have also shaped other approaches to thinking about risk assessment including PCCRARM (1997) and a recent publication by Krewski et al. (2007).

CURRENT CONCEPTS AND PRACTICES

EPA's statement of task for this committee (Appendix B) seeks a "scientific and technical review of EPA's current risk analysis concepts and practices." In addition, EPA invites the committee to develop "recommendations for improving" EPA's risk-analysis approaches, "taking into consideration past evaluations." At the outset, the committee approached its task in part by reviewing major National Research Council reports published since 1983. It also examined EPA risk-assessment activities in light of themes and trends in those reports. The discussion that follows highlights EPA's progress in many spheres and shortfalls and committee uncertainty about the nature and extent of progress.

The National Research Council reports and EPA documents arrayed in the timeline diagram in Figure 2-3 and the timeline table in Appendix C are the primary sources for this analysis. The implementation table in Appendix D isolates and highlights National Research Council recommendations on selected risk-assessment topics with relevant EPA responses as documented in a recent EPA *Staff Paper* (EPA 2004a), guideline documents, and other EPA sources; it also draws on a Government Accountability Office (GAO) study requested by Congress (GAO 2005).

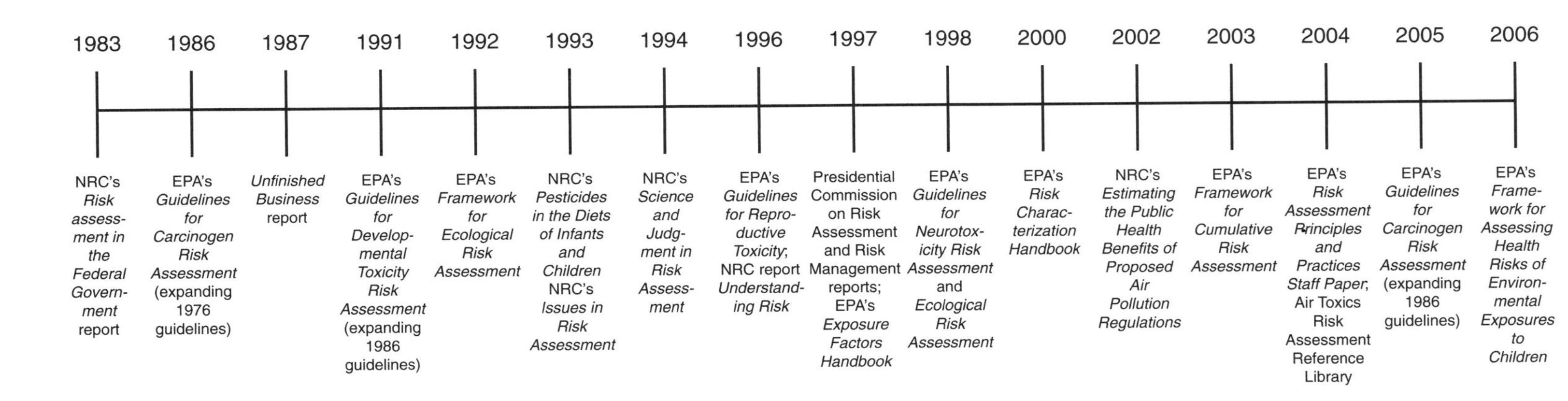

FIGURE 2-3 Timeline of major documentary milestones. Documents arrayed here represent major risk assessment reports presented in Tables C-1 and D-1 (see Appendixes C and D). Sources: NRC 1983, 1993a,b, 1994, 1996, 2002; EPA 1986a, 1987a, 1991, 1992b, 1996a, 1997c, 1998a,b, 2000b, 2003d, 2004a,b, 2005a, 2006; PCCRARM 1997.

Environmental Protection Agency Progress in Implementing National Research Council Recommendations

In general, as shown in Table D-1, National Research Council committees have recommended improvements related to a broad array of risk-assessment issues and activities. Most recommendations provide technical advice on scientific topics, such as cumulative risk, toxicity assessment, mode of action, and uncertainty analysis; but others address associated matters, such as peer review, guideline development, and principles like the conceptual distinction between risk assessment and risk management. EPA responses to the recommendations take several forms, including internal guidance memoranda and formal guidelines, handbooks and manuals, new programs, and standing committees to study identified risk-assessment topics.

Table D-1 shows that some recommendations have prompted complementary activities in various agency offices. For instance, the agency has both generic and program-specific guidance related to cumulative risk and aggregate exposure (Table D-1).[8] Agencywide guidance issued under the auspices of the Science Policy Council and Risk Assessment Forum includes a 1997 guidance memorandum and supplemental guidelines for chemical mixtures. Individual offices have undertaken separate projects to meet office-specific needs. Examples include, for the Office of Air and Radiation, the Integrated Air Toxics Strategy (64 Fed. Reg. 38705 [1991]), the TRIM model (EPA 2007a), and the Multiple Pathways of Exposure Model (EPA 2004b); for the Office of Pesticide Programs (OPP), the report *Guidance on Cumulative Risk Assessment of Pesticide Chemicals That Have a Common Mechanism of Toxicity* (EPA 2002a); and for the Office of Research and Development (ORD), the cumulative-risk components of the Human Health Research Strategy (EPA 2003a).

Table D-1 shows a long-standing emphasis on "risk characterization" in both National Research Council recommendations and EPA guidance memoranda, formal guidelines, and other documents (see Box 2-3). The 1994 National Research Council committee described risk characterization as involving integration of information developed in the hazard-identification, dose-response, and exposure analyses and "a full discussion of uncertainties associated with the estimates of risk" (NRC 1994, p. 27). The agency's risk-characterization guidance, including a handbook (EPA 2000b) devoted to the topic, was consistent with that recommendation in emphasizing "transparency" and "clarity" in explaining risk-assessment approaches and results, especially specifying strength and weaknesses of data and methods and identifying related uncertainties.

Citing 1994 National Research Council recommendations for greater attention to the use of defaults, EPA applies this general risk-characterization guidance to the specific subject of defaults in the proposed (EPA 1996b)[9] and final (EPA 2005a) cancer guidelines (Table D-1). Those documents articulate the scientific basis of five major defaults used in cancer risk assessment in the absence of scientific data. The *Staff Paper* (EPA 2004a) explains that the agency "invokes defaults only after the data are determined to be not usable at that point in the assessment" (EPA 2004a, p. 51), emphasizing that this is a "different approach from

[8]National Research Council recommendations are not by themselves responsible for EPA activities on topics covered by them. EPA's Science Advisory Board (SAB) and the International Life Sciences Institute (ILSI) Risk Science Institute have also provided recommendations on these issues. The burst of activity on cumulative risk and aggregate exposure, for example, reflects a confluence of such factors as new statutory requirements in the 1996 FQPA and advances in the state of the science.

[9]The 1996 proposal cited here and elsewhere (for example, Table D-1) represents an intermediate step in the evolution of EPA cancer principles from 1986 to 2005; also, although the guidelines were not completed for almost 10 years, the 1996 proposal documented contemporaneous EPA work on the 1994 National Research Council recommendations related to cancer risk assessment.

BOX 2-3 Agency Guidance on Risk Characterization: Attention to Uncertainty

A 1992 guidance memorandum reinforces principles enunciated in the 1983 Red Book and in EPA's 1986 risk-assessment guidelines and was a forerunner of later guidance documents.

> Highly reliable data are available for many aspects of an assessment. However, scientific uncertainty is a fact of life for the risk assessment process as a whole. . . . Scientists call for fully characterizing risk not to question the validity of the assessment, but to fully inform others about critical information in the assessment. . . . Even though risk characterization details limitations in an assessment, a balanced discussion of reliable conclusions and related uncertainties enhances, rather than detracts, from the overall credibility of each assessment [Reprinted in NRC 1994, Appendix B, pp. 352-353].

The *Risk Characterization Handbook* (EPA 2000b) instructs risk assessors to, among other things, "carry forward the key information from hazard identification, dose-response, and exposure assessment, using a combination of qualitative information, quantitative information, and information about uncertainties" (p. 24) and "describe the uncertainties inherent in the risk assessment and the default positions used to address these uncertainties or gaps in the assessment" (p. 21).

After highlighting the emphasis on "transparency" in EPA's 1995 risk-characterization policy (EPA 1995), the *Staff Paper* (EPA 2004a) notes that "one of the major comments on EPA risk assessment practices is that they do not characterize uncertainty and variability transparently enough" (p. 33). The statement of task for EPA (2004a) confirms that "this is an issue EPA is attempting to address" (p. 33). (See Box 2-4 for related peer-review commentary on one assessment.)

choosing defaults first and then using data to depart from them" (EPA 2004a, p. 51), as in the past. The committee found this framing of defaults problematic, as discussed at length in Chapter 6.

Table D-1 was instructive for the present committee's review of EPA risk-assessment concepts and practices called for in the statement of task. For example, GAO's survey reports broad-based approval in EPA of the program for developing risk-assessment guidelines in line with 1983 Red Book recommendations for inference guidelines (Table D-1). As new methods emerged, the agency revised and updated several of the original 1986 guidelines (on cancer, developmental toxicity, mixtures, and exposure assessment). The addition of new topics to the guideline library, such as neurotoxicity in 1998 and ecologic risk assessment in 1998, suggests that adding other new topics may be a useful way to implement recommendations in the present report.

EPA's response to recommendations from its Scientific Advisory Board (SAB) and the National Research Council for an enlarged peer-review program offers another model for the future. EPA's 1992 and 1994 peer-review policy memorandums (EPA 1992c, 1994) expanded peer review beyond statutory mandates[10] to "major scientifically and technically based work products related to Agency decisions" (EPA 2000b; Table D-1). The general objective of both the National Research Council recommendations and EPA's new policy was to add scientific expertise to the overall risk-assessment process. The expanded policy was intended to move assessments not then subject to peer review into the ambit of peer review. The calls for more peer review, like the call for more stakeholder participation, demonstrate concern about both the increasing complexity of risk assessment and the credibility of EPA assessments. However, EPA (2000b) acknowledges the need for upfront planning of the peer

[10]Section 109 of the CAA requires peer review of the criteria documents setting forth the scientific analyses underlying national ambient air quality standards; Section 6 of FIFRA requires peer review of identified pesticide actions. See also 70 Fed. Reg. 2664 [2005] (federal peer-review guidelines).

review to ensure it provides the appropriate insight and direction to the risk assessment. In that regard, the new framework proposed in this report may well require a different kind of peer review in which experience and expertise on decision theory, social sciences, and risk management may be required along with scientific expertise.

The enormous variety and scope of EPA risk-assessment responsibilities and activities preclude a detailed and full assessment by the present committee of risk-assessment practices in all parts of the agency. For example, the GAO survey (see GAO [2006], Table D-1) implies extensive use of the guidelines by agency risk assessors but does not provide information on the extent to which individual risk assessments (assessments of particular hazardous air pollutants, pesticides, or Superfund sites) follow some of or all of the principles enunciated in the guidelines. Similarly, even with the strong emphasis on identifying uncertainties, explaining defaults, and justifying science-policy choices as critical features of risk characterization in EPA guidance documents (see Table D-1 and Box 2-3), peer reviewers and other commenters recommend greater clarity and transparency in characterizing variability, uncertainty, and risk (GAO 2006; see Box 2-4 for one example). Those concerns raise questions about the extent to which guidance on risk characterization is fully used in practice, whether the guidance is adequate, and how to guide characterization during periods when science, practice, and expectations are evolving.

The 1994 National Research Council report called for explanation of the scientific basis of default options and identification of "criteria" for departure from defaults. *Guidelines for Carcinogen Risk Assessment* (EPA 2005a) includes as an appendix an extended discussion

BOX 2-4 Commentary on Risk Characterization for the Dioxin Reassessment

In a recent report (NRC 2006) on EPA's dioxin reassessment (EPA 2003b), the peer-review panel complimented some features of EPA's approach to scientific uncertainties in the assessment and then recommended that the agency "substantially revise the risk characterization section of Part III of the Reassessment to include a more comprehensive risk characterization and discussion of the uncertainties surrounding key assumptions and variables" (NRC 2006, p. 25).

For more than 20 years, EPA guidance documents have stressed displaying "all relevant information pertaining to the decision at hand" (EPA 1984, p. 14), fully informing others about "critical information from each stage of a risk assessment" (EPA 2000b, p. A-2), and the importance of transparency and "describing uncertainties inherent in risk assessment and default positions" (p. 21), among other things. See Box 2-3 and Table D-1 (section on risk characterization) for fuller statements and references. In view of this long-standing internal guidance emphasizing complete and transparent characterization in agency risk assessments, the need for "substantial improvement" in EPA's description of the scientific basis for *key elements* in this important assessment suggests inattention to principles enunciated in EPA guidance (NRC 2006, p. 9; emphasis in original):

> The Committee identified three areas that require substantial improvement in describing the scientific basis for EPA's dioxin risk assessment to support a scientifically robust risk characterization:
>
> - Justification of approaches to *dose-response modeling* for cancer and noncancer end points.
> - Transparency and clarity in *selection of key data sets* for analysis.
> - Transparency, thoroughness, and clarity in *quantitative uncertainty analysis.*

The calls for improved risk characterization in dioxin risk assessment by NRC (2006) illustrate the need for greater clarity and transparency that are often voiced in reviews of EPA risk assessments. Consistent with the statement of task, this report develops information and approaches for addressing these issues.

of the scientific basis of defaults and alternatives but does not provide criteria for invoking defaults (see Chapter 6).

Chapter 6 of the present report analyzes EPA implementation of selected recommendations regarding defaults in greater depth than in Table D-1. For example, Table 6-3 characterizes some EPA practices as implicit or "missing" defaults. As also shown in Table D-1, National Research Council committees have made various recommendations related to uncertainty analysis. However, as noted in Chapter 4, uncertainty analysis and characterization pose difficult technical issues, and in general related best practices have not been established. In the absence of guidelines on the appropriate degrees of detail, rigor, and sophistication needed in an uncertainty analysis for a given risk assessment, it is not surprising that expert advisory committees recommend technical improvements in this regard.[11] (See Box 2-5 on importance of implementation of guidelines.)

EPA and GAO comments on the Integrated Risk Information System (IRIS) (see Table D-1) may be instructive as to the outlook for the present committee's recommendations to the agency. The GAO report details numerous improvements in the IRIS process over the past 10 years. It also indicates that in 2005 EPA completed only eight IRIS reviews, falling "considerably short" of the recommended (and highly optimistic) goal of 50 each year (GAO 2006). GAO states that agency officials explained the shortfall in terms of such factors as risk-assessment complexity, resource limitations, and peer-review requirements.[12] Those factors will also be at play as the agency applies recommendations in this report to the current IRIS backlog and to new risk assessments for individual chemicals or sites.[13] Similarly, in reporting that 90% of its 2002 scientific and technical work products were peer-reviewed (Gilman 2003; Table D-1), the agency also tracks how the peer-review comments were addressed (EPA 2000c). In sum, Table D-1 identifies both EPA guidance responding to National Research Council recommendations and an impressive set of practices undertaken to improve agency risk assessments. However, the breadth and scope of EPA's risk-assessment agenda limit the table to a selected subset of current concepts and practices. Although the record demonstrates the extent to which National Research Council recommendations have been implemented on paper through guidelines and other guidance statements, the committee does not have detailed information on the extent to which the guidelines have been fully and effectively incorporated in practice. As EPA explained to GAO (in relation to IRIS), many factors could lead to partial implementation, including data availability, staff expertise and experience, resource constraints, adequate peer review, and the impact of statutory deadlines and legal frameworks on the risk-assessment process.

The Role of Policy

Each stage in the risk-assessment process calls for a series of choices, each with the potential to influence, and in some cases determine, the outcome of the risk assessment. As

[11]For example, one recent review "finds that EPA guidance concerning specific use of the Integrated Exposure Uptake Biokinetic (IEUBK) model and additional use of blood lead studies is incomplete. . . . The Office of Solid Waste and Emergency Response (OSWER) directive fails . . . to give adequate guidance about what to do when [data] and IEUBK model results disagree by a substantial margin" (NRC 2005a, p. 273).

[12]One published paper reports that in 2006 EPA added only two assessments to the IRIS database (Mills 2006).

[13]The impact of these factors on the high-profile IRIS program, which is based in the scientist-rich ORD, raises questions about the capacity of the agency as a whole, where many risk assessors have less experience than those in ORD, to expand its risk-assessment activities in line with recommendations set forth in this report. See Chapter 9.

BOX 2-5 Guideline Implementation and Risk-Assessment Impacts

As shown in Box 2-1, EPA's library of risk assessment guidelines covers a broad array of topics. In 1994, the NRC committee concluded that "the guidelines were generally consistent with the Red Book recommendations. . . . They include default options which are essentially policy judgments of how to accommodate uncertainties. They include various assumptions that are needed for assessing exposure and risk" (NRC 1994, p. 5).

Despite conformity with the Red Book, approval of peer reviewers (see peer-review history at the front of individual guidelines), and staff appreciation of the guideline documents (GAO 2006), concerns identified in EPA's *Staff Paper* (EPA 2004a) and the GAO report (GAO 2005) regarding EPA risk assessments (for example, overconservativism and underconservatism in risk estimates, use or nonuse of defaults, incomplete discussion of uncertainty, and delays in completing assessments) prompt questions about the extent to which the guidelines fulfill their intended function in individual assessments. That is, to what extent are problems associated with EPA risk assessments traceable to guideline content or use?

One question is related to the scientific adequacy and general utility of the guidelines themselves as a resource for assessors and managers; that is, do they provide information needed in a usable form? A second question is related to risk assessors' use or nonuse of the guidelines in any particular case; that is, do assessors and managers have the technical experience, scientific data, funding, and time to use the guidelines as intended? (See Box 2-4 for an example of incomplete attention to existing guidance.) Factors contributing to ineffective guidelines or guideline use may include

- Nonavailability of relevant data, risk-assessment methodology (for example, established defaults), or both.
- Complexity, lack of clarity, or infeasibility in the recommendations by the National Research Council and other bodies that advise the agency.
- Complexity, lack of clarity, or infeasibility in the related EPA guidelines.
- Optional vs mandatory wording in the guidelines.
- Individual or ad hoc policy overriding guideline policy.
- Lack of experience on the part of risk assessors.
- Management issues, such as lack of experience or oversight on the part of supervisors and decision-makers.

In view of EPA's pattern of developing guidelines to address previous National Research Council recommendations (Table D-1), understanding of factors that influence effective use of the guidelines by assessors and managers could be critical for effective implementation of recommendations in the present report.

developed more fully in Chapters 4-7, the data gaps and uncertainties inherent in the process generate the need for defaults and assumptions; in addition, alternative approaches to each assumption introduce the element of choice (NRC 1994, p. 27):

> Risk assessors might be faced with several scientifically plausible approaches (for example, choosing the most reliable dose-response model for extrapolation beyond the range of observable effects) with no definitive basis for distinguishing among them. The [Red Book] committee pointed out that selection of a particular approach under such circumstances involves what it called a science-policy choice. Science-policy choices are distinct from the policy choices associated with ultimate decision-making. . . . The science-policy choices that regulatory agencies make in carrying out risk assessments have considerable influence on the results.

However, it is critical that science-policy choices underlying risk-assessment guidelines be based on the need for consistency, reproducibility, and fairness.

Some choices are normal aspects of scientific endeavors, whether part of a regulatory process or not. For example, each stage of the risk-assessment process involves an initial survey of the scientific literature and relevant databases to identify and isolate studies pertinent to the pollutant or situation under review. The array includes information from many sources: reports in peer-reviewed journals, reports in the gray literature, personal communications about recent results not yet published, and the like. Some studies have been replicated or otherwise substantiated; others may have a questionable provenance. Judgments on those issues parallel judgments made in developing any scientific analysis. Continuing analysis involves reviewing each study for fundamental strengths and weaknesses, for example, quality-assurance issues, replicability, consistency with comparable studies, and peer-review status.

Other considerations are specific to the regulatory process. They include the relevance of any particular piece of evidence in the decision context (see Chapters 3 and 8), information submitted by stakeholders and other interested parties, applicability of relevant agency policies and guidelines, and factors that might compromise use of data for standard-setting purposes (for example, the presence of potential conflicts of interest in generating or censoring data).

It is easy to narrow the options by eliminating nonconforming studies. However, more than one study may meet basic scientific standards, and studies vary with respect to quality attributes. Benchmark dose (BMD) calculations for perchlorate offer an example, as described in a recent National Research Council peer-review report (NRC 2005b, p. 170):

> As part of its deliberations on the point of departure, the committee reviewed the BMD analyses conducted by EPA (2003c), the California Environmental Protection Agency (CalEPA 2004), and Crump and Goodman (2003) on the data from Greer et al. (2002). Overall these analyses used different models, approaches, parameters, response levels, and input data, so comparison of the results of the analyses is difficult.

The task, then, was to identify the "critical" study or studies for use in continuing the risk assessment (see, for example, EPA 2002b, 2004a, 2005a,b), which may involve choosing among or combining varied results from different scientifically adequate studies. When different scientists make different judgments—that is, different choices—among the alternative studies, related risk-assessment results may differ substantially (Box 2-6).

In addition to choosing one set of "hard" data over another where necessary, risk assessors identify uncertainties and unknowns at each stage in the process. In the hazard-identification stage, questions about the applicability to humans of findings in specific animal studies lead to uncertainty in the animal-to-human extrapolation, an assumption that data in those studies are predictive of adverse effects in humans under particular conditions of exposure. When relevant data are unavailable, other uncertainties lead to questions on other matters, such as the relevance of effects observed in studies on males to females, adults to children, and "healthy" workers to the general population. Similar uncertainties are important in all types of risk assessments.

The dose-response analysis almost invariably raises questions about the likelihood that effects observed at the generally higher doses used in animal studies (or under conditions of workplace exposures) would be observed at the generally lower doses expected in connection with environmental exposures. As shown for perchlorate, the number of choice points and the options at each point open the door to different reference dose (RfD) values, depending on the combination of choices made:

BOX 2-6 Choices and a Reference Dose Value for Perchlorate

In 2002, EPA issued a draft reference dose (RfD) for perchlorate, a contaminant found in public drinking-water supplies for more than 11 million people. After peer-review challenges to the scientific basis of EPA's proposed RfD, the National Research Council produced an independent analysis at the request of several agencies.

- EPA based the RfD on adverse effects in rats; the National Research Council committee chose a key biochemical event seen in healthy humans that would precede adverse effects as the basis of the RfD (NRC 2005b, pp. 14, 166).
- EPA used changes in brain morphometry, thyroid histopathology, and serum thyroid-hormone concentrations in rats (oral exposure) as the basis of its point of departure for the RfD calculation; the National Research Council committee recommended using inhibition of iodine uptake by the thyroid in a small group of exposed healthy humans, a nonadverse effect, as the basis of the point of departure (p. 168).
- EPA selected a "composite" uncertainty factor of 300 to account for animal-human differences, use of a lowest-observed-adverse-effect level, lack of chronic data, and other database gaps (p. 172). The National Research Council committee used a total uncertainty factor of 10 to account for interindividual variability (p. 178). This was consistent with the use of human data, and assumed that the point of departure was a no-observed-effect level.

EPA had proposed an RfD of 0.00003 mg/kg per day; committee recommendations would lead to an RfD of 0.0007 mg/kg per day (p. 178). In 2005, EPA responded to the National Research Council recommendation by issuing a new perchlorate RfD of 0.0007 mg/kg per day (EPA 2005c).

The analytic process for this chemical indicates that different scientific bodies can come to different risk conclusions, with a majority of the differences arising from different emphases placed on datasets and on how uncertainty and variability are viewed. Large-scale epidemiology studies can bring these variability and risk issues into sharper focus. For example, a recent large Centers for Disease Control and Prevention study found associations between relatively low perchlorate exposures and reduced thyroid function in sensitive populations of women (Blount et al. 2006). Further followup studies will provide insight as to whether the current RfD is adequate. Further analysis of CDC data suggest an interaction of perchlorate and tobacco smoking (perhaps via thiocyanate) to affect thyroid function (Steinmaus et al. 2007).

- Use of BMD or low dose for RfD calculation.
- Use of the lowest-observed-adverse-effect level or the no-observed-adverse-effect level.
- Use of the ED_{01}, ED_{05}, or ED_{10} to define the benchmark response.[14]
- For noncancer end points, an uncertainty factor of 1, 10, 100, 1000, or other.
- For carcinogens, a threshold or nonthreshold approach.

Exposure assessment can involve an even broader range of uncertainties and related choice points. Some are related to the fate and transport of the pollutant in the environment, others to data on and uncertainties about the metabolism, distribution, and fate of the chemical in the target population. In each case, chemical-specific data are rarely available on all the parameters critical for estimating expected exposures.

[14] ED_{01}, ED_{05}, or ED_{10} is the dose associated with either a 1%, 5%, or 10% increase in an adverse effect relative to the control response (EPA 2008).

As a result, exposure scenarios are just that—hypothetical situations based on combinations of measured and, where data are unavailable, modeled estimates of the form and amount of a chemical in the environment or human tissue. They often combine data specific to the chemical at issue and, where such data are not available, data on similar chemicals or on the same chemical in different conditions. After examining the database for answers to these questions, EPA risk assessors turn to assumptions and extrapolation to develop information for completing an assessment:

- In the absence of chemical-specific data, what data on what other chemicals best represent the chemical under study?
- In the absence of reliable measurements of exposure in the environment, which assumptions and models can be expected to provide reasonably valid estimates?
- In the absence of reliable measurements of tissue exposure in humans, which assumptions and models can be expected to provide reasonably valid estimates?
- Of several potentially vulnerable populations (for example, infants, children, the elderly, and pregnant women) with comparable exposure potential, which populations are the most sensitive and in need of protection under the standard?
- When and how should exposure assessment take account of cumulative or aggregate exposure?

Choices at those and other decision points shape predictions of risks to populations of interest and the credibility of the risk assessment itself.

Superimposed on those choices among candidate scientific studies, assumptions, models, and the like, policy choices are required as to which scientifically plausible assumptions and models to use in completing the assessment. The process is designed to accommodate discussion of the choices and the reasons for them. The Red Book paradigm and successor reports and EPA guidance documents stress the importance of characterizing risk by advising decision-makers and the public about uncertainties, assumptions, and choices made. A National Research Council report on EPA's dioxin reassessment illustrates the point (NRC 2006, p. 55):

> The impact of the choices made in the risk assessment process can be characterized by quantifying the impact of plausible alternative assumptions at critical steps. The risk estimates can be most fully characterized by performing probabilistic analyses when possible and by presenting the range of possible risk estimates rather than by reporting the single point estimates. Risk characterization should provide useful information to risk managers to help them understand the variability and uncertainty in the risk estimates.

Chapter 6 of the present report provides additional recommendations on developing alternative risk estimates in light of plausible alternatives to defaults. The Red Book points out that "risk characterization, the estimate of the magnitude of the public health problem, involves no additional scientific knowledge or concepts" (NRC 1983, p. 28). Rather, it calls for synthesizing information from the preceding analyses with special attention to identifying uncertainties and their impact on the assessment (see Chapter 4 of this report).

The Role of Time

Time is a major and rarely acknowledged influence in the nature and quality of environmental risk assessment in EPA. Some time factors are immediately obvious. The statutory deadlines for some regulatory decisions necessarily require completed risk assessments to meet the deadlines. When EPA fails to meet a standard-setting deadline, as often happens, regulated entities, advocacy groups, and other interested parties exercise their statutory right to bring "deadline" suits, which result in court orders to issue standards by a specified date. The result may bring closure to an assessment that has been languishing or lead to an assessment that meets the deadline but falls short of some scientific standards.

Such statutory requirements constitute advance notice of the need for specific risk assessments in specified timeframes and can lead to regular schedules for many assessments and related analyses. Examples of such requirements include the 5-year cycle for review and revision of the national ambient air quality standards (Section 109) and the 8-year deadline for maximum achievable control technology (MACT) standards for hazardous air pollutants (Section 112) under the CAA amendments of 1990. In 1996, Congress set new deadlines for pesticide actions under the FQPA, requiring the agency to reassess the risks of all existing pesticide food tolerances (standards) over a ten year period; that same year Congress enacted a new Safe Drinking Water Act requiring the agency to select five new contaminants each year for decisions on maximum contaminant levels (MCLs) for drinking water.

Several predictable but highly variable factors can upset the best-laid plans. The most obvious is the unavailability of scientifically reliable and context-relevant data and methods. Other situations can be cited. Some involve new research or monitoring data that identify issues that affect the assessment or information on the imminent appearance of new studies expected to make a substantial difference in the analysis; others involve emergency environmental problems or changes in political priorities that result in reassignment of resources and staff to other assessments. Undue political influence in the process can also result in delays (GAO 2008). And initial planning may have been inadequate with regard to what could reasonably be achieved with available data and resources and the corresponding setting of unreasonable expectations.

In some circumstances, EPA is faced with an abundance of data, especially on high-profile chemicals. Specifically, where chemicals have been studied for many years, multiple studies of comparable quality on a single chemical may yield different results, in some cases large differences in RfDs or risk and in other cases slight but critical differences—a situation that invites debate and controversy and may take years to resolve. In these circumstances, new studies and new data, while at the same time shedding light on assessments, can complicate reviews (Box 2-7). However, it is important to recognize the value of analyses that synthesize data across a number of different studies and end points, which can result in a more precise and defensible analysis.

In addition to recommending attention to previously unavailable new studies, almost every peer review recommends research that would improve the assessment. Recommendations of both types hold the prospect of reducing uncertainty and contributing to a more reliable risk assessment. Such recommendations also invite delay, require additional resources, and contribute to ambiguity as to whether the assessment is scientifically sufficient. Such delay can have significant impact on communities who are awaiting risk assessment results to make decisions regarding the safety of their neighborhoods where hazards may be present.

Iteration is an important feature of an adequate risk-assessment process and should be built into the planning. Addressing late-arising problems uncovered in discussion between assessors and managers will improve the assessment but may also delay its completion. Similarly, stakeholder and peer-review involvement brings many benefits but may extend the process. Changing administrations may also add to the time required.[15]

---

[15]EPA's recent dioxin reassessment and cancer guidelines are examples. Specifically, the dioxin report or parts of it were submitted for peer review on several occasions from 1992 to 2003, when a National Research Council

Box 2-7 Impact of New Studies

In 1997, concern about the effects of human exposure to mercury led Congress to request a National Research Council review of EPA's RfD for methyl mercury (MeHg). At the time, scientists were awaiting results from studies of three populations because the existing RfD was based on a 1987 study of 81 Iraqi children accidentally exposed in utero (NRC 2000a, p. 306). Noting that MeHg exposures in the Iraqi study population were not comparable with low-level chronic exposures expected in North American populations, the National Research Council committee recommended basing the RfD on new studies that were incomplete at the time of the 1997 Mercury Study Report to Congress (EPA 1997b).

A National Research Council committee recommended that EPA retain the 0.1-µg/kg per day RfD but replace the study used to set the RfD with new studies: "Since the establishment of the current RfD, results from the prospective studies in the Faroe Islands (Grandjean et al. 1997, 1998, 1999) and the Seychelles (Davidson et al. 1995a,b, 1998), as well as a peer-reviewed re-analysis of the New Zealand study (Crump et al. 1998) have added substantially to the body of knowledge concerning the developmental neurotoxic effects of chronic low-level exposure to MeHg" (NRC 2000a, p. 312).

Similarly, National Research Council recommendations on the long-running dioxin assessment expand the scope of the assessment: "EPA is encouraged to review newly available studies on the effects of TCDD on cardiovascular development in its risk assessment for noncancer end points" (NRC 2006, p. 174).

Perchlorate (Box 2-6) provides an example of how emerging data may inform risk after an assessment has been finalized.

The iterative nature of risk assessment and research ensures that new data will enter the process. The salutary effect of new data can also result in additional time for analysis and incorporation of data into the risk assessment.

In some ways, problems with timeliness are inherent in a decision-making environment that places a premium on "sound science" or "credible science." The nature of the conflict can be understood if it is recalled that the scientific process of seeking the truth, by design and to its credit, has no natural end point. In addition, the training of scientists, by design, and the embedded cultural traditions, such as requiring p values in tests of significance, instill values of prudence, replication, scientific debate, and peer review as prerequisites of a conclusion characterized as "sound science." This issue is discussed in more detail in Chapter 3.

INSTITUTIONAL ARRANGEMENTS FOR MANAGING THE PROCESS

Consideration of EPA's risk-assessment accomplishments and shortfalls and of the effects of policy and time leads to questions about institutional arrangements for "managing the process," the subtext of the Red Book. EPA has established an enormous array of programs for this purpose. The combination of people and programs reflects close attention to statutory requirements and advisory-body recommendations. That salutary orientation around diverse statutory requirements also leads to criticism of "apparent inconsistencies in risk assessment

panel undertook the most recent review. EPA's cancer guidelines were first published for comment and peer review in 1996; intermediate reviews took place before publication as final guidelines in 2005. Work began on both documents in the late 1980s. The development period included changes in the general approaches to risk assessment and specific new data and theories regarding cancer risk assessment and the toxicity of dioxin. In addition, several changes at the White House during this period led, at different times, to EPA decision-makers with different constituencies.

practices across EPA" (EPA 2004a, p. 14), which are traceable to statutory and managerial, as well as scientific, factors and to calls for greater coordination of agency programs.

Environmental Protection Agency Risk-Assessment Programs and Activities

EPA's major program offices have scientific responsibilities on the one hand and regulatory responsibilities on the other. For scientific data development and risk assessment, the agency relies on environmental professionals trained in diverse technical disciplines, such as chemistry, geology, toxicology, epidemiology, statistics, and communication. For risk management and regulatory decision-making, professionals in economics, engineering, law, and other fields work with agency policy-makers to shape regulatory decisions. As indicated in agency guidelines and other documents, assessors and managers have different roles but interact regularly throughout the process (EPA 1984, 2003d, 2004b; Table D-1, sections on "distinguishing linking risk assessment and risk management" and "problem formulation").

In addition to different statutes and scientists with expertise in many fields, EPA's risk-assessment work takes place in a variety of organizational and geographic locations and includes collaborative activities with numerous public and private scientific organizations. The result is a complex set of interactions that strengthen the agency's risk-assessment processes in the main, but the diversity of inputs also introduces drawbacks.

Each major program office manages several risk-assessment activities. For example, the Office of Water has programs for conducting health risk assessments under the Safe Drinking Water Act (SDWA) and ecologic risk assessments under the Clean Water Act. The Office of Air and Radiation conducts human health risk assessments for use in setting regulatory standards related to "criteria" pollutants (such as particulate matter [PM] and sulfur dioxide), in a different program "hazardous" pollutants (such as arsenic and mercury) from stationary sources, and in still another program pollutants from cars and other mobile sources. That office is also responsible for assessments related to stratospheric ozone depletion and acid rain. As evident in EPA's Science Inventory (EPA 2005d), other agency offices have comparably wide-ranging programs for a total set of activities that almost defies description. The diverse risk-assessment tasks impose demands for both breadth and quality in staffing and managing these activities.

Several offices have overarching responsibilities to help meet the demands. ORD conducts environmental research at more than 10 laboratories and centers around the country. The laboratories are organized around the basic units in the risk-assessment paradigm (for example, effects, exposure assessment, and risk characterization). ORD plans, conducts, and oversees most EPA risk assessments and risk-assessment-related research for the agency as a whole. In addition to its core program of fundamental research, a substantial portion is planned in collaboration with program and regional offices to address data needs for regulation. In keeping with congressional and agency guidance priorities, ORD-led multioffice research-planning teams coordinate planning and budgeting in line with data needs identified by program and regional offices. However, it is important to note that because EPA relies heavily on data in the published literature and these are not the studies conducted by EPA, there is no mechanism for developing the data necessary to address emerging issues, and this contributes to a scarcity of data on particular agents.

ORD scientists coordinate generic risk-assessment activities, such as guideline development and the reference-dose–reference-concentration (RfD-RfC) process, including management of the IRIS database. ORD also conducts individual chemical-specific assessments at the behest of program and regional offices and, variably, in collaboration with them.[16]

Some offices are staffed to meet particular needs. In keeping with its responsibilities to oversee the safety of pesticide products, OPP employs a highly specialized scientific staff to evaluate data related to testing and licensing requirements for new pesticides before they are marketed and to conduct risk assessments to set limits on the use of pesticides as appropriate. Because pesticides are toxic by definition, this office has special statutory authority to mandate testing procedures and require specific scientific data from pesticide manufacturers.

The authority to mandate data generation in that way is not generally available to other offices, which depend on ORD, the scientific literature, and outside contractors. One of the paradoxes of the risk-assessment process is that the same scientific uncertainties that hamper and complicate risk assessment stimulate the development of new data and methods. For example, scientific uncertainties and controversy related to standards under development for PM led to special funding for new research to reduce the uncertainties (see NRC 1998, 1999a, 2001a, 2004).

EPA regularly incorporates the expertise of external scientists into its risk-assessment activities. The agency has extensive long-term and ad hoc collaborative relationships with numerous risk-assessing entities in the public and private sectors. Public-sector partners include other federal entities, such as the National Toxicology Program, which is administratively housed in NIEHS; Argonne and other Department of Energy national laboratories; and the FDA National Center for Toxicological Research. Private-sector collaborators include the Health Effects Institute in Boston, ILSI, and the American Chemistry Council in Washington. EPA scientists also participate in numerous international programs, such as the UN International Programme on Chemical Safety (IPCS), of which the World Health Organization (WHO) is a partner. The IPCS Harmonization Project, which is designed to harmonize approaches to the assessment of risk, has been a particularly influential partner with EPA in advancing the practice of risk assessment.

EPA's ten regional offices have risk-assessment and regulatory activities corresponding to those in the major program offices but focused at the local level. They have diverse risk-assessment responsibilities. Scientists interact with EPA program offices in Washington, DC, and ORD risk-assessment centers and laboratories on the one hand and with nongovernment organizations and state, local, and tribal entities on the other. In some cases, the regional offices apply risk assessments or toxicity values (for example, RfD, RfC, or potency estimates from IRIS) developed elsewhere to regional problems; in other cases, they develop region-specific assessments. Through those interactions, state, local, and tribal information and perspectives become part of the process.

The diverse inputs to risk assessment in EPA are a natural outgrowth of the diverse environmental problems facing the nation and the agency and of the scientific complexities of the risk-assessment process. Several EPA activities, including risk-assessment guidelines and the RfD-RfC process, are designed to counteract the effects of compartmentalization by standardizing and unifying some of the diverse elements. In addition, the Office of the Science Advisor coordinates the work of two standing committees with agencywide, rather than program-specific, risk-assessing responsibilities. The Risk Assessment Forum was chartered in response to recommendations in the 1983 Red Book. Somewhat later, the agency set up a Risk Management Council composed of senior EPA risk managers with oversight

[16]In addition to the ORD laboratories, program and regional offices manage laboratories, such as that in Ann Arbor for the air program, that in Bay St. Louis for the pesticide program, and the National Enforcement Investigation Center in Colorado.

responsibilities for forum activities. Later, renamed and rechartered as the Science Policy Council, that group has enlarged its membership and responsibilities to address a variety of science-policy issues.

Risk Management: Regulations and Risk Assessment

EPA statutes lodge responsibility for regulatory decisions with the EPA administrator and the assistant administrators who head the program offices. All are political appointees who require Senate confirmation and generally change when the White House changes hands. In their roles as risk managers, those officials are responsible for using completed risk assessments with information from other disciplines to shape regulatory decisions. In addition, they and other risk managers provide oversight for the risk-assessment process from inception to conclusion.

As indicated above, the 1983 Red Book stressed the importance of a "conceptual distinction" (p. 7) between risk assessment and risk management but rejected the concept of "institutional separation" between the processes. EPA adheres to those principles in the sense that, although assessors and managers are colocated and interact regularly, assessors do not set standards and decision-makers do not conduct risk assessments.

Owing to the committee's statement of task, this chapter has focused on the evolution of *risk assessment* and related practices. The committee considers that the same degree of concern about uncertainty, variability, and inferences that has been applied to the assessment of risks should also be applied to the assessment of costs, but this was beyond the scope of this report. For example, economists on the administrator's planning, evaluation, and innovation staff provide information and analyses on costs and benefits for use in making regulatory decisions and for the regulatory impact analyses (RIAs) that accompany major regulatory actions. (The benefits are computed from the results of risk assessments.) In addition, many program and regional offices have units responsible for analysis of the economic benefits of proposed decisions and regulatory actions. ORD's National Risk Management Research Laboratory in Cincinnati conducts engineering research for use in developing and evaluating the technical feasibility of pollution-control methods used in formulating regulatory options. In accord with statutory directives, EPA program and regional offices interact with state and local offices on implementation and compliance issues, such as schedules, costs, feasibility, impacts, and enforcement.

Regarding regulation development, as indicated earlier, the Red Book emphasis on the "conceptual distinction" between risk assessment and risk management reflects the statutory dichotomy between information used in assessing risk and other kinds of information—"the public health, economic, social, political consequences of regulatory options" (Figure 2-1)—used with risk-assessment results to determine "agency decisions and actions." For example, in evaluating whether a pesticide poses an "unreasonable risk" to health or the environment, the pesticide law (FIFRA) calls for consideration of the economic, social, and environmental costs of using the pesticide. EPA "interprets this broad statutory language to mean that any significant benefits to public health through disease control or prevention, or through vector control, need to be considered in the suspension, cancellation, or denial of an application for registration or a determination of ineligibility for deregistration of a public health use of any pesticide that offers such benefits" (EPA 2007b). In the same vein, the 1996 amendments to the SDWA explicitly direct EPA to evaluate incremental benefits, costs, and risks associated with compliance with alternatives—a more specific delineation of nonscience considerations than in the original enactment.

Differences between the information base for risk assessment, which has science at its core, and that for the regulatory decision, which takes account of costs and other nonrisk factors, mean that regulatory decisions are not necessarily congruent with risk assessment. That is, concern about, for example, economic consequences or societal impacts may outweigh public-health or environmental concerns in such a way as to make a regulatory decision more or less protective than if the decision were based solely on the risk assessment. An additional asymmetry is that the uncertainties associated with cost and benefits are rarely considered although these uncertainties are often explicitly acknowledged in the risk assessment. The distinction between the SDWA's maximum contaminant level *goal* (MCLG) and the maximum contaminant *level* (MCL) illustrates the point: the MCLG for a carcinogenic contaminant may be zero, but costs and feasibility concerns may lead the agency to set a regulatory standard, the MCL, to allow a higher level of contamination (see Box 2-8).

Some statutes authorize a combination of risk assessment and "technology-based" processes in setting regulatory standards. Such standards as the SDWA's MCL illustrate the special case of "technology-based" standards for which a decision does not depend only on risk assessment. Rosenthal et al. (1992) explain that the SDWA calls for MCLGs, "which are concentrations at which no adverse human health effects are believed to occur." A health-based MCLG is not an enforceable limit. For enforcement purposes, the statute directs EPA to establish a MCL as close to the MCLG as "feasible with the use of the best technology, treatment techniques, and other means which the Administrator finds after examination for efficiency under field conditions . . . are available (taking costs into consideration)" (42 USC § 300g-1).

Other examples appear in the CAA. The 1990 amendments introduced a two-part scheme—part technology-based, part risk assessment—for 189 toxic pollutants regulated under Section 112 of the CAA. The first step directs EPA to identify major emitters of the

BOX 2-8 Arsenic in Drinking Water: Uncertainties and Standard-Setting

On January 22, 2001, EPA issued a pending standard of 10 µg/L as the maximum contaminant level of arsenic in drinking water. Although the scientific analysis underlying the proposal and the proposal itself had been peer-reviewed by both the EPA SAB (1995) and the National Research Council (1999b) and had gone through the public comment process, EPA on March 23, 2001, issued a notice delaying the effective date of the standard to address questions about the science supporting the rule and about the expected implementation costs for affected communities.

The National Research Council peer-review committee identified uncertainties and data gaps of several kinds (NRC 2001b):

> More research is needed on the possible association between arsenic exposure and cancers other than skin, bladder, and lung, as well as noncancer effects. . . . In addition, more information is needed on the variability in metabolism of arsenic among individuals, and the effect of that variability on an arsenic risk assessment. Laboratory and clinical research is also needed to define the mechanisms by which arsenic induces cancer to clarify the risks at lower doses [p. 10].

Nonetheless, the committee made it clear that data gaps and uncertainties do not disqualify the risk assessment for decision-making.

> There is a sound database on the carcinogenic effects of arsenic in humans that is adequate for the purposes of risk assessment. The subcommittee concludes that arsenic-induced internal (lung and bladder) cancers should continue to be the principal focus of arsenic risk assessment for regulatory decision making, as discussed and recommended in the 1999 NRC report [p. 10].

A final 10-µg/L standard was issued in 2002; EPA and Congress continue to study costs and technical issues associated with implementing the standard (Tiemann 2005).

pollutants among diverse source categories and requires that these sources use MACT within specified time limits. The second step takes risk into consideration: 8 years after promulgation of MACT standards to limit emissions of the 189 (later reduced to 187)[17] pollutants; EPA was required to evaluate the residual risk to the population and promulgate more stringent standards if necessary "to provide an ample margin of safety to protect public health" (1990 Amendments to the Clean Air Act, Title III, § 301 (d)(9)). The law specifies that for known, probable, or possible human carcinogens, the administrator is to promulgate revised standards if the MACT standards do not reduce the risk incurred by "the individual most exposed to emissions" from the source of pollution to less than one in a million. With the focus on the "individual most exposed," EPA models exposure with fine spatial resolution to characterize the maximum level of exposure associated with a toxic air pollutant. Chapter 4 reviews the current state of the science on variability in susceptibility to cancer, and Chapter 5 provides recommendations to EPA for considering this variability in risk assessments.

The CAA takes a different approach in setting national ambient air quality standards for criteria air pollutants (ozone, PM, carbon monoxide, sulfur dioxide, nitrogen oxides, and lead). Those standards are based solely on health criteria[18] without consideration of the cost and feasibility of compliance, which are reserved for later evaluation in developing state implementation plans. In this decision context, risk assessment plays a role in setting the NAAQS and in the RIAs generally used to evaluate control strategies for criteria air pollutants.

Strategic Planning, Priority-Setting, and Data Development

Scientifically informed strategic planning is critical. Reliable and relevant scientific data are major determinants of the quality of any risk assessment. As a result, the availability of such data strongly influences the agency's ability to improve its assessments in line with new methods, statutory directives, or advisory-body recommendations. In turn, the scientific quality and timeliness of reliable data depend in part on factors common to scientific work in general, such as the availability of methods and data needed to complete the assessment of any particular chemical. Near-term examples include emerging data and methods to understand modes of action that contribute to clarifying and reducing uncertainty in risk assessments. Another example is related to current studies of the use of new genomics and nanotechnology data and methods for environmental risk assessment.

In addition, and separate from state-of-the-science questions, data availability depends on congressional and White House subject-matter interests that determine budget priorities for annual and long-range data development. Examples include a 12-year congressional earmark for PM research and chemical-specific allocations or directives related to arsenic. At a different level, agencywide strategic planning, priority-setting, and budgeting processes determine how risk-assessment resources are allocated among EPA programs (for example, air vs water vs IRIS), entities (external grants vs EPA laboratories), practices (basic research vs routine monitoring), and prospective risk assessments (for example, dioxin vs arsenic vs a particular Superfund site).

Decisions on those issues are part of the annual planning and budgeting process, which involves scientists and managers with risk-assessment responsibilities in ORD laboratories

[17]The original list of hazardous air pollutants (HAPs) contained 189 compounds; however, caprolactam (see 61 Fed. Reg. 30816 [1996]) and methyl ethyl ketone (see 70 Fed. Reg. 75047 [2005]) were later delisted, reducing the number of HAPs to 187.

[18]The statute also calls for "secondary," or welfare standards to protect the environment and property.

and program and regional offices. The resulting budget and subject-matter priorities are crucial in the availability or nonavailability of relevant data for risk-assessment purposes and thus in the quality of agency risk assessments. Although changes in budget allocations and priorities have resulted in more funding in such fields as computational toxicology and nanotechnology and less funding for postdoctoral research fellowships and intramural and extramural research, the fact remains that, in real dollar terms, EPA's research and development funding is nearly unchanged since at least 1990, and has been steadily declining since fiscal year 2004 (Coull 2007). The resulting budget and subject-matter priorities also influence the availability and workload of scientists who have the risk-assessment experience needed to study issues raised in the statement of task.[19]

EXTRAMURAL INFLUENCES AND PARTICIPANTS

Executive Orders: Risk-Assessment Policy

As indicated above, congressional legislation determines the broad outlines of risk-assessment principles and practices. The White House influences the process through executive orders addressing diverse risk-assessment topics and activities. Executive orders directing EPA (and other agencies) to expand the scope of their risk-assessment programs to cover cumulative risks[20] and children's risks,[21] in combination with related congressional legislation, led to new emphases as to data collection and approaches to risk analysis.[22] Furthermore, such provisions as Section 3-301(a) in Executive Order 12898 on environmental justice are highly specific as to the kind of data required:

> Environmental health research, whenever practicable and appropriate, shall include diverse segments of the population in epidemiological and clinical studies, including segments at high risk from environmental hazards, low income populations, and workers who may be exposed to substantial environmental hazards.

Historically, Office of Management and Budget (OMB) oversight of EPA regulatory activities has focused on planning and budget, congressional directives and priorities, cost-benefit issues, and related administrative and accountability matters. In recent years, OMB has greatly expanded its involvement in risk-assessment practices to include governmentwide information-quality guidelines (67 Fed. Reg. 8452 [2002]), an "Information Quality Bulletin for Peer Review" (70 Fed. Reg. 2664 [2005]), a "Proposed Risk Assessment Bulletin" (OMB 2006), and a memorandum on "Updated Principles for Risk Analysis" (OMB/OSTP 2007). The present committee did not assess the impact of OMB oversight on EPA risk assessment.[23]

---

[19]Such advisory bodies as the National Research Council, the National Science Foundation, EPA's SAB, and EPA's Board of Scientific Counselors regularly review and comment on EPA's research priorities, both annual and for long-term strategic planning. See, for example, NRC 1998, 1999a, 2000b, 2001a, 2004; and www.EPA.gov/SAB.

[20]From Executive Order 12898 (February 11, 1994): "Environmental health analysis, whenever practicable and appropriate, shall identify multiple and cumulative exposures."

[21]From Executive Order 13045 (April 21, 1997): Federal agencies "shall make it a high priority to identify and assess environmental health risks and safety risks that may disproportionately affect children."

[22]Indeed, these executive orders led to the creation of the EPA Office of Environmental Justice and, later, the Office of Children's Health Protection.

[23]OMB and several government agencies asked the National Research Council to review the "Proposed Risk Assessment Bulletin." In its report (NRC 2007), the review committee lauds the goal of increasing the quality and objectivity of risk assessment in the federal government, but "concludes that the OMB bulletin is fundamentally flawed and recommends that it be withdrawn" (p. 6).

In sum, many factors—statutory requirements, the diverse array of environmental problems and agency programs, executive orders, OMB directives, and the vagaries of the risk-assessment process—give rise to risk-assessment practices and individual assessments that differ in form, information content, and analytic quality. Such diversity demands informed and experienced attention to managing the process.

Executive Orders: Regulatory Policy

Several executive orders illuminate the role of the White House in risk management and regulatory decision-making. Described as a "cornerstone of White House administrative policy" (OMB Watch 2002), Executive Order 12866 (October 4, 1993)[24] calls for each agency head to designate a regulatory-policy officer and outlines requirements related to risk assessment, cost-benefit analysis, performance-based regulatory standards, and other aspects of regulation development. A recent amendment, Executive Order 13422 (Jan. 18, 2007), requires the regulatory-policy officer to be a presidential appointee. The present committee did not assess the impact of those and other executive orders on EPA risk assessment.

Public Participation

EPA relies on information from the public in developing both general principles and risk assessments of individual chemicals. By law, EPA, like other federal agencies, is required to publish proposed regulations (including any underlying scientific analysis) in the *Federal Register*, invite public comments, and consider the comments in its final decision. EPA often follows that process for guidance documents that apply only internally (for example, risk-assessment guidelines) and for *preliminary* analyses used in rule-making. In addition, separately from the peer-review activities discussed above, the agency often convenes scientific experts to discuss strategic planning and research priorities and to introduce and develop background documents. Notice is given in the *Federal Register,* and the public is invited to observe and comment during the session.

Public meetings, workshops, and the notice and comment process are avenues for stakeholders to present risk-assessment-relevant information and opinion. One example is the Pesticide Program Dialogue Group, a forum established in 1995 for a diverse group of stakeholders to provide feedback on issues from nonanimal testing to endangered species to risk assessment. The group includes pesticide manufacturers, public-interest and advocacy groups, and trade associations. It is one of several groups on pesticide issues, with corresponding groups in other agency offices, such as those which involve air-program consultation with state and local air-pollution programs and waste-office consultation with responsible parties and community groups regarding Superfund sites. EPA regional offices work closely with the Indian tribes on selected issues. Thus, EPA expressly solicits information from interested and knowledgeable parties, whether scientists or nonscientists.

EPA's statement of task anticipates near-term and long-term improvements in risk assessment as a result of the present report. New approaches can be expected to require ad-

[24]Executive Order 12866 replaces and extends Executive Orders 12291 and 12498, issued during the Reagan administration. It directs federal regulatory agencies, including EPA, to "assess both the costs and the benefits of the intended regulation and, recognizing that some costs and benefits are difficult to quantify, propose or adopt a regulation only upon a reasoned determination that the benefits of the intended regulation justify its costs" [Sec. (b)(1)]. The order requires EPA to conduct a formal RIA for proposed regulations expected to impose economic costs in excess of $100 million per year.

BOX 2-9 Risk-Assessment Planning: Multiple Participants

The committee that produced *Understanding Risk* (NRC 1996) identified several criteria for judging success at the end of the process: getting the science right, getting the right science, getting the participation right, getting the right participation, and developing an accurate, balanced, and informative synthesis. As discussed below (Chapter 3), achieving those objectives depends in part on informed "planning and scoping" activities involving risk assessors, risk managers, and interested and affected parties. The emphasis on the "right" *participants* as well as the "right" science is important (McGarity 2004):

> There is little evidence that the *scientific information* that the agencies are currently using and disseminating is unreliable. Virtually all of the challenges that have been filed so far under the [2004 Information Quality Act] have involved disputes over interpretations, inferences, models and similar policy issues, and not the "soundness" of the underlying data.

justments of agency processes for allocating funds, scheduling research, expanding training, and other activities. New methods may also require enhanced peer review and expanded public participation to ensure that affected and interested parties in and outside the regulated community have an opportunity to contribute to new approaches and are prepared for change (see Box 2-9).

Peer Review, Quality Control, and Advisory Committees

Quality-control and peer-review procedures are particularly important when new approaches are introduced into the risk-assessment process. EPA uses several mechanisms to ensure the quality and relevance of laboratory and field data. In addition to general methods and guidelines, including uniform guidance applicable to all federal agencies, the major programs have program-specific methods related to, for example, air emissions, microbiologic contaminants, and underground storage tanks (EPA 2007c).

Similarly, EPA's peer-review program gives attention to new approaches and individual risk assessments. For example, a subcommittee of EPA's SAB monitored the development of EPA's first guidelines for ecologic risk assessment. Of course, assessments of individual chemicals based on new methods are subject to statutory requirements for peer review, such as the CAA requirement for review of the scientific basis of national ambient air quality standards and the FIFRA requirement for EPA's Scientific Advisory Panel (SAP) review of the scientific basis of some pesticide decisions. Other statutes require SAB review of a wide variety of analyses (see Box 2-10).[25]

Independent advisory committees that provide information and advice on special topics may contribute to new approaches. In addition to advisory committees required by statute,

[25]In response to recommendations from the EPA SAB and others (EPA 1992d), EPA peer-review policies issued in 1992 call for external review of scientific assessments not subject to statutory requirements. The processes were reinforced and augmented (and in some ways redefined) by OMB's 2002 governmentwide directive on peer review applicable to all federal agencies (67 Fed. Reg. 8452 [2002]). EPA risk assessments and underlying scientific analyses are also peer-reviewed when laboratory scientists, as well as those in program and regional offices, publish work developed for risk-assessment use in scholarly journals. That work includes individual laboratory or field studies on toxicology, epidemiology, and monitoring and subunits of risk assessment, such as hazard identification and exposure analysis.

BOX 2-10 After Peer Review

Peer review is not an end in itself. Ideally, peer review identifies deficiencies, suggests modifications, and otherwise leads the agency to improve a risk assessment to conform more fully with scientific standards and to guide decision-making and support regulatory standards. Two situations invite inquiry and attention because, while enhancing the assessment, they also cause delays and add costs to the risk-assessment process.

- Peer-review "spirals" involve repeated reviews that return assessments to the agency for further revision because the agency has not responded adequately to science-based recommendations in earlier reviews or because of science-policy debates or inadequacies in the peer-review process itself (GAO 2001). Recent examples include the reviews of dioxin and the cancer risk-assessment guidelines (see 68 Fed. Reg. 39086 [2003]; EPA 2005a; NRC 2006).
- Some assessments fail to reach closure or completion within a typical period after peer review. An example of such an unfinished assessment is that of dichloromethane (methylene chloride), which was peer-reviewed by the SAB in 1987; the health assessments remain in draft form (EPA 1987b,c), and the SAB comments have never been incorporated (EPA 2003e). The EPA assessment (EPA 1987b,c) at the time was regarded as a good example of the use of pharmacokinetic modeling. Specifically, the SAB review stated (EPA SAB 1988, p. 1) that "the Subcommittee concludes that the Addendum [EPA 1987c] was one of the best documents it has reviewed in terms of its clarity, coverage of the data and analysis of scientific issues. This document clearly demonstrates the potential utility of pharmacokinetic data in risk assessment. EPA should continue to use this approach in future risk assessments, whenever scientifically possible."

A confluence of factors may explain extended timeframes and unfinished assessments, including scientific complexity and controversy, a continually evolving database, and stakeholder and advocacy-group demands. Contributing factors in the case of dichloromethane were the absence of strong regulatory pressure for the assessment; the increasing importance of other chemicals, including trichloroethylene and tetrachloroethylene; and the replacement of dichloromethane with substitutes (L. Rhomberg, Gradient Corporation, Cambridge, MA, personal commun., May 31, 2007).

EPA is scheduled to update the IRIS value for dichloromethane in the middle of 2009 (Risk Policy Report 2007; 40 CFR Part 63 [2007]).

such as the SAB and SAP, EPA has chartered committees to provide advice on selected issues pertinent to risk assessment, such as research planning and priorities (the Board of Scientific Counselors), endocrine-disrupting chemicals (the National Committee on Endocrine Disrupting Chemicals and Toxic Substances), and children's health (the Children's Health Protection Advisory Committee) (www.EPA.gov).

International Organizations

EPA consults and collaborates with programs associated with the risk-assessment arms of numerous international organizations. EPA scientists sit on numerous international committees including the IPCS, the International Agency for Research on Cancer (IARC)/WHO, the International Commission on Radiological Protection, and the Intergovernmental Forum on Chemical Safety; participate in the writing of scholarly papers; and conduct risk-assessment training in conjunction with these international organizations. As with state and local regulatory bodies, EPA and these organizations share scientific data, exchange information on developments in risk assessment, and work to harmonize risk-assessment concepts and guidelines. Those interactions provide opportunities for EPA scientists to be alert to advances made in the organizations that will contribute to new approaches under way in EPA.

In sum, several mechanisms are available to inform and upgrade EPA risk-assessment processes. Beyond the basic procedures outlined above, complementary planning and oversight activities make it clear that the risk-assessment enterprise involves more than its basic scientific elements. Numerous overarching factors—tangible and intangible, scientific and nonscientific—shape the process and influence the quality of agency assessments.

CONCLUSIONS AND RECOMMENDATIONS

Congressional mandates give EPA a diverse set of risk-assessment and regulatory responsibilities. The process is informed by many factors, including congressional legislation, generic guidance, and advice from scientific advisory bodies, peer-review recommendations specific to individual risk assessments and guidelines, information from stakeholders and other interested parties, and the principle of comity with other government entities (state, local, and international) on risk-assessment issues. The result is a complex set of risk-assessment activities that have drawn high praise in many cases and sustained criticism in others. The process recommendations below identify institutional and management issues that require sustained attention by agency leadership. Except for the longer timeframe expected for *new* guidelines (see final recommendation), the committee contemplates implementation in the immediate and near future.

Conclusions

- Some deficiencies in current EPA risk-assessment practices can be attributed in part to the unavailability of relevant data and methods. Those limitations head the list of EPA concerns about implementing future recommendations for improvement (Appendix E). Implementing several of the recommendations in the present report will require *additional* data and methods related to each of the three analytic fields in the Red Book paradigm. In addition, *new* kinds of data or methods will be required to enable EPA to undertake analyses that are given new emphasis or recommended for the first time here.
- Although EPA has a 20-year history of issuing guidelines and other reports designed to implement recommendations for improvement offered by the National Research Council and other advisory bodies, moving from policy to practice has in some cases been incomplete or only partially effective (as to provisions put into practice) and in others uneven (as to use for all assessments in all parts of the agency, where applicable).
- Effective use of new methods and attention to new policies require instruction and training for both experienced risk assessors and newcomers. And putting new policies and methods into practice—that is, moving beyond policy documents—requires understanding and appreciation on the part of agency managers and decision-makers.
- Historically, guideline development in EPA has taken from as little as 3 years to more than 15 years (for example, the cancer guidelines were issued in 2005 after a 15-year development period). Improvements in risk assessment will involve issuing new guidelines, revising existing guidelines or issuing supplemental guidance, and implementing existing guidelines more effectively.

Recommendations

- The committee seconds the Government Accountability Office recommendation that the administrator of the Environmental Protection Agency direct agency offices to "more proactively identify the data most relevant to the current risk assessment needs, including the specific studies required and how those studies should be designed, and communicate those needs to the research community" (GAO 2006, p. 69). The committee recommends that the Environmental Protection Agency consider recommendations in the present report as part of that process.
- Putting recommendations from this report into practice will require additional staff in fields that are now lightly staffed (for example, epidemiology and quantitative uncertainty analysis) and new staff in fields that are generally understaffed relative to this report's emphasis on the social-science components of environmental decision-making (for example, psychology, sociology, economics, and decision theory).
- Agency leaders should give high priority to establishing and maintaining risk-assessment and decision-making training programs for scientists, managers responsible for risk-assessment activities, and other participants in the process. This reinforces the Government Accountability Office recommendation that the Administrator of the Environmental Protection Agency "ensure that risk assessors and risk managers have the skills needed to produce quality risk assessments by developing and implementing in-depth training" (GAO 2006, p. 69). A regular schedule of refresher courses is critical for such a program. This recommendation calls for training to ensure that all relevant managers and decision-makers are fully informed on risk-assessment principles and principles related to the other disciplines (such as economics and engineering) that, with risk assessment, influence regulatory decisions.
- To reduce the effects of the compartmentalization resulting from the Environmental Protection Agency's organization around diverse statutory mandates, the administrator can buttress the scientific talent brought to bear on improvement activities by revitalizing and expanding interoffice and interagency collaboration through existing structures (for example, the Risk Assessment Forum, the Science Policy Council, and the National Science and Technology Council Committee on Environment and Natural Resources) and by joining scientists from other agencies (for example, the National Institute of Environmental Health Sciences and the Food and Drug Administration) in these activities. This reinforces the Government Accountability Office recommendation that the administrator of the Environmental Protection Agency "develop a strategy to ensure that offices engage in early planning to identify and seek the expertise needed, both within the EPA workforce and from external subject matter experts" (GAO 2006, p. 69).
- The administrator of the Environmental Protection Agency should give special attention to expanding the scientific and decision-making core in the regional offices to ensure that they have the capacity to use improved risk-assessment methods and to meet their obligations for interaction with stakeholders, local agencies, and tribes.
- The Environmental Protection Agency should establish a tiered schedule for guideline implementation: (1) *immediate* and uniform use and oversight as to existing guidelines and risk-assessment policies (for example, 1-2 years), except where inapplicable; a *shorter-term* schedule for revision or updating of existing guidelines where appropriate (for example, 2-6 years); and a *longer-term* but definite schedule for development and issuance of new guidelines (for example, 6-15 years).

REFERENCES

Blount, B.C., J.L. Pirkle, J.D. Osterloh, L. Valentin-Blasini, K.L. Caldwell. 2006. Urinary perchlorate and thyroid hormone levels in adolescent and adult men and women living in the United States. Environ. Health Perspect. 114(12):1865-1871.

CalEPA (California Environmental Protection Agency). 2004. Public Health Goal for Chemicals in Drinking Water, Perchlorate. Office of Environmental Health Hazard Assessment, California Environmental Protection Agency [online]. Available: http://www.oehha.ca.gov/water/phg/pdf/finalperchlorate31204.pdf [accessed August 25, 2004].

Coull, B.C. 2007. Testimony of Dr. Bruce C. Coull, President of the U.S. Council of Environmental Deans and Directors, National Council for Science and the Environment, Before the Subcommittee on Energy and Environment, Committee on Science and Technology, U.S. House of Representatives. March 14, 2007.

Crump, K.S., T. Kjellström, A.M. Shipp, A. Silvers, and A. Stewart. 1998. Influence of prenatal mercury exposure upon scholastic and psychological test performance: Benchmark analysis of a New Zealand cohort. Risk Anal. 18(6):701-713.

Crump, K., and G. Goodman. 2003. Benchmark Analysis for the Perchlorate Inhibition of Thyroidal Radioiodine Uptake Utilizing a model for the Observed Dependence of Uptake and Inhibition on Iodine Excretion. Prepared for J. Gibbs, Kerr-McGee Corporation. January 24, 2003. (Presentation at the Fifth Meeting on Assess the Health Implications of Perchlorate Ingestion, July 29-30, 2004, Washington, DC.)

Davidson, P.W., G.J. Myers, C. Cox, C. Shamlaye, O. Choisy, J. Sloane-Reeves, E. Cernchiari, D.O. Marsh, M. Berlin, M. Tanner, and T.W. Clarkson. 1995a. Neurodevelopmental test selection, administration, and performance in the main Seychelles child development study. Neurotoxicology 16(4):665-676.

Davidson, P.W., G.J. Myers, C. Cox, C.F. Shamlaye, D.O. Marsh, M.A. Tanner, M. Berlin, J. Sloane-Reeves, E. Cernichiari, O. Choisy, A. Choi, and T.W. Clarkson. 1995b. Longitudinal neurodevelopmental study of Seychellois children following in utero exposure to methylmercury from maternal fish ingestion: Outcomes at 19 and 29 months. Neurotoxicology 16(4):677-688.

Davidson, P.W., G.J. Myers, C. Cox, C. Axtell, C. Shamlaye, J. Sloane-Reeves, E. Cernichiari, L. Needham, A. Choi, Y. Wang, M. Berlin, and T.W. Clarkson. 1998. Effects of prenatal and postnatal methylmercury exposure from fish consumption on neurodevelopment: Outcomes at 66 months of age in the Seychelles child development study. JAMA 280(8):701-707.

EPA (U.S. Environmental Protection Agency). 1984. Risk Assessment and Management: Framework for Decision Making. EPA 600/9-85-002. Office of the Administrator, U.S. Environmental Protection Agency, Washington, DC. December 1984.

EPA (U.S. Environmental Protection Agency). 1986a. Guidelines for Carcinogen Risk Assessment. EPA/630/R-00/004. Risk Assessment Forum, U.S. Environmental Protection Agency, Washington, DC. September 1986 [online]. Available: http://www.epa.gov/ncea/raf/car2sab/guidelines_1986.pdf [accessed Jan. 7, 2008].

EPA (U.S. Environmental Protection Agency). 1986b. Guidelines for Mutagenicity Risk Assessment. EPA/630/R-98/003. Risk Assessment Forum, U.S. Environmental Protection Agency, Washington, DC. September 1986 [online]. Available: http://www.epa.gov/osa/mmoaframework/pdfs/MUTAGEN2.PDF [accessed Jan. 7, 2008].

EPA (U.S. Environmental Protection Agency). 1986c. Guidelines for Health Risk Assessment of Chemical Mixtures. EPA/630/R-98/002. Risk Assessment Forum, U.S. Environmental Protection Agency, Washington, DC. September 1986 [online]. Available: http://www.epa.gov/ncea/raf/pdfs/chem_mix/chemmix_1986.pdf [accessed Jan. 7, 2008].

EPA (U.S. Environmental Protection Agency). 1987a. Unfinished Business: A Comparative Assessment of Environmental Problems. Office of Policy Analysis, Office of Policy Planning and Evaluation, U.S. Environmental Protection Agency, Washington, DC.

EPA (U.S. Environmental Protection Agency). 1987b. Technical Analysis of New Methods and Data Regarding Dichloromethane: Pharmacokinetics, Mechanism of Action and Epidemiology. External Review Draft. EPA/600/8-87/029A. PB-87-228557/XAB. Office of Health and Environmental Assessment, U.S. Environmental Protection Agency, Washington, DC. June 1, 1987.

EPA (U.S. Environmental Protection Agency). 1987c. Update to the Health Assessment Document and Addendum for Dichloromethane (Methylene Chloride): Pharmacokinetics, Mechanism of Action and Epidemiology. Review Draft. EPA/600/8-87/030A. PB-87-228565/XAB. Office of Health and Environmental Assessment, U.S. Environmental Protection Agency, Washington, DC. July 1, 1987.

EPA (U.S. Environmental Protection Agency). 1991. Guidelines for Developmental Toxicity Risk Assessment. EPA/600/FR-91/001. Risk Assessment Forum, U.S. Environmental Protection Agency, Washington, DC. December 1991 [online]. Available: http://www.epa.gov/NCEA/raf/pdfs/devtox.pdf [accessed Jan. 10, 2008].

EPA (U.S. Environmental Protection Agency). 1992a. Guidelines for Exposure Assessment. EPA600Z-92/001. Risk Assessment Forum, U.S. Environmental Protection Agency, Washington, DC.

EPA (U.S. Environmental Protection Agency). 1992b. Framework for Ecological Risk Assessment. EPA/63-R-92/001. Risk Assessment Forum, U.S. Environmental Protection Agency, Washington, DC.

EPA (U.S. Environmental Protection Agency). 1992c. Guidance on Risk Characterization for Risk Managers and Risk Assessors. Memorandum to Assistant Administrators, and Regional Administrators, from F. Henry Habicht, Deputy Administrator, Office of the Administrator, Washington, DC. February 26, 1992 [online]. Available: http://www.epa.gov/oswer/riskassessment/pdf/habicht.pdf [accessed Oct. 10, 2007].

EPA (U.S. Environmental Protection Agency). 1992d. Safeguarding the Future: Credible Science, Credible Decisions. The Report of an Expert Panel on the Role of Science at EPA. EPA/600/9-91/050. U.S. Environmental Protection Agency, Washington, DC. March 1992.

EPA (U.S. Environmental Protection Agency). 1994. Peer Review and Peer Involvement at the U.S. Environmental Protection Agency. Memorandum to Assistant Administrators, General Counsel, Inspector General, Associate Administrators, Regional Administrators, and Staff Office Directors, from Carol M. Browner, Administrator, U.S. Environmental Protection Agency. June 7, 1994 [online]. Available: http://www.epa.gov/osa/spc/pdfs/perevmem.pdf [accessed Oct. 16, 2007].

EPA (U.S. Environmental Protection Agency). 1995. Guidance for Risk Characterization. Science Policy Council, U.S. Environmental Protection Agency, Washington, DC. February 1995 [online]. Available: http://www.epa.gov/osa/spc/pdfs/rcguide.pdf [accessed Jan. 7, 2008].

EPA (U.S. Environmental Protection Agency). 1996a. Guidelines for Reproductive Toxicity Risk Assessment. EPA/630/R-96/009. Risk Assessment Forum, U.S. Environmental Protection Agency, Washington, DC. October 1996 [online]. Available: http://www.epa.gov/ncea/raf/pdfs/repro51.pdf [accessed Jan. 10, 2008].

EPA (U.S. Environmental Protection Agency). 1996b. Proposed Guidelines for Carcinogen Risk Assessment. EPA/600/P-92/003C. Office of Research and Development, U.S. Environmental Protection Agency, Washington, DC. April 1996 [online]. Available: http://www.epa.gov/NCEA/raf/pdfs/propcra_1996.pdf [accessed Jan. 7, 2007].

EPA (U.S. Environmental Protection Agency). 1997a. Guiding Principles for Monte Carlo Analysis. EPA/630/R-97/001. Risk Assessment Forum, U.S. Environmental Protection Agency, Washington, DC. March 1997 [online]. Available: http://www.epa.gov/ncea/raf/montecar.pdf [accessed Jan. 7, 2008].

EPA (Environmental Protection Agency). 1997b. Mercury Study Report to Congress, Volume 1. Executive Summary. EPA-452/R-97-003. Office of Air Planning and Standards and Office of Research and Development, U.S. Environmental protection Agency, Washington, DC. December 1997 [online]. Available: http://www.epa.gov/ttn/oarpg/t3/reports/volume1.pdf [accessed Jan. 16, 2008].

EPA (U.S. Environmental Protection Agency). 1997c. Exposure Factors Handbook. National Center for Environmental Assessment, Office of Research and Development, U.S. Environmental Protection Agency, Washington, DC. August 1997 [online]. Available: http://www.epa.gov/ncea/efh/report.html [accessed Aug. 5, 2008].

EPA (U.S. Environmental Protection Agency). 1998a. Guidelines for Ecological Risk Assessment. EPA/630/R-95/002F. Risk Assessment Forum, U.S. Environmental Protection Agency, Washington, DC. April 1998 [online]. Available: http://oaspub.epa.gov/eims/eimscomm.getfile?p_download_id=36512 [accessed Feb. 9, 2007].

EPA (U.S. Environmental Protection Agency). 1998b. Guidelines for Neurotoxicity Risk Assessment. EPA/630/R-95/001F. Risk Assessment Forum, U.S. Environmental Protection Agency, Washington, DC. April 1998 [online]. Available: http://www.epa.gov/ncea/raf/pdfs/neurotox.pdf [accessed Jan. 10, 2008].

EPA (U.S. Environmental Protection Agency). 2000a. Supplementary Guidance for Conducting Health Risk Assessment of Chemical Mixtures. EPA/630/R-00/002. Risk Assessment Forum, U.S. Environmental Protection Agency, Washington, DC. August 2000 [online]. Available: http://www.epa.gov/ncea/raf/pdfs/chem_mix/chem_mix_08_2001.pdf [accessed Jan. 7, 2008].

EPA (U.S. Environmental Protection Agency). 2000b. Risk Characterization Handbook. EPA 100-B-00-002. Office of Science Policy, Office of Research and Development, U.S. Environmental Protection Agency, Washington, DC. December 2000 [online]. Available: http://www.epa.gov/OSA/spc/pdfs/rchandbk.pdf [accessed Jan. 7, 2008].

EPA (U.S. Environmental Protection Agency). 2000c. Peer Review Handbook, 2nd Ed. EPA 100-B-00-001. Science Policy Council, U.S. Environmental Protection Agency, Washington, DC. December 2000 [online]. Available: http://www.epa.gov/osa/spc/pdfs/prhandbk.pdf [accessed May 16, 2008].

EPA (U.S. Environmental Protection Agency). 2002a. Guidance on Cumulative Risk Assessment of Pesticide Chemicals That Have a Common Mechanism of Toxicity. Office of Pesticide Programs, U.S. Environmental Protection Agency, Washington, DC. January 14, 2002 [online]. Available: http://www.epa.gov/pesticides/trac/science/cumulative_guidance.pdf [accessed Jan. 7, 2008].

EPA (U.S. Environmental Protection Agency). 2002b. Guidelines for Ensuring and Maximizing the Quality, Utility and Integrity of Information Disseminated by the Environmental Protection Agency. EPA/260R-02-008. Office of Environmental Information, U.S. Environmental Protection Agency, Washington, DC. October 2002 [online]. Available: http://www.epa.gov/QUALITY/informationguidelines/documents/EPA_InfoQualityGuidelines.pdf [accessed Feb. 9, 2007].

EPA (U.S. Environmental Protection Agency). 2003a. Human Health Research Strategy. EPA/600/R-02/050. Office of Research and Development, U.S. Environmental Protection Agency, Washington, DC [online]. Available: http://www.epa.gov/nheerl/humanhealth/HHRS_final_web.pdf [accessed July 31, 2008].

EPA (U.S. Environmental Protection Agency). 2003b. Exposure and Human Health Reassessment of 2,3,7,8-Tetrachlorodibenzo-*p*-Dioxin (TCDD) and Related Compounds. NAS Review Draft. National Center for Environmental Assessment, Office of Research and Development, U.S. Environmental Protection Agency, Washington, DC. December 2003 [online]. Available: http://www.epa.gov/NCEA/pdfs/dioxin/nas-review/ [accessed Jan. 9, 2008].

EPA (U.S. Environmental Protection Agency). 2003c. Disposition of Comments and Recommendations for Revisions to "Perchlorate Environmental Contamination: Toxicological Review and Risk Characterization, External Review Draft (January16, 2002)." National Center for Environmental Assessment, Risk Assessment Forum, U.S. Environmental Protection Agency, Washington, DC [online]. Available: http://cfpub2.epa.gov/ncea/cfm/recordisplay.cfm?deid=72117 [accessed Jan 4, 2008].

EPA (U.S. Environmental Protection Agency). 2003d. Framework for Cumulative Risk Assessment. EPA/600/P-02/001F. National Center for Environmental Assessment, Risk Assessment Forum, U.S. Environmental Protection Agency, Washington, DC [online]. Available: http://cfpub.epa.gov/ncea/cfm/recordisplay.cfm?deid=54944 [accessed Jan. 4, 2008].

EPA (U.S. Environmental Protection Agency). 2003e. Dichloromethane. (CASRN 75-09-2). Integrated Risk Information System, U.S. Environmental Protection Agency, Washington, DC [online]. Available: http://www.epa.gov/NCEA/iris/subst/0070.htm [accessed Jan. 10, 2008].

EPA (U.S. Environmental Protection Agency). 2004a. Risk Assessment Principles and Practices: Staff Paper. EPA/100/B-04/001. Office of the Science Advisor, U.S. Environmental Protection Agency, Washington, DC. March 2004 [online]. Available: http://www.epa.gov/osa/pdfs/ratf-final.pdf [accessed Jan. 9, 2008].

EPA (U.S. Environmental Protection Agency). 2004b. Risk Assessment and Modeling-Air Toxics Risk Assessment Library, Vol.1. Technical Resources Manual, Part III. Human Health Risk Assessment: Multipathway. EPA-453-K-04-001A. Office of Air Quality Planning and Standards, U.S. Environmental Protection Agency, Research Triangle Park, NC.

EPA (U.S. Environmental Protection Agency). 2005a. Guidelines for Carcinogen Risk Assessment. EPA/630/P-03/001F. Risk Assessment Forum, U.S. Environmental Protection Agency, Washington, DC. March 2005 [online]. Available: http://cfpub.epa.gov/ncea/cfm/recordisplay.cfm?deid=116283 [accessed Feb. 7, 2007].

EPA (U.S. Environmental Protection Agency). 2005b. Supplemental Guidance for Assessing Susceptibility for Early-Life Exposures to Carcinogens. EPA/630/R-03/003F. Risk Assessment Forum, U.S. Environmental Protection Agency, Washington, DC. March 2005 [online]. Available: http://cfpub.epa.gov/ncea/cfm/recordisplay.cfm?deid=160003 [accessed Jan. 4, 2008].

EPA (U.S. Environmental Protection Agency). 2005c. Perchlorate and Perchlorate Salts: Reference Dose for Chronic Oral Exposure. Integrated Risk Information System, U.S. Environmental Protection Agency, Washington, DC [online]. Available: http://www.epa.gov/iris/subst/1007.htm [accessed Jan. 10, 2008].

EPA (U.S. Environmental Protection Agency). 2005d. Science Inventory (SI) Database. U.S. Environmental Protection Agency, Washington, DC [online]. Available: http://cfpub.epa.gov/si/ [accessed Jan. 11, 2008].

EPA (U.S. Environmental Protection Agency). 2006. Framework for Assessing Health Risks of Environmental Exposures to Children (External Review Draft). EPA/600/R-05/093A. National Center for Environmental Assessment, Office of Research and Development, U.S. Environmental Protection Agency, Washington, DC. March 2006 [online]. Available: http://cfpub.epa.gov/ncea/cfm/recordisplay.cfm?deid=150263 [accessed Oct. 11, 2007].

EPA (U.S. Environmental Protection Agency). 2007a. Total Risk Integrated Methodology (TRIM) – General Information. Office of Air and Radiation, U.S. Environmental Protection Agency, Washington, DC [online]. Available: http://www.epa.gov/ttn/fera/trim_gen.html [accessed Feb. 22, 2008].

EPA (U.S. Environmental Protection Agency). 2007b. Explanation of Statutory Framework for Risk Benefit Balancing for Public Health Pesticides. Public Health Issues, Office of Pesticides, U.S. Environmental Protection Agency, Washington, DC [online]. Available: http://earth1.epa.gov/pesticides/health/risk-benefit.htm [accessed Jan. 11, 2008].

EPA (U.S. Environmental Protection Agency). 2007c. Environmental Test Methods and Guidelines. Information Sources, U.S. Environmental Protection Agency, Washington, DC [online]. Available: http://www.epa.gov/Standards.html [accessed Jan. 11, 2007].

EPA (U.S. Environmental Protection Agency). 2008. IRIS Glossary. Integrated Risk Information System National Center for Environmental Assessment, U.S. Environmental Protection Agency, Washington, DC [online]. Available: http://www.epa.gov/ncea/iris/help_gloss.htm#e [accessed Feb. 28, 2008].

EPA SAB (U.S. Environmental Protection Agency Science Advisory Board). 1988. Assess Health Effects Associated with Dichloromethane (Methylene Chloride). Final report. SAB-EHC-88-013. PB-89-108997/XAB. U.S. Environmental Protection Agency, Science Advisory Board, Washington, DC. May 9, 1988 [online]. Available: http://yosemite.epa.gov/sab/sabproduct.nsf/0708B8BE9D8693968525732800 5CF44F/$File/DICHLOROMETHANE+++++++EHC-88-013_88013_5-22-1995_264.pdf [accessed Jan. 11, 2008].

EPA SAB (U.S. Environmental Protection Agency Science Advisory Board). 1995. SAB Review of Issues Related to the Regulation of Arsenic in Drinking Water. EPA-SAB-DWC-95-015. U.S. Environmental Protection Agency Science Advisory Board, Washington, DC. July 1995 [online]. Available: http://yosemite.epa.gov/sab/sabproduct.nsf/D3435DF2A691B8328525719B006A500E/$File/dwc95015.pdf [accessed July 31, 2008].

GAO (Government Accountability Office). 2001. EPA's Science Advisory Board Panels. Improved Policies and Procedures Needed to Ensure Independence and Balance. Report to the Ranking Minority Member, Committee on Government Reform, House of Representatives. GAO-01-536. Washington, DC: Government Accountability Office. June 2001 [online]. Available: http://www.gao.gov/new.items/d01536.pdf [accessed July 31, 2008].

GAO (U.S. Government Accountability Office). 2005. Chemical Regulation: Options Exist to Improve EPA's Ability to Assess Health Risks and Manage Its Chemical Review Program. GAO-05-458. Washington, DC: U.S. Government Accountability Office [online]. Available: http://www.gao.gov/new.items/d05458.pdf [accessed Jan. 10, 2008].

GAO (U.S. Government Accountability Office). 2006. Human Health Risk Assessment: EPA Has Taken Steps to Strengthen Its Process, but Improvements Needed in Planning, Data Development, and Training. GAO-06-595. Washington, DC: U.S. Government Accountability Office [online]. Available: http://www.gao.gov/new.items/d06595.pdf [accessed Jan. 10, 2008].

GAO (U.S. General Accountability Office). 2008. Chemical Assessments: Low Productivity and New Interagency Review Process Limit the Usefulness and Credibility of EPA's Integrated Risk Information System. GAO-08-440. Washington, DC: U.S. General Accountability Office. March 2008 [online]. Available: http://www.gao.gov/new.items/d08440.pdf. [accessed June 11, 2008].

Gilman, P. 2003. Statement of Paul Gilman, Assistant Administrator for Research and Development and EPA Science Advisor, U.S. Environmental Protection Agency, before the Committee on Transportation and Infrastructure, Subcommittee on Water Resources and the Environment, U.S. House of Representatives, March 5, 2003 [online]. Available: http://www.epa.gov/ocir/hearings/testimony/108_2003_2004/2003_0305_pg.pdf [accessed February 9, 2007].

Greer, M.A., G. Goodman, R.C. Pleus, and S.E. Greer. 2002. Health effects assessment for environmental perchlorate contamination: The dose response for inhibition of thyroidal radioiodine uptake in humans. Environ. Health Perspect. 110(9):927-937.

Grandjean, P., P. Weihe, R.F. White, F. Debes, S. Araki, K. Yokoyama, K. Murata, N. Sørensen, R. Dahl, and P.J. Jørgensen. 1997. Cognitive deficit in 7-year-old children with prenatal exposure to methylmercury. Neurotoxicol. Teratol. 19(6):417-428.

Grandjean, P., P. Weihe, R.F. White, and F. Debes. 1998. Cognitive performance of children prenatally exposed to "safe" levels of methylmercury. Environ. Res. 77(2):165-172.

Grandjean, P., E. Budtz-Jørgensen, R.F. White, P.J. Jørgensen, P. Weihe, F. Debes, and N. Keiding. 1999. Methylmercury exposure biomarkers as indicators of neurotoxicity in children aged 7 years. Am. J. Epidemiol. 150(3):301-305.

IOM (Institute for Medicine). 1997. Dietary Reference Intakes for Calcium, Phosphorus, Magnesium, Vitamin D, and Fluoride. Washington, DC: National Academy Press.

IOM (Institute for Medicine). 1998. Dietary Reference Intakes: A Risk Assessment Model for Establishing Upper Intake Levels for Nutrients. Washington, DC: National Academy Press.

IOM (Institute for Medicine). 2003. Dietary Reference Intakes: Applications in Dietary Planning. Washington, DC: The National Academies Press.

Krewski, D., V. Hogan, M.C. Turner, P.L. Zeman, I. McDowell, N. Edwards, and J. Losos. 2007. An integrated framework for risk management and population health. Hum. Ecol. Risk Asses. 13(6):1288-1312.

McGarity, T.O. 2004. Our science is sound science and their science is junk science: Science-based strategies for avoiding accountability and responsibility for risk-producing products and activities. Kans. Law. Rev. 52(4):897-937.

Mills, A. 2006. IRIS from the inside. Risk Anal. 26(6):1409-1410.

NRC (National Research Council). 1983. Risk Assessment in the Federal Government: Managing the Process. Washington, DC: National Academy Press.

NRC (National Research Council). 1993a. Issues in Risk Assessment. Washington, DC: National Academy Press.

NRC (National Research Council). 1993b. Pesticides in the Diets of Infants and Children. Washington, DC: National Academy Press.

NRC (National Research Council). 1994. Science and Judgment in Risk Assessment. Washington, DC: National Academy Press.

NRC (National Research Council). 1996. Understanding Risk: Informing Decisions in a Democratic Society. Washington, DC: National Academy Press.

NRC (National Research Council). 1998. Research Priorities for Airborne Particulate Matter: I. Immediate Priorities and a Long-Range Research Portfolio. Washington, DC: National Academy Press.

NRC (National Research Council). 1999a. Research Priorities for Airborne Particulate Matter: II. Evaluating Research Progress and Updating Portfolio. Washington, DC: National Academy Press.

NRC (National Research Council). 1999b. Arsenic in Drinking Water. Washington DC: National Academy Press.

NRC (National Research Council). 2000a. Toxicological Effects of Methyl Mercury. Washington, DC: National Academy Press.

NRC (National Research Council). 2000b. Strengthening Science at the U.S. Environmental Protection Agency. Washington, DC: National Academy Press.

NRC (National Research Council). 2001a. Research Priorities for Airborne Particulate Matter: III. Early Research Progress. Washington, DC: National Academy Press.

NRC (National Research Council). 2001b. Arsenic in Drinking Water: 2001 Update. Washington, DC: National Academy Press.

NRC (National Research Council). 2002. Estimating the Public Health Benefits of Proposed Air Pollution Regulations. Washington, DC: National Academy Press.

NRC (National Research Council). 2004. Research Priorities for Airborne Particulate Matter: IV. Continuing Research Progress. Washington, DC: The National Academies Press.

NRC (National Research Council). 2005a. Superfund and Mining Megasites/Lessons from Coeur D'Alene River Basin. Washington, DC: The National Academies Press.

NRC (National Research Council). 2005b. Health Implications of Perchlorate Ingestion. Washington, DC: The National Academies Press.

NRC (National Research Council). 2006. Health Risks from Dioxin and Related Compounds/Evaluation of the EPA Reassessment. Washington, DC: The National Academies Press.

NRC (National Research Council). 2007. Scientific Review of the Proposed Risk Assessment Bulletin from the Office of Management and Budget. Washington, DC: The National Academies Press.

OMB (Office of Management and Budget). 2006. Proposed Risk Assessment Bulletin. Office of Management and Budget, Washington, DC. January 9, 2006 [online]. Available: http://www.whitehouse.gov/omb/inforeg/proposed_risk_assessment_bulletin_010906.pdf [accessed Jan. 4, 2008].

OMB/OSTP (Office of Management and Budget/Office of Science and Technology Policy). 2007. Updated Principles for Risk Analysis. Memorandum for the Heads of Executive Departments and Agencies, from Susan E. Dudley, Administrator, Office of Information and Regulatory Affairs, Office of Management and Budget, and Sharon L. Hays, Associate Director and Deputy Director for Science, Office of Science and Technology Policy, Washington, DC. September 19, 2007 [online]. Available: http://www.whitehouse.gov/omb/memoranda/fy2007/m07-24.pdf [accessed Jan. 4, 2008].

OMB Watch. 2002. Executive Order 12866. OMB Watch. February 10, 2002 [online]. Available: http://www.ombwatch.org/article/articleview/180/1/67 [accessed Jan. 11, 2008].

Parkin. 2007. Foundations and Frameworks for Microbial Risk Assessments. Presentation at the 4th Meeting on Improving Risk Analysis Approaches Used By the U.S. EPA, April 17, 2007, Washington DC.

PCCRARM (Presidential/Congressional Commission on Risk Assessment and Risk Management). 1997. Framework for Environmental Health Risk Management - Final Report, Vol. 1. [online]. Available: http://www.riskworld.com/nreports/1997/risk-rpt/pdf/EPAJAN.PDF [accessed Jan. 4, 2008].

Risk Policy Report. 2007. EPA Says Future Studies May Force Tighter Toxics Limits for Solvents. Inside EPA's Risk Policy Report. 14(19):14. May 8, 2007.

Rosenthal, A., G.M. Gray, and J.D. Graham. 1992. Legislating acceptable cancer risk from exposure to toxic chemicals. Ecol. Law Q. 19(2):269-362.

Steinmaus, C., M.D. Miller, R. Howd. 2007. Impact of smoking and thiocyanate on perchlorate and thyroid hormone associations in the 2001-2002 national health and nutrition examination survey. Environ. Health Perspect. 115:1333-1338.

Suter, G., T. Vermeire, W. Munns, and J. Sekizawa 2001. Framework for the Integration of Health and Ecological Risk Assessment. Chapter 2 in Integrated Risk Assessment Report. Prepared for the WHO/UNEP/ILO International Programme on Chemical Safety. WHO/IPCS/IRA/01/12. The International Programme on Chemical Safety, World Health Organization. December 2001 [online]. Available: http://www.who.int/ipcs/publications/en/ch_2.pdf [accessed July 30, 2008].

Tiemann, M. 2005. Arsenic in Drinking Water: Regulatory Developments and Issues. CRS Report for Congress 05-RS-20672a. Washington, DC: Congressional Research Service. October 20, 2005 [online]. Available: http://assets.opencrs.com/rpts/RS20672_20051020.pdf [accessed Jan. 7, 2008].

WHO (World Health Organization). 2006. A Model for Establishing Upper Levels of Intake for Nutrients and Related Substances. Report of a Joint FAO/WHO Technical Workshop on Risk Assessment, May 2-6, 2005, Geneva, Switzerland. World Health Organization, Food and Agriculture Organization of the United Nations. January 13, 2006 [online]. Available: http://www.who.int/ipcs/highlights/full_report.pdf [accessed July 30, 2008].

3

The Design of Risk Assessments

RISK ASSESSMENT AS A DESIGN CHALLENGE

Risk assessment is sometimes used to describe a process and sometimes to describe the product of a process. The dual use can create confusion, but it also serves as a reminder that the task of improving risk analysis necessarily requires attention both to desirable qualities of the process and to desirable qualities of the product. Given that there are inevitable constraints on efforts to assess risk and multiple objectives to be met, the selection of appropriate elements of *process* and the specification of required elements of the final *product* constitute a complex design challenge.

Well-designed risk-assessment processes create products that serve the needs of a community of consumers, including risk managers, community and industrial stakeholders, risk assessors themselves, and ultimately the public. Multiple interpretations of the word *design* apply to our presentation. One of the primary goals of design reflects the overall utility of a product to its end users. A second key aspect of design is the assurance of technical quality. Many of the technical aspects of quality may not be apparent to end users, but they are important prerequisites that provide the foundation for the quality of a decision-support product. Finding the appropriate mix of technical quality and utility, given constraints, is the essence of design of a decision-support product.

The Decision-Making Environment and the Importance of Process

Many decision-making situations involving matters of public heath and environmental risk have five common elements: the desire to use the best scientific methods and evidence in informing decisions, uncertainty that limits the ability to characterize both the magnitude of the problem and the corresponding benefits of proposed interventions, a need for timeliness in decision-making that precludes resolving important uncertainties before decisions are required, the presence of some sort of tradeoff among disparate adverse outcomes (which may be health, ecologic, or economic outcomes, each affecting a different set of stakeholders), and the reality that, because of the inherent complexity of the systems being managed and the long-term implications of many decisions (such as cancer latency, changes in the structure of ecosystems, or multiple simultaneous sources of exposure), there will be little or no short-term feedback as to whether the desired outcome has been achieved by the decisions.

The combination of uncertainty in the scientific data and assumptions (the "inputs") and inability to validate assessment results directly or to isolate and evaluate the impact of a resulting decision (the "outputs") creates a situation in which decision-makers, the scientific community, the public, industry and other stakeholders have little choice but to rely on the overall quality of the many *processes* used in the conduct of risk assessment to provide some assurance that the assessment is aligned with societal goals.

Those challenging properties of the decision-making environment may be considered particularly acute for many health and environmental decisions, but they are by no means new to decision-makers generally. The academic discipline of decision analysis under uncertainty, among others, has a rich literature on which to draw for methods and findings (Morgan et al. 1990; Clemen 1996; Raiffa 1997). The importance of attention to *process* is entirely compatible with the theory of the management sciences that defines a good decision under uncertainty as one that uses the most appropriate processes and methods to assemble and interpret evidence, to apply the decision-maker's values properly, and to make timely choices with available resources rather than defining a good decision only according to its (apparent) outcomes. This attention to process is also compatible with arguments for the inclusion of more deliberative approaches to assessment and decision-making. As such, the most appropriate processes and methods in a given situation may be an appropriate balance of deliberative and analytic methods, as advocated in NRC (1996).

Risk Assessment as a Decision-Support Product

The process of risk assessment involves generation of a number of individual products that are combined to form a final product (which is often referred to as "the risk assessment"). The final product of a risk assessment process is most often understood to be a report. The present committee suggests that the product of a risk assessment should be considered to include not only the report but various subproducts, such as computational models and other information that is assembled during the process. The subproducts have different uses and serve a variety of audiences. For example, a computational model with a user-friendly interface may be at least as valuable in informing decision-making as the technical report most often associated with the term *risk assessment*. In addition, such subproducts as dose-response assessments typically have value that transcends a particular decision-support application and may be used in thousands of future decision-support situations. It is also useful to consider that risk assessments and individual subproducts experience a life cycle (consisting, for example, of conception, design, development, testing, use, maintenance, obsolescence, and replacement) that should be explicitly recognized.

The products of risk assessment may be thought of as, among other things, communication products. Their value lies in their contribution to the objectives of the decision-making function, including their effects on the primary decision-maker and other interested parties who participate in the decision or otherwise use the information that the products convey. Although the effort expended in the process is largely scientific, the critical final process in risk assessment is ultimately communication.

The Quality of Risk Assessment Includes Both Process and Product Attributes

The decision-making environment associated with health and environmental risk management compels the various users of risk assessment to value and scrutinize the assessment process. In addition, risk assessment is understood to result in a set of final products whose specific attributes are critical for meeting their objectives. In a sense, it may be neither possible nor appropriate to separate the process from the product. The situation is somewhat analogous to that of other products whose quality is more readily scrutinized with respect to the process that is used rather than through scrutiny of detectable qualities of the final product. For example, the safety aspects of the quality of complex engineered systems, medical devices, and foods are increasingly scrutinized with respect to the quality of the process that generates and maintains them rather than judged solely on the basis of measurable qualities of the final product. Similarly, the final products of a risk assessment have a mixture of detectable and undetectable qualities, and both the final product and the underlying process must be considered in judging the overall quality.

Given the demands of health and environmental decision-making, perhaps the most appropriate element of quality in risk-assessment products is captured in their ability to improve the capacity of decision-makers to make informed decisions in the presence of substantial, inevitable and irreducible uncertainty. A secondary but surely important quality is the ability of the assessment products to improve other stakeholders' understanding and to foster and support the broader public interests in the quality of the decision-making process (for example, fairness, transparency, and efficiency). Those attributes are difficult to measure, and some elements of quality often cannot be judged until some time after the completion of the risk assessment.

Formative and Iterative Design of Risk Assessments

For the committee's purposes, the term *design* implies adopting a user-centered perspective to craft both an assessment process and a decision-support product that achieves the objectives of supporting high-quality decision-making while working within inevitable constraints. Accordingly, an important part of the early design process is the understanding and weighing of all the objectives, recognition of constraints, and explicit acknowledgment of the need for tradeoffs.

Design will inevitably occur throughout the risk-assessment process, and flexibility and iteration will be important aspects of the overall process design. Like any complex product designed in a complex environment, the process and product may need to be redesigned as objectives and constraints inevitably change and in response to new knowledge. While recognizing the iterative nature of risk-assessment planning, the committee strongly encourages increasing attention to design in the formative stages of a risk assessment. Such a shift in attention is recognized by the Environmental Protection Agency (EPA 2004a). It is also captured in guidance documents for ecologic risk assessment and cumulative risk assessment (EPA 1992, 1998, 2003). In those applications, EPA has adopted two tasks labeled *planning and scoping* and *problem formulation*. The two tasks are examples of early design activities, and the committee believes that they should be formalized, applied more consistently in risk-assessment activities, and, perhaps most important, result in concrete outputs detailing the rationale and findings of the early design process. The tasks are described in more detail later in this chapter.

DESIGN CONSIDERATIONS: OBJECTIVES, CONSTRAINTS, AND TRADEOFFS

As in any complex design problem, the process of design is intended to find the best solution to achieve multiple simultaneous and competing objectives while satisfying constraints on the process or the end product. As decision-support and communication products for use in public decision-making, risk-assessment products inherit objectives from their parent domains of science and public policy. The objectives are not always compatible and, considered individually, would influence the design in different and sometimes opposing directions. In addition, general constraints on the process (such as resources and time) require that tradeoffs be made in pursuit of the objectives.

The candidate objectives of risk assessment can, for present purposes, be separated into three categories, which are related to the inputs to the process (including evidentiary and participatory aspects), the process that transforms the inputs into risk-assessment products, and the impact of the products on decision-making. The objectives described below are examples that might be considered by EPA in designing risk-assessment processes and products; clearly, it is the responsibility of EPA to interpret its mandate to choose and weigh the relative importance of different objectives.

Objectives Related to Inputs

Use of the Best Scientific Evidence and Methods

A core aspect of health and environmental risk assessment is the universal desire to make use of the best scientific methods and the highest-quality evidence. Pursuit of that objective would lead EPA to acquire and interpret evidence by using established, trusted, and formal methods. The specifics underlying the notion of the "best science" are, not surprisingly, highly contested. Many attributes might define "best," and different parties will place considerably different weights on them. Even though the objective, simply stated, is superficially clear and uncontroversial, some aspects of the implementation are necessarily complex and controversial. In addition, pursuit of the best scientific understanding is inevitably resource-intensive and time-intensive, and this leads to conflict with other objectives and with constraints on resources.

Inclusiveness of Scope

For various reasons, human health risk assessment has traditionally focused on single cause-effect pathways that involve a single chemical and single identified adverse effect. The narrowness of scope is frequently questioned with respect to both its scientific merits and its relevance to decision contexts of considerably greater scope. The scope of consideration in health and environmental risk management would ideally be as large as possible. It can be argued that any limitation in scope constitutes a simplification of reality that must be recognized and justified because important parts of the total cause-effect network may have been missed. A narrow scope has the potential to distort the external validity of the conclusions and the associated decisions they support and thus to limit their applicability to the "real world."

From a decision-support perspective, limitations in scope might create what is seen as highly imbalanced information support, supporting a particular concern with voluminous technical analysis while other concerns of great relevance to stakeholders (which cannot be readily dismissed on purely scientific grounds) remain largely or completely unaddressed and explained only by the chosen scope of the risk assessment. For example, in situations where stakeholders are concerned about exposure from both food and water pathways, the provision of an elaborate risk assessment for waterborne exposure while providing only a cursory review for foodborne exposure may appear to be imbalanced with respect to the information needs. A somewhat more simplified risk assessment that includes both pathways may be preferable, if the foodborne pathway cannot be dismissed on strong grounds. Here, the objective of broadening of scope may compete with the desire to perform the "best" risk assessment on a single pathway.

The desire to broaden the scope of human health risk assessment appears to be shared by EPA. Table 3-1 illustrates the expansion of the scope (in both risk assessment and decision-making) to which EPA aspires, at least as far as can be inferred from its guidance for cumulative risk assessment. Some of the "new" characteristics are current practice in ecologic risk assessment.

A critical dimension of scope (and a theme of Chapter 8 of this report) is the explicit inclusion of the various possible mitigation options that might be considered to reduce the risk that is being assessed. The scope would be expanded so that the assessment would provide not only estimates of existing risk but estimates of risk reduction associated with a variety of changes in the risk-generating system. To provide more complete information to the decision-maker, the decision-support products would ideally include (or be reasonably integrated with) estimates of the associated costs and any countervailing risks associated with the proposed mitigation options, as might be presented, for example, in a remedial action report under Superfund or in assessments that inform pesticide registration decisions.

Additional elements of scope derive from the desire to support decision-makers other than EPA's internal risk managers. The often-advocated goal of supporting local decision-makers, communities, and industrial stakeholders in a participatory decision-making model suggests the need for more customized decision-support tools on the basis of the nuanced information needs and value foci of other decision-makers. This implies either that the scope of the risk assessment increases to include those diverse needs and values or that separate assessments are conducted with different scopes and end points considered (with the associated problems of compatibility).

The concept of extended decision support can be taken further to support the broad array of decisions that EPA may not be directly involved in but ultimately is interested in their being risk-based, particularly for preventive risk management. Product and process-

TABLE 3-1 Transition in EPA Human Health Risk-Assessment Characteristics According to EPA (1997)

Old	New
Single end point	Multiple end points
Single source	Multiple sources
Single pathway	Multiple pathways
Single route of exposure	Multiple routes of exposure
Central decision-making	Community-based decision-making
Command and control	Flexibility in achieving goals
One-size-fits-all response	Case-specific responses
Single-medium-focused	Multiple-media-focused
Single-stressor risk reduction	Holistic reduction of risk

Source: EPA 1997.

development decisions that are made every day around the world and have short-term and long-term effects on human health and the environment may be the most important class of external decisions that would ideally be increasingly risk-informed. This class of decisions includes decisions based on life-cycle analysis and various related approaches with similar goals, in which risks are ideally reduced by design of energy and material flows in advance rather than by end-of-pipe mitigation strategies. Some of these preventative strategies may benefit from risk-assessment components (like dose-response information, or quantification of common exposure scenarios) without the need for an entire risk assessment to be completed. This might suggest that risk-assessment products be designed, prepared, and disseminated in a modular fashion to allow for the individual components to be used and reused by third parties making different types of decisions.

Inclusiveness of Input

A process that considers a broader evidence base and uses diverse methods to reach conclusions is generally preferred to one that is limited to a narrower evidence base or a narrower selection of methods. Breadth can be achieved by considering input from different academic disciplines and by including traditional knowledge and a variety of deliberative methods of arriving at conclusions about what can be considered to be "known." The ideal becomes problematic when disciplinary biases rightly or wrongly determine that input from some other sources of information lacks sufficient *validity*—according to criteria that are idiosyncratic in each discipline—to be included as reliable input into a given analysis. Breadth can be seen as a potential threat to the integrity of the evidence base and of the conclusions derived from it. Because there is no universal standard for inclusion and weighing of evidence among disciplines (and often even within a discipline), resolution of the competing ideals of breadth and integrity of evidence requires careful attention to process.

Integrity of Science-Policy Assumptions

As a primary theme of both the Red Book (NRC 1983) and *Science and Judgment in Risk Assessment* (NRC 1994) and continuing in the present report, the careful application of science-policy assumptions (or "defaults") is critical for the integrity of the risk-assessment process. The use of defaults is necessary to complete risk assessments in the presence of substantial uncertainties and the embedded policy choices can have profound impacts on the risk-assessment findings and the associated decision-making functions.

In addition to the science-policy assumptions that are easily recognizable, the process should take account of the presence of key subjective elements in evidence-gathering and integration that can influence the results of risk assessment. They may include a number of standard practices or conventions that are not normally recognized as elements of science policy.

Objectives Related to Process

Inclusiveness in Process

Decision-making processes ideally are inclusive with respect to the participation and deliberation of affected and interested parties. In pursuit of that objective, risk-assessment processes would be structured to accommodate the needs of diverse stakeholders, including accepting their input at appropriate points, ensuring fairness in the influence of various aspects of the design of the risk-assessment process and products (for example, input into its scope and access to information), fostering their desired level of understanding of the process, and meeting their specific information needs.

Transparency

It is both a scientific and a policy-making objective that the process of conducting a risk assessment and the risk-assessment products themselves be transparent. Transparency is a requirement that is always present, but it is rarely defined in operational terms. Some strict interpretations of transparency are akin to requirements for scientific reproducibility: that enough information is provided for a skilled analyst to be able to follow all the reasoning and independently reproduce the results. Transparency in risk-assessment models could be interpreted to mean that the computer code is entirely in the public domain (but may be executable only on specified computers) or to suggest that the models be publicly available to be downloaded, complete with a user guide, and to be able to be run by individual interested users who lack advanced computer skills. In other interpretations, transparency would require that simplified versions of documents be produced to increase the number and diversity of parties that could follow the main arguments and understand the overall process of analysis and its conclusions. Given the lack of specificity in the operational definition of transparency, some effort is required during the early design period to achieve agreement among risk assessors and those seeking or responsible for ensuring transparency on the attributes that are sought and how they will be implemented.

Compliance with Statutes and Administrative Law Requirements

Some risk-assessment activities must comply with a variety of requirements imposed on federal policy-making activities, with the level of requirements depending on the risk assessment and the statutes that govern them. The nature and impact of these requirements is reviewed by NRC (2007). For example, EPA and other federal agencies are required by law to provide opportunity for public comment on proposed regulations and to take comments into account in making decisions. Some statutes have requirements for stakeholder participation in various aspects of the risk-assessment and rule-making processes; others require peer review of particular categories of risk assessment. Other statutory provisions call for EPA Science Advisory Board meetings to be open to the public and for agency records to be made available to the public through the Freedom of Information Act.[1] The administrative requirements regarding the risk-assessment process generally increase effort in the process, add costs, and affect the schedule. However, good practice would suggest that many of the required elements (such as peer review and stakeholder consultation) would often be included even if they were not required by statute or other administrative requirements.

[1]As outlined in Chapter 2, the organic statutes administered by EPA include substantive standards and criteria bearing on risk-assessment activities specific to different EPA programs (such as those involving air and water). In addition, program-specific and agencywide guidelines detail principles and practices related specifically to the risk assessment process (Table D-1).

Objectives Related to Impact on Decision-Making

Consideration of Uncertainty and Its Impacts

A shared ideal in science and decision-making is that uncertainties in evidence be fully exposed and described. The task of confronting the implications of uncertainty is ultimately the domain of the risk manager, so it is important that key sources of uncertainties be described individually and in the context of their collective impact on the conclusions of the risk assessment. When the set of decision-maker options is known, an uncertainty analysis can be most profitably directed toward describing the impact of uncertainty on the consideration of these options.

A difficult challenge in risk assessment is determining the best way to communicate the nature and magnitude of uncertainties. Analysis and judgment are required for focusing the discussion of uncertainty on important sources and describing the impacts of uncertainty in a manner that is relevant to the decision-making process. There are many potential uses of information about uncertainty for risk managers, including choices to delay or to expedite decision-making or to invest in research to reduce uncertainties. Assessing and communicating the utility of investing in additional information (such as conducting or considering more studies or gathering or formally eliciting expert input) are among the most challenging aspects of risk assessment. Formal and less formal methods for assessing the value of information are discussed below.

Control of 'Iatrogenic Risk' in the Decision-Making Process

There are a number of ways in which the process of assessing and managing risk can lead to an increase in risk—analogous to the notion of iatrogenic risk in medicine (risk "caused by the doctor"). In the same way that a delay in diagnosis by a physician can increase risk to the patient, delays in the process of assessing risks may increase overall exposure to risk when decisions are delayed. In the presence of low risk, the increased risk may also come from the prolonged stress of being in a state of uncertainty with regards to health. The design of a risk-assessment process should balance the pursuit of individual attributes of technical quality in the assessment and the competing attribute of timeliness of input into decision-making.

The critical process of triage, like other resource-allocation decisions in health care, must balance the needs of individual patients with those of others seeking attention. An overburdensome process of assessing individual risks can result in a lack of attention to other risks that deserve the attention of both risk assessment and risk management. Design must consider not only the needs of the individual assessment but the institutional role in simultaneously assessing and managing many other risks. Thus, the design of risk assessments should provide flexibility with respect to resource demands to foster balance in the management of multiple risks across the organization.

The health-care analogy is readily extended to the issue of risk-risk tradeoffs. Physicians routinely consider side effects of their treatment decisions. They also need to consider the impacts of decisions that patients themselves make in response to information about risks. In the same way, health and environmental risk-assessment and risk-management processes need to consider the *complete impact* of risk-assessment products and decisions given their inevitable potential to inadvertently contribute to increased risk. Ideally, the design of a risk assessment takes into account foreseeable consequences of decisions, including substitution risks (for example, replacement of one source of hazard with another of similar, greater, or unknown risk or diversion of waste from one waste stream to another), side effects of risk controls (for example, increase in risks due to disinfection byproducts in an effort to control microbial hazards or development of resistance in pests, microorganisms, and invasive species), and other potential adverse outcomes associated with decisions taken by EPA or foreseeable decisions that might be taken by other stakeholders. It is also possible to extend the analogy to post-market surveillance for medicine to suggest that decisions based on risk assessments be monitored for the potential for unanticipated impacts (or the absence of anticipated impacts).

ENVIRONMENTAL PROTECTION AGENCY'S CURRENT GUIDANCE RELATED TO RISK-ASSESSMENT DESIGN

The 1983 Red Book described the four key stages in the risk-assessment process as hazard identification, exposure assessment, dose-response assessment, and risk characterization (see Figure 3-1). In the intervening years, *planning and scoping* (a deliberative process that assists decision-makers in defining a risk-related problem) and *problem formulation* (a technically oriented process that assists assessors in operationally structuring the assessment) have emerged as additional distinct but related stages in both the human health and ecologic risk-assessment paradigms (EPA 1992, 1998, 2003, 2004a).

Not all decisions require or are amenable to the results of a risk assessment. Decision-makers must first consciously identify risk assessment as an appropriate decision-support

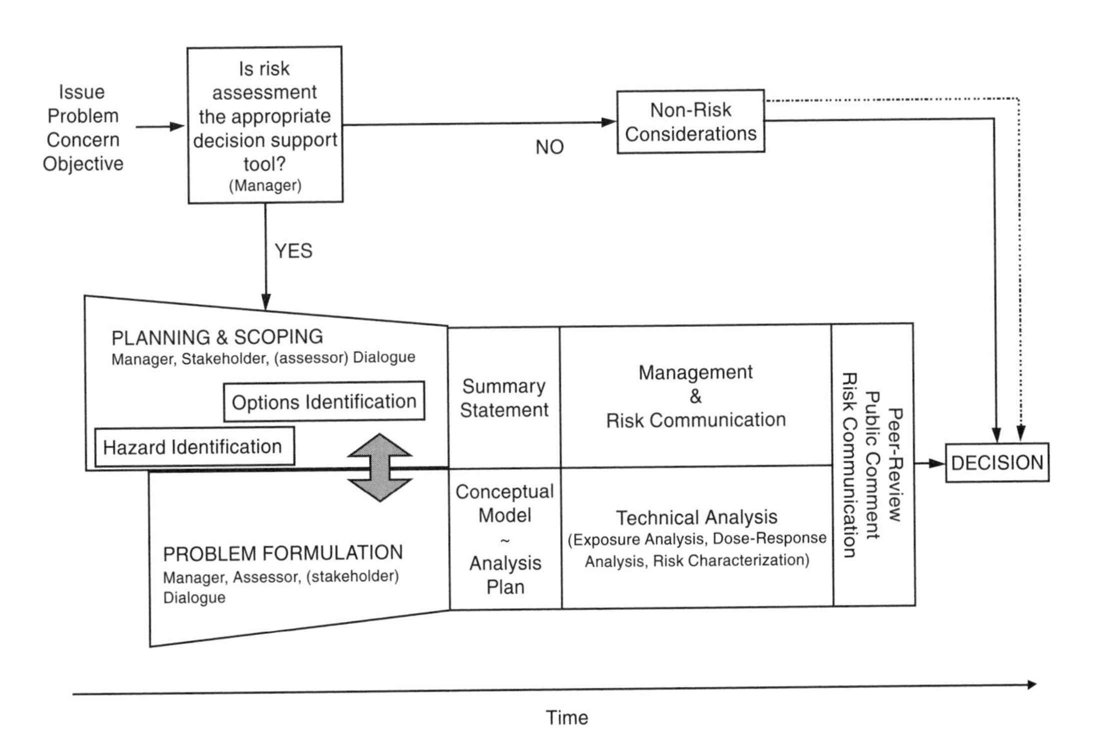

FIGURE 3-1 Schematic representation of the formative stages of risk-assessment design. Dotted line in figure denotes that decisions informed by risk assessment will be influenced by nonrisk considerations. Source: Adapted from EPA 1998, 2003.

tool. If risk assessment is not selected as a tool, the decision-maker can be guided by a host of other, nonrisk-related considerations. Clearly, even decisions that are informed by the results of a risk assessment will be influenced by the same nonrisk-related considerations (as indicated by the dotted connection in Figure 3-1).

Here, *planning and scoping* is used as described by EPA (2003, 2004a), and *problem formulation* is used as described by EPA (1998, 2003, 2004a). Planning and scoping are considered to constitute primarily a discussion between decision-makers (risk managers) and stakeholders in which assessors have a supporting role, and problem formulation involves a discussion between decision-makers and assessors (and technically oriented stakeholders) to develop a detailed technical design for the assessment that reflects the broad conceptual design developed in the scoping stage.

As illustrated in Figure 3-1, planning and scoping determine which hazards and risk-mitigation options are of concern for the assessment and set boundaries for the assessment (that is, its purpose, structure, content, and so on). Box 3-1 lists some of the specific issues related to scope that may be discussed during this stage. Once planning and scoping are under way, problem formulation begins and runs in parallel with them. Discussions during this stage focus primarily on methodologic issues of the desired assessment, as illustrated in Box 3-2. It is important to note that communication between the two, *now parallel* stages, needs to occur for the assessment to be useful. The overarching purpose of the two critical, but often underused, stages of the risk-assessment process is to provide a clearer and more explicit connection between the decision-making context and the risk assessment that will inform the decision-maker. It also makes more explicit the relative roles of the decision-maker, stakeholders, and the risk assessor (EPA 2003, 2004a).

Planning and Scoping

In 1989, EPA's guidance for Superfund provided several pages of guidance specific to the planning and scoping of a human health risk assessment (EPA 1989). Because assessment of complex ecologic systems challenged both decision-makers and assessors, it was

BOX 3-1 Selected Elements of Scope Considered During Planning and Scoping

- Spatial and temporal scope options
- Direct hazards and stressors
- Mitigation-related hazards and stressors
- Sources
- Source-mitigation options
- Environmental exposure pathways
- Exposure-mitigation options
- Individual intake pathways
- Individual intake mitigations
- At-risk populations
- Populations at mitigation-related risk
- Direct adverse health outcomes
- Mitigation-related adverse health outcomes

BOX 3-2 Selected Methodologic Considerations in Problem Formulation

- Hazard-identification methods
- Stressor-characterization methods
- Source-characterization models and methods
- Environmental transport and fate models and methods
- Computational methods
- Uncertainty-characterization methods
- Intake and internal-dose models
- Dose-response models and methods
- Health-outcome measurement (risk measurement) methods
- Integrated cost-benefit methods
- Transparency, dissemination, and peer-review methods

the ecologic risk-assessment community that ultimately championed the need to define the scope of a risk assessment and the need for discussion between decision-makers, assessors, and interested parties from the outset of an assessment. The need to scope an assessment and the need for assessors and managers to interact were discussed briefly in EPA's 1992 framework for ecologic risk assessment (EPA 1992). NRC (1993) advocated for the integration of ecologic risks into the 1983 Red Book paradigm, and expressed a need to extend this paradigm to include the need for interaction between risk assessment and management at the early stages of a risk assessment, based on experience in ecologic assessment. In 1996, a National Research Council committee commented on the importance of planning from the beginning of a risk assessment (NRC 1996). In 1998, EPA released its guidance for ecologic risk assessment, which superseded the 1992 framework document and provided a greatly expanded discussion of scoping and of the roles of assessors and decision-makers; it also drew a clear distinction between the goals and content of the planning and scoping stage and the problem-formulation stage. More recently, EPA has further articulated how critical planning and scoping are for the conduct of a successful risk assessment and has provided detailed guidance for their conduct (EPA 2003, 2004a). During planning and scoping, a team of decision-makers, stakeholders, and risk assessors identifies the issue (or concern, problem, or objective) to be assessed and establishes the goals, breadth, depth, and focus of the assessment. Once the decision to use a risk assessment has been made, this stage becomes critical for developing a common understanding of why the risk assessment is being conducted, the boundaries of the assessment (for example, time, space, regulatory options, and impacts), the quantity and quality of data needed to answer the assessment questions, and how decision-makers will use and communicate the results. During this stage, decision-makers charged with protecting health and the environment, in the context of other competing interests, can identify the kinds of information they need to reach their decisions, risk assessors can ensure that science is used effectively to inform decision-makers' concerns, and stakeholders can bring a sense of realism and purpose to the assessment. This stage is a focal point for stakeholder involvement in the risk-assessment process and the point at which risk communication should begin (EPA 2003). The relevance of risk-assessment results to decision-making can be enhanced by the up-front involvement of decision-makers and stakeholders in setting goals, defining options, and defining the scope and complexity of an assessment (Suter et al. 2003). Together, all can evaluate whether the assessment will help to address the identified problems (EPA 2004b).

While a common plan for the risk assessment is one of the goals of these stages, reaching consensus on all aspects of the scope and conduct of a risk assessment among decision-makers and stakeholders representing diverse interests, will not always be feasible. In addition, it is not necessarily in the public interest to delay the risk assessment where consensus is difficult to achieve. The process requires a balance among the competing values of deliberative input into a risk assessment, timeliness in the risk assessment process, and the resource burden associated with these early stages.

Early Identification of Decision-Making Options

As discussed later in this chapter and further in Chapter 8, the utility of a risk assessment is greatly enhanced when it is constructed and carried out in the context of a clear set of options under consideration by the decision-maker. Figure 3-1 explicitly includes identification of options as a critical element of planning and scoping. Although present EPA guidance (for example, on ecologic risk assessment, cumulative risk assessment, and air toxics) does not contain exact language calling for the explicit identification of decision-making options during the planning and scoping stage, it does allow preliminary consideration of regulatory or other management options. Existing EPA risk-assessment frameworks unquestionably contemplate consideration of *options as they are related to decision-making,* with plenty of interpretive room for arraying options if that is desired by or available to decision-makers and risk managers. For example (EPA 1998, p. 10), "risk assessors and risk managers both consider the potential value of conducting a risk assessment to address identified problems. Their discussion explores what is known about the degree of risk, *what management options are available to mitigate or prevent it*, and the value of conducting a risk assessment compared with other ways of learning about and addressing environmental concerns" (emphasis added). Not every issue faced by a risk manager will necessarily lend itself to "arraying options." Some complex problems may also best be addressed by completing a thorough assessment of health risks and vulnerable populations prior to considering necessary control options. The Clean Air Act has used this approach to reduce air pollution concentrations over the past four decades. In the management of contaminated sediment, for example, it may be possible to examine the tradeoffs between various options, such as removal vs monitored natural recovery or capping versus hot-spot removal; in the case of soil contaminants for which no practical treatment options exist, options may be limited to various degrees of soil removal; and there may be instances in which the regulatory environment is so prescriptive as to preclude all but a few stipulated options.

Although the planning and scoping stage is primarily deliberative, in that it involves extensive discussion between decision-makers and stakeholders and to a smaller extent with risk assessors, it is expected to produce tangible products that are critical for the performance of a credible and useful risk assessment (EPA 2003, 2004a). The primary product is a statement, with explanation, of why the assessment is being performed and what it will include and exclude (that is, how comprehensive it will be). Other products may be descriptions of those involved and their roles (for example, technical, legal, or stakeholder advisers), key agreements made and understandings reached among those involved, the resources (such as budgets, staff, data, and models) required by or available to the assessment, and the schedule to be followed (including provision for timely and adequate internal and independent external peer review). A statement (Box 3-3) often summarizes the end result of the planning

BOX 3-3 Planning and Scoping: An Example Summary Statement

"Air toxics emissions may be causing increased long-term inhalation health risk (both cancer and noncancer concerns) to people in the immediate vicinity of Acme Refining Company. A modeling risk assessment will be performed to evaluate potential long-term human health impacts of inhalation exposures to all air toxics emitted by the facility. Inhalation risks for populations within 50 km of the Acme property boundary will be assessed under residential exposure conditions. Noninhalation pathways will not be assessed for either human or ecological receptors" (EPA 2004a, Chapter 5).

and scoping process, describing the specific concerns that the risk assessment will address and generally what will be included in its purview. The problem-formulation stage, whose specific products are a conceptual model and an analysis plan, develops the specific technical details for the assessment laid out during planning and scoping.

Problem Formulation

The extension of the concept of "problem formulation" to human health risk assessment first emerged during a 1991 National Research Council–sponsored risk-assessment workshop where the absence of such an activity in health risk assessment and the criticality of its use for ecologic risk assessment were discussed (NRC 1993). In 1992, EPA published *Framework for Ecological Risk Assessment* as the first statement of principles for ecologic risk assessments, including a further articulation of the concept of problem formulation (EPA 1992). The concept reached fruition in the agency's 1998 *Guidelines for Ecological Risk Assessment*, which superseded the 1992 framework document (EPA 1998). Those documents describe methods for conducting conventional single-species, chemical-based risk assessments and techniques for assessing risk to ecosystems from multiple exposures (or stressors) and multiple effects (or end points) (EPA 1991). For several reasons, ecologic risk assessments in the United States have generally placed a greater emphasis on problem formulation than have human-health risk assessments (Moore and Biddinger 1996). But by emphasizing completion of problem formulation early, the ecologic-risk framework provides a clear procedural advantage over the existing human health risk framework in achieving an assessment that can be used to inform a management decision. The advantage is derived from having decision-makers and stakeholders as active participants from the beginning of an assessment rather than passively awaiting receipt of the results.

The problem-formulation stage sketches out the technical implications and decisions that are implied by the discussions that occur among decision-makers and stakeholders during planning and scoping so that risk assessors can proceed with the technical aspects of the assessment in a manner consistent with the decision context. This stage translates the results of the planning and scoping stage into two critical products: a conceptual model that explicitly identifies the stressors, sources, receptors, exposure pathways, and potential adverse human health effects that the risk assessment will evaluate and an analysis plan (or work plan) that outlines the analytic and interpretive approaches that will be used in the risk assessment. The general concern and approach articulated in the summary statement developed during scoping are given greater detail in a study-specific conceptual model. The model comprises both graphic illustrations (see Figure 3-2) and narrative descriptions that explicitly identify

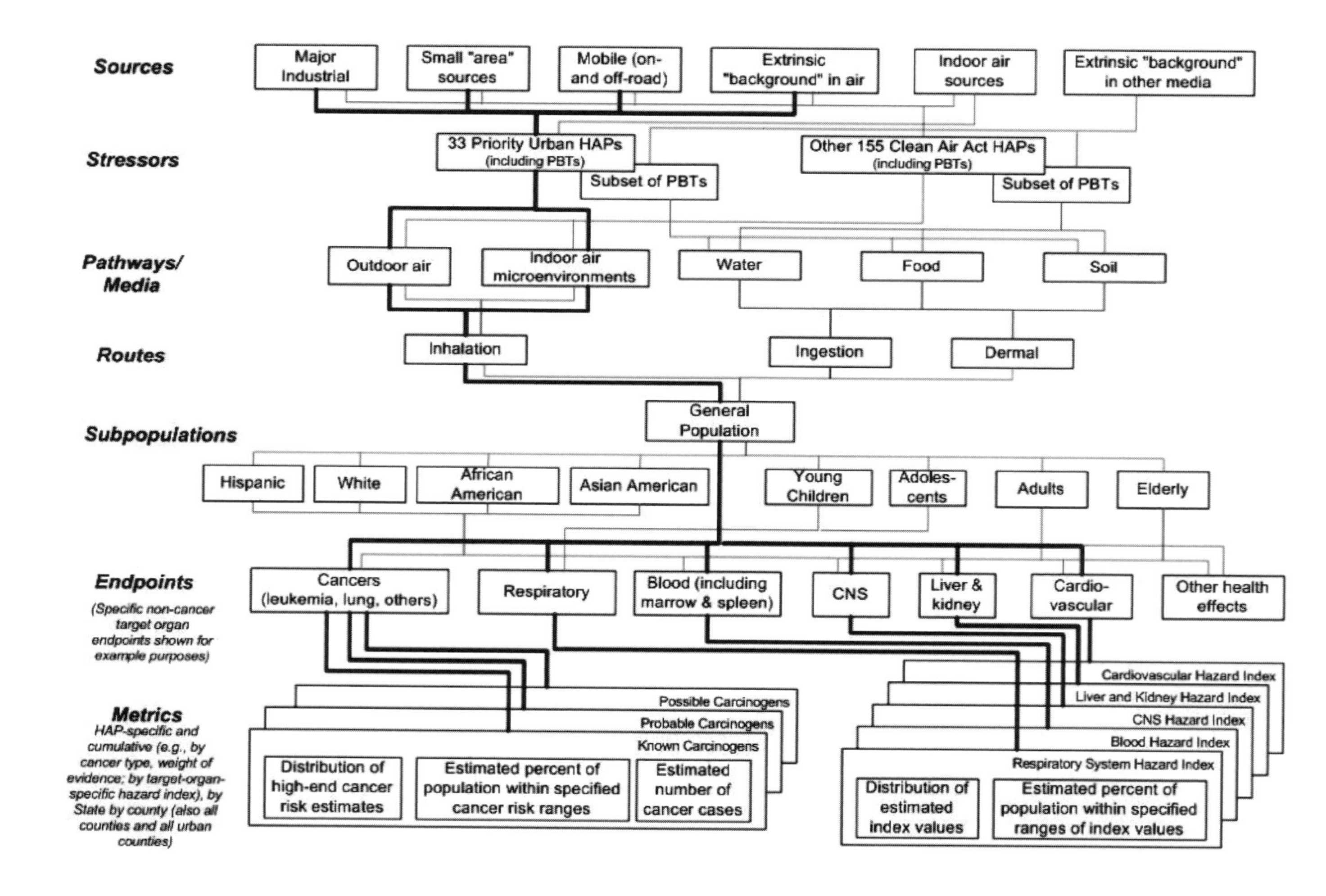

FIGURE 3-2 Illustration of the scope of a risk assessment, indicating both pathways considered (bold lines) and pathways not considered. Source: EPA 2004a.

sources, contaminants of concern (stressors), exposure pathways, potential receptors, and adverse human health effects that the risk assessment is going to evaluate.

> The review of the conceptual model led to significant savings in the application of the model for calculating air dispersion, exposure and risk estimation. More than a third of the possible analyses were shown to be unnecessary to address the problem formulated in the planning and scoping discussion [EPA 2002, p. E-6].

For important risk assessments, particularly controversial or precedent-setting ones, it may be advisable that the scientific and technical credibility of the conceptual model be examined with a peer-review process. Although the conceptual model serves as a guide for determining what types, amount, and quality of data are needed for the assessment to address the issues and concerns of interest to decision-makers, the analysis plan matches each element of the conceptual model with the analytic approach that the assessors initially intend to use to develop data or otherwise represent that element. Box 3-4 lists some of the major elements of an analysis plan (from EPA 2004a).

BOX 3-4 Major Elements of an Analysis Plan

Sources	How will information on the sources in the analysis (e.g., source location, important release parameters) be obtained and analyzed?
Pollutants	How will chemicals of potential concern (COPC) be confirmed and their emissions values be estimated?
Exposure pathways	How will the identified exposure pathways be assessed? How will ambient concentrations be estimated?
Exposed populations(s)	How will exposures to populations of interest be characterized? How will their exposure concentrations be estimated? What will be the temporal resolution? What sensitive subpopulations may be affected?
End points	How will information on the toxicity of the COPC be obtained (what are the data sources)? What risk metrics will be derived for the risk characterization?

In addressing the above aspects of the analysis, the plan should also clearly describe the following:

- How will *quality* be ensured in each step (e.g., what will be included in the quality assurance/quality control plans)?
- How will *uncertainty* and *variability* in the results be assessed?
- How will all stages of the assessment be *documented*?
- Who are the *participants* and what are their *roles* and *responsibilities* in the various activities?
- What is the *schedule* for each step (including milestones)?
- What are the *resources* (e.g., time, money, personnel) being allocated for each step?

Source: EPA 2004a.

Recognition of the Need to Strengthen the Use of Formative Design Stages

The specific nature and needs of the decision environment are often neglected in risk assessment, if there is no systematic approach (Crawford-Brown 1999). It is increasingly clear that even "the highest-quality risk assessment is worthless if it does not address the needs of the decision-maker" (Suter 2006, p. 4). EPA guidance documents make it evident that the agency recognizes, at least in theory, that "[planning and scoping] may be the most important step in the risk assessment process" and that "without adequate [planning and scoping], most risk assessments will not succeed in providing the type of information that risk management needs to make a well founded decision" (EPA 2004a, p. 5-9). Similar ideas were also expressed in a report on EPA risk-assessment practices by GAO (2006). EPA has also observed that many of the shortcomings or failures of ecologic risk assessments can be traced to a weakness in or lack of problem formulation (CENR 1999).

Both the planning and scoping and problem-formulation stages are necessary to ensure that the form and content of a risk assessment are determined by the nature of the decision to be supported. Both stages offer opportunities to reach some level of consensus on how to proceed (for example, with respect to regulatory context and objectives, scientific objectives, data needs, or reasonably expected limitations) in an assessment so that its results will be useful and informative to decision-makers. Those stages also offer excellent opportunities to give risk communication an early and pivotal role in the overall risk-assessment process rather than allowing it to become an afterthought. Although both planning and scoping and problem formulation can be challenging and time-consuming, the time and effort are usually well spent and have been shown to result in risk assessments that are more useful to and better accepted by decision-makers (EPA 2002, 2003, 2004a).

The incorporation of those stages in the risk-assessment process is, however, still inconsistent. For example, both stages are missing from EPA's new cancer guidelines and from the current Office of Solid Waste and Emergency Response risk-assessment protocol for combustion facilities (EPA 2005a,b). Thus, although the stages are now widely acknowledged, at least conceptually, as critical for the success of a risk assessment (particularly for complex, controversial, or precedent-setting assessments) and guidance for their conduct is available, the question remains as to whether EPA or other public agencies, the regulated community, or their contractors are taking full advantage of them to focus, refine, and improve human health and ecologic risk-assessment efforts. The question warrants attention in that continued inattention to the importance of planning and scoping and of problem formulation can be expected to yield human health risk assessments (by EPA and others) that fail to reach their full potential in providing support to decision-makers and others seeking solutions to environmental and health concerns.

INCORPORATING VALUE-OF-INFORMATION PRINCIPLES IN FORMATIVE AND ITERATIVE DESIGN

Scylla and Charybdis:[2] Navigating the Twin Hazards of Uncertainty and Delay in Decision Support

The combination of the magnitude and the practical irreducibility of key uncertainties and their impact on decision-making constitute the core challenge in efforts to achieve a

[2]Scylla and Charybdis are two sea monsters of Greek mythology. They were located on opposite sides of a narrow strait such that they posed an inescapable choice in that avoiding Scylla (a six-headed monster) required passing too closely to Charybdis (presenting a whirlpool hazard) and vice versa.

robust yet practical approach to public-health and environmental risk management. The combination of uncertainty and the prospect of delaying important decisions constitute a key hazard in navigating the difficult waters of health and environmental decision-making. To an extent, it can be argued that the conflict is inherent in a decision-making environment that, while valuing a timely decision, places a large premium on the often-repeated yet ill-defined goal that the decision be "scientific," "based on sound science," or "based on the best available science." The nature of the conflict can be understood by recalling that the scientific process of seeking the truth, by design and to its great credit, has no natural end point. Also by design, the training of scientists and such embedded traditions as applying tests of statistical significance instill the value of prudence and the "due-process" tasks of peer review, replication, and scientific debate before a conclusion can be said to be based in science. The idea that there are risks (for example, prolonged exposure to a hazard, or stress in the community awaiting an assessment of health risks) that may be associated with waiting for a particular study to be completed or for a scientific consensus to emerge is not readily incorporated into the standard scientific paradigm.

The lack of established "stopping criteria" in science contributes to the conflict wherein any attempt to put an end to or otherwise constrain scientific inquiry and debate to meet regulatory or legal deadlines or, perhaps most problematically, to achieve an abstract notion of *timeliness* can lead to the accusation that the corresponding decision is "unscientific." In pursuing the goal of timely decision-making, there is an inherent conflict between meeting the requirements associated with the goal of *knowing* and the requirements associated with the more pragmatic goal of *deciding*.

Protection of the public and protection of the scientific knowledge base from Type 1 errors (that is, avoiding false positives) are not equivalent goals. That fact, somewhat obvious when considered carefully, is a fundamental source of tension that is not sufficiently acknowledged or confronted directly in risk assessment, risk communication, and risk management practices. Navigating (as opposed to resolving) the conflict between those goals is best addressed through its careful consideration by both risk managers and risk assessors in the formative and iterative design of risk assessment. To confront that challenge, risk managers must see themselves as managing uncertainty and delay as well as managing risk. Managing under uncertainty requires diverse strategies that address different aspects of the overall decision-making process, including investments to collect, store, and manage information; investments to improve the knowledge base, that is, to generate new knowledge; formalization of the processes used to collect, use, and process information; formalization of processes to calculate and communicate uncertainty; adjustment of the risk-assessment process to mitigate the practical impact of the uncertainty on the analytic process; adjustment of the decision-making process to accommodate the consideration of the uncertainty; and adjustment of the timing of decision-making in both directions—to delay or to expedite—when uncertainty is acknowledged to be sufficiently great.

It is important to note that the day-to-day work of uncertainty management should not be considered the sole domain of analytic experts. It is primarily the responsibility of the risk manager to prescribe and implement appropriate accommodations in the overall decision-making process if the analytic efforts aimed at supporting decisions under great uncertainty are to have the desired impact and to ensure that risks associated with delays in decision-making are balanced by the likelihood and magnitude of any benefit that is believed to be associated with proposed enhancements of the knowledge base or with the process of risk assessment. Choosing a strategy involves important tradeoffs because any strategy to deal with uncertainty will be incomplete and imperfect. The committee believes that one of the dominant pathways to improving risk analysis involves correctly matching the uncertainty-management strategy to the particular demands and resources of the decision-making environments in and outside EPA. This issue is discussed in greater detail in Chapter 4. Ideally, the matching process would be expanded to consider the many other decision-makers that make use of EPA's analytic products.

Value of Information: What Makes Information Valuable?

A fundamental aspect of decision-making under uncertainty involves the inevitable choice between making an immediate decision with the information and analysis available and delaying the decision while, for example, more raw information is collected, a more refined analytic product is prepared, or consultations with affected parties are conducted. Even if delay is not the primary concern, the direct and indirect costs of acquiring the information will often need to be considered.

As the most generic analytic framework for valuing information in the context of decisions, value-of-information (VOI) analysis provides a set of methods for optimizing efforts and resources to gather, to process, and to apply information to help decision-makers achieve their objectives. The application of VOI analysis is illustrated schematically in Figure 3-3.

The Process of Quantitative Value-of-Information Analysis

The decision-theoretic process to quantitatively value information begins with analyzing the best option available to the decision-maker in a certain state of uncertainty. This serves as a baseline scenario with respect to information available to the decision-maker. The process then systematically considers when and how the decision-maker's preferred option might be changed if the decision-maker was able to incorporate additional information into the decision that was not available in the baseline scenario. This new information is expected to either eliminate or reduce the extent of a source of uncertainty.

In VOI analysis, the decision-maker is assumed to change the preferred option only when there would be a change in the net expected benefits. Accordingly, in addition to consideration of how likely it is that the preferred decision would change, the process measures how much of an increase in benefit would be expected given the additional information. The net (or expected) value of gathering information to resolve or reduce uncertainty is calculated by weighing the increase in benefits associated with each potential outcome of the information collected by the probability of each outcome. This weighing process includes assigning the value zero (that is, representing no increase in benefits) for situations where the information gathered does not change the decision-maker's preferred option.

A critical part of understanding the concept of VOI analysis is to differentiate scientific and decision-analytic perspectives on the value of information. In research proposals and in the literature, scientists often describe proposed studies as valuable with respect to enhancing the overall knowledge base, perhaps with a suggestion that it will inform important decisions. Conversely, the decision-analytic notion of VOI is entirely decision-centric. In a VOI analysis, an information source is valued solely on the basis of the probability and magnitude of its potential impacts on a specific decision at a specific time with a specific state of prior knowledge. Therefore, it is a common and expected result of VOI analysis to estimate that an information source, which may otherwise be considered valuable as a general scientific matter, has little or no value in support of a particular decision. This happens when the specific decision is not sensitive to the resolution of the uncertainty that the information source addresses. Considering this situation in Figure 3-3, the arrow indicator, which denotes that option C is preferred given currently available information, would not be moved much by this source of new information.

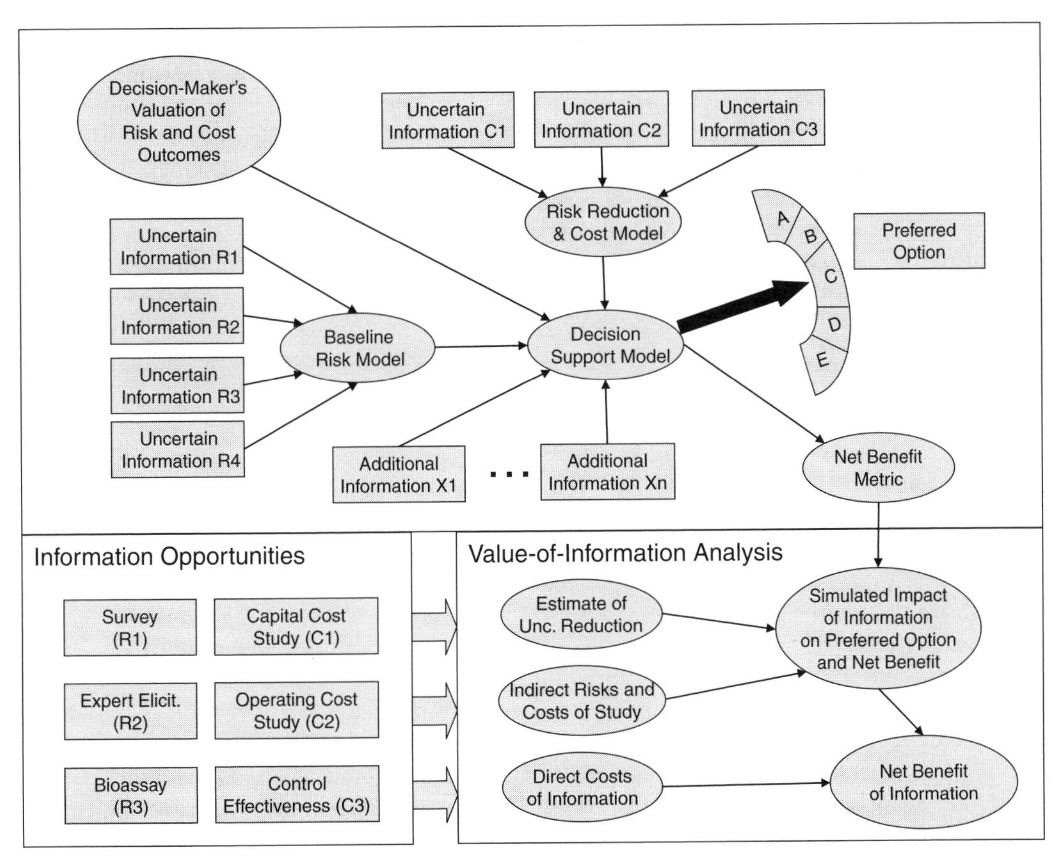

FIGURE 3-3 Schematic of the application of value-of-information analysis to assess the impacts of additional studies in a specific decision context. Information opportunities that address uncertainties in the baseline model are considered with respect to the changes they would have on the decision-maker's preferred decision option and the associated change in net benefits. The analysis may also consider any direct costs (for example, financial) and indirect costs (for example, the health or economic impacts of delayed decision-making) associated with the information opportunity. The valuation of information is ultimately driven by the decision-maker's values with respect to the distribution of risks and costs, including any costs associated with delayed decisions.

Experience in the Application of Value-of-Information Methods

The applications of VOI methods in environmental health decision-making might be characterized as sporadic and somewhat academic (Yokota and Thompson 2004). In the academic literature, there has been a considerable interest in the use of VOI techniques to evaluate various activities within toxicity testing (Lave and Omenn 1986; Lave et al. 1988; Taylor et al. 1993; Yokota et al. 2004). Recently Hattis and Lynch (2007) applied a VOI framework to assess the expected effect of improved human pharmacodynamic or pharmacokinetic variability information on doses deemed to be protective for noncancer effects. VOI methods have been employed to estimate value of sampling information in the context of environmental remediation (Dakins et al. 1996), and in an assessment of information value in the context of alternate control policies for source water protection in a watershed impacted by agricultural runoff (Borisova et al. 2005). Other applications can be found in the value of improved exposure information in the case of drycleaning operations (Thompson and Evans 1997), and the value of genetic screening options related to prevention of beryllium disease (Bartel et al. 2000).

There is evidence of sporadic interest and research aimed at employing VOI methods at EPA. For example, Messner and Murphy (2005) present an analysis of VOI about the quality of source water in the context of decisions about investments in drinking water treatment plants. In other applications, EPA staff and contractors have applied VOI principles in assessing the value of environmental information systems and human exposure information across a class of regulatory decisions (IEc 2000; Koines 2005).

Prospects for Formal Value-of-Information Analysis at EPA

VOI analysis has a number of benefits in support of decision-making compared with the more common scientific characterization of the potential value of a study. The intuitive and idiosyncratic views of individual scientists and decision-makers tend to place high value on information from their own discipline while diminishing the value of information from other disciplines. Scientists from all disciplines may devalue information that is not scientifically *interesting* (for example, that would not be publishable in a scientific journal) even if it substantially reduces a critical uncertainty in a risk assessment and the knowledge has considerable potential to affect the decision-maker's choice of the best option. In contrast, VOI analysis could provide both a more context-specific and a more objective assessment of the decision-centric value of a piece of information or, by extension, the value of an information system to a class of decisions that might use it. Despite the potential benefits, it is important to note that a VOI analysis is not considered to be generally *superior* to the use of expert scientific judgment about the importance of a scientific investigation; rather, it answers a much narrower question about the importance of a study for the outcome of a *specific decision* and is not appropriate as a general measure of the scientific merit and broader utility of a study.

For example, in the context of some specific decision, a VOI analysis might place great value on a small survey to estimate the fraction of businesses using a near-obsolete technology and very little value on a large, well-designed, and broadly important scientific study when considering only the narrow purposes of the specific decision at hand. The decision-maker's preferences for options (perhaps in choosing among options B, C, and D in Figure 3-3) may be very sensitive to the level of uncertainty in risk reductions and the costs that would be imposed on businesses by a decision that would, for example, forbid the continued use of the older technology. In both the risk estimation and the cost estimation, the number of such businesses may be an important consideration in this particular decision context. Conversely, a scientific study that would contribute to the understanding of the risk and may reduce the overall uncertainty in a broadly desirable and scientifically rigorous way may not be able to add information that changes the relative desirability of the specific options enough to change the decision-maker's preferred choice. Clearly, there are many other scenarios in which scientific investigation is precisely what is required to differentiate adequately among available options.

Despite the intellectual appeal of the formal VOI analytic framework and the ever-present need for a robust means of assessing information value, the formal VOI paradigm imposes a number of challenges that limit its practical and widespread use in the near term. The use of the formal VOI framework in environmental health applications has been extensively reviewed by Yokota and Thompson (2004). One of their findings relates to the somewhat academic status of VOI in this field:

> Rigorous VOI analyses provide opportunities to evaluate strategies to collect information to improve EHRM [environmental health risk-management] decisions. This review of the methodology and applications shows that advances in computing tools allow analysts to tackle problems with greater complexity, although the literature still lacks "real" applications, probably due to a number of barriers. These barriers include the lack of guidance from EPA and others on criteria for standardizing EHRM risk and decision analyses, the lack of consensus on values to use for health outcomes, the lack of default distributions for frequently used inputs, and inexperience of risk managers and communicators with using probabilistic risk results.

There are important considerations in addition to the barriers expressed above.

- VOI computation can be technically challenging, particularly when one is trying to evaluate imperfect information, which is almost always the relevant case.
- Its analytic formality does not lend itself to being combined with the more common deliberative approaches of determining the potential value of information.
- The approach presumes that the analyst can fully describe the change in a decision-maker's choices in response to new information. This condition is not very realistic (or at least is rarely the case) and is particularly problematic when the decision-making process is not rule-driven or whenever the VOI analyst is forced to speculate as to the behavior of the decision-maker in response to new information.
- The impact of the new information must be characterized with respect to the resulting change in a probability distribution that describes the current level of uncertainty, which may not be formally characterized as a probability distribution.
- Very few technical or policy analysts or decision-makers have had any exposure to this type of analysis, suggesting a considerable burden of training.
- The "value" assigned in a VOI analysis is itself, ultimately, an uncertain quantity.

A key challenge for uncertainty management in EPA and elsewhere is the need to design the risk assessment to support decisions with respect to an explicit array of candidate options that the decision-maker is likely to consider. Without these options, it is not possible to assert a formal decision-centric valuation of information; indeed, in this case, a formal VOI analysis cannot even be attempted. A key potential side effect is the perpetuation of "incomplete" risk assessments. The perpetuation side effect is a natural result in the absence of a well-characterized decision-support context, including a concrete array of decision options, because there will always be a scientific rationale, as opposed to a decision-centric rationale, to continue to gather information, perform or review new studies, and to improve technical aspects of a risk assessment.

The committee recognizes both the advantages of VOI analysis for risk assessment and risk management as well as the presence of continuing barriers to the use of formal and computational VOI analysis in EPA. As a result, there is likely to be only a small proportion of risk assessments and decision contexts that meet the criteria where a formal VOI is possible (for example, having clear decision rules and prior estimates of uncertainty) and for which the stakes are high enough to make a VOI analysis cost-effective.

Alternative VOI Methods for Diverse Decision-Making Contexts

As an alternative that is applicable to a larger proportion of decision contexts, the committee believes that EPA would benefit from developing and applying a structured but less quantitative method for assessing the value of new information that captures the essential reasoning embodied in VOI analysis. The essential reasoning in the formal VOI approach is based on the explicit characterization of a *direct causal link* between a specific source of new information, the predicted change in the behavior of a decision-maker given this new information, and the resulting *improvement with respect to the decision-maker's objectives that can be expected in the presence of the additional knowledge*. Essentially, the process of valuation would involve the presentation of a qualitative or semiquantitative argument (as opposed to formal computation) that describes the causal relationship between the knowledge that might come from the considered source of information and the potential for improved decision outcomes. The process could also consider the potential for risk in delaying the decision until the information is available and is adequately incorporated into the decision-support products (either risk assessments or cost assessments). An example of the development and application of a structured semiquantitative VOI method, including a discussion of the complementary role of these methods, can be found in Hammitt and Cave (1991).

Valuing Methodologic and Procedural Improvements in Risk-Assessment Design

Earlier in this chapter, the committee described the rationale for placing a great premium on aspects of *process* in risk assessment. When all the combinations of choices of scope and technical, consultative, and quality-control methods are considered with the variations in the intensity of their application, it could be argued that there are an uncountable number of ways in which a risk assessment could be constructed. Such flexibility is generally welcome and has the potential to make risk assessment relevant to the broadest possible array of applications, but it can be problematic.

The essentially deliberative process of matching opportunities to enhance the risk-assessment process with the objectives of achieving high-quality decision support may be facilitated by using a decision-centric evaluation model that characterizes the impact of any proposed enhancements to the risk assessment—and its manifestation in the form of a risk-assessment product with corresponding attributes—on the desired objectives of the decision-making function. The committee encourages the development of such an evaluation framework for *methodologic improvements* in risk assessment that instills some of the concepts of decision-analytic value of information. A schematic of such an evaluation model is illustrated in Figure 3-4.

The proposed evaluation framework would expand the consideration of the casual relationship between risk-assessment activity and the quality of decision-making in two respects. It would be structured to assist in the relative valuation of the many attributes of risk-assessment processes and products that need to be considered in the formative and iterative design process. By relaxing the formality of the VOI approach, it could include a broader set of decision-making objectives—such as transparency, timeliness, integration with other decision inputs, and compatibility with stakeholder participation—that are less tangible and quantifiable but nonetheless critically important in determining the overall decision-support *value* of a given activity or effort.

An important aspect of instilling the benefits that are analogous to VOI analysis will be in drawing explicit causal linkages, even if expressed qualitatively, between risk-assessment design options and the ultimate impact on the decision-making environment. In this way, the potential for the "value-of-methods" approach is limited in an analogous way by one of the barriers in the formal VOI approach. In VOI analysis, the analyst must know the decision-maker's valuation of risk assessment or other quantitative outcomes in sufficient detail as to predict a change in the decision-maker's behavior in response to new information (that is,

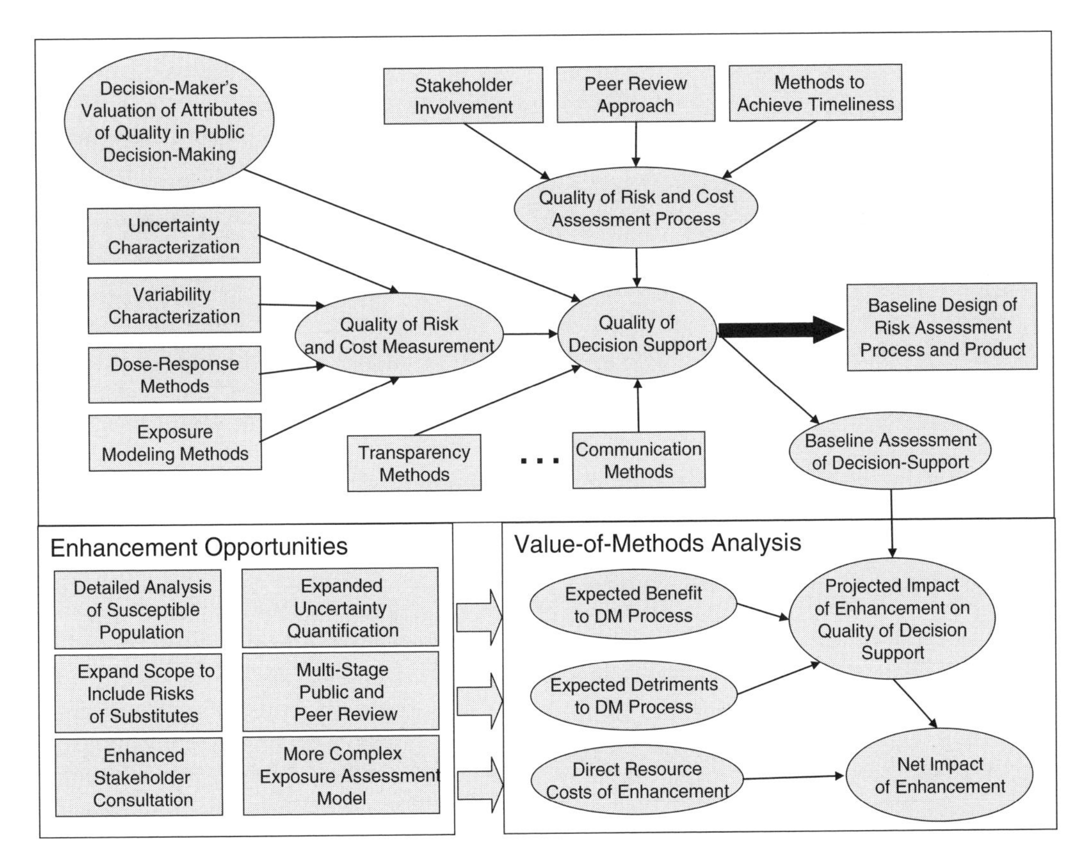

FIGURE 3-4 Schematic of an analysis of the value of various methodologic opportunities (or "value of methods" analysis) to enhance the risk-assessment process and products. The structure mimics the standard VOI approach, but focuses on different impacts. In contrast with VOI analysis, the valuation of these opportunities is derived from the value system that specifies the desirable attributes of the overall process of public-health and environmental decision-making. Whereas VOI analysis considers the impact of information on the decision outcome (the "ends"), this type of analysis would consider the impact of diverse risk-assessment methods on the overall quality of decision support (the "means").

predicting their choice among available options, or their choice in setting a single number within a continuum). In the value-of-methods approach, the analyst who is contemplating the value of a particular risk-assessment method (for example, in choosing among a qualitative, quantitative scenario-based, or fully probabilistic characterization of uncertainty) requires some way to characterize the change in the decision-support environment that corresponds to each of these alternative methods. Further, the analyst would need to know how much the different changes in the decision support environment are valued based on the capacity of the decision-making process to take advantage of the method, and the institutional values of the desirable qualities of decision-making. In order to remove this potential barrier, this expression of the valued attributes of decision support would be made highly context specific (for example, having very different objectives for community-level decision support as compared to a national standard-setting process) and would be agreed to and documented in the formative stages of risk-assessment design.

**Weight-of-Evidence and Hazard Classification:
An Example of a Value-of-Methods Question**

The phrase *weight of evidence* (WOE) is used by EPA and other scientific bodies to describe the strength of the scientific inferences that can be drawn from a given body of evidence. In its most common applications in EPA, WOE is used to characterize the hazardous (toxic or carcinogenic) properties of chemicals on the basis of an integrated analysis of all relevant observational and experimental data. It is increasingly used to describe the strength of evidence supporting particular modes of (toxic) action (MOAs) and dose-response relationships. Because scientific evidence used in WOE evaluations varies greatly among chemicals and other hazardous agents in type, quantity, and quality, it is not possible to describe the WOE evaluation in other than relatively general terms. It is thus not unexpected that WOE judgments in particular cases can vary among experts and that consensus is sometimes difficult to achieve.

Perhaps the most formal WOE activity undertaken in EPA concerns the classification of carcinogens. The weighing of evidence from epidemiology and experimental studies pertaining to specific chemicals or chemical mixtures that may be carcinogenic involves substantial agency resources and can lead to controversy and extended debate.

One distinction made in EPA carcinogen classification is whether the available evidence is sufficient to establish causality for humans (that is, whether a substance can be labeled as a "known" human carcinogen) or falls short and indicates that the agent is a "likely" human carcinogen. Causal relationships can be more straightforward to establish in well-done clinical and (in animals) experimental studies, but an individual observational (epidemiology) study typically can establish only a statistical association. A larger body of epidemiologic evidence can be sufficient to rule out bias and confounding with sufficient confidence to support a causal relationship; with experimental evidence, it may be sufficient to establish causality in humans. The weighing of such evidence can be controversial, so such institutions as EPA, the International Agency for Research on Cancer (IARC), the Institute of Medicine (IOM), and the National Toxicology Program (NTP) have developed practices and classification schemes to aid the process of reaching conclusions about the overall evidence. NTP, IOM, and IARC convene expert bodies to undertake WOE analyses of carcinogenicity data; EPA relies on peer review by expert groups, such as its Science Advisory Board, to vet staff findings on carcinogenicity evidence.

The committee notes that in some cases there does not appear to be substantial value in the agency's making distinctions between certain carcinogenicity classifications. Whether a chemical is "carcinogenic in humans" or "likely to be carcinogenic in humans" generally has no important influence on the ultimate quantification of risk and the use of risk estimates in decision-making. In many regulatory contexts, known human carcinogens may be treated no differently from "likely" human carcinogens: risks are estimated for all substances for which there is sufficiently convincing evidence of carcinogenicity, irrespective of whether human causality has been established, and the risk estimates are not adjusted according to the WOE classification.

As a result, once the available evidence, either epidemiologic or experimental, is judged sufficient to establish that a given finding of toxicity or carcinogenicity is potentially relevant to humans, there may not be the need for further distinctions in classification, except in some circumstances as a communication tool. Unless clear reasons are brought forward at some stage, such as in the formative design stage of risk assessment, to support the need for such a definitive human causality assessment, the committee sees no reason for the agency to spend time and resources to fine-tune the hazard classification in order to settle the question of whether the agent is a likely or known cause of the effect in humans.

However, the systematic consideration of evidence in WOE analyses remains important as a matter of good scientific practice. Thus, whether the accumulated evidence is sufficient to consider a substance potentially hazardous to humans or is sufficient to support a given MOA requires a weighing of individual studies and pieces of evidence, and this practice should continue. The committee recommends that the agency remain mindful of cases in which fine distinctions have little or no impact on the overall use of risk information.

WOE classification provides an example of distinctions between the formal VOI analysis and the less formal value-of-methods analysis. The fact that these finer distinctions in WOE classification are not used further in risk assessment or in any apparent decision rule used by EPA suggests placing no value on the exercise to seek these distinctions, when the potential benefit is viewed purely from a formal VOI analysis perspective (as illustrated in Figure 3-3). But a WOE classification that distinguishes known from likely carcinogens may be deemed by EPA to be required in support of other values associated with risk assessment practice (for example, using a "good scientific practices" argument, or as the basis for a simplified means of communication of the epistemic status of a claim of carcinogenicity). WOE is an example of how EPA may benefit from a structured characterization (as described above and illustrated in Figure 3-4) of the exact role of a resource-intensive method in supporting the broader goals of public-health and environmental decision-making, which would include, among many other aspects, the use of good scientific practices and consideration of good communication practices. The method would require a more explicit valuation of important attributes of quality in decision support.

CONCLUSIONS

- The nature of health and environmental risk management places great demands on both the processes and the products of risk assessment. In reviewing the history and many objectives of risk assessment, the committee finds that a more aggressive formative design stage is critical for the future success of risk assessment. The design should reflect the many objectives of the decision-making function and maintain this focus throughout the life cycle of the assessment.
- The key role of design in risk assessment is captured in current EPA guidance for ecologic risk assessment and cumulative human health risk assessment and embodied in the tasks of planning and scoping and problem formulation.
- A key design consideration for risk assessment lies in the potential for a poorly designed risk-assessment process to contribute to increased risk by a number of pathways. These include the potential to contribute to excessive delays in decision-making, to divert assessment and management attention from competing hazardous concerns, to contribute to ill-informed substitution of one risk for another, and to create barriers to inclusion or acceptance of risk assessments by various stakeholders.
- Decisions to invest in additional information to support a risk assessment are standard and important in risk management. The investment can be in the form of direct costs, resource costs, or delay. Standard scientific rationales for asserting that a study is important may be misleading when considered from a purely decision-centric perspective. The committee acknowledges the potential for a key beneficial role of VOI analysis in providing an objective measure of the potential impact of new information on a particular decision. A number of barriers to application of formal VOI methods limit its general applicability. However, the

underlying structure of VOI analysis in expressing an explicit causal link between information, decision-maker behavior, and decision-making objectives is broadly applicable. It can be extended to guiding a number of design decisions at the formative and later stages of risk-assessment design. A value-of-methods analysis would provide an approach for considering the impact of opportunities, in the form of specific activities or methods, to enhance a risk assessment with respect to the overall quality of decision support and for considering any costs associated with the activity or methods. The approach could be applied to assess the value of current or proposed risk-assessment activities, for example, in weighing the value of advanced methods of uncertainty analysis, weight-of-evidence methods, or the development of complex computational models. The approach could also be applied to assess the benefit of procedural methods, such as stakeholder consultations, more intensive peer reviews, or methods to achieve greater transparency.

RECOMMENDATIONS

- The committee recommends that EPA strengthen its commitment to risk-assessment planning. That can be achieved by formally including the requirement for formative and iterative design of risk assessments that is user-centric and maintains focus on informing decisions.
- The committee recommends formalizing and implementing planning and scoping and problem formulation in human health risk assessment and ensuring their continued and intensive application in ecologic risk assessment. Important elements of formalization would include specification of concrete documentary and related communication products that would be expected as the outcomes of these formative design stages, and consideration of the feasibility and benefits of explicitly arraying decision-making options as early as possible in the process in order to focus the analytic tasks in the risk-assessment process.
- The committee recommends that EPA design risk assessments with due consideration of the potential for risk-assessment processes to contribute to unintended consequences, such as delays in risk-based decision-making that may prolong exposure to risk, diversion of attention away from other important risks within EPA's mandate, and the potential for uninformed risk-risk substitutions.
- The committee recommends that EPA consider the adoption of formal VOI methods for highly quantified and well-structured decision-making problems, particularly those with very high stakes, clear decision rules, and the possibility of substantial risks associated with delays in decision-making. For the great majority of decisions that are not readily amenable to formal VOI analysis, the committee recommends that EPA develop a structured evaluation method that exploits, in a less quantitative fashion than formal VOI analysis, a causal understanding of the impact of new information in specific decision-making situations. The committee further recommends that EPA consider an extension of the structured evaluation method, conceptually related to VOI analysis, to assess the potential value of diverse methodologic options in risk assessment with respect to improving the overall quality of decision support.

REFERENCES

Bartel, S.M., R.A. Ponce, T.K. Takaro, R.O. Zerbe, G.S. Omenn, and E. M. Faustman. 2000. Risk estimation and value-of-information analysis for three proposed genetic screening programs for chronic beryllium disease prevention. Risk Anal. 20(1):87-100.

Borisova, T., J. Shortle, R.D. Horan, and D. Abler. 2005. Value of information for water quality management. Water Resour. Res. 41, W06004, doi:10.1029/2004WR003576.

CENR (Committee on Environment and Natural Resources). 1999. Ecological Risk Assessment in the Federal Government. CENR/5-99/001. Committee on Environment and Natural Resources, National Science and Technology Council, Washington, DC.

Clemen, R.T. 1996. Making Hard Decisions: An Introduction to Decision Analysis, 2nd Ed. Boston: Duxbury Press.

Crawford-Brown, D.J. 1999. Risk-Based Environmental Decisions: Methods and Culture. New York: Kluwer.

Dakins, M.E., J.E. Toll, M.J. Small, and K.P. Brand. 1996. Risk-based environmental remediation: Bayesian Monte Carlo Analysis and the expected value of sample information. Risk Anal. 16(1):67-79.

EPA (U.S. Environmental Protection Agency). 1989. Risk Assessment Guidance for Superfund, Vol. 1. Human Health Evaluation Manual Part A. EPA/540/1-89/002. Office of Emergency and Remedial Response, U.S. Environmental Protection Agency, Washington, DC. December 1989 [online]. Available: http://rais.ornl.gov/homepage/HHEMA.pdf [accessed Jan. 11, 2008].

EPA (U.S. Environmental Protection Agency). 1991. Ecological Assessment of Superfund Sites: An Overview. EPA 9345.0-05I. Office of Solid Waste and Emergency Response, U.S. Environmental Protection Agency, Washington, DC. ECO Update 1(2) [online]. Available: http://www.epa.gov/swerrims/riskassessment/ecoup/pdf/v1no2.pdf [accessed Jan. 11, 2008].

EPA (U.S. Environmental Protection Agency). 1992. Framework for Ecological Risk Assessment. EPA/63-R-92/001. Risk Assessment Forum, U.S. Environmental Protection Agency, Washington, DC.

EPA (U.S. Environmental Protection Agency). 1997. Guidance on Cumulative Risk Assessment, Part 1 Planning and Scoping. Science Policy Council, U.S. Environmental Protection Agency, Washington, DC. July 3, 1997 [online]. Available: http://www.epa.gov/brownfields/html-doc/cumrisk2.htm [accessed Jan. 14, 2008].

EPA (U.S. Environmental Protection Agency). 1998. Guidelines for Ecological Risk Assessment. EPA/630/R-95/002F. Risk Assessment Forum, U.S. Environmental Protection Agency, Washington, DC. April 1998 [online]. Available: http://oaspub.epa.gov/eims/eimscomm.getfile?p_download_id=36512 [accessed Feb. 9, 2007].

EPA (U.S. Environmental Protection Agency). 2002. Lessons Learned on Planning and Scoping for Environmental Risk Assessments. Science Policy Council Steering Committee, U.S. Environmental Protection Agency, Washington, DC. January 2002 [online]. Available: http://www.epa.gov/OSA/spc/pdfs/handbook.pdf [accessed Jan. 11, 2008].

EPA (U.S. Environmental Protection Agency). 2003. Framework for Cumulative Risk Assessment. EPA/600/P-02/001F. National Center for Environmental Assessment, Risk Assessment Forum, U.S. Environmental Protection Agency, Washington, DC [online]. Available: http://cfpub.epa.gov/ncea/cfm/recordisplay.cfm?deid=54944 [accessed Jan. 4, 2008].

EPA (U.S. Environmental Protection Agency). 2004a. Risk Assessment and Modeling-Air Toxics Risk Assessment Library, Vol.1. Technical Resources Manual. EPA-453-K-04-001A. Office of Air Quality Planning and Standards, U.S. Environmental Protection Agency, Research Triangle Park, NC.

EPA (U.S. Environmental Protection Agency). 2004b. Risk Assessment Principles and Practices: Staff Paper. EPA/100/B-04/001. Office of the Science Advisor, U.S. Environmental Protection Agency, Washington, DC. March 2004 [online]. Available: http://www.epa.gov/osa/pdfs/ratf-final.pdf [accessed Jan. 9, 2008].

EPA (U.S. Environmental Protection Agency). 2005a. Guidelines for Carcinogen Risk Assessment. EPA/630/P-03/001F. Risk Assessment Forum, U.S. Environmental Protection Agency, Washington, DC. March 2005 [online]. Available: http://cfpub.epa.gov/ncea/cfm/recordisplay.cfm?deid=116283 [accessed Feb. 7, 2007].

EPA (U.S. Environmental Protection Agency). 2005b. Human Health Risk Assessment Protocol for Hazardous Waste Combustion Facilities. EPA530-R-05-006. Office of Solid Waste and Emergency Response, U.S. Environmental Protection Agency, Washington, DC [online]. Available: http://www.weblakes.com/hh_protocol.html [accessed Jan. 11, 2008].

GAO (U.S. Government Accountability Office). 2006. Human Health Risk Assessment: EPA Has Taken Steps to Strengthen Its Process, but Improvements Needed in Planning, Data Development, and Training. GAO-06-595. Washington, DC: U.S. Government Accountability Office [online]. Available: http://www.gao.gov/new.items/d06595.pdf [accessed Jan. 10, 2008].

Hammitt, J.K., and J.A.K. Cave. 1991. Research Planning for Food Safety: A Value of Information Approach. R-3946-ASPE/NCTR. RAND Publication Series [online]. Available: http://www.rand.org/pubs/reports/2007/R3946.pdf [accessed Jan. 11, 2008].

Hattis, D., and M.K. Lynch. 2007. Empirically observed distributions of pharmacokinetic and pharmacodynamic variability in humans—Implications for the derivation of single point component uncertainty factors providing equivalent protection as existing RfDs. Pp. 69-93 in Toxicokinetics in Risk Assessment, J.C. Lipscomb, and E.V. Ohanian, eds. New York: Informa Healthcare.

IEc (Industrial Economics, Inc.). 2000. Economic Value of Improved Exposure Information, Review Draft. EPA Contract Number: GS-10F-0224J. Industrial Economics, Inc., Cambridge, MA.

Koines, A. 2005. Big Decisions: Initial Results of Benefit-Cost Analysis. Presentation at EPA's Environmental Information Symposium 2005: Supporting Decisions to Achieve Environmental Results, November 30, 2005, Las Vegas, NV [online]. Available: http://www.epa.gov/oei/proceedings/2005/pdfs/koines.pdf [accessed Aug. 2, 2008].

Lave, L.B., and G.S. Omenn. 1986. Cost-effectiveness of short-term tests for carcinogenicity. Nature 324(6092):29-34.

Lave, L.B., F.K. Ennever, H.S. Rosenkranz, and G.S. Omenn. 1988. Information value of the rodent bioassay. Nature 336(6200):631-633.

Messner, M. and T.B. Murphy. 2005. Reducing Risk of Waterborne Illness in Public Water Systems: The Value of Information in Determining the Optimal Treatment Plan. Poster presentation at EPA Science Forum 2005: Collaborative Science for Environmental Solutions, May 16-18, 2005, Washington, DC [online]. Available: http://www.epa.gov/sciforum/2005/pdfs/oeiposter/messner_michael_illness.pdf [accessed Aug. 5, 2008].

Moore, D.R.J., and G.R. Biddinger. 1996. The interaction between risk assessors and risk managers during the problem formulation phase. Environ. Toxicol. Chem. 14(12):2013-2014.

Morgan, M.G., M. Henrion, and M. Small. 1990. Uncertainty: A Guide to Dealing with Uncertainty in Quantitative Risk and Policy Analysis. Cambridge, MA: Cambridge University Press.

NRC (National Research Council). 1983. Risk Assessment in the Federal Government: Managing the Process. Washington, DC: National Academy Press.

NRC (National Research Council). 1993. Issues in Risk Assessment. Washington, DC: National Academy Press.

NRC (National Research Council). 1994. Science and Judgment in Risk Assessment. Washington, DC: National Academy Press.

NRC (National Research Council). 1996. Understanding Risk: Informing Decision in a Democratic Society. Washington, DC: National Academy Press.

NRC (National Research Council). 2007. Scientific Review of the Proposed Risk Assessment Bulletin from the Office of Management and Budget. Washington, DC: The National Academies Press.

Raiffa, H. 1997. Decision Analysis: Introductory Lectures on Choices under Uncertainty. New York: McGraw-Hill.

Suter II, G.W. 2006. Ecological Risk Assessment, 2nd Ed. Boca Raton, FL: CRC Press.

Suter II, G.W., S.B. Norton, and L.W. Barnthouse. 2003. The evolution of frameworks for ecological risk assessment from the Red Book ancestor. Hum. Ecol. Risk Assess. 9(5):1349-1360.

Taylor, A.C., J.S. Evans, and T.E. McKone. 1993. The value of animal test information in environmental control decisions. Risk Anal. 13(4):403-412.

Thompson, K.M., and J.S. Evans. 1997. The value of improved national exposure information for perchloroethylene (Perc): A case study for dry cleaners. Risk Anal. 17(2): 253-271.

Yokota, F., and K.M. Thompson. 2004. Value of information analysis in environmental health risk management decisions: Past, present, and future. Risk Anal. 24(3):635-650.

Yokota, F., G. Gray, J.K. Hammitt, and K.M. Thompson. 2004. Tiered chemical testing: A value of information approach. Risk Anal. 24(6):1625-1639.

4

Uncertainty and Variability: The Recurring and Recalcitrant Elements of Risk Assessment

INTRODUCTION TO THE ISSUES AND TERMINOLOGY

Characterizing uncertainty and variability is key to the human health risk-assessment process, which must engage the best available science in the presence of uncertainties and difficult-to-characterize variability to inform risk-management decisions. Many of the topics in the committee's statement of task (Appendix B) address in some way the treatment of uncertainty or variability in risk analysis. Some of those topics have existed since the early days of environmental risk assessment. For example, *Risk Assessment in the Federal Government: Managing the Process* (NRC 1983), referred to as the Red Book, addressed the use of inference guidelines or default assumptions. *Science and Judgment in Risk Assessment* (NRC 1994) provided recommendations on defaults, use of quantitative methods for uncertainty propagation, and variability in exposure and susceptibility. The role of expert elicitation in uncertainty analysis has been considered in other fields for decades, although it has only been examined and used in select recent cases by the Environmental Protection Agency (EPA). Other topics identified in the committee's charge whose improvement requires new consideration of the best approaches for addressing uncertainty and variability include the cumulative exposures to contaminant mixtures involving multiple sources, exposure pathways, and routes; biologically relevant modes of action for estimating dose-response relationships; models of environmental transport and fate, exposure, physiologically based pharmacokinetics, and dose-response relationships; and linking of ecologic risk-analysis methods to human health risk analysis.

Much has been written that addresses the taxonomy of uncertainty and variability and the need and options for addressing them separately (Finkel 1990; Morgan et al. 1990; EPA 1997a,b; Cullen and Frey 1999; Krupnick et al. 2006). There are also several useful guidelines on the mechanics of uncertainty analysis. However, there is an absence of guidelines on the appropriate degree of detail, rigor, and sophistication needed in an uncertainty or variability analysis for a given risk assessment. The committee finds this to be a critical issue. In presentations to the committee (Kavlock 2006; Zenick 2006) and recent evaluations of emerging scientific advances (NRC 2006a, 2007a,b), there is the promise of improved capacity for assessing risks posed by new chemicals and risks to sensitive populations that are left unaddressed by current methods. The reach and depth of risk assessment are sure to improve with expanding computer tools, additional biomonitoring data, and new toxicology techniques. But such advances will bring new challenges and an increased need for wisdom and creativity in addressing uncertainty and variability. New guidelines on uncertainty analysis (NRC 2007c) can help enormously in the transition, facilitating the introduction of the new knowledge and techniques into agency assessments.

Characterizing each stage in the risk assessment process—from environmental release to exposure to health effect (Figure 4-1)—poses analytic challenges and includes dimensions of uncertainty and variability. Consider trying to understand the possible dose received by individuals and, on the average, by a population from the application of a pesticide. The extent of release during pesticide application may not be well characterized. Once the pesticide is released, the exposure pathways leading to an individual's exposure are complex and difficult to understand and model. Some of the released substance may be transformed in the environment to a more or less toxic substance. The resulting overall exposure of the community near where the pesticide is released can vary substantially among individuals by age, geographic location, activity patterns, eating habits, and socioeconomic status. Thus, there can be considerable uncertainty and variability in how much pesticide is received. Those factors make it difficult to establish reliable exposure estimates for use in a risk assessment, and they illustrate how the characterization of exposure with a single number can be misleading. Understanding the dose-response relationship—the relationship between the dose and risk boxes in Figure 4-1—is as complex and similarly involves issues of uncertainty and variability. Quantifying the relationship between chemical exposure and the probability of an adverse health effect is often complicated by the need to extrapolate results from high doses to lower doses relevant to the population of interest and from animal studies to humans. Finally, there are interindividual differences in susceptibility that are often difficult to portray with confidence. Those issues can delay the completion of a risk assessment (for decades in the case of dioxin) or undermine confidence in the public and those who use risk assessments to inform and support their decisions.

Discussions of uncertainty and variability involve specific terminology. To avoid confusion, the committee defines in Box 4-1 key terms as it has used them.

The importance of evaluating uncertainty and variability in risk assessments has long been acknowledged in EPA documents (EPA 1989a, 1992, 1997a,b, 2002a, 2004a, 2006a) and National Research Council reports (NRC 1983, 1994). From the Red Book framework and the committee's emphasis on the need to consider risk management options in the design of risk assessments (Chapters 3 and 8), it is evident that risk assessors must establish procedures that build confidence in the risk assessment and its results. EPA builds confidence in its risk assessments by ensuring that the assessment process handles uncertainty and variability in ways that are predictable, scientifically defensible, consistent with the agency's statutory mission, and responsive to the needs of decision-makers (NRC 1994). For example, several environmental statutes speak directly to the issue of protecting susceptible and highly exposed people (EPA 2002a, 2005c, 2006a). EPA has accordingly developed risk-assessment practices for implementing these statutes, although, as noted below and in Chapter 5, the overall treatment of uncertainty and variability in risk assessments can be insufficient. Box 4-2 provides examples of why uncertainty and variability are important to risk assessment.

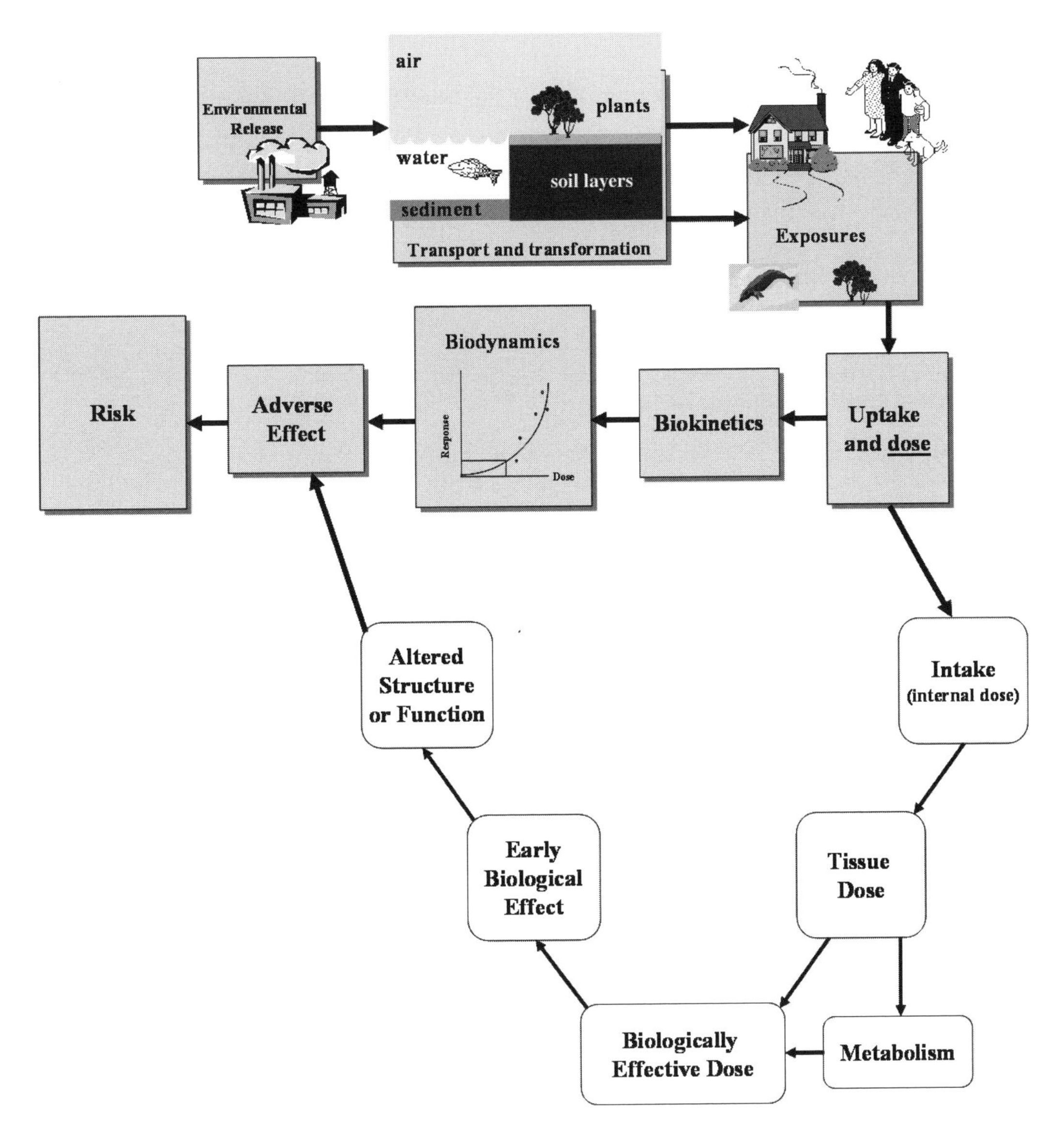

FIGURE 4-1 Illustration of key components evaluated in human health risk assessment, tracking pollutants from environmental release to health effects.

In the sections below, the committee first reviews approaches to address uncertainty and variability and comments on whether and how the approaches have been applied to EPA risk assessments. The committee then focuses on uncertainty and variability as applied to each of the stages of the risk-assessment process (as illustrated in Figure 4-1, which expands beyond the four steps from the Red Book to consider subcomponents of risk assessment). The chapter concludes by articulating principles for uncertainty and variability analysis, leaving detailed recommendations on specific aspects of the risk-assessment process to Chapters 5 through 7. The committee notes that elements of exposure assessment are not addressed extensively

BOX 4-1 Terminology Related to Uncertainty and Variability[a]

Accuracy: Closeness of a measured or computed value to its "true" value, where the "true" value is obtained with perfect information. Owing to the natural heterogeneity and stochastic nature of many biologic and environmental systems, the "true" value may exist as a distribution rather than a discrete value.

Analytic model: A mathematical model that can be solved in closed form. For example, some model algorithms that are based on relatively simple differential equations can be solved analytically to provide a single solution.

Bias: A systematic distortion of a model result or value due to measurement technique or model structure or assumption.

Computational model: A model that is expressed in formal mathematics with equations, statistical relationships, or a combination of the two and that may or may not have a closed-form representation. Values, judgment, and tacit knowledge are inevitably embedded in the structure, assumptions, and default parameters, but computational models are inherently quantitative, relating phenomena through mathematical relationships and producing numerical results.

Deterministic model: A model that provides a single solution for the stated variables. This type of model does not explicitly simulate the effects of uncertainty or variability, as changes in model outputs are due solely to changes in model components.

Domain (spatial and temporal): The limits of space and time that are specified in a risk assessment or risk-assessment component.

Empirical model: A model that has a structure based on experience or experimentation and does not necessarily have a structure informed by a causal theory of the modeled process. This type of model can be used to develop relationships that are useful for forecasting and describing trends in behavior but may not necessarily be mechanistically relevant. Empirical dose-response models can be derived from experimental or epidemiologic observations.

Expert elicitation: A process for obtaining expert opinions about uncertain quantities and probabilities. Typically, structured interviews and questionnaires are used in such elicitation. Expert elicitation may include "coaching" techniques to help the expert to conceptualize, visualize, and quantify the quantity or understanding being sought.

Model: A simplification of reality that is constructed to gain insights into select attributes of a particular physical, biologic, economic, or social system. Mathematical models express the simplification in quantitative terms.

---

[a]Compiled or adapted from NRC (2007d) and IPCS (2004).

in further chapters, as compared with other steps in the risk-assessment process, given our judgment that previous reports had sufficiently addressed many key elements of exposure assessment and that the exposure-assessment methods that EPA has developed and used in recent risk assessments generally reflect good technical practice, other than the overarching issues related to uncertainty and variability analysis and decisions about the appropriate analytic scope for the decision context.

Parameters: Terms in a model that determine the specific model form. For computational models, these terms are fixed during a model run or simulation, and they define the model output. They can be changed in different runs as a method of conducting sensitivity analysis or to achieve calibration goals.

Precision: The quality of a measurement that is reproducible in amount or performance. Measurements can be precise in that they are reproducible but can be inaccurate and differ from "true" values when biases exist. In risk-assessment outcomes and other forms of quantitative information, *precision* refers specifically to variation among a set of quantitative estimates of outcomes.

Reliability: The confidence that (potential) users should have in a quantitative assessment and in the information derived from it. Reliability is related to both precision and accuracy.

Sensitivity: The degree to which the outputs of a quantitative assessment are affected by changes in selected input parameters or assumptions.

Stochastic model: A model that involves random variables (see definition of *variable* below).

Susceptibility: The capacity to be affected. Variation in risk reflects susceptibility. A person can be at greater or less risk relative to the person in the population who is at median risk because of such characteristics as age, sex, genetic attributes, socioeconomic status, prior exposure to harmful agents, and stress.

Variable: In mathematics, a variable is used to represent a quantity that has the potential to change. In the physical sciences and engineering, a variable is a quantity whose value may vary over the course of an experiment (including simulations), across samples, or during the operation of a system. In statistics, a random variable is one whose observed outcomes may be considered outcomes of a stochastic or random experiment. Their probability distributions can be estimated from observations. Generally, when a variable is fixed to take on a particular value for a computation, it is referred to as a parameter.

Variability: Variability refers to true differences in attributes due to heterogeneity or diversity. Variability is usually not reducible by further measurement or study, although it can be better characterized.

Vulnerability: The intrinsic predisposition of an exposed element (person, community, population, or ecologic entity) to suffer harm from external stresses and perturbations; it is based on variations in disease susceptibility, psychological and social factors, exposures, and adaptive measures to anticipate and reduce future harm, and to recover from an insult.

Uncertainty: Lack or incompleteness of information. Quantitative uncertainty analysis attempts to analyze and describe the degree to which a calculated value may differ from the true value; it sometimes uses probability distributions. Uncertainty depends on the quality, quantity, and relevance of data and on the reliability and relevance of models and assumptions.

UNCERTAINTY IN RISK ASSESSMENT

Uncertainty is foremost among the recurring themes in risk assessment. In quantitative assessments, *uncertainty* refers to lack of information, incomplete information, or incorrect information. Uncertainty in a risk assessment depends on the quantity, quality, and relevance of data and on the reliability and relevance of models and inferences used to fill data gaps.

BOX 4-2 Some Reasons Why It Is Important to Quantify Uncertainty and Variability

Uncertainty

- Characterizing uncertainty in risk informs the affected public about the range of possible risks from an exposure that they may be experiencing. Risk estimates sometimes diverge widely.
- Characterizing the uncertainty in risk associated with a given decision informs the decision-maker about the range of potential risks that result from the decision. That helps in evaluating any decision alternative on the basis of the possible risks, including the most likely and the worst ones; it also informs the public.
- Mathematically, it is often not possible to understand what may occur on average without understanding what the possibilities are and how probable they are.
- The value of new research or alternative research strategies can be assessed by considering how much the research is expected to reduce the overall uncertainty in the risk estimate and how the reduction in uncertainty leads to different decision options.
- Although the committee is not aware of any research to prove it, there is a strong sense among risk assessors that acknowledging uncertainty adds to the credibility and transparency of the decision-making process.

Variability

- Assessing variability in risk enables the development of risk-management options that focus on the people at greatest risk rather than on population averages. For example, the risk from exposures to particular vehicle emissions varies in a population and can be much higher in those close to roadways than the population average. That has implications for zoning and school-siting decisions.
- Understanding how the population may vary in risk can facilitate understanding of the shape of the dose-response curve (see Chapter 5). Greater use of genetic markers for factors contributing to variability can support this effort.
- It is often not possible to estimate an average population risk without knowing how risk varies among individuals in the population.
- On the basis of understanding how different exposures may affect risk, people might alter their own level of risk, for example, by filtering their drinking water or eating fewer helpings of swordfish (which is high in methyl mercury).
- The aims of environmental justice are furthered when it becomes clear that some community groups are at greater risk than the overall group and policy initiatives are undertaken to rectify the imbalance.

For example, the quantity, quality, and relevance of data on dietary habits and a pesticide's fate and transport will affect the uncertainty of parameter values used to assess population variability in the consumption of the pesticide in food and drinking water. The assumptions and scenarios applied to address a lack of data on how frequently a person eats a particular food affect the mean and variance of the intake and the resulting risk distribution. It is the risk assessor's job to communicate not only the nature and likelihood of possible harm but the uncertainty in the assessment. One of the more significant types of uncertainties in EPA risk assessments can be characterized as "unknown unknowns"—factors that the assessor is not aware of. These uncertainties cannot be captured by standard quantitative uncertainty analyses, but can only be addressed with an interactive approach that allows timely and effective detection, analysis, and correction.

EPA's practices in uncertainty analysis are reviewed below. The discussion of practice begins by considering EPA's use of defaults. An expanded treatment of uncertainty beyond defaults requires additional techniques. Specific analytic techniques that EPA has used or could use in these contexts are discussed below, including Monte Carlo analysis for quantitative uncertainty analysis, expert elicitation, methods for addressing model uncertainty, and addressing uncertainty in risk comparisons. In parallel, the conduct of assessments (including uncertainty analysis) that are appropriate in complexity for risk-management decisions is discussed with considerations for uncertainty analyses used to support risk-risk, risk-benefit, and cost-benefit comparisons and tradeoffs.

The Environmental Protection Agency's Use of Available Methods for Addressing Uncertainty

EPA's treatment of uncertainty is evident both in its guidance documents and from a review of important risk assessments that it has conducted (EPA 1986, 1989a,b, 1997a,b,c, 2001, 2004a, 2005b). The agency's guidance follows in large part from recommendations in the Red Book (NRC 1983) and other National Research Council reports (for example, NRC 1994, 1996).

Use of Defaults

As described in the Red Book, because of large inherent uncertainties, human health risk assessment "requires judgments to be made when the available information is incomplete" (NRC 1983, p. 48). To ensure that the judgments are consistent, explicit, and not unduly influenced by risk-management considerations, the Red Book recommended that so-called "inference guidelines," commonly referred to as defaults, be developed independently of any particular risk assessment (p. 51). *Science and Judgment in Risk Assessment* (NRC 1994) reaffirmed the use of defaults as a means of facilitating the completion of risk assessments. EPA often relies on default assumptions when "the chemical- and/or site-specific data are unavailable (i.e., when there are data gaps) or insufficient to estimate parameters or resolve paradigms . . . to continue with the risk assessment" (EPA 2004a, p. 51). Defaults which are the focus of controversy and debate are often needed to complete cancer-hazard identification and dose-response assessment. Because of their importance and the need to address some of the above concerns, the committee devotes Chapter 6 to default assumptions. Consideration is given to how risk assessments can use emerging methods to characterize uncertainties more explicitly while conveying the information needed to inform near-term risk-management decisions.

Some approaches based on defaults lead to confusion about levels of uncertainty. For example, EPA estimates cancer risk from the results of animal studies based on default assumptions and then applies likelihood methods to fit models to tumor data and characterizes the dose-response relationship with the lower 95% confidence bound typically on a dose that causes a 10% tumor response beyond background (see Chapter 5). In the past, it estimated the upper 95% confidence bound in the linear term in the multistage polynomial, that is, the "cancer potency." It usually does not show the opposite bound or other points in the distribution. EPA's approach is reasonable, but it can lead to misunderstanding when the bounds on the final risk calculations are overinterpreted, for example, when bounds are discussed as characterizing the full range of uncertainty in the assessment. When a new study shows a higher upper bound on the potency or a lower bound on the risk-specific dose, it may appear that uncertainty has increased with further study. From a strictly Bayesian perspective, additional information can never increase uncertainty if the underlying distributional structure of uncertainty is correctly specified. However, when mischaracterized and misunderstood, the framework for defaults used by EPA can make it appear that uncertainty is increasing. For example, suppose that there was an epidemiologic study of the effects of an environmental contaminant, and suppose that the degree of overall uncertainty is incorrectly characterized by the parameter uncertainty in fitting a dose-response slope to the results of that single study. If a second study caused EPA to select an alternative value for the dose-response slope, the risk estimate would change. The uncertainty *conditional* on one or the other causal model may or may not change. Chapters 5 and 6 suggest approaches to establishment of defaults and uncertainty characterization that may encourage research that could reduce key uncertainties.

Quantitative Uncertainty Analysis

In a quantitative uncertainty analysis (QUA), both uncertainty and variability in different components of the assessment (emissions, transport, exposure, pharmacokinetics, and dose-response relationship) are combined by using an uncertainty-propagation method, such as Monte Carlo simulation, with two-stage Monte Carlo analysis utilized to separate uncertainty and variability to the extent possible. This approach has been referred to as probabilistic risk assessment, but the committee prefers to avoid this term because of its association with fault-tree analysis in engineering. The use of the term QUA to encompass variability as well as uncertainty is awkward, but we use this term going forward to be consistent with its usage elsewhere.

In the federal government, an early user of QUA was the Nuclear Regulatory Commission. In the mid-1970s, the Nuclear Regulatory Commission used QUA that involved considerable use of expert judgment to characterize the likelihood of nuclear reactor failure (USNRC 1975). QUA became more commonly used in EPA in the late 1980s. EPA has since been encouraging the use of QUA in many programs, and the computational methods required have become more readily available and practicable.

An example of the evolution of the use of QUA in EPA is its risk-assessment guidance for Superfund. The 1989 *Risk Assessment Guidance for Superfund* (RAGS), Volume 1 (EPA 1989a) and supporting guidance describe a point-estimate (single-value) approach to risk assessment. The output of the risk equation is a point estimate that could be a central-tendency exposure estimate of risk (for example, the mean or median risk) or reasonable-maximum-exposure (RME) estimate of risk (for example, the risk expected if the RME occurred), depending on the input values used in the risk equation. But RAGS, Volume 3, Part A (EPA 2001) describes a probabilistic approach that uses probability distributions for one or more variables in a risk equation to characterize variability and uncertainty quantitatively.

The common practice of choosing high percentile values (ensuring one-sided confidence) for multiple uncertain variables provides results that are probably above the median but still at an unknown percentile of the risk distribution (EPA 2002a). QUA techniques, such as those in RAGS, Volume 3, can address this issue in part, but a few major concerns regarding their use in EPA remain. First, they require training to be used appropriately. Second, even if they are used appropriately, their outputs may not be easily understood by decision-makers. So training is recommended not only for risk assessors but for risk managers (see recommendations in Chapter 2). Third and perhaps most important, in many contexts, the data may not be available to characterize all input distributions fully, in which case the assessment either involves subjective judgments or systematically omits key uncertainties. For formal QUA to be most informative, the treatment of uncertainty should, to the extent feasible, be homologous among components of the risk assessment (exposure, dose, and dose-response relationship).

The differential treatment of uncertainty among components of a risk assessment makes the communication of overall uncertainty difficult and sometimes misleading. For example, in EPA's regulatory impact analysis for the Clean Air Interstate Rule (EPA 2005c), formal probabilistic uncertainty analysis was conducted with the Monte Carlo method, but this considered only sampling variability in epidemiologic studies used for dose-response functions and in valuation studies. EPA used expert elicitation for a more comprehensive characterization of dose-response relationship uncertainty, but this was not integrated into a single output distribution. Within the quantitative uncertainty analysis, emissions and fate and transport modeling outputs were assumed to be known with no uncertainty. Although EPA explicitly acknowledged the omitted uncertainty in a qualitative discussion, it was not addressed quantitatively. The 95% confidence intervals reported did not reflect the actual confidence level, because the important uncertainties in other components were not included. The training mentioned above therefore should not only be related to the mechanical aspects of software packages but address issues of interpretability and the goal of treating uncertainty consistently among all components of risk assessment.

An earlier National Research Council committee (NRC 2002) and the EPA SAB (2004) also raised concerns about the inconsistent approach to uncertainty characterization. However, it is important to recognize that there are some uncertainties in environmental and health risk assessments that defy quantification (even by expert elicitation) (IPCS 2006; NRC 2007d) and that inconsistency in approach will be an issue to grapple with in risk characterization for some time to come. The call for homologous treatment of uncertainty should not be read as a call for "least-common-denominator" uncertainty analysis, in which the difficulty of characterizing uncertainty in one dimension of the analysis leads to the omission of formal uncertainty analysis in other components.

Use of Expert Judgment[1]

It often happens in practice that empirical evidence on some components of a risk assessment is insufficient to establish uncertainty bounds and evidence on other components captures only a fraction of the total uncertainty. When large uncertainties result from a combination of lack of data and lack of conceptual understanding (for example, a mechanism of action at low dose), some regulatory agencies have relied on expert judgment to fill the gaps or establish default assumptions. Expert judgment involves asking a set of carefully selected experts a series of questions related to a specific array of potential outcomes and usually providing them with extensive briefing material, training activities, and calibration exercises to help in the determination of confidence intervals. Formal expert judgment has been used in risk analysis since the 1975 Reactor Safety Study (USNRC 1975), and there are multiple examples in the academic literature (Spetzler and von Holstein 1975; Evans et al. 1994; Budnitz et al. 1998; IEc 2006). EPA applications have been more limited, perhaps in part because of institutional and statutory constraints, but interest is growing in the agency. The 2005 *Guidelines for Carcinogen Risk Assessment* (EPA 2005b, p. 3-32) state that "these cancer guidelines are flexible enough to accommodate the use of expert elicitation to characterize cancer risks, as a complement to the methods presented in the cancer guidelines." A recent study of health effects of particulate matter used expert elicitation to characterize uncertainties in the concentration-response function for mortality from fine particulate matter (IEc 2006).

Expert elicitation can provide interesting and potentially valuable information, but some

[1]Expert judgment is analogous to the term expert elicitation.

critical issues remain to be addressed. It is unclear precisely how EPA can use this information in its risk assessments. For example, in its regulatory impact analysis of the National Ambient Air Quality Standard for $PM_{2.5}$ (particulate matter no larger than 2.5 μm in aerodynamic diameter), EPA did not use the outputs of the expert elicitation to determine the confidence interval for the concentration-response function for uncertainty propagation but instead calculated alternative risk estimates corresponding to each individual expert's judgment with no weighting or combining of judgments (EPA 2006b). It is unclear how that type of information can be used productively by a risk manager, inasmuch as it does not convey any sense of the likelihood of various values, although seeing the range and commonality of judgments of individual experts may be enlightening. Formally combining the judgments can obscure the degree of their heterogeneity, and there are important methodologic debates on the merits of weighing expert opinions on the basis of their performance on calibration exercises (Evans et al. 1994; Budnitz et al. 1998). Two other problems are the need to combine incompatible judgments or models and the technical issue of training and calibration when there is a fundamental lack of knowledge and no opportunity for direct observation of the phenomenon being estimated (for example, the risk of a particular disease at an environmental dose). Although methods have been developed to address various biases in expert elicitation, expert mischaracterization is still expected (NRC 1996; Cullen and Small 2004). Some findings about judgment in the face of uncertainty that can apply to experts are provided in Box 4-3. Other practical issues are the cost of and time required for expert elicitation, management of conflict of interest, and the need for a substantial evidence base on which the experts can draw to make expert elicitation useful.

Given all of those limitations, there are few settings in which expert elicitation is likely to provide information necessary for discriminating among risk-management options. The committee suggests that expert elicitation be kept in the portfolio of uncertainty-characterization

BOX 4-3 Cognitive Tendencies That Affect Expert Judgment

Availability: The tendency to assign greater probability to commonly encountered or frequently mentioned events.

Anchoring and adjustment: The tendency to be over-influenced by the first information seen or provided in an initial problem formulation.

Representativeness: The tendency to judge an event by reference to another that in the eye of the expert resembles it, even in the absence of relevant information.

Disqualification: The tendency to ignore data or strongly discount evidence that contradicts strongly held convictions.

Belief in "law of small numbers": The tendency of scientists to believe small samples from a population to be more representative than is justified.

Overconfidence: The tendency of experts to overestimate the probability that their answers are correct.

Source: Adapted from NRC 1996; Cullen and Small 2004.

options available to EPA but that it be used only when necessary for decision-making and when evidence to support its use is available. The general concept of determining the level of sophistication in uncertainty analysis (which could include expert elicitation or complex QUA) based on decision-making needs is outlined in more detail below.

Level of Uncertainty Analysis Needed

The discussion of the variety of ways in which EPA has dealt with uncertainty—from defaults to standard QUA to expert elicitation—raises the question of the level of analysis that is needed in any given problem. A careful assessment of when a detailed assessment of uncertainty is needed may avoid putting additional analytic burdens on EPA staff or limiting the ability of EPA staff to complete timely assessments. Formal QUA is not necessary and not recommended for all risk assessments. For example, for a risk assessment conducted to inform a choice among various control strategies, if a simple (but informative and comprehensive) evaluation of uncertainties reveals that the choice is robust with respect to key uncertainties, there is no need for a more formal treatment of uncertainty. More complex characterization of uncertainty is necessary only to the extent that it is needed to inform specific risk-management decisions. It is important to address the extent and nature of uncertainty analysis needed in the planning and scoping phase of a risk assessment (see Chapter 3).

For many problems, an initial sensitivity analysis can help determine those parameters whose uncertainty might most impact a decision and thus require a more detailed uncertainty analysis. One valuable approach involves utilizing tornado diagrams, in which individual parameters are permitted to vary while all other uncertain parameters are held fixed. The output of this exercise provides a graphical plot of parameters that have the largest influence on the final risk calculation. This both provides a visual representation of the sensitivity analysis, helpful for communication to risk managers and other stakeholders, and determines the subset of parameters that could be carried forward in more sophisticated QUA.

"Tiers" or "levels" of sophistication in QUA in risk assessment have been discussed. Paté-Cornell (1996) proposed six levels ranging from level 0 (hazard detection and failure-mode identification) to level 5 (QUA with multiple risk curves reflecting variability at different levels of uncertainty). Similarly, in its draft report on the treatment of uncertainty in exposure assessment, the International Programme on Chemical Safety (IPCS 2006) has proposed four tiers for addressing uncertainty and variability in exposure assessment, from the use of default assumptions to sophisticated QUA. The IPCS tiers are shown in Box 4-4.

BOX 4-4 Levels of Uncertainty Analysis

Tier 0: Default assumptions—single value of result.
Tier 1: Qualitative but systematic identification and characterization of uncertainty.
Tier 2: Quantitative evaluation of uncertainty making use of bounding values, interval analysis, and sensitivity analysis.
Tier 3: Probabilistic assessment with single or multiple outcome distributions reflecting uncertainty and variability.

Source: IPCS 2006.

The committee does not endorse any specific ranking approaches but favors the up-front consideration of levels of sophistication in uncertainty analyses and notes that there is a continuum of approaches rather than a number of discrete options. The characterization of uncertainty and variability in a risk assessment should be planned and managed and matched to the needs of the stakeholders involved in risk-informed decisions. In evaluating the tradeoff between the higher level of effort needed to conduct a more sophisticated analysis and the need to make timely decisions, EPA should take into account both the level of technical sophistication needed to identify the optimal course of action and the negative impacts that will result if the optimal course of action is incorrectly identified. If a relatively simple analysis of uncertainty (for example, a nonprobabilistic assessment of bounds) is sufficient to identify one course of action as clearly better than all the others, there is no need for further elucidation. In contrast, when the best choice is not so clear and the consequences of a wrong choice would be serious, EPA can proceed in an iterative manner, making the analysis more and more sophisticated until the optimal choice *is* sufficiently clear. In so doing, EPA should be mindful that one of the greatest costs of more sophisticated analysis can be the time involved, during which populations may continue to be exposed to an agent or costs may be incurred unnecessarily. Related to these issues, in planning the uncertainty analysis and interpreting lower-tier uncertainty analyses, it is preferable to have up-front agreement on terms of reference. For example, calls for "central tendencies," "best estimates," or "plausible" upper or lower bounds of risk are of little value if these terms are not clearly defined.

EPA has an opportunity and responsibility to develop guidelines for uncertainty analysis both to define terms of reference and to offer insight into appropriate tailoring of sophistication and level of practice to individual risk-management decisions. EPA has limited resources and should not be expected to treat all issues using a single approach or process. The tiered approach to uncertainty analysis provides EPA the opportunity to match the degree of sophistication in uncertainty analysis to the level of concern for a specific risk problem and to the decision-making needs to address that problem. Lower-tier uncertainty analysis methods can be used in a screening step to determine whether the information is adequate to make decisions and to identify situations in which more intensive quantitative methods would be necessary.

Special Concerns about Uncertainty Analysis for Risk or Cost-Benefit Tradeoffs

In making risk comparisons or cost-benefit determinations, consistency in addressing uncertainty in the risks, costs, and benefits being compared is particularly important, and fuller descriptions of uncertainty than provided by an upper confidence limit are also important. The approaches described above are typically applied to develop confidence bounds and a probability distribution for a single risk. Although assessors commonly analyze one risk at a time, many assessments are done to support analyses of various options for controlling a hazard. They can involve considering more than one uncertain quantity at the same time with respect to

- Which of several risks deserves higher priority.
- The net risk of an environmental control action (reduction in risk less any increases in risk because of substitution or risk transfer).
- The net benefits of an action (reduction in risk less any costs incurred).
- The total benefits of an action (the monetized reduction in risk in light of the baseline level of risk even if costs are ignored).

Two issues make uncertainty analyses for risk-risk and risk-benefit or cost comparisons more informative but also more difficult to do properly than single-item QUA. First, uncertainty in multiple risks means that simply stating that one risk is or is not larger than another risk, or that the benefits are or are not larger than the costs, is not a well-formulated comparison; the key is to determine the probability that one risk is larger or one action is preferable. Second, there is the question of how large the uncertainty is when comparing multiple with individual risks (Finkel 1995b). If the uncertainties in each of the items being compared are related, the uncertainty in the comparison can be less than that in an individual risk. But usually the uncertainties will be independent and not related. For example, uncertainty in risk based on estimating exposure and addressing toxicologic information will generally be completely independent of cost estimates for reducing the risk, which may be based on consumer and producer behavior.

As a result, uncertainties in a comparison can exceed the uncertainty in items being compared, an important issue that has implications in developing and using risk estimates. Box 4-5 provides a simple but informative example about comparing two uncertain quantities. These quantities are risks, but they could be any measurable quantities of interest. The examples include a comparison of discrete and continuous probabilities. This simple example reveals the need to address confidence intervals both when assessing risk and when comparing risk.

This discussion illustrates that statements regarding risk comparisons, or costs vs benefits, would be made better in probabilistic than in deterministic terms. The question "Do the benefits exceed the costs?" can be given an unequivocal yes answer only if virtually all possible values of the net benefit distribution are positive. This does not necessarily imply that EPA must utilize sophisticated QUA whenever risk-risk or benefit-cost comparisons are required. An iterative approach as proposed earlier can allow for a determination of whether benefits clearly exceed costs (or vice versa) using a relatively simple analysis of uncertainty, or whether more detailed analyses would be required to make this comparison interpretable. These efforts would benefit from EPA guidance on uncertainty and the concept of statistical significance as applied to cost-benefit and risk comparison analyses, with a specific emphasis on the use of a tiered uncertainty analysis approach in this context.

Model Uncertainty

One of the dimensions of uncertainty that is difficult to capture quantitatively (or even qualitatively) involves model uncertainty. The National Research Council (NRC 2007d) noted that there is a range of options for performing model-uncertainty analysis. One computationally intense option is to represent all model uncertainties probabilistically, including the uncertainties associated with a choice between alternative models or alternative model assumptions. Another option is to use a scenario or sensitivity assessment that might consider model results for a small number of plausible cases. A third option is to address uncertainty with default parameters and a "default model such that there is no explicit quantification of model uncertainty." The first option has the problem of demanding detailed probabilistic analyses among one or more models that include potentially large numbers of parameters whose uncertainties must be estimated, often with little information. Such problems are compounded when models are linked into a highly complex system. In the second option noted above, when scenario assessment and sensitivity analysis are used to evaluate model uncertainty without making explicit use of probability, such a deterministic approach is easy to implement and understand but typically does not include what is known about each scenario's likelihood. In many situations, some combination of these first two approaches is

BOX 4-5 Examples of Uncertainties for Comparisons of Discrete and Continuous Possibilities

Example 1: Discrete

Consider two quantities, A and B—they could be two disparate risks being compared, a "target" risk and an "offsetting" risk, or a benefit estimate (A) and the corresponding cost estimate (B). In any case, we are fairly confident (80%) that A has the value 20, but believe with 10% probability each that we might have over- or underestimated A by a factor of 2 (that is, A can be 10 with probability 0.1, or 40 with probability 0.1). Similarly, we are fairly confident (80%) that B has the value 15, but with 10% probability it could be a factor of 3 higher or lower.

Given the 3 possible discrete values of A, and the 3 possible values of B, there are 9 possible true values of the ratio (A/B), as given in the following table. Assigning A and B as independent random variables with the marginal distributions specified, for example, P(A=10)=P(A=40)=0.10 and P(A=20)=0.80, leads immediately to the joint distribution specified below since the joint distribution of independent random variables is the product of their marginal distributions.

Ratio of A to B for different values and probabilities of A and B

	Value of A [prob(A					
	10 (10%)		20 (80%)		40 (10%)	
Value of B [prob(B)]	A/B	prob(A/B)	A/B	prob(A/B)	A/B	prob(A/B)
5 (10%)	2	1%	4	8%	8	1%
15 (80%)	0.67	8%	1.3	64%	2.7	8%
45 (10%)	0.22	1%	0.44	8%	0.89	1%

In this case, although the highest possible value of A differs from its lowest possible value by a factor of 4, and the extreme values of B differ from each other by a factor of 9, the ratio A/B can be as low as 0.22 or as high as 8, a factor of 36 difference. The uncertainty in the comparison exceeds the uncertainty in either quantity. A is "probably" greater than B, but for four of the nine possibilities, with a total likelihood of 18%, B is in fact greater than A.

Example 2: Continuous

Now suppose A and B are both lognormally distributed, and each have the exact same PDF but are uncorrelated with one another. Assume that the median value is 10, and the logarithmic standard deviation is 1.0986 (a geometric standard deviation of exactly e1.0986, or 3). In this case, the PDF for (A/B) has an exact solution: it too is lognormal, with a median of 1.0 (the median of A divided by the median of B), and a logarithmic standard deviation of 1.554 (which is the square root of the sum of [1.09862 plus 1.09862]).

In this case, we could say that on the basis of median values, A and B are equal, but that statement would be highly uncertain. In fact, there is a 5 percent chance that (A/B) is equal to 12.9 or larger[a], and a corresponding chance that (A/B) is equal to 0.078 or smaller. Note that while the 90th percentile width for A alone spans a factor of 37, as does the 90th percentile width for B alone, the ratio is even more uncertain: (12.9) divided by (0.078) equals 165.

Even though the typical values of the two risks are "equal," it would be incorrect to report that they are equal (or that the net benefit is zero, or that the substitution risk cancels out the primary risk). In fact, this analysis tells us that we cannot confidently determine which quantity is greater, which is quite different from being able to pronounce them as equal.

[a]This number is equal to the median (1) times exp[(1.554)(1.645)], the upper 95th percentile point.

appropriate. The balance between detailed probabilistic modeling and scenario and sensitivity evaluation is determined by the purpose of the model and the specific needs of a given risk assessment—another matter that would benefit from guidance.

Finally, with respect to the third option of default modeling, the National Research Council (NRC 2007d, pp. 26-27) observed that models of natural systems are necessarily never complete and that in regulatory modeling "assumptions and defaults are unavoidable as there is never a complete data set to develop a model." It also noted that the fundamental uncertainties and limitations, although "critical to understand when using environmental models . . . do not constitute reasons why modeling should not be performed. When done in a manner that makes effective use of existing science and that is understandable to stakeholders and the public, models can be very effective for assessing and choosing amongst environmental regulatory activities and communicating with decision-makers and the public." The present committee agrees.

Committee Observations Regarding the Treatment of Uncertainty

Although EPA has developed methods for addressing parameter uncertainty, particularly for exposure assessment, the remaining challenge is to address uncertainties that are difficult to capture with probability distributions and to provide guidance for the level of detail needed to capture and communicate key uncertainties. Many decision-makers tend to believe that with sufficient resources, science and technology will provide an obvious and cost-effective solution to the problems of protecting human health and the environment. In reality, however, there are many sources of uncertainty, and many uncertainties cannot be reduced or even quantified (see Box 4-6 for a discussion of model and parameter uncertainty). The committee's review of uncertainty reveals that developing quantitative risk estimates in the face of substantial uncertainty and appropriately characterizing the degree of confidence in the results are recurring challenges in risk assessment that must be addressed over the coming decade.

As noted above, there are different strategies (or levels of sophistication) for addressing uncertainty. Regardless of which level is selected, it is important to provide the decision-maker with information to distinguish reducible from irreducible uncertainty, to separate individual variability from true scientific uncertainty, to address margins of safety, and to consider benefits, costs, and comparable risks when identifying and evaluating options. To make risk assessment consistent with such an approach, EPA should incorporate formal and transparent treatment of uncertainties in each component of the risk-characterization process and develop guidelines to advise assessors on how to proceed.

The methods of addressing uncertainty vary widely in their implementation, their expected formality, and their cost and time requirements. The options for uncertainty analysis vary considerably in their ability to be understood by decision-makers and other parties. Although it is not stressed in the technical literature on uncertainty analysis, it is worth remembering that the product of risk assessment is in the end primarily a communication product (see Chapter 3). Therefore, perhaps the most appropriate measure of quality in the uncertainty analysis is whether it improves the capacity of the primary decision-maker to make informed decisions in the presence of substantial, inevitable, and irreducible uncertainty. Another important measure of quality is whether it improves the understanding of other stakeholders and thus fosters and supports the broader public interests in the decision-making process. The choice of methods of expressing uncertainty is important and is clearly a design problem that requires careful attention to objectives.

BOX 4-6 Expressing and Distinguishing Model and Parameter Uncertainty

Choosing which uncertainties to leave unaddressed and which to express and deciding how best to express them can be daunting tasks. As a simple example of expressing uncertainty, consider two distinct sources of uncertainty in generating an estimate of risk.

- *Fundamental causal uncertainty*: uncertainty about the existence of critical cause-effect relationships, for example, uncertainty about whether a particular compound causes cancer.
- *Uncertainty* in the strength of the causal relationship: the degree to which the cause results in the effect, for example, how much cancer is caused by a given dose of the compound.

The latter uncertainty is typically more easily expressed than the former in quantitative terms, with a probability distribution. But it should be noted that there are quantitative aspects for the casual uncertainty (hazard) in that there are statistical thresholds around positive findings from toxicity experiments. The two types of uncertainty can be addressed in a cause-effect model that takes on a value of zero to represent the lack of existence of a causal relationship and nonzero values to characterize the strength of the relationship. With such a representation, the outcomes of the overall model can have a multimodal distribution in which some finite probability at zero represents no causal relationship and a range of nonzero values represents the uncertainty in the strength of the relationship. That could be made more complex while allowing different mathematical forms to represent different possible ways that the effect is caused, for example, whether the compound causes the effect by a mechanism that is linear or nonlinear at low doses.

It is often difficult to assign probabilities to different mathematical relationships. As an alternative, causal scenarios could be used, with each scenario representing distinct theories of causality. In the example here, one scenario would be no causal relationship, another would be a linear dose-response relationship, and a third would be a nonlinear dose-response relationship. Each scenario would have a corresponding conditional uncertainty analysis. Each model would be assumed true, and the likely range of model values in it could be derived. In this scenario approach, the individual uncertainty analyses are much simpler and may be more widely applied and understood. However, decision-making that is directed toward reducing important sources of uncertainty may be misguided by a focus on readily quantifiable uncertainties (for example, How much water is consumed by specific subpopulations?) when the global uncertainty may well be dominated by causal uncertainties whose collective impact is not quantified (for example, Are children disproportionately sensitive to the contaminant? Which of many possible adverse effects does the contaminant cause? Is exposure by inhalation an important contributor to total risk?). Efforts to measure a subset of readily quantifiable uncertainties when fundamental causal uncertainties dominate the overall uncertainty may therefore not be justifiable.

VARIABILITY AND VULNERABILITY IN RISK ASSESSMENT

There are important variations among individuals in a population with respect to susceptibility and exposure. Many of the statistical techniques and general concepts described above in relation to uncertainty analysis are applicable to variability analysis. For example, probabilistic approaches, such as Monte Carlo methods, can be used to propagate variability throughout all components of a risk assessment, expert elicitation can be used to characterize various percentiles in a distribution, and the level of analytic sophistication should be matched to the problem at hand. But the key difference between uncertainty analysis and variability analysis is that variability can only be better characterized, not reduced, so it often must be addressed with strategies different from those used to address uncertainty. For example, the strategy that a policy-maker uses to address uncertainty about whether a rodent carcinogen is a human carcinogen differs from the strategy to address the variability in cancer susceptibility between children and adults. The latter is a case where the variability can be represented by a probability distribution, but likely a mixed (bimodal) distribution rather than a standard normal distribution. This section briefly describes key concepts and methods, EPA's treatment of variability in general, and the basis of the committee's recommendations related to variability in each component of risk assessment.

People differ in susceptibility to the toxic effects of a given chemical exposure because of such factors as genetics, lifestyle, predisposition to diseases and other medical conditions, and other chemical exposures that influence underlying toxic processes. Examples of factors that affect susceptibility are shown in Table 4-1 along with some estimates of increased

TABLE 4-1 Examples of Factors Affecting Susceptibility to Effects of Environmental Toxicants

Ratio of Sensitive Case to "Normal"		Reference
	Genetic	
10:1	"While the risk of cancer following irradiation may be elevated up to 100-fold in some heritable cancer disorders a single best estimate of a 10-fold increase in risk is appropriate for the purposes of modeling radiological impact."	ICRP 1998; Tawn 2000
>10:1	Wilson's heterozygotes (about 1% of population) and copper sensitivity	NRC 2000
	Predisposing exposures	
20:1	Greater sensitivity to arsenic-induced lung cancer in smokers than in nonsmokers.	CDHS 1990
10-20:1	Greater sensitivity to lung cancer due to radon in smokers than in nonsmokers.	ATSDR 1992
20-100:1	Suggestive evidence that low-iodide female smokers are much more sensitive to perchlorate-induced thyroid hormone disruption than "normal" adults.	Blount et al. 2006
10-30:1	Liver-cancer risk from aflatoxin in those with vs without hepatitis.	Wu-Williams et al. 1992
	Physiologic and Pharmacokinetic	
>10:1	Difference in sensitivity to 4-aminobiphenyl (median vs upper 2 percentile of population) due to physiologic and pharmacokinetic differences (modeled).	Bois et al. 1995
	Lifestage	
5-10:1	Breast-cancer risk. Radiation exposure of pubescent girls and those before first completed pregnancy vs younger girls.	Bhatia et al. 1996
	Stochastic	
100:1	Estimated with two-stage clonal model. Increased liver-cancer risk due to stochastic effects (in 0.1% of population compared with median).	Heidenreich 2005
	Overall	
50:1	Modeled heterogeneity in cancer risk—95th percentile compared with median—from age-specific incidence curves for two most common human tumors (lung and colorectal).	Finkel 1995a, 2002
2-110:1	Differences between median vs 98th percentile in noncancer effects at site of contact, responses differ with end point and toxicant.	Hattis et al. 1999

sensitivity that have been reported in the literature. The factors are similar to effect modifiers in epidemiology, in that they modify the effect of another factor on a disease. The first column in Table 4-1 should be interpreted with caution, as there are notable differences in the percentiles used to characterize the size of the susceptible population. Susceptibility factors are broadly considered to include any factor that increases (or decreases) the response of an individual to a dose relative to a typical individual in the population. The distribution of disease in a population can result not only from differences in susceptibility but from disproportionate distributions of exposures of individuals and subgroups in a population. Taken together, variations in disease susceptibility and exposure potential give rise to potentially important variations in vulnerability to the effects of environmental chemicals. Figure 4-2 illustrates how variations in exposure result in variations in risk. Individuals may be more vulnerable than others because they have or are exposed to

- Factors that increase biologic sensitivity or reduce resilience to exposures (such as age, pre-existing disease, and genetics).
- Prior or concurrent exposures to substances that increase a person's susceptibility to the effects of additional exposures.
- Factors that contribute to greater potential for exposure, including personal behavior patterns, the built environment, and modified environmental conditions in locations where time is spent (such as community, home, work, and school).
- Social and economic factors that may influence exposure and biologic responses.

Variability can be more important when independent susceptibility factors can interact to increase susceptibility. For example, genetic and other predisposing conditions interact in ultraviolet-radiation-induced melanoma. Low DNA-repair capacity itself measured in lymphocytes was not observed to increase the risk of melanoma, but statistically significant interactions and large increases in the risk of melanoma were observed in people with low DNA-repair capacity *and* either low tanning capacity or dysplastic nevi (Landi et al. 2002).

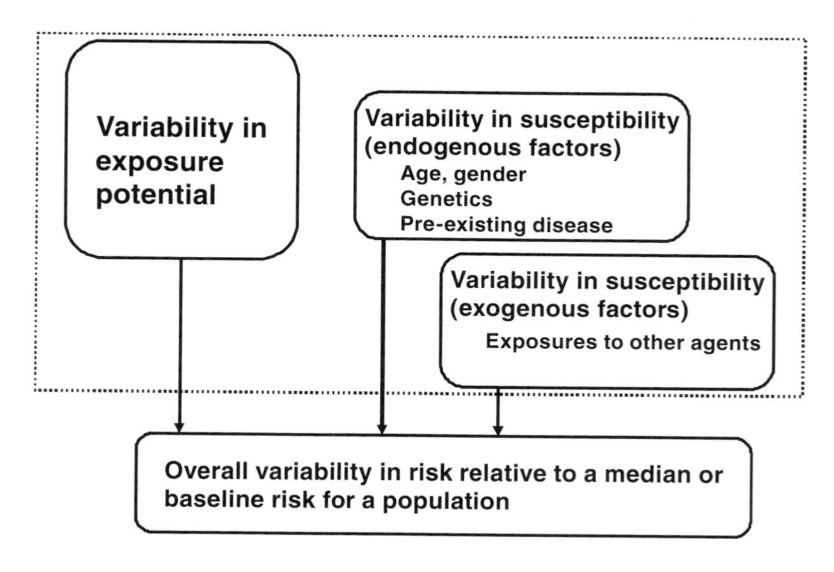

FIGURE 4-2 Factors contributing to variability in risk in the population.

Alcohol consumption, obesity, and diabetes can affect the expression of metabolizing enzymes, such as CYP2E1, whose expression is also under the influence of genetic factors (Ingelman-Sundberg et al. 1993, 1994; Micu et al. 2003; Sexton and Hattis 2007). Interactions are expected to be common but unknown in many diseases caused or exacerbated by environmental chemicals.

Environmental Protection Agency's Approach to Variability in Health-Effects Assessments

EPA's approach to variability assessment is described in its recent *Risk Assessment Principles and Practices: Staff Paper* (EPA 2004a) and guidelines. The staff paper emphasizes that EPA focuses on characterizing variability in exposure, particularly high-end exposures, using as an example the maximally exposed individual in its hazardous air pollutant program. The committee observes that over the last several years some EPA programs have advanced considerably in their efforts to characterize variability in exposure. However, variability in susceptibility and vulnerability has received less detailed evaluation in most EPA health-effects assessments, although there are notable exceptions such as lead, ozone, and sulfur oxides. EPA efforts are considered and options for further improvements presented below.

To address variability in vulnerability to noncancer end points, EPA assumes population-threshold dose-response behavior and assigns uncertainty (adjustment) factors. EPA also endorses such an approach for low-dose nonlinear cancer end points but has been inconsistent in whether and how it is applied. For human-to-human variability in noncancer end points, the default "uncertainty" factor is typically 10, but it can be reduced or increased with sufficient supporting data often by partitioning it into pharmacokinetic and pharmacodynamic factors. The agency has done that with a few assessments based on human data. Only six cases in the Integrated Risk Information System (IRIS) database rely on human occupational data; of these, three had a human intraspecies factor of 10, two had a factor of 3, and one, beryllium, had a factor of 1 because it was assumed that the most sensitive group was included in the occupational study. Thus, in all but four cases in IRIS, a default human intraspecies factor of 10 was assumed, but 10 was the highest value assumed in all cases (EPA 2007a).

The 2005 *Guidelines for Carcinogen Risk Assessment* (EPA 2005b) recognize a number of the factors in Table 4-1 as contributing to cancer susceptibility. Indeed, the guidelines call for the derivation of "separate estimates for susceptible populations and life stages so that these risks can be explicitly characterized" (p. 3-27). The guidelines also lay out a number of reasons why risk estimates derived from occupational studies may not be representative of the general population, including the healthy-worker effect, lack of representation of some subpopulations (for example, fetuses and the young), and underrepresentation of others (for example, women). Guidance in addressing the generalizability of risk estimates derived from occupational studies to the general population is not provided. Similarly the 2005 guidelines point out that animal studies are conducted in relatively homogeneous groups, in contrast with the heterogeneous human population to which the study results are applied. To address variability in susceptibility, the 2005 guidelines (EPA 2005b) call for

- Development of a separate risk estimate for those who are susceptible "when there is an epidemiologic study or animal bioassay that reports quantitative results for susceptible individuals" (p. 3-28).
- Adjustment of the general population estimate for susceptible individuals based on risk-related parameters, for example, pharmacokinetic modeling using pharmacokinetic parameters corresponding to susceptible groups compared with the general population.

- Use of general information in the absence of agent-specific information about early life-stage susceptibility as outlined in *Supplemental Guidance for Assessing Susceptibility from Early-Life Exposure to Carcinogens* (EPA 2005a) and whatever updates follow.

Committee Observations and Comments on Environmental Protection Agency's Approach to Variability

The guidelines provide a useful starting point, but given the agency's limited experience in implementing the 2005 guidelines it is unclear how EPA practice will develop to account for variability. The committee has some concerns based on the guideline language and recent EPA assessments and draft guidance (EPA 2004a, 2005a,b).

With regard to life stages, the 2005 guidelines note that in nature susceptibility differs among various life stages, and the committee agrees that this should be given formal consideration. In an example of late and early life-stage susceptibilities, repair of ultraviolet-damaged DNA declines at 1% per year in subjects 20-60 years old (Grossman 1997), but misrepair in those overexposed when very young has a much longer time to be manifested as cancer. The 2005 guidelines and supplemental guidance that developed generic factors for early-life susceptibility was a step in the right direction. The supplemental guidance provides weighting factors for exposures to mutagenic compounds in the early postnatal and juvenile period. However, *in utero* periods and nonmutagenic chemicals were not covered, and in practice EPA treats the prenatal period as devoid of sensitivity to carcinogenicity, although it has funded research to explore this issue (Hattis et al. 2004, 2005). That stands in contrast with the language in the 2005 guidelines: "Exposures that are of concern extend from conception through adolescence and also include pre-conception exposures of both parents" (EPA 2005b, p. 1-16). EPA needs methods for explicitly considering in cancer risk assessment *in utero* exposure and chemicals that do not meet the threshold of evidence that the agency is considering for judging whether a chemical has a mutagenic mode of action (EPA 2005b). Special attention should be given to hormonally active compounds and genotoxic chemicals that do not meet the threshold of evidence requirements.

The committee encourages EPA to quantify more explicitly variations in exposure and in dose-response relationships. The tiered approach to variability assessment discussed in the 2005 guidelines, with multiple risk descriptions for different susceptible subgroups, is a step in the right direction but falls short of what is needed. The guidelines embrace a default of no variability in the absence of chemical-specific evidence to the contrary. When there is evidence, the focus is on differences between groups. It is important at a minimum to address people who fall into groups that have identified susceptibility. But the guidelines adopt the rather narrow view that variation comes solely from the identified factors that are used to "group" people (for example, a polymorphism) and that are established as important for the chemical under study but not other factors, such as age, ethnic group, socioeconomic status, or other attributes that affect individuals and only incidentally make them part of a new "group." But it will also be important to describe and estimate variability among individuals and the extent of individual differences.

Thus, there is a need for a nonzero default to address the variation in the population expected in the absence of chemical-specific data. The reliance on agent-specific data for all but the early-life assessments of susceptibility is problematic. Because of lack of data, formally addressing variability in cancer risk assessment is feasible only for the most data-rich compounds. That echoes the concern raised earlier about the need to develop more simplified approaches for uncertainty analysis that are tailored to the problems under study: more generalized approaches must be developed to address variability in cancer risk to avoid analyses in which uncharacterized sources of variability are implicitly presumed to have zero effect on individual and population risk. In Chapter 5, the committee proposes an alternative framework for both cancer and noncancer end points that accounts more explicitly for variations in susceptibility and background disease processes and that includes approaches for compounds without substantial data. The framework provides the needed quantitative descriptions of variability in risk for both cancer and noncancer end points.

UNCERTAINTY AND VARIABILITY IN SPECIFIC COMPONENTS OF RISK ASSESSMENT

Each component of a risk assessment includes uncertainty and variability, some explicitly characterized and some unidentified. For each component, current approaches used by EPA to characterize uncertainty and variability are discussed below, and potential improvements are considered.

Hazard Identification

Hazard assessment makes a classification regarding toxicity, for example, whether a chemical is "carcinogenic to humans" or "likely to be" (EPA 2005b), is a neurotoxicant (EPA 1998), or is a potential reproductive hazard (EPA 1996). This gives rise to both quantitative and qualitative uncertainties in hazard characterization. Hazard-identification activities at EPA and other agencies (such as the International Agency for Research on Cancer) focus on protocols for making consistent and transparent classifications but not on a formal treatment of uncertainty. In contrast with the other components of risk assessment, the hazard-identification stage often involves uncertainty about the existence of critical cause-effect relationships that lead to categorically distinct classifications. This type of uncertainty is distinct from uncertainty about such factors as dose-response or exposure-source relationships that have an inherent confidence interval. In this case, one element of an uncertainty analysis involves the issue of misclassification, that is, assigning the wrong outcome to a substance. EPA and the International Agency for Research on Cancer (IARC) have relied on weight-of-evidence classifications (IARC: 1, 2A, 2B, 3, and 4; EPA: "likely to be carcinogenic to humans") to express uncertainty in hazard classifications. Because hazard assessment typically involves a statement or classification regarding the potential for harm, the uncertainty in hazard is not captured well by probability distributions. A formal analysis of hazard uncertainty often requires expert elicitation and discrete probability to communicate uncertainty. Another option is the use of fuzzy sets (Zadeh 1965) or possibility theory (Dubois and Prade 2001), which is a special case of fuzzy set theory. Fuzzy sets and possibility theory were introduced to represent and manipulate data that have "membership" uncertainty. An element of a fuzzy set, such as a toxic characteristic, has a grade of membership, for example, membership in the set "carcinogen" or "not carcinogen." The grade of membership is different in concept from probability. Membership is a quantitative noncommittal measure of imperfect knowledge. The advantage of these methods is that they can characterize nonrandom uncertainties arising from vagueness or incomplete information and give an approximate estimate of the uncertainties. The limitations of fuzzy methods are that they: (1) cannot provide a precise estimate of uncertainty but only an approximate estimation, (2) might not be applicable to situations involving uncertainty resulting from random sampling error, and (3) create difficulties in communicating because set membership or possibilities do not necessarily add to 1. The committee does not endorse any of these specific methods to address uncertainty in hazard assessment but notes in Chapter 3 the need to consider the impact on the overall use of risk information in the fine distinctions between labels describing uncertainty in the weight-of-evidence classification (for example, known vs likely).

Emissions

The first key step in linking pollutant sources to impact in risk assessments, particularly those used to discriminate among various control options, involves characterizing emissions by relevant sources both under baseline conditions and with implementation of controls. In a few situations (for example, in evaluating sulfur dioxide emissions from power plants in the Acid Rain Program), continuous monitoring data are readily available and can be used to characterize baseline emissions with little uncertainty and to characterize the benefits of controls with relatively low uncertainty. But in most cases, there are few source-specific emission measurements, so risk assessors must rely on interpretations based on limited data and emission models.

For example, EPA provides emission factors for stationary sources through the AP-42 database (EPA 2007b). Typically, information on source configuration, fuel composition, control technologies, and other items is used to determine an emission factor based on extrapolation from a limited number of field measurements and known characteristics of the fuel and technology. Uncertainty is included through an emission-factor quality rating, scaled from A to E, that is not quantitatively interpretable and conflates uncertainty and variability. For example, an emission-factor quality rating of A (excellent) is awarded when data are taken from many randomly selected facilities in the source category. But the degree of uncertainty related to measurement techniques is ignored, and the variability among facilities is not carried forward to the overall risk characterization. Because information on variability is not retained and uncertainty is not quantified, EPA treats emission estimates in effect as known quantities in risk assessments. That leads to multiple problems, including mischaracterization of total uncertainty or variability in the assessment and an inability to determine whether improvements in emission estimation are necessary to inform risk-management decisions better (that is, within a value-of-information context). More generally, the AP-42 database has many entries that have not been updated in decades, and this raises the question of whether the emission factors accurately capture current technologies (and adds an unacknowledged source of uncertainty). A final issue is the difficulty of estimating how emissions will change once a risk-management decision is applied; this requires an assessment of the performance of the regulated parties with regard to compliance and noncompliance.

Many risk assessments in EPA use emission models other than those found in AP-42, but most emission estimates suffer from similar issues related to limitations of validation and unacknowledged uncertainty and variability. For example, traffic emissions are characterized with models, such as MOBILE6, in which the estimates are derived from traffic-flow data and calibrated with dynamometer studies on specific vehicles. However, that may not represent true driving-cycle conditions, and some pollutants (such as particulate matter) may be more uncertain than others. In spite of the potentially larger uncertainties associated with emission models, in such analyses as the regulatory impact analysis of nonroad diesel emissions (EPA 2004b), the benefits of controls are presented with up to six significant digits of precision, and no uncertainty is incorporated into the benefits analysis; indeed, in a table titled "Primary Sources of Uncertainty in Benefits Analysis" (EPA 2004b, Table 9A-17), emissions are not even mentioned as a source of uncertainty. EPA and other practitioners should take care to present data with an appropriate number of significant figures, no greater than the smallest number of significant figures reasonably available in the input data, and should formally address emissions as a key source of uncertainty.

For emission characterization, the committee sees an important opportunity for EPA to address variability and uncertainty about emissions explicitly and quantitatively. It will require EPA to evaluate existing models to characterize the uncertainty and variability of individual emission estimates better. The committee recognizes that site-specific emissions data on many situations are lacking and this results in continued reliance on emission models, but it encourages EPA to pursue emission-evaluation studies when plausible and to make more regular refinements in emission-model structures.

Transport, Fate, and Exposure Assessment

Exposure assessment is the process of measuring and modeling the magnitude, frequency, and duration of contact between the potentially harmful agent and a target population, including the size and characteristics of that population (IPCS 2000; Zartarian et al. 2005). For risk assessments, exposure assessment should characterize the sources, routes, pathways, and the attendant uncertainties linking source to dose. It is common for assessors to pose exposure scenarios to define plausible pathways for human contact. Recognition of the multiple possible exposure pathways highlights the importance of a multimedia, multipathway exposure framework. In a multipathway exposure framework, the omission of key exposure pathways (potentially due to data limitations) can contribute to an exposure assessment uncertainty that is often difficult to formally quantify.

Given the framework of exposure assessment in the context of risk assessment, critical inputs include emissions data (described above), fate and transport models to characterize environmental concentrations (both indoors and outdoors), and methods for estimating human exposure given assumed or estimated concentrations. It is also necessary to relate exposure to intake and intake to dose. Further analytic efforts related to modeling human dose are considered later.

The number of transport, fate, and exposure models in active use in EPA or elsewhere is too large to evaluate them individually or to make general statements about their utility and reliability (see the Council for Regulatory Environmental Modeling Web site for a current list [EPA 2008]). Transport, fate, and exposure models can vary substantially in their level of detail, geographic scope, and geographic resolution. Some models are based on environmental parameters that are "archetypal" and provide values that are typical of some regions or populations but not representative of any specific geographic area. These models are used to understand the likely behavior of pollutants as a function of basic chemical properties (Mackay 2001; McKone and MacLeod 2004) and are typically used for comparative assessments of pollutants and for interpreting how partitioning properties and degradability determine transport and fate. Site-specific models apply to releases at specific locations and often track pollutant transport with much more spatial and temporal detail than regional mass-balance models. They are used in a broad array of decision-support activities, including screening-level assessments; setting goals for air emissions, water quality, and soil-cleanup standards; assessing the regional and global fate of persistent organic chemicals; and assessing life-cycle impacts.

There have been many more performance evaluations of transport, fate, and exposure models than of emission models (see, for example, Cowan et al. 1995; Fenner et al. 2005). Although their reliability can vary widely among chemicals considered and the spatial and temporal scale of application, a large literature, methods, and software are available to characterize their uncertainty and sensitivity when they are used in risk assessments.

A critical insight that should be recognized by EPA and other practitioners is that there is no "ideal" transport, fate, or exposure model that can be used under all circumstances.

Some models may be considered to have greater fidelity than others, given the degree to which they capture theoretical constructs and have been evaluated against field measurements, but this does not necessarily imply that the more detailed model should be used under all circumstances. A model with lower resolution (and more uncertainty) but more timely outputs may have greater utility in some decision contexts, especially if the uncertainty can be reasonably characterized to determine its influence on the decision process. Similarly, a model that is highly uncertain with respect to maximum individual exposure but can characterize population-average exposures well may be suitable if the risk management decision is driven by the latter. That reinforces a recurring theme of this report regarding the selection of the appropriate risk-assessment methods in light of the competing demands and constraints described in Chapter 3.

With respect to human exposure modeling, EPA has placed increasing emphasis over the last 25 years on quantitative characterization of uncertainty and variability in its exposure assessments. Exposure assessments and exposure models have evolved from simple assessments that addressed only conditions of maximum exposure to assessments that focus explicitly on exposure variation in a population with a quantitative uncertainty analysis. For example, EPA guidelines for exposure assessment issued in 1992 (EPA 1992) called for both high-end and central-tendency estimates for the population. The high end was considered as what could occur for the 90th percentile or higher of exposed people, and the central tendency might represent an exposure near the median or mean of the distribution of exposed people. Through the 1990s, there was increasing emphasis on an explicit and quantitative characterization of the distinction between interindividual variability and uncertainty in exposure assessments. There was also growing interest in and use of probabilistic simulation methods, such as those based on Monte Carlo or closely related methods, as the basis of estimation of differences in exposure among individuals or, in some cases, of the uncertainty associated with any particular exposure estimate. That effort has been aided by a number of comprehensive studies in the United States and Europe that have used individual personal monitoring in conjunction with ambient and indoor measurements (Wallace et al. 1987; Özkaynak et al. 1996; Kousa et al. 2001, 2002a,b). Expanded use of biomonintoring will provide an opportunity both to evaluate and expand the characterization of exposure variability in human populations.

The committee anticipates expanded efforts by EPA to quantify uncertainty in exposure estimates and to separate uncertainty and population variability in these estimates. Decisions about controlling exposures are typically based on protecting a particular group of people, such as a population or a highly exposed subpopulation (for example, children), because different individuals have different exposures (NRC 1994). The transparency afforded by probabilistic characterization and separation of uncertainty and variability in exposure assessment offers potential benefits for increasing common understanding as a basis of greater convergence in methodology (IPCS 2006).

To date, however, probabilistic exposure assessments have focused on the uncertainty and variability associated with variables in an exposure-assessment model. Missing from the EPA process are guidelines for addressing how model uncertainty and data limitations affect overall uncertainty in exposure assessment. In particular, probabilistic methods have provided estimates of exposure to a compound at the 99th percentile of variability in the population, for example, but have often not considered how model uncertainty affects the reliability of the estimated percentiles. That is an important subject for improvement in future efforts. EPA should also strive for continual enhancement of databases used in exposure modeling, focusing attention on evaluation (that is, personal exposure measurements vs predicted exposures) and applicability to subpopulations of interest. Such documents as the *Exposure Factors Handbook* (EPA 1997d) provide crucial data for such analyses and should be regularly revised to reflect recommended improvements.

Dose Assessment

Assessment of doses of chemicals in the human population relies on a wide array of tools and techniques with varied applications in risk assessment. Monitoring and modeling approaches are used for dose assessment, and important uncertainties and variability are linked to them. Many of the above conclusions for exposure assessment are applicable to dose assessment, but with the recognition that there will be greater variability in doses than exposures across the population as well as greater uncertainty in characterizing those doses.

For monitoring, there have been limited but important efforts in recent years to develop comprehensive databases of tissue burdens of chemicals in representative samples of the human population (for example, the National Health and Nutrition Examination Survey [NHANES], the Center for Health Assessment of Mothers and Children of Salinas, the National Children's Study). There are also efforts to conduct systematic biomonitoring programs in the European Union and in California. Biomonitoring data can provide valuable insight into the degree of variability in internal doses in the population, and analyses of these data can help to determine factors that contribute to dose variability or that modify the exposure-dose relationship. But there are limits to how much variability can be assessed from these data. For example, NHANES is a database of representative samples for the entire U.S. population, but does not capture any geographic subgroups. A discussion of the limitations of NHANES can be found in NRC (2006a). Even with these emerging biomonitoring data, it is still a challenge to assess the contribution of a single source or set of sources to measures of internal dose, which can limit the risk management applicability of these data. In addition there is the challenge of interpreting what the biomonitoring data mean in terms of potential risk to human health (NRC 2006a). Issues related to the value of data obtained through biomonitoring programs are considered in more detail in Chapter 7 in the context of cumulative risk assessment.

Dose modeling is commonly based on physiologically-based pharmacokinetic (PBPK) models. PBPK models are used as a means of addressing species, route, and dose-dependent differences in the ratio of tissue-specific dose to applied dose and thus serve as an alternative to default assumptions for extrapolation that link dose to outcome. PBPK models may address some of the uncertainty associated with extrapolating dose-response data from an animal model to humans, but they often fail to fully capture variability of pharmacokinetics and dose in human populations. Toxicologic research can be used to suggest the structure of PBPK models. And sensitive subpopulations or differing senstivities within the population might be described in terms of some attributes through pharmacokinetic modeling (see Chapter 5, 4-aminobiphenyl case study).

A number of issues related to uncertainty and variability in pharmacokinetic models were addressed in a 2006 workshop (EPA 2006a; Barton et al. 2007). Because the present committee determined that that was a timely and comprehensive review of issues, key findings of the workshop are summarized here. The 2006 workshop considered both short-term and long-term goals for incorporating uncertainty and variability into PBPK models. In particular, Barton et al. (2007) reported the following short-term goals: multidisciplinary teams to integrate deterministic and nondeterministic statistical models; broader use of sensitivity analyses, including those of structural and global (rather than local) parameter changes; and enhanced transparency and reproducibility through more complete documentation of model structures and parameter values, the results of sensitivity and other analyses, and supporting, discrepant, or excluded data. The longer-term needs reported by Barton et al. (2007) included theoretical and practical methodologic improvements for nondeterministic and statistical modeling; better methods for evaluating alternative model structures; peer-reviewed databases of parameters and covariates and their distributions; expanded coverage of PBPK models for chemicals with different properties; and training and reference materials, such as cases studies, tutorials, bibliographies and glossaries, model repositories, and enhanced software.

Many recent examples of PBPK models applied in toxicology have been for volatile organic chemicals and have used similar structures. PBPK models are needed for a broader array of chemical species (for example, from low to high volatility and low to high log Kow[2]). Methods for comparing alternative model structures rapidly with available data would facilitate testing of new structural ideas, provide perspective on model uncertainty, and help to address chemicals on which data are sparse. Ultimately, the recognition that models of various degrees of complexity may all describe the available data reasonably will encourage the acquisition of data to differentiate between competing models.

Mode of Action and Dose-Response Models

Many of the most substantial issues related to both uncertainty and variability can be seen in the realm of dose-response assessment for both cancer and noncancer end points. Historically, risk assessments for carcinogenic end points have been conducted very differently from noncancer risk assessments. In reviewing the issue of mode of action, the committee recognized a clear and important need for a consistent and unified approach in dose-response modeling. For carcinogens, it has generally been assumed that there is no threshold of effect, and risk assessments have focused on quantifying their potency, which is the low-dose slope of the dose-response relationship. For noncancer risk assessment, the prevailing assumption has been that homeostatic and other repair mechanisms in the body result in a population threshold or low-dose nonlinearity that leads to inconsequential risk at low doses, and risk assessments have focused on defining the reference dose or concentration that is sufficiently below the threshold or threshold-like dose to be deemed safe ("likely to be without an appreciable risk of deleterious effects") (EPA 2002b, p. 4-4). Noncancer risk assessments simply compare observed or predicted doses with the reference dose to yield a qualitative conclusion about the likelihood of harm.

The committee finds substantial deficiencies in both approaches with respect to core concepts and the treatment of uncertainty and variability. Cancer risk assessments often provide estimates of the population burden of disease or fraction of the population likely to be above a defined risk level. But there is no explicit treatment of uncertainty associated with such factors as interspecies extrapolation, high-dose to low-dose extrapolation, and the limitations of dose-response studies to capture all relevant information. Moreover, there is essentially no consideration of variations in the population in susceptibility and vulnerability other than consideration of the increased susceptibility of infants and children. The noncancer risk-assessment paradigm remains one of defining a reference value with no formal quantification of how disease incidence varies with exposure. Human heterogeneity is accommodated with a "default" factor, and it is often unclear when the evidence is sufficient to deviate from such defaults. The structure of the reference dose also omits any formal quantification of uncer-

[2]Kow is the octanol-water partition coefficient or the ratio of the concentration of a chemical in octanol and in water at equilibrium and at a specified temperature.

tainty. And the current approach does not address compounds for which thresholds are not apparent (for example, fine particulate matter and lead) or not expected (for example, in the case of background additivity). To address the issue of improving dose-response modeling, both from the perspective of uncertainty and variability characterization and in the context of new information on mode of action, the committee has developed a unified and consistent approach to dose-response modeling (Chapter 5).

Beyond toxicologic studies of chemicals, there are multiple examples where uncertainty and variability have been more explicitly treated. For example, two National Research Council reports prepared by the Committee on Biological Effects of Ionizing Radiation (NRC 1999, 2006b) have provided examples for addressing dose-response uncertainty for ionizing radiation. Both the BEIR VI report dealing with radon (NRC 1999) and the BEIR VII report dealing with low linear energy transfer (LET) ionizing radiation (NRC 2006b) provided a quantitative analysis of the uncertainties associated with estimates of radiation cancer risks.

More generally, epidemiologic studies provide enhanced mechanisms for characterizing uncertainty and variability, sometimes providing information that is more relevant for human health risk assessment than dose-response relationships derived by extrapolating laboratory-animal data to humans. Emerging disciplines such as health tracking, molecular epidemiology, and social epidemiology provide opportunities to improve resolution in linking exposure to disease, which may enhance the ability of epidemiologists to uncover both main effects and effect modifiers, providing greater insight about human heterogeneity in response. A more detailed discussion of the role of these emerging epidemiologic disciplines from the perspective of cumulative risk assessment is provided in Chapter 7.

An additional consideration in the treatment of uncertainty and variability in dose-response modeling is related to approaches to combine information across multiple publications, especially in the context of epidemiologic evidence. Various meta-analytic techniques have been employed both to provide pooled central estimates with uncertainty bounds and to evaluate factors that could explain variability in findings across studies (Bell et al. 2005; Ito et al. 2005; Levy et al. 2005). While these approaches will not be applicable in most contexts, because they require a sufficiently large body of epidemiologic literature to allow for pooled analyses, these methods can be utilized to reduce uncertainty associated with selection of a single epidemiologic study for a dose-response function, to characterize uncertainty associated with application of a pooled estimate to a specific setting, and to determine factors that contribute to variability in dose-response functions. EPA should consider these and other meta-analytic techniques, especially for risk management applications tied to specific geographic areas.

PRINCIPLES FOR ADDRESSING UNCERTAINTY AND VARIABILITY

EPA and policy analysts are not constrained by a lack of methods for conducting uncertainty analysis but can be paralyzed by the absence of guidance on what levels of detail and rigor are needed for a particular risk assessment. That creates situations that splinter the parties involved into those who favor application of the most sophisticated methods to all cases and those who would rather ignore uncertainty completely and simply rely on point estimates of parameters and defaults for all models. But risk assessment often requires something in between. To confront the issue, EPA should develop guidance for conducting and establishing the level of detail in uncertainty and variability analyses that is required for various risk assessments. To foster optimal treatment of variability in its assessments, the agency could develop general guidelines or further supplemental guidance to its health-effects

BOX 4-7 Recommended Principles for Uncertainty and Variability Analysis

1. Risk assessments should provide a quantitative, or at least qualitative, description of uncertainty and variability consistent with available data. The information required to conduct detailed uncertainty analyses may not be available in many situations.
2. In addition to characterizing the full population at risk, attention should be directed to vulnerable individuals and subpopulations that may be particularly susceptible or more highly exposed.
3. The depth, extent, and detail of the uncertainty and variability analyses should be commensurate with the importance and nature of the decision to be informed by the risk assessment and with what is valued in a decision. This may best be achieved by early engagement of assessors, managers, and stakeholders in the nature and objectives of the risk assessment and terms of reference (which must be clearly defined).
4. The risk assessment should compile or otherwise characterize the types, sources, extent, and magnitude of variability and substantial uncertainties associated with the assessment. To the extent feasible, there should be homologous treatment of uncertainties among the different components of a risk assessment and among different policy options being compared.
5. To maximize public understanding of and participation in risk-related decision-making, a risk assessment should explain the basis and results of the uncertainty analysis with sufficient clarity to be understood by the public and decision-makers. The uncertainty assessment should not be a significant source of delay in the release of an assessment.
6. Uncertainty and variability should be kept conceptually separate in the risk characterization.

(for example, EPA 2005a) and exposure guidance used in its various programs. To support the effort, the committee offers the principles presented in Box 4-7.

The principles in Box 4-7 are consistent with and expand on the "Principles for Risk Analysis" originally established in 1995, noted as useful by the National Research Council (NRC 2007c), and recently re-released by the Office of Management and Budget and the Office of Science and Technology Policy (OMB/OSTP 2007). They are derived from the more detailed discussions above. In particular, they are based on the following issues.

- Qualitative thinking about uncertainty that reveals that despite the uncertainty, one can have confidence in which risk-management option to pick and not need to quantify further.
- A need to ensure that uncertainty and variability are addressed by ensuring that the risk is not underestimated.
- Characterization of a variety of risks and their corresponding confidence intervals.

Depending on the risk-management options, a quantitative treatment of uncertainty and variability may be needed to differentiate among the options for making an informed decision. Uncertainty analysis is important for both data-rich and data-poor situations, but confidence in the analysis will vary according to the amount of information available.

Because resources are limited in EPA, it is important to match the level of effort to the extent to which a more detailed analysis may influence an important decision. If an uncertainty analysis will not substantially influence outcomes of importance to the decision-maker, resources should not be expended on a detailed uncertainty analysis (for example, two-dimensional Monte Carlo analysis). In developing guidance for uncertainty analysis, EPA first should develop guidelines that "screen out" risk assessments that focus on risks that do not warrant the use of substantial analytic resources. Second, the guidelines should describe the level of detail that is warranted for "important" risk assessments. Third, the analysis should be tailored to the decision-rule outcome by addressing what is valued in a decision; for example, if the decision-maker is interested only in the 5% most-exposed or most at-risk members of a population, there is little value in structuring an uncertainty analysis that focuses on uncertainty and variability in the full population.

The risk assessor should consider the uncertainties and variabilities that accrue in all stages of the risk assessment—in emissions or environmental concentration data, fate and exposure assessment, dose and mechanism of action, and dose-response relationship. It is important to identify the largest sources of uncertainty and variability and to determine the extent to which there is value in focusing on other components. This approach should be based on a value-of-information (VOI) strategy even when resources for a fully quantitative VOI analysis are limited (see discussion in Chapter 3). For example, when uncertainty gives rise to risk estimates that are spread across one or more key decision points, such as a range that includes acceptable and unacceptable levels of risk, then there is value in addressing uncertainty in other components when this information provides more insight on whether one choice of action for reducing risk is better than another.

When the goal of a risk assessment is to discriminate among various options, the uncertainty analysis supporting the evaluation should be tailored to provide sufficient resolution to make the discriminations (to the extent that it can). It is important to distinguish when and how to engage an uncertainty analysis to characterize one-sided confidence (confidence that the risk does not exceed X or confidence that all or most individuals are protected from harm, and so on) or richer descriptions of the uncertainty (for example, two-sided confidence bounds, or the full distribution). Depending on the options being considered, a fuller description may be needed to understand tradeoffs. When a "safe" level of risk is being established, without consideration of costs or countervailing risks, a single-sided (bounding) risk estimate or lower-bound acceptable dose may be sufficient.

RECOMMENDATIONS

This chapter addressed the need to consider uncertainty and variability in an interpretable and consistent manner among all components of a risk assessment and to communicate them in the overall risk characterization. The committee focused on more detailed and transparent methods for addressing uncertainty and variability, on specific aspects of uncertainty and variability in key computational steps of risk assessment, and on approaches to help EPA to decide what level of detail to use in characterizing uncertainty and variability to support risk-management decisions and public involvement in the process. The committee recognizes that EPA has the technical capability to do two-stage Monte Carlo and other very detailed and computationally intensive analyses of uncertainty and variability. But such analyses are not necessary in all decision contexts, given that transparency and timeliness are also desirable attributes of a risk assessment, and given that some decisions can be made with less complex analyses. The question is not often about better ways to do these analyses, but about developing a better understanding of when to do these analyses.

To address those issues, the committee provides the following recommendations:

- EPA should develop a process to address and communicate the uncertainty and variability that are parts of any risk assessment. In particular, this process should encourage risk assessments to characterize and communicate uncertainty and variability in all key computational steps of risk assessment—emissions, fate-and-transport modeling, exposure assessment, dose assessment, dose-response assessment, and risk characterization.

- EPA should develop guidance to help analysts determine the appropriate level of detail needed in uncertainty and variability analyses to support decision-making. The principles of uncertainty and variability analysis above provide a starting point for development of this guidance, which should include approaches both for analysis and communication
- In the short term, EPA should adopt a "tiered" approach for selecting the level of detail used in uncertainty and variability assessment. A discussion of the level of detail used for uncertainty analysis and variability assessment should be an explicit part of the problem formulation and planning and scoping.
- In the short term, EPA should develop guidelines that define key terms of reference used in the presentation of uncertainty and variability, such as *central tendency*, *average*, *expected*, *upper bound*, and *plausible upper bound*. In addition, because risk-risk and benefit-cost comparisons pose unique analytic challenges, guidelines could provide insight into and advice on uncertainty characterizations to support risk decision-making in these contexts.
- Improving characterization of uncertainty and variability in risk assessment comes at a cost, and additional resources and training of risk assessors and risk managers will be required. In the short term, EPA should build the capacity to provide guidance to address and implement the principles of uncertainty and variability analysis.

REFERENCES

ATSDR (Agency for Toxic Substances and Disease Registry). 1992. Case Studies in Environmental Medicine: Radon Toxicity. U.S. Department of Health and Human Services, Public Health Service, Agency for Toxic Substances and Disease Registry, Atlanta, GA.

Barton, H.A., W.A. Chiu, R. Woodrow Setzer, M.E. Andersen, A.J. Bailer, F.Y. Bois, R.S. Dewoskin, S. Hays, G. Johanson, N. Jones, G. Loizou, R.C. MacPhail, C.J. Portier, M. Spendiff, and Y.M. Tan. 2007. Characterizing uncertainty and variability in physiologically based pharmacokinetic models: State of the science and needs for research and implementation. Toxicol. Sci. 99(2):395-402.

Bell, M.L., F. Dominici, and J.M. Samet. 2005. A meta-analysis of time-series studies of ozone and mortality with comparison to the National Morbidity, Mortality and Air Pollution Study. Epidemiology 16(4):436-445.

Bhatia, S., L.L. Robison, O. Oberlin, M. Greenberg, G. Bunin, F. Fossati-Bellani, and A.T. Meadows. 1996. Breast cancer and other second neoplasms after childhood Hodgkin's disease. N. Engl. J. Med. 334(12):745-751.

Blount, B.C., J.L. Pirkle, J.D. Osterloh, L. Valentin-Blasini, and K.L. Caldwell. 2006. Urinary perchlorate and thyroid hormone levels in adolescent and adult men and women living in the United States. Environ. Health Perspect. 114(12):1865-1871.

Bois, F.Y., G. Krowech, and L. Zeise. 1995. Modeling human interindividual variability in metabolism and risk: The example of 4-aminobiphenyl. Risk Anal. 15(2):205-213.

Budnitz, R.J., G. Apostolakis, D.M. Boore, L.S. Cluff, K.J. Coppersmith, C.A. Cornell, and P.A. Morris. 1998. Use of technical expert panels: Applications to probabilistic seismic hazard analysis. Risk Anal. 18(4):463-469.

CDHS (California Department of Health Services). 1990. Report to the Air Resources Board on Inorganic Arsenic. Part B. Health Effects of Inorganic Arsenic. Air Toxicology and Epidemiology Section. Hazard Identification and Risk Assessment Branch. Department of Health Services. Berkeley, CA.

Cowan, C.E., D. Mackay, T.C.J. Feijtel, D. Van De Meent, A. Di Guardo, J. Davies, and N. Mackay, eds. 1995. The Multi-Media Fate Model: A Vital Tool for Predicting the Fate of Chemicals. Pensacola, FL: Society of Environmental Toxicology and Chemistry.

Cullen, A.C., and H.C. Frey. 1999. The Use of Probabilistic Techniques in Exposure Assessment: A Handbook for Dealing with Variability and Uncertainty in Models and Inputs. New York: Plenum Press.

Cullen, A.C., and M.J. Small. 2004. Uncertain risk: The role and limits of quantitative analysis. Pp. 163-212 in Risk Analysis and Society: An Interdisciplinary Characterization of the Field, T. McDaniels, and M.J. Small, eds. Cambridge, UK: Cambridge University Press.

Dubois, D., and H. Prade. 2001. Possibility theory, probability theory and multiple-valued logics: A clarification. Ann. Math. Artif. Intell. 32(1-4):35-66.

EPA (U.S. Environmental Protection Agency). 1986. Guidelines for Carcinogen Risk Assessment. EPA/630/R-00/004. Risk Assessment Forum, U.S. Environmental Protection Agency, Washington, DC. September 1986 [online]. Available: http://www.epa.gov/ncea/raf/car2sab/guidelines_1986.pdf [accessed Jan. 7, 2008].

EPA (U.S. Environmental Protection Agency). 1989a. Risk Assessment Guidance for Superfund, Vol. 1. Human Health Evaluation Manual Part A. EPA/540/1-89/002. Office of Emergency and Remedial Response, U.S. Environmental Protection Agency, Washington, DC. December 1989 [online]. Available: http://rais.ornl.gov/homepage/HHEMA.pdf [accessed Jan. 11, 2008].

EPA (U.S. Environmental Protection Agency). 1989b. Interim Procedures for Estimating Risks Associated with Exposures to Mixtures of Chlorinated Dibenzo-p-Dioxins and Dibenzofurans (CDDs and CDFs): 1989 Update. EPA/625/3-89/016. Risk Assessment Forum, U.S. Environmental Protection Agency, Washington, DC.

EPA (U.S. Environmental Protection Agency). 1992. Guidelines for Exposure Assessment. EPA600Z-92/001. Risk Assessment Forum, U.S. Environmental Protection Agency, Washington, DC [online]. Available: http://cfpub.epa.gov/ncea/raf/recordisplay.cfm?deid=15263 [accessed Jan. 14, 2008].

EPA (U.S. Environmental Protection Agency). 1996. Guidelines for Reproductive Toxicity Risk Assessment. EPA/630/R-96/009. Risk Assessment Forum, U.S. Environmental Protection Agency, Washington, DC. October 1996 [online]. Available: http://www.epa.gov/ncea/raf/pdfs/repro51.pdf [accessed Jan. 10, 2008].

EPA (U.S. Environmental Protection Agency). 1997a. Guiding Principles for Monte Carlo Analysis. EPA/630/R-97/001. Risk Assessment Forum, U.S. Environmental Protection Agency, Washington, DC. March 1997 [online]. Available: http://www.epa.gov/ncea/raf/montecar.pdf [accessed Jan. 7, 2008].

EPA (U.S. Environmental Protection Agency). 1997b. Policy for Use of Probabilistic Analysis in Risk Assessment at the U.S. Environmental Protection Agency. Science Policy Council, U.S. Environmental Protection Agency, Washington, DC. May 15, 1997 [online]. Available: http://www.epa.gov/osp/spc/probpol.htm [accessed Jan. 15, 2008].

EPA (U.S. Environmental Protection Agency). 1997c. Guidance on Cumulative Risk Assessment, Part 1. Planning and Scoping. Science Policy Council, U.S. Environmental Protection Agency, Washington, DC. July 3, 1997 [online]. Available: http://www.epa.gov/brownfields/html-doc/cumrisk2.htm [accessed Jan. 14, 2008].

EPA (U.S. Environmental Protection Agency). 1997d. Exposure Factors Handbook. National Center for Environmental Assessment, Office of Research and Development, U.S. Environmental Protection Agency, Washington, DC. August 1997 [online]. Available: http://www.epa.gov/ncea/efh/report.html [accessed Aug. 5, 2008].

EPA (U.S. Environmental Protection Agency). 1998. Guidelines for Neurotoxicity Risk Assessment. EPA/630/R-95/001F. Risk Assessment Forum, U.S. Environmental Protection Agency, Washington, DC. April 1998 [online]. Available: http://www.epa.gov/ncea/raf/pdfs/neurotox.pdf [accessed Jan. 10, 2008].

EPA (U.S. Environmental Protection Agency). 2001. Risk Assessment Guidance for Superfund (RAGS): Vol. 3 - Part A: Process for Conducting Probabilistic Risk Assessment. EPA 540-R-02-002. Office of Emergency and Remedial Response, U.S. Environmental Protection Agency, Washington, DC. December 2001. http://www.epa.gov/oswer/riskassessment/rags3a/ [accessed Jan. 14, 2008].

EPA (U.S. Environmental Protection Agency). 2002a. Calculating Upper Confidence Limits for Exposure Point Concentrations at Hazardous Waste Sites. OSWER 9285.6-10. Office of Emergency and Remedial Response, U.S. Environmental Protection Agency, Washington, DC. December 2002 [online]. Available: http://www.hanford.gov/dqo/training/ucl.pdf [accessed Jan. 14, 2008].

EPA (U.S. Environmental Protection Agency). 2002b. A Review of the Reference Dose and Reference Concentration Processes. Final report. EPA/630/P-02/002F. Risk Assessment Forum, U.S. Environmental Protection Agency, Washington, DC. December 2002 [online]. Available: http://www.epa.gov/iris/RFD_FINAL%5B1%5D.pdf [accessed Jan. 14, 2008].

EPA (U.S. Environmental Protection Agency). 2004a. Risk Assessment Principles and Practices: Staff Paper. EPA/100/B-04/001. Office of the Science Advisor, U.S. Environmental Protection Agency, Washington, DC. March 2004 [online]. Available: http://www.epa.gov/osa/pdfs/ratf-final.pdf [accessed Jan. 9, 2008].

EPA (U.S. Environmental Protection Agency). 2004b. Final Regulatory Analysis: Control of Emissions from Nonroad Diesel Engines. EPA420-R-04-007. Office of Transportation and Air Quality, U.S. Environmental Protection Agency. May 2004 [online]. Available: http://www.epa.gov/nonroad-diesel/2004fr/420r04007a.pdf [accessed Jan. 14, 2008].

EPA (U.S. Environmental Protection Agency). 2005a. Supplemental Guidance for Assessing Susceptibility from Early-Life Exposures to Carcinogens. EPA/630/R-03/003F. Risk Assessment Forum, U.S. Environmental Protection Agency, Washington, DC. March 2005 [online]. Available: http://cfpub.epa.gov/ncea/cfm/recordisplay.cfm?deid=160003 [accessed Jan. 4, 2008].

EPA (U.S. Environmental Protection Agency). 2005b. Guidelines for Carcinogen Risk Assessment. EPA/630/P-03/001F. Risk Assessment Forum, U.S. Environmental Protection Agency, Washington, DC. March 2005 [online]. Available: http://cfpub.epa.gov/ncea/cfm/recordisplay.cfm?deid=116283 [accessed Jan. 15, 2008].

EPA (U.S. Environmental Protection Agency). 2005c. Regulatory Impact Analysis for the Final Clean Air Interstate Rule. EPA-452/R-05-002. Air Quality Strategies and Standards Division, Emission, Monitoring, and Analysis Division and Clean Air Markets Division, Office of Air and Radiation, U.S. Environmental Protection Agency. March 2005 [online]. Available: http://www.epa.gov/CAIR/pdfs/finaltech08.pdf [accessed Jan. 14, 2008].

EPA (U.S. Environmental Protection Agency). 2006a. International Workshop on Uncertainty and Variability in Physiologically Based Pharmacokinetic (PBPK) Models, October 31-November 2, 2006, Research Triangle Park, NC [online]. Available: http://www.epa.gov/ncct/uvpkm/ [accessed Jan. 15, 2008].

EPA (U.S. Environmental Protection Agency). 2006b. Regulatory Impact Analysis (RIA) of the 2006 National Ambient Air Quality Standards for Fine Particle Pollution. Air Quality Strategies and Standards Division, Office of Air and Radiation, U.S. Environmental Protection Agency. October 6, 2006 [online]. Available: http://www.epa.gov/ttn/ecas/regdata/RIAs/Executive%20Summary.pdf [accessed Nov. 17, 2008].

EPA (U.S. Environmental Protection Agency). 2007a. Integrated Risk Information System (IRIS). Office of Research and Development, U.S. Environmental Protection Agency, Washington, DC [online]. Available http://www.epa.gov/iris/ [accessed Jan. 15, 2008].

EPA (U.S. Environmental Protection Agency). 2007b. Emissions Factors & AP 42. Clearinghouse for Inventories and Emissions Factors, Technology Transfer Network, U.S. Environmental Protection Agency [online]. Available: http://www.epa.gov/ttn/chief/ap42/index.html [accessed Jan. 15, 2008].

EPA (U.S. Environmental Protection Agency). 2008. EPA's Council for Regulatory Environmental Modeling (CREM). Office of the Science Advisor, U.S. Environmental Protection Agency. October 23, 2008 [online]. Available: http://www.epa.gov/crem/ [accessed Nov. 20, 2008].

EPA SAB (U.S. Environmental Protection Agency Science Advisory Board). 2004. EPA's Multipmedia Multipathway and Multireceptor Risk Assessment (3MRA) Modeling System: A Review by the 3MRA Review Panel of the EPA Science Advisory Board.. EPA-SAB-05-003. U.S. Environmental Protection Agency, Science Advisory Board, Washington, DC. October 22, 2004 [online]. Available: http://yosemite.epa.gov/sab/sabproduct.nsf/99390EFBFC255AE885256FFE00579745/$File/SAB-05-003_unsigned.pdf [accessed Sept. 9, 2008].

Evans, J. S., G.M. Gray, R.L. Sielken, A.E. Smith, C. Valdez-Flores, and J.D. Graham. 1994. Use of probabilistic expert judgment in uncertainty analysis of carcinogenic potency. Regul. Toxicol. Pharmacol. 20(1):15-36.

Fenner, K., M. Scheringer, M. MacLeod, M. Matthies, T. McKone, M. Stroebe, A. Beyer, M. Bonnell, A.C. Le Gall, J. Klasmeier, D. Mackay, D. van de Meent, D. Pennington, B. Scharenberg, N. Suzuki, and F. Wania. 2005. Comparing estimates of persistence and long-range transport potential among multimedia models. Environ. Sci. Technol. 39(7):1932-1942.

Finkel, A.M. 1990. Confronting Uncertainty in Risk Management: A Guide for Decision Makers. Washington, DC: Resources for the Future.

Finkel, A.M. 1995a. A quantitative estimate of the variations in human susceptibility to cancer and its implications for risk management. Pp. 297-328 in Low-Dose Extrapolation of Cancer Risks: Issues and Perspectives, S.S. Olin, W. Farland, C. Park, L. Rhomberg, R. Scheuplein, and T. Starr, eds. Washington, DC: ILSI Press.

Finkel, A.M. 1995b. Toward less misleading comparisons of uncertain risks: The example of aflatoxin and alar. Environ. Health Perspect. 103(4):376-385.

Finkel, A.M. 2002. The joy before cooking: Preparing ourselves to write a risk research recipe. Hum. Ecol. Risk Assess. 8(6):1203-1221.

Greer, M.A., G. Goodman, R.C. Pleus, and S.E. Greer. 2002. Health effects assessment for environmental perchlorate contamination: The dose response for inhibition of thyroidal radioiodine uptake in humans. Environ. Health Perspect. 110(9):927-937.

Grossman, L. 1997. Epidemiology of ultraviolet-DNA repair capacity and human cancer. Environ. Health Perspect. 105(Suppl. 4):927-930.

Hattis, D., P. Banati, and R. Goble. 1999. Distributions of individual susceptibility among humans for toxic effects: How much protection does the traditional tenfold factor provide for what fraction of which kinds of chemicals and effects? Ann. NY Acad. Sci. 895:286-316.

Hattis, D., R. Goble, A. Russ, M. Chu, and J. Ericson. 2004. Age-related differences in susceptibility to carcinogenesis: A quantitative analysis of empirical animal bioassay data. Environ. Health Perspect. 112(11):1152-1158.

Hattis, D., R. Goble, and M. Chu. 2005. Age-related differences in susceptibility to carcinogenesis. II. Approaches for application and uncertainty analyses for individual genetically acting carcinogens. Environ. Health Perspect. 113(4):509-516.

Heidenreich, W.F. 2005. Heterogeneity of cancer risk due to stochastic effects. Risk Anal. 25(6):1589-1594.

ICRP (International Commission on Radiological Protection). 1998. Genetic Susceptibility to Cancer. ICRP Publication 79. Annals of the ICPR 28(1-2). New York: Pergamon.

IEc (Industrial Economics, Inc). 2006. Expanded Expert Judgment Assessment of the Concentration-Response Relationship Between PM2.5 Exposure and Mortality. Prepared for the Office of Air Quality Planning and Standards, U.S. Environmental Protection Agency, Research Triangle Park, NC, by Industrial Economics Inc., Cambridge, MA. September, 2006 [online]. Available: http://www.epa.gov/ttn/ecas/regdata/Uncertainty/pm_ee_report.pdf [accessed Jan. 14, 2008].

Ingelman-Sundberg, M., I. Johannson, H. Yin, Y. Terelius, E. Eliasson, P. Clot, and E. Albano. 1993. Ethanol-inducible cytochrome P4502E1: Genetic polymorphism, regulation, and possible role in the etiology of alcohol-induced liver disease. Alcohol 10(6):447-452.

Ingelman-Sundberg, M., M.J. Ronis, K.O. Lindros, E. Eliasson, and A. Zhukov. 1994. Ethanol-inducible cytochrome P4502E1: Regulation, enzymology and molecular biology. Alcohol Suppl. 2:131-139.

IPCS (International Programme on Chemical Safety). 2000. Human exposure and dose modeling. Part 6 in Human Exposure Assessment. Environmental Health Criteria 214. Geneva: World Health Organization [online]. Available: http://www.inchem.org/documents/ehc/ehc/ehc214.htm#PartNumber:6 [accessed Jan. 15, 2008].

IPCS (International Programme on Chemical Safety). 2004. IPCS Risk Assessment Terminology Part 1: IPCS/OECD Key Generic Terms used in Chemical Hazard/Risk Assessment and Part 2: IPCS Glossary of Key Exposure Assessment Terminology. Geneva: World Health Organization [online]. Available: http://www.who.int/ipcs/methods/harmonization/areas/ipcsterminologyparts1and2.pdf [accessed Jan. 15, 2008].

IPCS (International Programme on Chemical Safety). 2006. Draft Guidance Document on Characterizing and Communicating Uncertainty of Exposure Assessment, Draft for Public Review. IPCS Project on the Harmonization of Approaches to the Assessment of Risk from Exposure to Chemicals. Geneva: World Health Organization [online]. Available: http://www.who.int/ipcs/methods/harmonization/areas/draftundertainty.pdf [accessed Jan. 15, 2008].

Ito, K., S.F. DeLeon, and M. Lippmann. 2005. Associations between ozone and daily mortality: Analysis and meta-analysis. Epidemiology 16(4):446-457.

Kavlock, R. 2006. Computational Toxicology: New Approaches to Improve Environmental Health Protection. Presentation on the 1st Meeting on Improving Risk Analysis Approaches Used by the U.S. EPA, November 20, 2006, Washington, DC.

Kousa, A., C. Monn, T. Totko, S. Alm, L. Oglesby, and M.J. Jantunen. 2001. Personal exposures to NO_2 in the EXPOLIS study: Relation to residential indoor, outdoor, and workplace concentrations in Basel, Helsinki, and Prague. Atmos. Environ. 35(20):3405-3412.

Kousa, A., J. Kukkonen, A. Karppinen, P. Aarnio, and T. Koskentalo. 2002a. A model for evaluating the population exposure to ambient air pollution in an urban area. Atmos. Environ. 36(13):2109-2119.

Kousa, A., L. Oglesby, K. Koistinen, N. Kunzli, and M. Jantunen. 2002b. Exposure chain of urban air $PM_{2.5}$-associations between ambient fixed site, residential outdoor, indoor, workplace, and personal exposures in four European cities in EXPOLIS study. Atmos. Environ. 36(18):3031-3039.

Krupnick, A., R. Morgenstern, M. Batz, P. Nelsen, D. Burtraw, J.S. Shih, and M. McWilliams. 2006. Not a Sure Thing: Making Regulatory Choices under Uncertainty. Washington, DC: Resources for the Future. February 2006 [online]. Available: http://www.rff.org/rff/Documents/RFF-Rpt-RegulatoryChoices.pdf [accessed Nov. 22, 2006].

Landi, M.T., A. Baccarelli, R.E. Tarone, A.Pesatori, M.A. Tucker, M. Hedayati, and L. Grossman. 2002. DNA repair, dysplastic nevi, and sunlight sensitivity in the development of cutaneous malignant melanoma. J. Natl. Cancer Inst. 94(2):94-101.

Levy, J.I., S.M. Chemerynski, and J.A. Sarnat. 2005. Ozone exposure and mortality: An empiric Bayes meta-regression analysis. Epidemiology 16(4):458-468.

Mackay, D. 2001. Multimedia Environmental Models: The Fugacity Approach, 2nd Ed. Boca Raton: Lewis.

McKone, T.E., and M. MacLeod. 2004. Tracking multiple pathways of human exposure to persistent multimedia pollutants: Regional, continental, and global scale models. Annu. Rev. Environ. Resour. 28:463-492.

Micu, A.L., S. Miksys, E.M. Sellers, D.R. Koop, and R.F. Tyndale. 2003. Rat hepatic CYP2E1 is induced by very low nicotine doses: An investigation of induction, time course, dose response, and mechanism. J. Pharmacol. Exp. Ther. 306(3):941-947.

Morgan, M.G., M. Henrion, and M. Small. 1990. Uncertainty: A Guide to Dealing with Uncertainty in Quantitative Risk and Policy Analysis. Cambridge: Cambridge University Press.

NRC (National Research Council). 1983. Risk Assessment in the Federal Government: Managing the Process. Washington, DC: National Academy Press.

NRC (National Research Council). 1994. Science and Judgment in Risk Assessment. Washington, DC: National Academy Press.

NRC (National Research Council). 1996. Understanding Risk: Informing Decisions in a Democratic Society. Washington, DC: National Academy Press.

NRC (National Research Council) 1999. Health Effects of Exposure to Radon BEIR VI. Washington, DC: National Academy Press.

NRC (National Research Council). 2000. Copper in Drinking Water. Washington, DC: National Academy Press.

NRC (National Research Council). 2002. Estimating the Public Health Benefits of Proposed Air Pollution Regulations. Washington, DC: The National Academies Press.

NRC (National Research Council). 2006a. Human Biomonitoring of Environmental Chemicals. Washington, DC: The National Academies Press.

NRC (National Research Council). 2006b. Health Risks from Exposures to Low Levels of Ionizing Radiation BEIR VII. Washington, DC: The National Academies Press.

NRC (National Research Council). 2007a. Applications of Toxicogenomic Technologies to Predictive Toxicology and Risk Assessment. Washington, DC: The National Academies Press.

NRC (National Research Council). 2007b. Toxicity Testing in the Twenty-First Century: A Vision and a Strategy. Washington, DC: The National Academies Press.

NRC (National Research Council). 2007c. Scientific Review of the Proposed Risk Assessment Bulletin from the Office of Management and Budget. Washington, DC: The National Academies Press.

NRC (National Research Council). 2007d. Models in Environmental Regulatory Decision Making. Washington, DC: The National Academies Press.

OMB/OSTP (Office of Management and Budget/Office of Science and Technology Policy). 2007. Updated Principles for Risk Analysis. Memorandum for the Heads of Executive Departments and Agencies, from Susan E. Dudley, Administrator, Office of Information and Regulatory Affairs, Office of Management and Budget, and Sharon L. Hays, Associate Director and Deputy Director for Science, Office of Science and Technology Policy, Washington, DC. September 19, 2007 [online]. Available: http://www.whitehouse.gov/omb/memoranda/fy2007/m07-24.pdf [accessed Jan. 4, 2008].

Özkaynak, H., J. Xue, J. Spengler, L. Wallace, E. Pellizari, and P. Jenkins. 1996. Personal exposure to airborne particles and metals: Results from the particle TEAM study in Riverside, California. J. Expo. Anal. Environ. Epidemiol. 6(1):57-78.

Paté-Cornell, M.E. 1996. Uncertainties in risk analysis: Six levels of treatment. Reliab. Eng. Syst. Safe. 54(2):95-111.

Sexton, K., and D. Hattis. 2007. Assessing cumulative health risks from exposure to environmental mixtures: Three fundamental questions. Environ. Health Perspect. 115(5):825-832.

Spetzler, C.S., and C.S. von Holstein. 1975. Probability encoding in decision analysis. Manage. Sci. 22(3):340-358.

Tawn, E.J. 2000. Book Reviews: Genetic Susceptibility to Cancer (1998) and Genetic Heterogeneity in the Population and its Implications for Radiation Risk (1999). J. Radiol. Prot. 20:89-92.

USNRC (U.S. Nuclear Regulatory Commission). 1975. The Reactor Safety Study: An Assessment of Accident Risk in U.S. Commercial Nuclear Power Plants. WASH-1400. NUREG-75/014. U.S. Nuclear Regulatory Commission, Washington, DC. October 1975 [online]. Available: http://www.osti.gov/energycitations/servlets/purl/7134131-wKhXcG/7134131.PDF [accessed Jan. 15, 2008].

Wallace, L.A., E.D. Pellizzari, T.D. Hartwell, C. Sparacino, and R. Whitmore. 1987. TEAM (Total Exposure Assessment Methodology) study: Personal exposures to toxic substances in air, drinking water, and breath of 400 residents of New Jersey, North Carolina, and North Dakota. Environ. Res. 43(2):290-307.

Wu-Williams, A.H., L. Zeise, and D. Thomas. 1992. Risk assessment for aflatoxin B1: A modeling approach. Risk Anal. 12(4):559-567.

Zadeh, L.A. 1965. Fuzzy sets. Inform. Control 8(3):338-353.

Zartarian, V., T. Bahadori, and T. McKone. 2005. Adoption of an official ISEA glossary. J. Expo. Anal. Environ. Epidemiol. 15(1):1-5.

Zenick, H. 2006. Maturation of Risk Assessment: Attributable Risk as a More Holistic Approach. Presentation on the 1st Meeting on Improving Risk Analysis Approaches Used by the U.S. EPA, November 20, 2006, Washington, DC.

5

Toward a Unified Approach to Dose-Response Assessment

THE NEED FOR AN IMPROVED DOSE-RESPONSE FRAMEWORK

Introduction to the Problem

As described in Chapter 4, one of the urgent challenges to risk assessment is the evaluation of hazard and risk in a manner that is faithful to the underlying science, is consistent among chemicals, accounts adequately for variability and uncertainty, does not impose artificial distinctions among health end points, and provides information that is maximally useful for risk characterization and risk management. There have been efforts to harmonize dose-response methods for cancer and noncancer end points, but, as discussed below, criticisms have been raised regarding the validity of dose-response assessments for risk characterizations and management and regarding the treatment of uncertainty and variability in human sensitivity. This chapter examines the science governing dose-response assessment for a variety of end points (cancer and noncancer) and develops an integrative framework that provides conceptual and methodologic approaches for cancer and noncancer assessments.

Current Framework

Dose-response assessments for carcinogenic end points have been conducted very differently from noncancer assessments. For carcinogens, it has been assumed that there is no threshold of effect, and dose-response assessments have focused on quantifying the risk at low doses. The current Environmental Protection Agency (EPA) approach derives a "point of departure" (POD), such as the lower bound on the dose that results in an excess risk of 10% based on fitting of a dose-response model to animal bioassay data (EPA 2000a). After adjustment for animal-human differences in the dose metric, risk is assumed to decrease linearly with doses below the POD for carcinogens that are direct mutagens or are associated with large human body burdens (EPA 2005a). The population burden of disease or the population risk at a given exposure is estimated. In practice, EPA carcinogen assessments do not account for differences among humans in cancer susceptibility other than from possible early-life susceptibility (see Chapter 4).

For noncancer end points, it is assumed that homeostatic and defense mechanisms lead to a dose threshold[1] (that is, there is low-dose nonlinearity), below which effects do not occur or are extremely unlikely. For these agents, risk assessments have focused on defining the reference dose (RfD) or reference concentration (RfC), a putative quantity that is "likely to be without an appreciable risk of deleterious effects" (EPA 2002a, p. 4-4). The "hazard quotient" (the ratio of the environmental exposure to the RfD or RfC) and the "hazard index" (HI, the sum of hazard quotients of chemicals to which a person is exposed that affect the same target organ or operate by the same mechanism of action) (EPA 2000b) are sometimes used as indicators of the likelihood of harm. An HI less than unity is generally understood as being indicative of lack of appreciable risk, and a value over unity indicates some increased risk. The larger the HI, the greater the risk, but the index is not related to the likelihood of adverse effect except in qualitative terms: "the HI cannot be translated to a probability that adverse effects will occur, and is not likely to be proportional to risk" (EPA 2006a). Thus, current RfD-based risk characterizations do not provide information on the fraction of the population adversely affected by a given dose or on any other direct measure of risk (EPA 2000a). That deficiency is present whether the dose is above the RfD (in which case the risk may be treated as nonzero but is not quantified) or below the RfD (in which case the risk can be treated as "unappreciable" or zero even though with some unquantified probability it is not zero).

As in cancer dose-response assessment, the RfD is also derived from a POD, which could be a no-observed-adverse-effect level (NOAEL) or a benchmark dose (BMD). However, instead of extrapolating to a low-dose risk, the POD is divided by "uncertainty factors" to adjust for animal-human differences, human-human differences in susceptibility, and other factors (for example, data gaps or study duration). In a variant of the RfD approach to noncancer or low-dose nonlinear cancer risk assessment, the agency calculates a "margin of exposure" (MOE), the ratio of a NOAEL or POD to a projected environmental exposure (EPA 2000a, 2005b). The MOE is compared with the product of uncertainty factors; an MOE greater than the product is considered to be without appreciable risk or "of low concern," and an MOE smaller than the product reflects a potential health concern (EPA 2000b). MOEs and RfDs are defined for durations of exposure (for example, acute, subchronic, and chronic) and may be defined for specific life stages (for example, developmental) (EPA 2002a).

Recent refinements in risk-assessment methods in EPA have used mode-of-action (MOA)[2] evaluations in dose-response assessment. EPA's *Guidelines for Carcinogen Risk Assessment* (2005b) state that if a compound is determined to be "DNA reactive and [to] have direct mutagenic activity" or to have high human exposures or body burdens "near doses associated with key precursor events" (EPA 2005b, p. 3-21), a no-threshold approach is applied; risk below the POD is assumed to decrease linearly with dose. For carcinogens with sufficient MOA data to conclude nonlinearity at low doses, such as those acting through a cytotoxic MOA, the RfD approach outlined above for noncancer end points is applied (EPA 2005b),

[1]More recent noncancer guidelines have abandoned the term *threshold*, noting the difficulty of empirically distinguishing dose-response relationships with true biologic thresholds from ones that are nonlinear at low doses (EPA 2005b, p. 3-24).

[2]Following EPA 2005b (p. 1-10), the MOA is defined as "a sequence of key events and processes, starting with interaction of an agent with a cell, proceeding through operational and anatomical changes, and resulting" in the adverse effect. "A 'key event' is an empirically observable precursor step that is itself a necessary element of the mode of action or is a biologically based marker for such an element."

except when there is adequate evidence to support mechanistic modeling (there has been no such case).

Another refinement in dose-response assessment has been the derivation of the RfD or low-dose cancer risk from a POD that is calculated using BMD methodology (EPA 2000a). In noncancer risk assessment, this approach has the advantage of making better use of the dose-response evidence available from bioassays than do calculations based on NOAELs. It also provides additional quantitative insight into the risk presented in the bioassay at the POD because for quantal end points the POD is defined in terms of a given risk for the animals in the study.

EPA's treatment of noncancer and low-dose nonlinear cancer end points is a major step by the agency in an overall strategy to harmonize cancer and noncancer approaches to dose-response assessment. Other aspects of this harmonization for the different end points include consideration of the same cross-species factors (EPA 2006b), and the same pharmacokinetic adjustments. EPA staff have also explored for noncancer end points dose-response modeling that results in probabilistic descriptions (for example, for acrolein, Woodruff et al. 2007) and that could be readily integrated into benefits evaluation (for thyroid-disrupting chemicals, Axelrad et al. 2005). But these approaches have not found their way into agency practice.

Scientific, Technical, and Operational Problems with the Current Approach

The committee recognizes EPA's efforts to examine and refine dose-response assessment methodology and practice and the agency's work to clarify its approaches and practices in guidelines and other documents (for example, EPA 2000a, 2002b, 2004, 2005b). A number of improvements over the last decade can be noted, such as the movement toward using MOA determinations and the application of BMD methods. However, the current framework has important structural problems, some of which have been exacerbated by recent decisions. Figure 5-1 presents an outline of the current framework for dose-response assessment and risk characterization in EPA and some major limitations in the framework, which are discussed below.

Potential Low-Dose Linearity for Noncancer and "Nonlinear" Cancer End Points

Thresholds are assumed for noncarcinogens and for carcinogens believed to operate through an MOA considered nonlinear at low doses. The rationale is that at levels below the threshold dose, clearance pathways, cellular defenses, and repair processes have been thought to minimize damage so that disease does not result. However, as illustrated in Figure 5-2, threshold determinations should not be made in isolation, inasmuch as other chemical exposures and biologic factors that influence the same adverse effect can modify the dose-response relationship at low doses and should therefore be considered.

Nonlinear Cancer End Points

The current determination of "nonlinearity" based on MOA assessment is a reasonable approach to introduce scientific evidence on MOA into cancer dose-response assessment. However, some omissions in this overall approach for low-dose nonlinear carcinogens could yield inaccurate and misleading assessments. For example, the current EPA practice of determining "nonlinear" MOAs does not account for mechanistic factors that can create linearity at low dose. The dose-response relationship can be linear at a low dose when an exposure contributes to an existing disease process (Crump et al. 1976, Lutz 1990). Effects

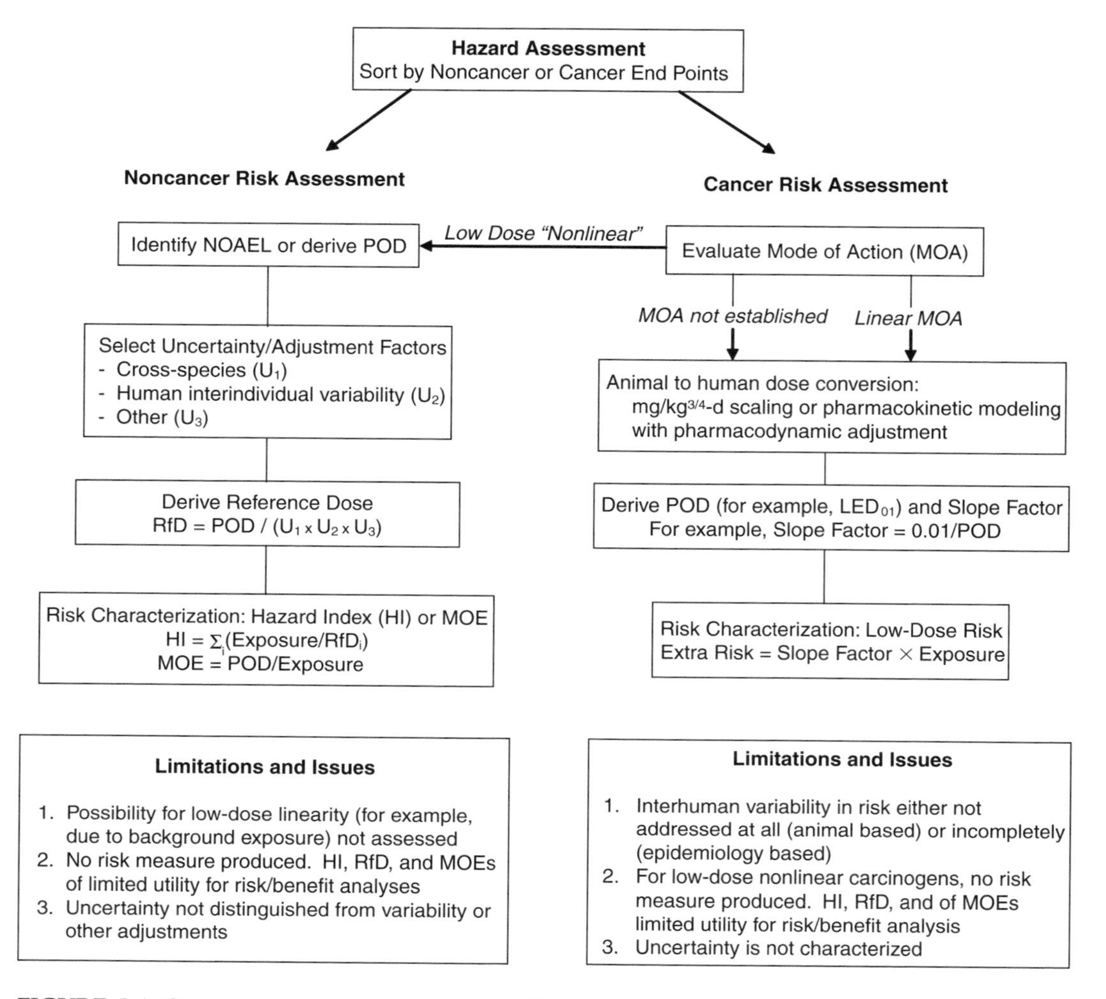

FIGURE 5-1 Current approach to noncancer and cancer dose-response assessment.

of exposures that add to background processes and background endogenous and exogenous exposures can lack a threshold if a baseline level of dysfunction occurs without the toxicant and the toxicant adds to or augments the background process. Thus, even small doses may have a relevant biologic effect. That may be difficult to measure because of background noise in the system but may be addressed through dose-response modeling procedures. Human variability with respect to the individual thresholds for a nongenotoxic cancer mechanism can result in linear dose-response relationships in the population (Lutz 2001).

In the laboratory, nonlinear dose-response processes—for example, cytotoxicity, impaired immune function and tumor surveillance, DNA methylation, endocrine disruption, and modulation of cell cycles—may be found to cause cancer in test animals. However, given the high prevalence of those background processes, given cancer as an end point, and given the multitude of chemical exposures and high variability in human susceptibility, the results may still be manifested as low-dose linear dose-response relationships in the human population (Lutz 2001). The possibility of low-dose linearity due to background is acknowledged

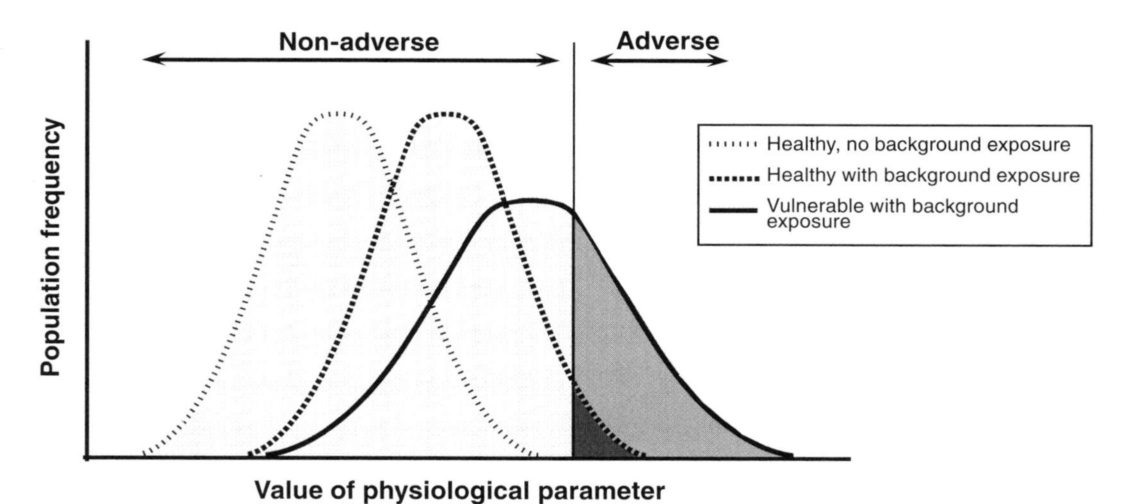

FIGURE 5-2 Value of physiologic parameter for three hypothetical populations, illustrating that population responses depend on a milieu of endogenous and exogenous exposures and on vulnerability of population due to health status and other biologic factors. Source: Adapted from Woodruff et al. 2007. Reprinted with permission; copyright 2007, *Environmental Health Perspectives*.

in the EPA (2005b) *Guidelines for Carcinogen Risk Assessment* to a limited degree—for chemicals with high body burdens or high exposures—but has not been addressed in EPA assessments. And EPA practices do not call for systematic evaluation of endogenous and exogenous exposures or mechanisms that can lead to linearity.

By segregating cancer and noncancer risk assessment, the current framework tends to place undue focus on "complete" carcinogens, ignoring contributions to ongoing carcinogenesis processes and the multifactorial nature of cancer. Chemicals that may increase human cancer risk by contributing to an underlying process are handled essentially as noncarcinogens even though they may be integral to the carcinogenic process. The dichotomy increases the burden of judging which chemicals are carcinogens rather than accepting the variety of carcinogenic MOAs and incorporating them into a comprehensive risk assessment.

Noncancer End Points

Similarly, noncarcinogens can exhibit low-dose linearity, for example, when there is considerable interindividual variability in susceptibility and each individual has his or her own threshold, especially when an underlying disease (such as cardiopulmonary disease) can interact with the toxicant (such as particulate matter [PM] or ozone). Schwartz et al. (2002) made the argument for the absence of a population threshold for mortality effects of PM. Other factors that support nonthreshold dose-response relationships for noncarcinogens include

- The observation of dose-response relationships with no apparent thresholds for subtle, common adverse end points, such as IQ loss or neurobehavioral deficits associated with lead or methylmercury exposures—an observation that continues to be made even as investigators probe for effects at smaller exposures (Axelrad et al. 2007). Those effects occur at lower doses than frank toxicity and are expected to become a more common basis of dose-response assessment as increasingly subtle end points are studied with more sensitive tests (for example, tests based on -omics) or epidemiologically.

- The fact that in receptor-mediated events, even at very low doses a chemical can occupy receptor sites and theoretically perturb cell function (such as signal transduction or gene expression) or predispose the cell to other toxicants that bind to or modulate the receptor system (such as organochlorines and the aryl hydrocarbon receptor or endocrine disruptors and hormonal binding sites) (Brouwer et al. 1999; Jeong et al. 2008).
- The observation that exposures that perturb or accelerate background endogenous disease processes and add to background endogenous and exogenous exposures may not show evidence of a threshold, as described above ("Nonlinear Cancer End Points").

There are multiple toxicants (for example, PM and lead) for which low-dose linear concentration-response functions rather than thresholds have been derived for noncancer end points. The current EPA framework treats them as exceptions (implicitly if not explicitly) and does not provide methods and practices for readily assessing the dose-response relationship for cases in which thresholds are not apparent or not expected, for example, because of background additivity. As discussed in this chapter, for critical end points driving the risk characterization at low doses, such cases may be common, and a new framework and practice are needed.

Another problem posed by the current noncancer framework is that the term *uncertainty factors* is applied to the adjustments made to calculate the RfD to address species differences, human variability, data gaps, study duration, and other issues. The term engenders misunderstanding: groups unfamiliar with the underlying logic and science of RfD derivation can take it to mean that the factors are simply added on for safety or because of a lack of knowledge or confidence in the process. That may lead some to think that the true behavior of the phenomenon being described may be best reflected in the unadjusted value and that these factors create an RfD that is highly conservative. But the factors are used to adjust for differences in individual human sensitivities, for humans' generally greater sensitivity than test animals' on a milligrams-per-kilogram basis, for the fact that chemicals typically induce harm at lower doses with longer exposures, and so on. At times, the factors have been termed *safety factors*, which is especially problematic given that they cover variability and uncertainty and are not meant as a guarantee of safety.

The Need for Evaluation of Background Exposures and Predisposing Disease Processes

Dose-response assessments for noncancer and nonlinear cancer end points are generally performed without regard to exposure to other chemicals that affect the same pathologic processes or the extent of pre-existing disease in the population. The need to address chemicals that have "a common mechanism of toxicity" in a cumulative risk assessment has been established for pesticides under the Food Quality Protection Act (FQPA) of 1996 (EPA 2002b, p. 6). EPA (2002b) provides a useful example, but it was driven principally by the explicit requirements of the FQPA, and few noncarcinogens are evaluated in this way. Furthermore, dose additivity has been observed at relatively low doses for various endocrine-related toxicities with similar and dissimilar mechanisms of action (for example, Gray et al. 2001; Wolf et al. 2004; Crofton et al. 2005; Hass et al. 2007; Metzdorff et al. 2007). Dosing animals with two chemicals that have different MOAs at their NOAELs resulted in a significant adverse response, which suggested dose additivity (as when two chemicals at subthreshold doses lead to an effect). In practice, a common implicit assumption is effect additivity—two subthreshold doses yield a nonresponse because neither produces a response on its own.

Consideration of chemicals that have a common MOA has not included how endogenous and other chemicals, not the direct subjects of testing and evaluation by regulatory agencies, affect the human dose-response relationship. The recent EPA draft dibutyl phthalate (DBP) assessment is an example in which there was an opportunity to consider cumulative exposure to the various agents that can contribute to the antiandrogen syndrome seen with phthalates, but the impact of even other phthalates on the DBP dose-response relationship was not taken into account in setting the draft RfD (EPA 2006c). In the application of such an assessment, DBP exposures above the RfD would be treated as posing some undefined extra degree of risk and DBP exposures below the RfD would, without further guidance from the agency, potentially be treated as riskfree without regard to the presence of other antiandrogen exposures.

Risk-Assessment Outcomes Needed for Risk Evaluation and Benefit Analysis

The end products of noncancer (and nonlinear cancer) assessments in the current paradigm (exposure-effect quotients that qualitatively indicate potential risk—MOEs, RfDs, and RfCs, Figure 5-1) are inadequate for benefit-cost analyses or for comparative risk analyses. MOEs and RfDs as currently defined do not provide a basis for formally quantifying the magnitude of harm at various exposure levels. Therefore, the committee finds the 2005 *Guidelines for Carcinogen Risk Assessment* movement toward RfDs and away from an expression of risk posed by nonlinear carcinogens problematic. Similarly, although noncancer risk assessment has moved to a BMD framework that makes better use of evidence than an approach based on NOAELs and lowest observed-adverse-effect levels (LOAELs), the paradigm remains one of defining an RfD or RfC without any sense of the degree of population risk reduction that would be found in moving from one dose to another dose. A probabilistic approach to noncancer assessment, similar to how cancer risks are expressed, would be much more useful in risk-benefit analysis and decision-making. The current threshold-nonthreshold dichotomy creates an inconsistent approach for bringing toxicology and risk science into the decision-making process.

That paradigm has other unintended consequences. For example, the linear-extrapolation exercise for carcinogens and lack of consideration of linearity for noncarcinogens and "nonlinear" carcinogens create a high bar of evidence for carcinogen identification and reduce the consideration of the possibility of noncancer end points for carcinogens. More generally, the many noncancer health end points are generally given little weight in benefit-cost analyses or other analytically driven decision frameworks in part because of the nature of the resulting qualitative risk characterization.

In the general case in which an intervention reduces exposures from above the RfD to below the RfD, it is particularly unfortunate to fail to quantify this benefit. It might be possible, through economic valuation (willingness-to-pay or contingent-valuation) studies, to estimate the benefits of moving N members of the population from exposure above the RfD to exposure below the RfD, but it would be more straightforward and intelligible to directly estimate the benefits of such an exposure and risk reduction. The current approach also does not address the benefits of lowering exposures that are already below the RfD or the benefits of lowering exposures from above the RfD to an exposure level that is still above the RfD, both of which, if understood to be associated with a nonzero probability of harm, also need valuation. The framework described below provides a means of generating the data needed for such analyses.

Limitations of the Current Approach for Low-Dose Linear Cancer End Points

EPA assumes that the linear default approach for dose-response assessment provides "an upper-bound calculation of potential risk at low doses," which is "thought to be public-health protective at low doses for the range of human variation" (EPA 2005b, p. A-9). EPA (2005b) noted that the National Research Council reports (NRC 1993, 1994) generally discussed the variability in human susceptibility to carcinogens and that EPA and other agencies were conducting research on the issue. The committee finds that although the precise degree of human variability is not known, the upper statistical bound derived from fits to animal data does not address human variation, as discussed below. Further, with few exceptions (EPA 2001a), the current practice embeds an implicit assumption that it is zero. This is not credible and is increasingly unwarranted as more and more studies document the substantial interindividual variation in the human population (see Chapter 4).

According to EPA, "the linear default procedure adequately accounts for human variation unless there is case-specific information for a given agent or mode of action that indicates a particularly susceptible subpopulation or lifestage, in which case the special information will be used" (EPA 2005b, p. A-9). That implies that in general the linear-extrapolation procedure will overestimate the risk to an extent that will account for the underestimation bias related to the omission of human heterogeneity. EPA provides no evidence to support that assumption and in essence establishes a default (no variability in susceptibility) that is unsubstantiated (see Chapter 6 for discussion of "missing" defaults). There are three main steps in deriving human cancer risk from animal bioassay data: adjusting animal doses to equivalent human doses, deriving the POD by fitting a mathematical model to the data, and linearly extrapolating from the POD to lower doses. The default animal-to-human adjustment is based on metabolic differences due to the roughly 200- to 2,000-fold differences in body sizes and is set at a median value without accounting for the large qualitative uncertainty, in any particular application, of the humans being more sensitive than the animal or vice versa. The lower bound on the POD merely accounts for the uncertainty in the model fitted to data from the fairly homogeneous animals used in studies. If the true dose-response relationship for an agent is indeed linear, the statistical lower confidence limit (for example, the BMD lower confidence limit [BMDL]) associated with a POD (for example, the BMD) provides a small increment of "conservatism"—typically not more than a factor of 2 (Subramaniam et al. 2006). That is highly unlikely to account for variation in susceptibility in cancer in a large exposed human population (see Chapter 4). If, instead, the true dose-response relationship is nonlinear, treating it as linear might introduce enough "conservatism" to offset the underestimation of risk in people of above-average susceptibility, but the degree to which the high-dose-based estimate is in error would preferably be analyzed separately. The practice of assuming no human variation in response to compounds for which linearity is applied is simplistic and inconsistent with the manner in which noncancer assessments are conducted. Many factors can cause the cancer response to be highly variable in the population, including age, sex, genetic polymorphisms, endogenous disease processes, lifestyle, and coexposure to other xenobiotics common in the human environment (see "Variability and Vulnerability in Risk Assessment" in Chapter 4). Some of those factors, especially pharmacokinetics and early age, are beginning to be considered in a few cancer risk assessments, but much more emphasis needs to be placed on describing the ranges of susceptibility and risk.

Other Limitations of the Current Approach

One cross-cutting issue for all end points is the degree to which dose-response characterization is done in data-poor cases. Often, a compound on which information is sparse is not addressed in a quantitative risk assessment and operationally can be treated as though it posed no risk of regulatory importance. That is unlikely to describe the situation adequately or to be helpful in setting research priorities. An approach to that problem is described in Chapter 6.

In addition, any analysis must grapple with the best approach for integrating data from multiple studies and on multiple end points. There has been a tendency in risk assessment to pick a single dataset with which to describe risk, in part because it leads to straightforward rationales that are easy to explain, understand, and communicate. However, the direction toward better understanding of uncertainty, human variability, and more accurate assessment necessarily involves increasing complexity and integration of evidence from disparate sources. It also may involve constructing dose-response relationships based on evidence from a variety of study types (such as cancer bioassays and in vitro studies). Also, a given exposure to a particular chemical may affect multiple end points, and a risk description based on one tumor site or effect may fall short of conveying the overall risk posed by the substance.

In summary, the committee finds multiple scientific and operational limitations in the current approach for both cancer and noncancer risk assessments. The following section describes a means for addressing many of the issues by developing a unified framework for toxicity assessment that incorporates variability and uncertainty more completely and provides quantitative risk information on cancer and noncancer end points alike.

A UNIFIED FRAMEWORK AND APPROACH FOR DOSE-RESPONSE ASSESSMENT

The committee finds that the underlying science is more consistent with a new conceptual framework for dose-response modeling and recommends that the agency adopt a unified framework. Figure 5-3a illustrates the underlying dose-response principles for the framework, which includes background processes and exposures in considering risks on the individual and population scales. Figure 5-3a shows that an individual's risk from exposure to an environmental chemical is determined by the chemical itself, by concurrent background exposures to other environmental and endogenous chemicals that affect toxicity pathways and disease processes, and by the individual's biologic susceptibility due to genetic, lifestyle, health, and other factors. How the population responds to chemical insults depends on individual responses, which vary among individuals.

Clearly, background exposures and biologic susceptibility factors differ substantially between animals and humans, and there can be more confidence in dose-response descriptions that consider and account for background exposure and biologic susceptibility of populations for which risks are being estimated. Figure 5-3b provides a depiction of individual and population risk that formally takes these factors into account. The shape of the population dose-response relationship at low doses is inferred from an understanding of individual dose-response relationships, which in turn are based on consideration of background exposure and biologic susceptibility on human heterogeneity. An upper bound on the population dose-response relationship would be derived to express uncertainty in the population dose-response relationship. For compounds whose effects show a linear dose-response relationship, this upper bound is not the same as the familiar upper bound derived by fitting dose-response models to animal bioassay data. The latter upper bound measures only a very small aspect of uncertainty: that due to sampling variability and the statistical fit to animal data. Here, the committee envisions a more comprehensive description of uncertainty that accounts for other aspects, such as uncertainty in cross-species extrapolation. The dose of the environmental chemical that poses, say, a risk above background ("extra risk") of 10^{-5} in a population, could be described by a probability distribution that reflects

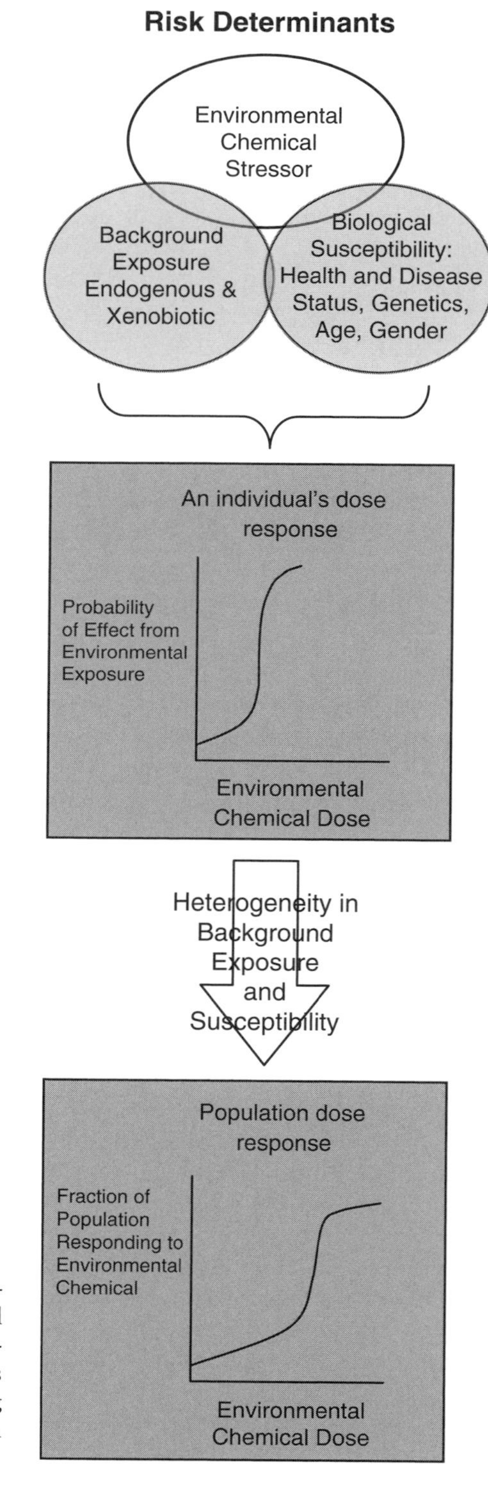

FIGURE 5-3a New conceptual framework for dose-response assessment. Risk posed by environmental chemical is determined from individual's biologic make-up, health status, and other endogenous and exogenous exposures that affect toxic process; differences among humans in these factors affect shape of population dose-response curve.

Risk Depiction: Based on data, defaults, and other inferences

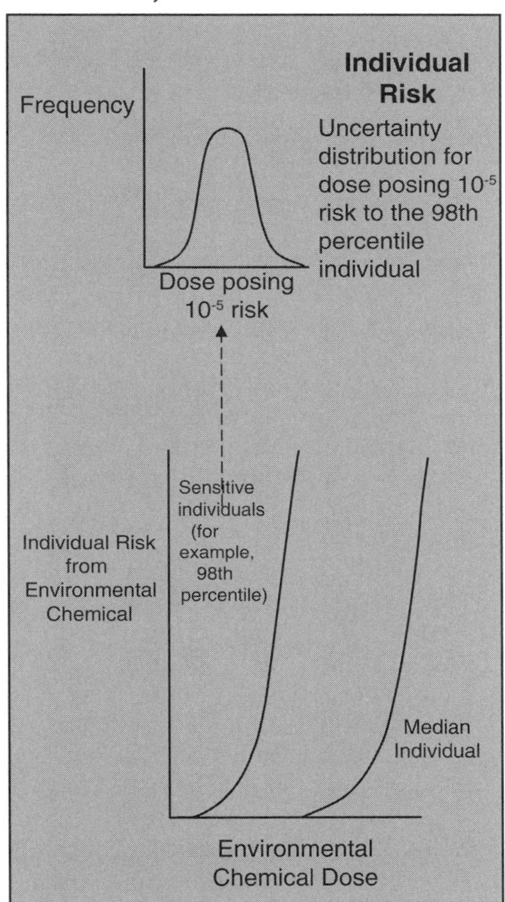

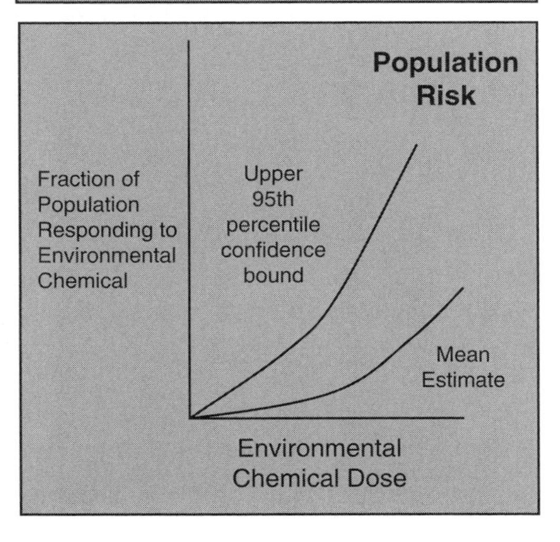

FIGURE 5-3b Risk estimation and description under the new conceptual framework for dose-response assessment. Risk estimates are based on inferences made from human, animal, MOA, and other data and understanding of possible background processes and exposures. Ideally, population dose-response relationship and uncertainty (represented by upper 95% bound) and dose-response relationships for sensitive members of population are described. (As explained in text, upper 95% confidence bound on risk is not same as upper-bound estimate generated in current cancer risk assessments.) Mean estimate of population risk can be derived from understanding of individual risk.

the uncertainty. Ideally, risk would be estimated for sensitive as well as typical individuals, and uncertainty in those estimates would also be described.

One important outcome of the new approach is the redefinition of the RfD as a risk-specific dose rather than as a dichotomous risk–unappreciable risk descriptor. The redefinition is described further below.

Characteristics of the Dose-Response Framework

The dose-response framework envisioned includes the following features:

- *Dose-response characterizations that use the spectrum of evidence from human, animal, mechanistic, and other relevant studies.* Whole-animal dose-response studies will continue to play a central role in establishing PODs for most chemicals, but information on human heterogeneity, background exposures, and disease processes and data from mechanistic in vitro and in vivo studies will be critical in selecting the approach to the dose-response analysis. Some information used in the dose-response derivation will be chemical-specific. In the absence of reliable chemical-specific information on human variability, interspecies differences, and other components of the analysis, generalizations and defaults based on evidence from other chemicals and end points and theoretical considerations may be used. Clearly, this presents challenges associated with selection of data sources, data synthesis, and model uncertainty.
- *The goal of providing a probabilistic characterization of harm*, such as a description of the form "at dose D, R fraction of the population would be anticipated to suffer harm with a confidence interval of R_L-R_H." For example, a summary statement of risk may be that at an air concentration of 0.05 ppm (= D), 1/10,000 (= R) of the population are likely to be affected with a 95% confidence interval (CI) of 5/100,000-3/10,000 of the population. That general form can be made more specific to particular outcomes and MOAs. For example, as described later in this chapter, for agents unlikely to have a threshold even at the individual level (such as mutagenic carcinogens), each person is assumed to be at a finite risk, and one can also make statements about individual risk. A summary statement may be given like that above with a further description that the 95th percentile individual at a dose of 0.05 ppm may face a risk of 1/1,000 (with a CI of 5/10,000-3/1,000). Thus, for a population uniformly exposed to a compound at 0.05 ppm, the characterization would indicate the distribution of risk among individuals (with variability driven by differences in background exposures and biologic susceptibility), in this example, with 5% of individuals having estimated risks above 1/1,000 (with associated confidence bounds). The key attribute of the characterization would be a quantitative and probabilistic characterization of harm for each critical end point. A similar position for probabilistic expression of noncancer risk has been advocated by the EPA Science Advisory Board (EPA SAB 2002). Multiple end points of varied severity would be considered. In many cases, new research or well-justified default approaches will be needed to attain this level of refinement in noncancer dose-response analysis.
- *Explicit consideration of human heterogeneity in response,* for both cancer and noncancer end points, that is distinguished from uncertainty. This variability assessment would consider susceptibility due to age, sex, health status, genetic makeup, and other factors. Uncertainty in human variability estimates would be described, preferably quantitatively. The rigor of this characterization would be commensurate with the needs of the assessment (see Chapters 3 and 4).
- *Treatment of uncertainty aimed at characterizing the most important types of uncertainties* for both cancer and noncancer end points. This could involve formal quantification

following probabilistic approaches that are consistent with recommendations about the use of default assumptions in Chapter 6. It could also include sensitivity analyses or qualitative characterizations if they would provide a better description of uncertainty or are commensurate with the needs of the assessment.

- *Evaluation of background exposure and susceptibility in order to select modeling approach.* The assessment of "background exposure" and "background disease processes" would involve characterization of other chemicals or nonchemical stressors that influence the same general pathologic processes as the chemical under evaluation. Such consideration should aid the evaluation of the shape of the dose-response relationship, including the potential for low-dose linearity and high-risk subpopulations and hence appropriate methodologic approaches for the dose-response analysis. Background exposures and susceptibility factors can result in linear low-dose-response relationships that would otherwise be considered low-dose nonlinear on the basis of MOA alone.
- *Use of distributions instead of "uncertainty factors,"* as the science and data develop and are found to provide a sufficient basis for doing so. For example, research is going on to develop uncertainty distributions for the pharmacokinetic (PK) and pharmacodynamic (PD) components of the interspecies and intraspecies human uncertainty factors (for example, Hattis and Lynch 2007). Data-driven adjustment factors developed by such bodies as the World Health Organization's International Program on Chemical Safety (IPCS 2005) are being expanded to probabilistic descriptions on the basis of information from the pharmaceutical sector and emerging from the biologic sciences. It will be a challenge to overcome some of the data limitations for developing those approaches. For example, many studies use small numbers of human subjects, so the sensitive individuals in the population may not be characterized quantitatively by distributions derived from these studies, particularly if the true human distribution is multimodal. Approaches are needed to address that issue. The formal incorporation of variability due to polymorphisms, aging, endogenous disease status, exposure, and other factors will probably prove to be complex and challenging. Later in this chapter, examples are given of an approach for developing and using an intrahuman variability adjustment and distribution for cancer risk derivations. It may sometimes be preferable to use single-value "uncertainty factors," either out of necessity or reflecting science-policy choices (see Chapter 6). Their use would preferably be accompanied by a qualitative description of the associated uncertainty in their application.

The term *uncertainty factors* can be problematic because it connotes only one aspect of the function of the factors. As the default distributions are developed, a better, more specific label for them would be preferable (for example, *human variability distribution*) to reflect their content more appropriately (for example, accounting for human heterogeneity). This would lessen the opportunity for transferring to the new default distributions the misunderstanding commonly associated with use of "uncertainty factors," as described earlier.

- *Descriptions of sensitive individuals or subpopulations.* The assessment would characterize individuals and subgroups according to whether they have coexposures to key nonchemical stressors, specific polymorphisms influencing metabolism or DNA repair, pre-existing or endogenous disease processes, high background endogenous or exogenous exposures, and other determinants of increased susceptibility.
- *Approaches and resulting assessments that are transparent and understandable by the public and by risk managers.* This may require alternative presentations of the characterization of risk to suit the needs of specific decisions.

Risk-Specific Definition of the Reference Dose

This framework facilitates a redefinition of the RfD and RfC in terms of a risk-specific dose and confidence level, as outlined in Box 5-1. Although Box 5-1 focuses on a risk-specific definition of the RfD, the framework developed in this chapter can be used to estimate risk at any dose, not just the RfD; for example, the risk and confidence bounds around the risk could be reported for continuous exposure to an air concentration of 1 part-per-billion. This redefinition will facilitate an understanding of the benefits of lowering exposure in valuation exercises for environmental decision-making.

An RfD defined in that manner can be used as RfDs have always been used in aiding risk-management decisions, but it has additional beneficial features. It presents a dose above which risks may be increased above a standard criterion or de minimis risk and below which risks are considered insignificant or minimal but not necessarily zero. It is analogous to the presentation of cancer risks to risk managers with the understanding that the bright-line risk-specific dose is based on a previously agreed on de minimis or acceptable level of risk inasmuch as zero risk cannot be assumed. However, rather than being an expression of the line between possible harm and safety, the newly defined RfD can be interpreted in terms of population risk. Managers can then weigh alternative options in terms of the percentage of the population that is above or below the de minimis risk-specific dose; this also enables a quantitative estimate of benefits for different risk-management options. An example of this approach is provided for a thyroid disrupting compound by Axelrad et al. (2005).

The de minimis risk for the RfD could depend on the nature of the health outcome (that is, a subtle, precursor effect, a mild effect, or a severe effect) and the subpopulation; for example, the RfD could be based on a 1 in 1,000 risk for a minimally adverse response in a sensitive subpopulation (Hattis et al. 2002).

As is the case for linear cancer end points, multiple risk-specific doses could be provided in the Integrated Risk Information System and in the various risk characterizations that EPA produces to aid environmental decision-making. Different risk-management decisions may call for different acceptable risks, and this redefinition would provide risk managers a means of considering the population risk associated with exposures resulting from specific control strategies. The doses related to different target risks could be distinguished from RfDs and RfCs with names like *risk-specific dose* to avoid confusion. The confidence values associated with these risk-specific doses should be included in any database with the risk targets to ensure that this key information is not lost. Over the years of experience with cancer—a severe effect with a relatively long latent period—an acceptable risk range has been adopted that is used in risk-management decisions. Such experience will accrue for other health end points.

BOX 5-1 A Risk-Specific Reference Dose

For quantal effects, the RfD can be defined to be the dose that corresponds to a particular risk specified to be de minimis (for example, 1 in 100,000) at a defined confidence level (for example, 95%) for the toxicity end point of concern. It can be derived by applying human variability and other adjustment factors (for example, for interspecies differences) represented by distributions rather than default uncertainty factors.

Conceptual Models

Approaches to describe dose-response relationships in probabilistic terms depend on how one conceives the underlying biologic processes and how they contribute to an individual's dose-response relationship, the nature of human variability, and the degree to which the processes may be independent of background exposures and processes. This is illustrated in three example prototypical conceptual models:

1. *Nonlinear individual response, low-dose linear population response with background dependence.* As discussed above, low-dose linearity can arise when the dose-response curves for individuals in the population are nonlinear or even have thresholds but the exposure to the chemical in question adds to prevalent background exposures that are contributing to current disease. The dose-response relationship would be determined to a great extent by human variability and background exposure. In Figure 5-4, each individual's dose-response relationship can be characterized by a threshold dose-response function with zero risk up to a particular dose and then sharply increasing risk with increasing dose above it. A collection of the threshold dose-response functions for a number of individuals is displayed on the left side of the figure. The proportion of individuals in the population whose threshold is exceeded by a particular dose is displayed on the right side.

2. *Low-dose nonlinear individual and population response, low-dose response independent of background.* This is the dose-response conceptual model currently in use for noncancer end points. For these dose-response relationships, the fraction of the human population responding drops to inconsequential levels at low doses. At very low doses, the threshold dose for toxicity is not exceeded in individuals, or the risk is infinitesimal. The same is true for the population, with the shape of its dose-response relationship determined by the variability in individuals' thresholds, as illustrated in Figure 5-5.

Clearly, there are many compounds and end points for which available compound-specific data are not sufficient to describe probabilistic dose-response relationships for nonlinear end points adequately. For some chemicals, default distributions may be constructed on the basis of known chemical and physiologic properties for chemicals considered representative for this purpose. Some default adjustment factors could be specific for some types of chemicals. Examples of how default distributions may be derived to support the derivation of risk for this conceptual model are given below; the committee cites these examples not to endorse particular distributions or specific results but to provide an example of a low-dose nonlinear dose-response modeling approach.

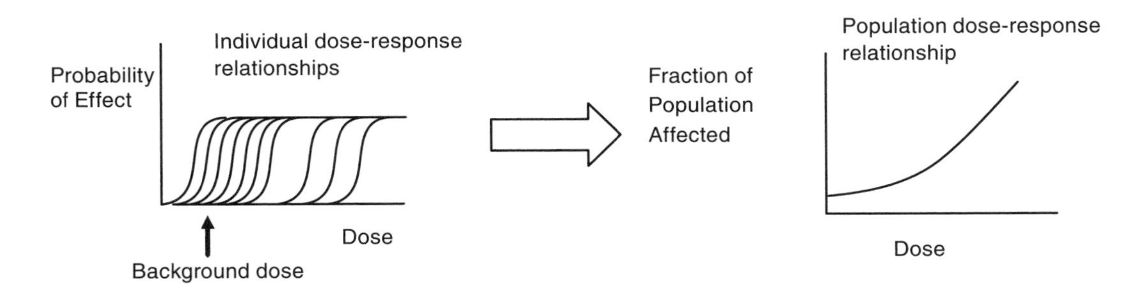

FIGURE 5-4 Linear low-dose response in the population dose-response relationship resulting from background xenobiotic and endogenous exposures and variable susceptibility in the population.

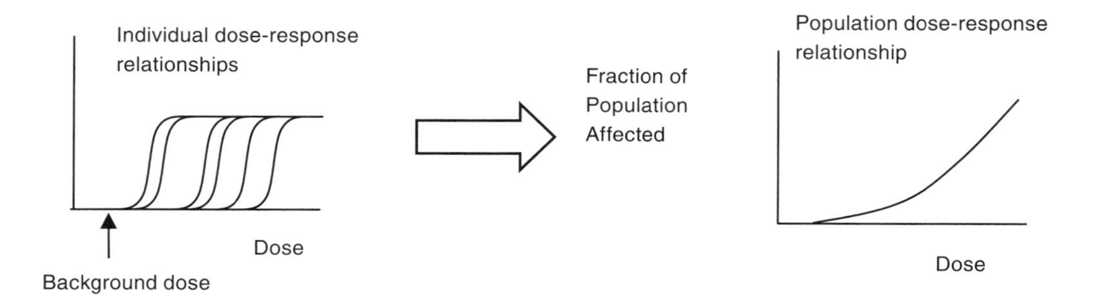

FIGURE 5-5 Nonlinear or threshold low-dose response relationships for individuals and populations.

3. *Low-dose linear individual and population dose-response.* For this conceptual model, both individual risk and population risk have no threshold and are linear at low doses, as illustrated in Figure 5-6. Note that *low-dose linear* means that at low doses "added risk" (above background) increases linearly with increasing dose; it does not mean that the dose-response relationship is linear throughout the dose range between zero dose and high doses. A possible approach for deriving linear cancer dose-response relationships and estimating risk for individuals at different quantiles and for the population is described below for this conceptual model illustrated in Figure 5-6.

To the extent that uncertainty in cross-species and other adjustments can be ascertained, rough quantitative estimates of uncertainty may be provided and incorporated into the characterization of the dose-response relationship. The upper confidence bound on the population dose-response curve in Figure 5-6 depicts the uncertainty in the model fit to data, as well as in the other adjustments.

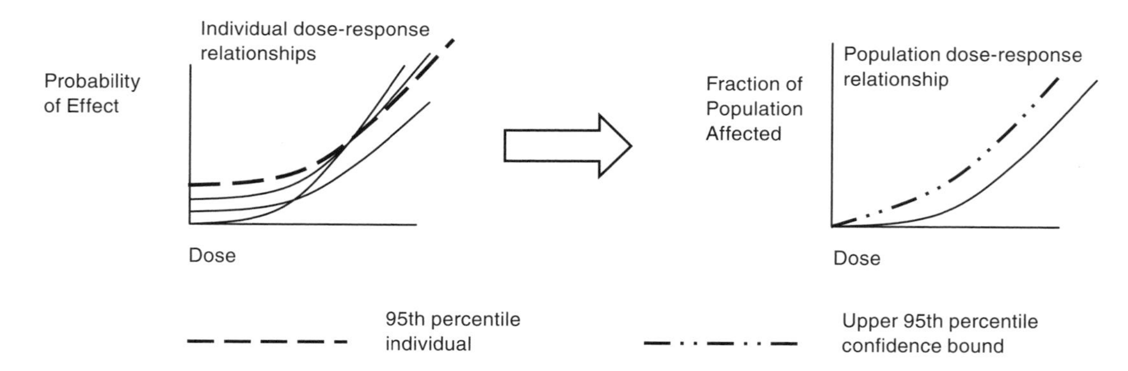

FIGURE 5-6 Linear low-dose response models for individuals and population. Individual dose-response relationships may cross. Thus, individual at the 95th percentile at one dose (dashed line in graph on left) may not be same individual at another dose. From uncertainty estimates for assessment components, upper 95th percentile estimate for population dose-response relationship can be derived (dashed line in graph on right).

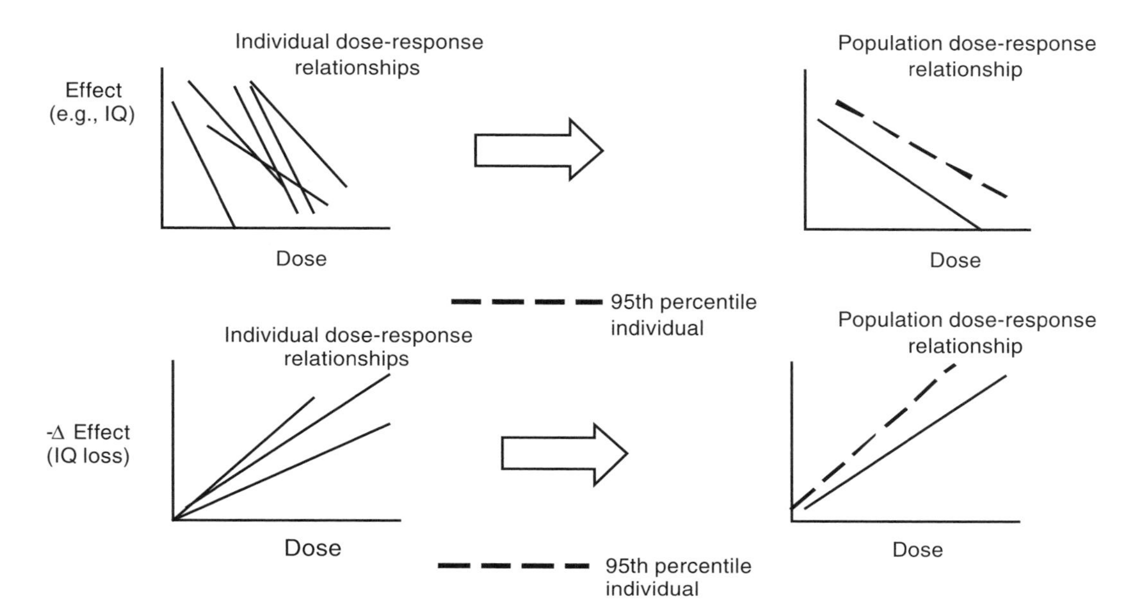

FIGURE 5-7 Dose-response relationships involving a continuous effect variable.

Low-dose linear dose-response relationships can also involve continuous-effect variables, such as decreasing IQ, illustrated in Figure 5-7. As the exposure increases, IQ decreases potentially shifting the entire population distribution in the direction of decreased function, as may occur with methylmercury (Axelrad et al. 2007).

General Approach to Dose-Response Assessment

The general approach, illustrated in Figure 5-8, involves consideration of MOA, background exposures, and possible vulnerable populations in selecting a conceptual model and methods for dose-response analysis.

Data Assembly and End-Point Assessment

The process begins, as is done currently, with review of the peer-reviewed scientific literature to assemble health-effects data for identifying end points of concern. The review emphasizes end points that are of greatest concern to populations exposed through environmental media. Thus, for chemicals with robust datasets, there is little focus on severe effects at high doses other than as indicators, for example, of possible target organs, route specificity, and dose-dependent pharmacokinetics. An exception is the plausible scenario, in which, for example, acute high-dose exposures occur from chemical terrorism or accidental releases.

One important aspect of dataset selection for dose-response estimation is the consideration of target organ (site) concordance between animals and humans. A toxic effect may be preferentially expressed in an animal model in a tissue that is particularly vulnerable because of unique features of metabolism in the tissue, the particular hormonal influences on the tissue, or the rates of aging, damage, and repair in the tissue, and other factors. In some cases, the target organ in a rodent species, such as the forestomach or Zymbal gland, may not have an exact human counterpart. However, the presence of carcinogenic action

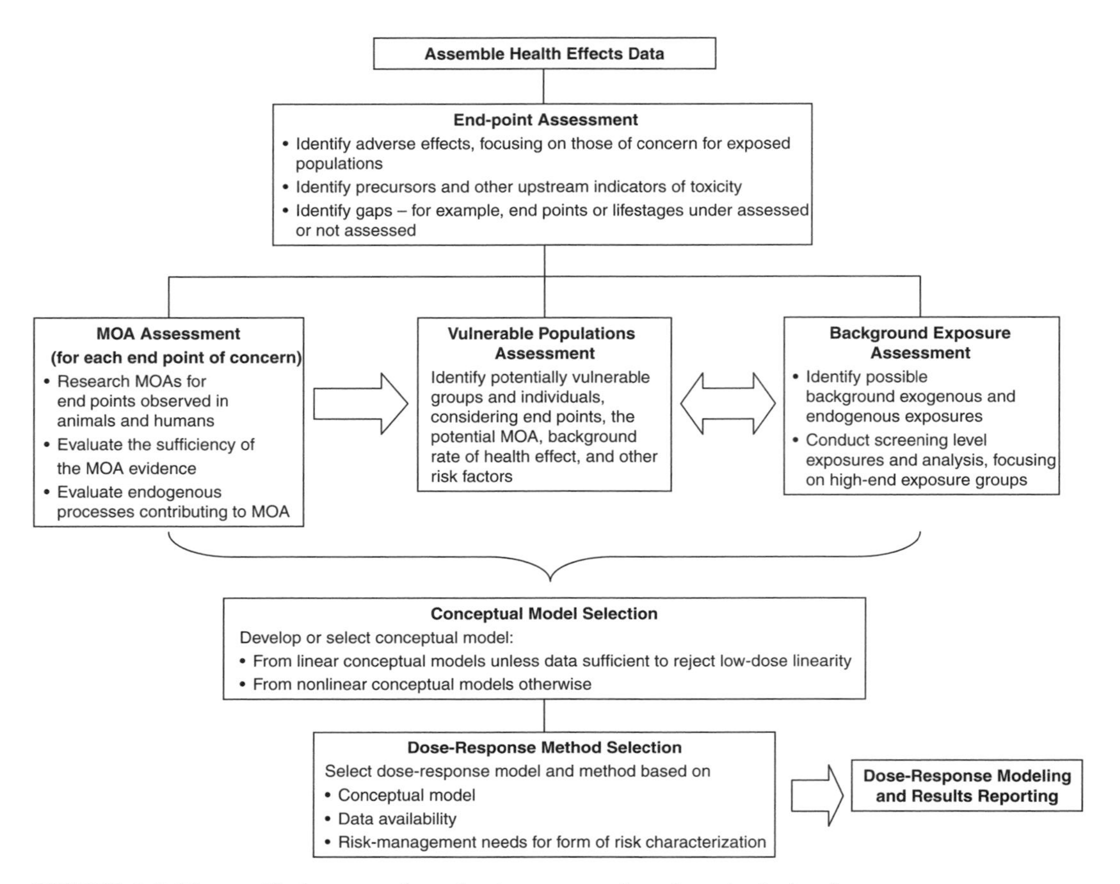

FIGURE 5-8 New unified process for selecting approach and methods for dose-response assessment for cancer and noncancer end points involves evaluation of background exposure and population vulnerability to ascertain potential for linearity in dose-response relationship at low doses and to ascertain vulnerable populations for possible assessment.

in tissues for which there is no correspondence in humans or that may be regulated differently in humans does not mean that the toxicity or tumor finding in animals is irrelevant. That the rodent tissue is sensitive to the toxicant signifies that the toxicant MOAs operate in a mammalian system that has characteristics in common with similar or even not obviously related tissues in humans or human subpopulations. Because epidemiologic studies are often limited in their ability to explore outcomes related to workplace or environmental exposures, it is typically impossible to rule out the relevance of an effect seen in a particular rodent tissue unless there is detailed mechanistic information on why humans would not be affected (IARC 2006). The finding that the high sensitivity of the rat Zymbal gland to benzene tumorigenesis occurs via an MOA (clastogenesis) similar to that which produces benzene-induced bone marrow toxicity and cancer in humans (Angelosanto et al. 1996) is an indication that a tissue that is specific to the rat can still provide important hazard and potency information related to human risk. In general, tissues that are responsive to a toxicant should be considered relevant to human risk assessment unless mechanistic information demonstrates that the processes occurring in the tissues could not occur in humans.

Mode-of-Action Assessment

The MOA evaluation explores what is known or hypothesized about the key events after chemical exposures that lead to the toxicity of a compound, including metabolic activation and detoxification, initial interactions with critical cellular targets (for example, covalent binding with protein or DNA, peroxidation of lipids and proteins, DNA methylation, and receptor binding), altered cellular processes (for example, apoptosis, gene expression, and signal transduction), and other types of biochemical perturbation that may involve defense mechanisms or be considered precursor events. Background or endogenous processes that might act in concert with those events would also be considered. Any MOA information that might be helpful in understanding dose-response relationships at both high and low doses would be considered, including dose-dependent nonlinearities in metabolic processes, depletion of cellular defenses, potential to outpace repair processes, induction of enzymes by repeat dosing, additivity and interaction with background disease processes, and additivity of the chemical and its metabolites with other chemical exposures.

The MOA assessment brings mechanistic information to bear on the dose-response assessment. However, the available data will often be too limited to explain how a chemical or its metabolites act to produce an effect. In such cases, default assumptions will apply; below possible defaults are presented in the context of conceptual models. Chapter 6 provides further recommendations and guidance on developing and applying defaults.

Precautionary lessons on the use of MOA data in dose-response assessment are presented by way of the following examples. As the first example, findings of rodent liver cancer have been hypothesized to be of limited or no human relevance for chemicals that are agonists for the peroxisome proliferator activated receptor α (PPARα), a hormone receptor involved in energy homeostasis (Klaunig et al. 2003). Notably findings of rodent liver cancer for di(2-ethylhexyl)phthalate were found by the International Agency for Research on Cancer (IARC 2000) not to be relevant to humans because peroxisome proliferation was demonstrated in mice and rats, but not in human hepatocyte cultures or livers of nonhuman primates exposed to DEHP. However, findings of liver cancer at a higher incidence in PPARα-null than wild-type mice (Ito et al. 2007) call into question this conclusion. Second, MOA assessment has recently been introduced as a way to determine whether a carcinogen has greater sensitivity early in life. Following EPA (2005c), a factor is to be applied when exposure occurs in early life to account for the greater sensitivity during this period, but only for chemicals with established mutagenic MOAs. These guidelines (EPA 2005c) raise the question of what constitutes a mutagenic MOA. It can be difficult to establish how a chemical with some genotoxic activity may induce a mutation (for example, direct vs indirect effect), how to translate findings from one biologic system or age group to another, and how effects are produced when a chemical induces cancer by multiple MOAs, as many carcinogens are likely to do. The practice is inconsistent with the EPA approach to low-dose extrapolation in its cancer risk-assessment guidance: when the MOA is uncertain, the default position is to assume a low-dose linear extrapolation (EPA 2005b, p. 3-21).

The "M" factor described later in this chapter is introduced to modify the dose-response slope at low doses to address the case of multiple MOAs or other aspects that can be different between high and low dose. The MOA assessment would inform the selection of M.

Background and Vulnerability Assessments

A critical aspect of the new approach is the determination that, whether addressing cancer or noncancer end points, dose-response models should fully address both intersubject variability and background disease processes and exposures. How those factors may "linearize" dose-response relationships, which would otherwise be low-dose nonlinear relationships on the basis of MOA, should be considered explicitly. The committee recommends that two systematic reviews be included as components of EPA dose-response assessments. The first is an assessment of background exposures to xenobiotics (for example, in pharmaceuticals, food, and environmental media) and endogenous chemicals that may affect the processes by which the chemical produces toxicity and may result in low-dose linearity. The second is an assessment of human vulnerability that identifies underlying disease processes in the population to which the chemical in question may be adding and that suggests groups of sensitive individuals and their characteristics. Those issues are considered further below in terms of how they may affect the choice of conceptual model used in dose-response analysis.

To facilitate this step of the dose-response assessment process, the committee provides an initial set of diagnostic questions that address whether background considerations are key factors:

- What is known or suspected to be the chemical's MOA?
- What underlying degenerative or disease processes might the toxicant affect or otherwise interact with?
- What are the background incidences and population distributions of these processes?
- Are there identified sensitive populations?
- Have the underlying processes been characterized in humans with markers of susceptibility and precursor effect?
- What known and probable factors can affect the underlying processes and thus potentially modulate adverse health outcomes of exposure to the toxicant?
- What are the levels of human-to-human and age-dependent variability and uncertainty with respect to background degenerative and disease processes, and how do they interact with the toxicant's MOA?
- What environmental contaminants in air, drinking water, food or in consumer products (for example, foods, pharmaceuticals, cosmetics) or endogenous chemicals (for example, natural hormones) are similar to the chemical in question?
- Could they potentially operate by MOAs similar to that of the chemical in question?
- What chemicals might operate by a different MOA but have the potential to affect the same toxic process as the chemical under study?
- How might the endogenous and exogenous background components vary among individuals? Can subgroups with particularly high exposures be identified?
- Is there a potential for people with high background exposures to have health conditions that predispose them to the critical end points or diseases caused by the chemical under study?

Questions, like those above, are essential to ask when conducting chemical risk assessment, whether using the unified framework or current approaches. These questions help identify potential data sources for understanding inter-human variability in response and the extent to which a chemical may pose risks at low doses, and the limits in that understanding. EPA's draft risk assessment for trichloroethylene (TCE) (EPA 2001a; NRC 2006a) took a step in this direction by considering how differences in metabolism, disease, and other factors contribute to human variability in response to TCE, and how other factors may alter its metabolism. EPA's draft dioxin risk assessment considered the impact of background and cumulative exposure to dioxin-like compounds and the potential impact on low-dose response (EPA 2004; NRC 2006b). The unified framework formalizes the incorporation of this type of information into human-health risk assessments, through background and vulnerability assessments and the subsequent selection of a conceptual model for dose-response assessment.

A Pictorial to Aid Vulnerability Assessment

Many factors can affect susceptibility to a chemical, including host genetics, disease status, sex, age, functional reserve, capability of defense mechanisms (for example, glutathione status), capability of repair mechanisms, activity of the immune system, and coexposure to other xenobiotics. Figure 5-9 is an aid to explore how the disease process may be influenced by numerous biochemical processes and risk factors. Someone who is not very vulnerable may have no or few risk factors, whereas someone who is vulnerable may have many or far greater exposure to one or several of them. Figure 5-9 portrays a hypothetical population vulnerability distribution, with the X-axis representing "functional decline," a continuous variable that is an indicator of vulnerability. For example, the indicator of functional decline for asthma could be reduced airway responsiveness. People who have generally lower levels of risk factors and disease precursors will be on the left side of the population distribution in Figure 5-9. Moving to the right will be people who experience a loss of function but are not symptomatic. With further loss of function, as may occur in people who have additional or greater exposure to risk factors, biomarker levels are higher and approach their threshold for symptoms and disease. Stressors that may be innocuous in healthy people may be life-threatening in those who are susceptible. For example, exposure to low concentrations of an infectious agent may cause clinical infection only rarely in the average person, but those whose lung clearance and immune function are compromised may develop pneumonia at a higher frequency and, when afflicted, may have a greater risk of death.

Figure 5-9 illustrates a hypothetical situation in which the population depicted is ex-

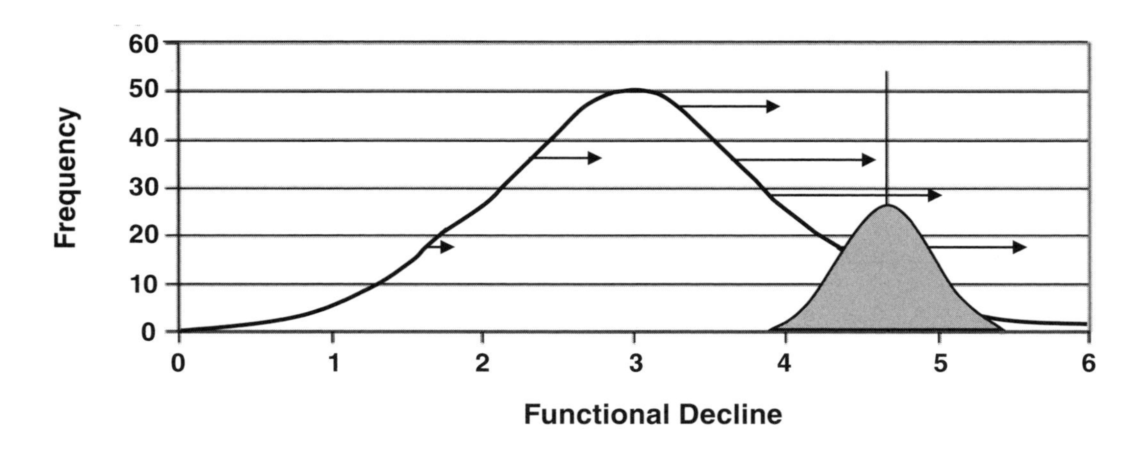

FIGURE 5-9 Population vulnerability distribution. Arrows represent hypothetical response to same toxicant dose for people at given level of functional decline unrelated to any particular toxicant. Vertical line represents presumed threshold between overt adverse and nonadverse effect in median person. Shaded area straddling line represents distribution of thresholds in population.

posed to a toxicant. The vertical line represents the theoretical threshold to elicit an adverse clinical effect in the median person. The threshold will not be the same in everyone, so it is represented in the figure as a normal distribution. The arrows represent the magnitude of toxicant effect in response to a given dose in people who are at a given level of functional decline. In this example, people who are more vulnerable are both closer to their threshold and more responsive to a given toxicant dose (represented by the larger arrows). Toxicant exposure will shift the vulnerability distribution to the right and make it more skewed, as indicated by the size of the arrows. Here, as in epidemiology, functional decline or baseline health status might be thought of as an effect modifier of the risk of interest. Sensitivity differs because the more vulnerable, on the one hand, have less functional reserve and cellular defense and, on the other hand, may have a greater number of processes that could contribute to disease (for example, less responsive airways, less pulmonary clearance, poorer immune surveillance, or impaired cardiac function). Low-functioning people can be at greater risk not only because they can be near the threshold but because they can have a greater response per unit dose.

Low doses cause a small shift, and even a very low dose may push a few people over their threshold. If the background level of clinical effect is high (for example, 1% of people have the disease) and there is considerable baseline variability, many people would be expected to be vulnerable to a toxicant-induced increase in the disease. In the case of rare diseases or effects (for example, affecting 1 per 100,000), few people are expected to be just shy of the threshold, and it would take a larger dose of toxicant to produce the same increase in effect as in the high-background case. The diagnostic questions listed above may help the risk assessor to understand the characteristics of the population vulnerability distribution and the potential for low-dose exposures to push some in the population over their threshold.

Selection of a Conceptual Model

Based on the background exposure, MOA, and vulnerability assessments, a decision is made as to the general approach to the dose-response analysis. It involves a selection of conceptual models for individual and population dose-response relationships. To guide this decision, the committee has developed examples of prototypical conceptual models, described earlier and summarized in Figure 5-10.

Consideration of background exposures and processes is critical for the determination of likelihood of low-dose linearity in the population dose-response relationship. Conceptual models 1 and 3 are illustrations of low-dose linearity in population response. The committee recommends that agents be considered as low-dose nonlinear, as in conceptual model 2, only if

- Biologic additivity is not a significant response modifier, for example, there are very low background rates of health end points or damage processes in the population in general, or relevant to the chemical's known or possible MOAs.
- Chemical additivity is not a significant response modifier, that is,
  - the totality of exposure to the toxicant and other agents (exogenous and endogenous) is unlikely to cause the adverse affect, or
  - the toxicant's contribution is so inconsequential that it will not promote the related ongoing toxic processes.

To illustrate the criteria, consider the case of ambient xenon. At high levels, say 70% (mixed with 30% oxygen), xenon is an analgesic and induces a hypnotic effect, and at high

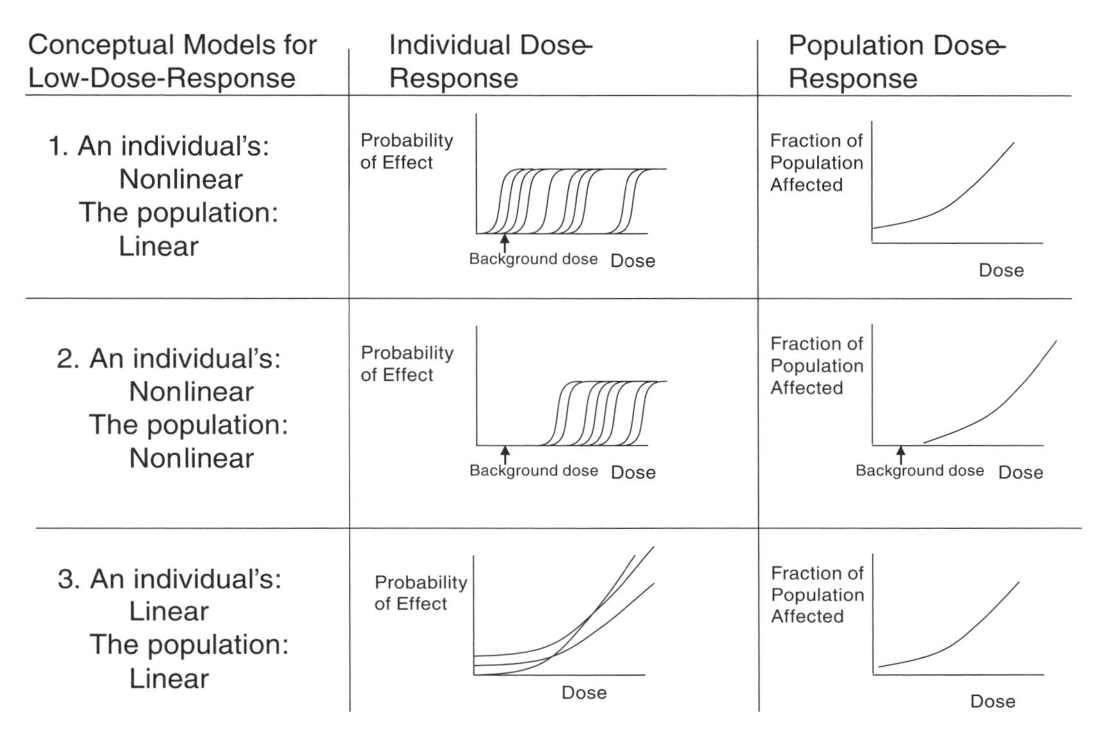

FIGURE 5-10 Examples of conceptual models to describe individual and population dose-response relationships.

levels, xenon displaces oxygen. The MOA for xenon's anesthetic action is unknown but is believed to be electrophysiological in nature, like the other volatile anesthetics. Xenon is ubiquitous in air, at quite a low concentration (0.0000087%). If asked to do a risk assessment for environmental levels of xenon, should a linear or nonlinear approach be applied?

While the MOA is unknown, the number of individuals in the general population with analgesia by xenon relevant MOAs will be restricted to those undergoing surgery, and so the first criterion is met. The totality of exposure to xenon and other volatile anesthetics is not producing anesthesia in the general population. Also, at 0.0000087% xenon's contribution to even those undergoing anesthesia would be inconsequential, as would the degree of oxygen displacement. Thus both criteria point to a threshold approach for the xenon analysis.

Carbon monoxide also impairs blood oxygenation. Its average ambient concentration, expressed as carboxyhemoglobin levels in blood (COHb), is 0.5% COHb. This concentration is less than an order of magnitude below the COHb concentration where effects are observed in human subjects: 2-6% COHb has been associated with increased angina symptoms in those with coronary artery disease. Even in apparently healthy subjects, COHb levels as low as 5% are seen to affect maximal exercise time and the maximal exercise level. Furthermore, concentrations of carbon monoxide in air can fluctuate diurnally, geographically, and by activity (for example, driving). Thus in evaluating the risk of carbon monoxide exposure, both of the above criteria indicate a linear approach should be considered: coronary heart disease is common and increased carbon monoxide exposures will likely contribute to ongoing toxic processes.

The recommendation to consider background exposure and vulnerability in deciding between linear and low-dose nonlinear approaches applies even to agents that, when tested in isolation in rodent models, appear to have a threshold and whose MOAs (in the absence of consideration of background and human heterogeneity) would otherwise suggest a threshold. Approaches and guidelines for conducting vulnerability and background assessments will be needed, as will guidelines for conducting the assessments and selecting conceptual models.

Selection of Method for Dose-Response Analysis

The approach to the analysis depends on the conceptual model, the data available for the analysis, and risk-management needs. If, for example, data are sparse and available only from animal studies and low-dose linearity is ruled out, the analysis may proceed by using default distributions for adjustment factors and using methods like those described in the next section. If there is a relatively high endogenous or exogenous background exposure to the same and related chemicals or vulnerability can be substantial and highly variable (perhaps in particularly sensitive subgroups), the analysis may proceed by a linear default or incorporate distributional information specific to the particular chemical or circumstance being analyzed.

The following section suggests approaches to dose-response analyses for a variety of toxic mechanisms and interactions with background processes and exposures. The general assumption in working through the examples provided is that variability distributions are unimodal: people who are at an extreme for a particular parameter are not numerous enough to constitute a subpopulation that should be analyzed separately. However, for any given parameter (for example, respiratory function, immunoglobulin E status, blood pressure, xenobiotic-metabolizing capacity, or DNA repair), a multimodal distribution may exist and be influential enough to create a multimodal distribution of risk at a given dose.

Unique subpopulations can be addressed as special cases within the framework. Figure 5-11 depicts such a case, showing that the dose-response relationship for sensitive people has very little overlap with that for the typical person. If the sensitive people constitute a distinct group either because of their numbers or because of identifiable characteristics—such as ethnicity, genetic polymorphism, functional or health status, or disease—they should be considered for separate treatment in the overall risk assessment. An example of a generally susceptible well-defined group is asthmatics, with respect to their response to irritant gases emitted from rocket engines (NRC 1998a). Analysis of dose-response functions of asthmatic subjects indicated sensitivity to hydrochloric acid potentially 3 times greater, to nitrogen dioxide 10 times greater, and to nitric acid 20 times greater than healthy individuals, respectively. The committee reviewing the data considered that a multimodal distribution that includes the variance and distributional form within each mode was needed for full characterization of the range of sensitivity to those irritants. Issues of threshold and background additivity can be analyzed separately for each mode to determine whether low-dose linearity assumptions are appropriate for one or more subpopulations. While consideration of susceptible subpopulations has been included in a number of environmental risk assessments (for example, NRC 2000 [copper and Wilson's disease heterozygotes]; EPA 2001b [methylmercury effects on developing children]), the level of consideration and incorporation in EPA assessments could be much improved. The conceptual framework and committee recommendations in this chapter support qualitative and quantitative improvements.

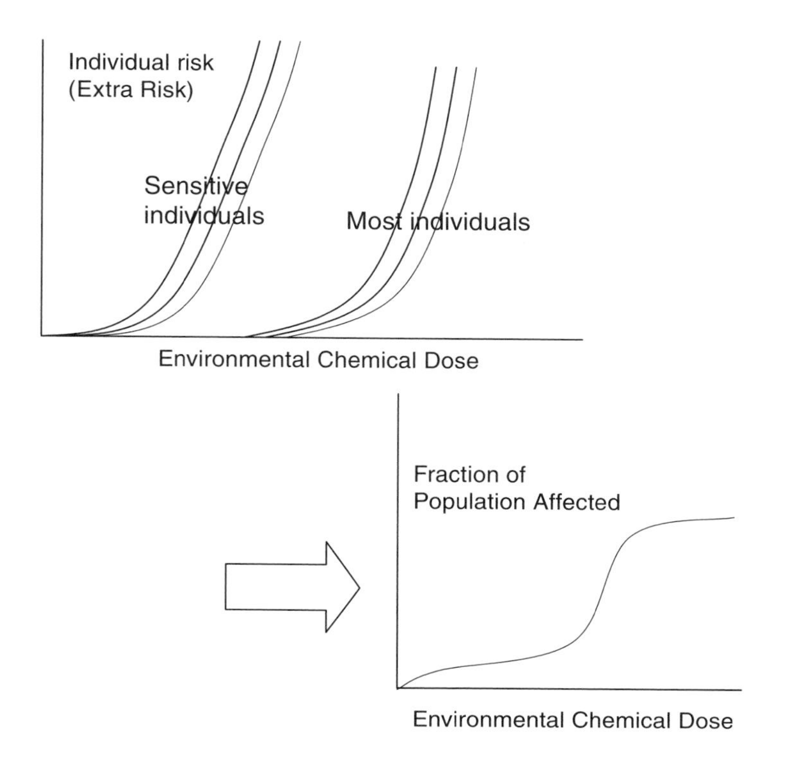

FIGURE 5-11 Widely differing sensitivity can create a bimodal distribution of risk.

CASE STUDIES AND POSSIBLE MODELING APPROACHES

This section provides case studies and possible methods for dose-response analysis for the three example conceptual models, as outlined in Figure 5-12. Methods take into account the nature of the data available. Some methods are "bottom up" in that the dose-response relationship is constructed from components. An example is given for how human variability in asthmatic response might be inferred from gene polymorphisms and might lead to a description of the population dose-response relationship for asthma. Other methods are "top down" in that the dose-response relationship at low doses is derived by fitting exposure-response models to observations from epidemiologic or animal studies.

Conceptual Model 1: Low-Dose Linear Dose-Response Relationship Due to Heterogeneous Individual Thresholds and High Background

Particulate-Matter Case Study

Fine PM ($PM_{2.5}$) belongs to a family of pollutants (including ozone) with noncancer end points for which the evidence points to a linear or other nonthreshold population response at low doses. For those agents, exposed individuals have different thresholds, and full characterization of the distribution of thresholds in the population (in this case based on epidemiologic evidence) is informative for a population concentration-response function. Numerous factors contribute to the distribution of the thresholds, as explained later.

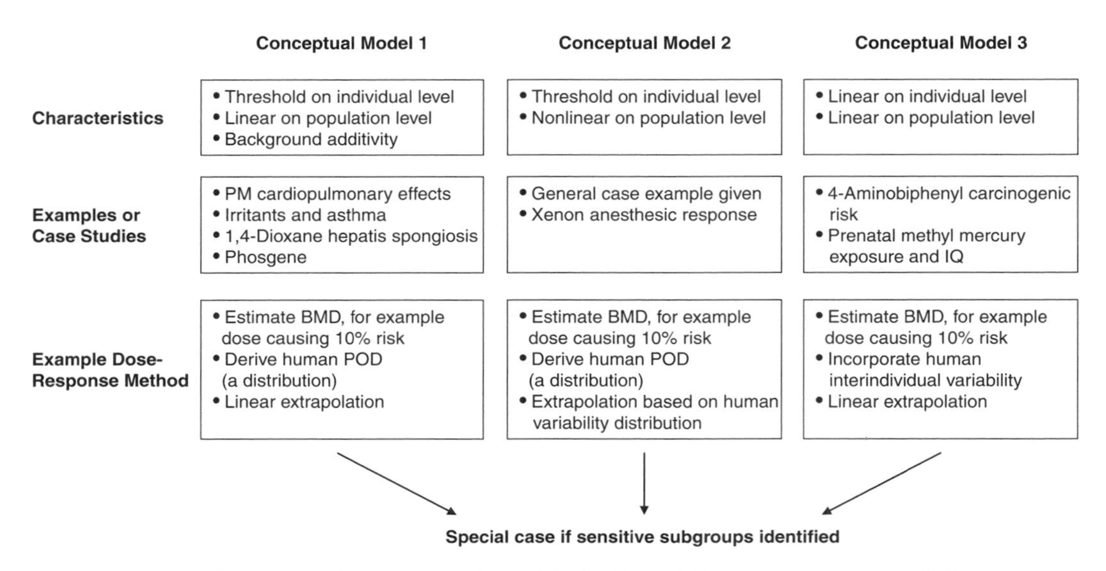

FIGURE 5-12 Three example conceptual models lead to different descriptions of dose-response relationships at individual or population levels. These are illustrated in the case studies. For each conceptual model, there may be a sensitive subgroup that should be addressed with separate dose-response analysis.

Furthermore, $PM_{2.5}$ is an example of pollutants that have numerous sources of exposure, so any analysis of a given source of $PM_{2.5}$ takes place against a background that may already be above a threshold for numerous people.

This case illustrates two dose-response issues that are of particular interest to the committee's framework for dose-response assessment:

- How concentration-response functions are developed throughout the range of observed exposures, taking into account potential nonlinearities and population thresholds.
- How human heterogeneity in response has been quantified and formally addressed both to understand sensitive subpopulations and to determine the distribution of individual thresholds to understand low-dose effects better.

How concentration-response functions are determined outside the range of observed exposures is not addressed. The available epidemiologic evidence for $PM_{2.5}$ analyses has involved fairly low-level exposures, and extrapolation below the level of observation to any great degree is less important than for compounds for which evidence is derived from animal bioassays or occupational (high dose) epidemiology.

The $PM_{2.5}$ dose-response assessment entails the construction, from epidemiologic observations, of a concentration-response function spanning all observed levels of exposure. Such a function could be used to determine directly the proportion of people whose thresholds were exceeded by a given concentration (as described above), if concentration-response functions were developed all the way down to the lowest observed exposure (ideally, approaching nonanthropogenic background). However, in a benefit-cost analysis framework, the question of the slope of the concentration-response curve near nonanthropogenic background is irrelevant because any feasible control strategies involve incremental exposure reductions and some residual exposure. For the $PM_{2.5}$ case, an important outcome of the assessment for risk management is the difference in the proportions of people adversely affected between a precontrol and a postcontrol scenario. Thus, the analysis has focused on risks in regions of the dose-response curve in which control options are relevant.

Some investigators have used statistical techniques to investigate whether any nonlinearities (including population thresholds) were present in $PM_{2.5}$ concentration-response functions in the range of observed data. For time-series studies looking at mortality and morbidity end points, the statistical methods used have included generalized additive models (Schwartz et al. 2002) and penalized regression splines (Samoli et al. 2005). Other studies have evaluated the questions of thresholds and nonlinearities explicitly by fitting piecewise linear concentration-response functions with defined knot points and then using model averaging based on the posterior probabilities of the various candidate models (Schwartz et al. 2008). Regardless of the approach, any of these techniques allow the explicit consideration of nonlinearities in concentration-response functions, including the possibility of population thresholds. However, these approaches are clearly applicable only to epidemiologic evidence, in which there are observations at a sufficient number of magnitudes of exposure to infer the shape of the concentration-response function empirically rather than on the basis of prior hypotheses about functional form. It is also most relevant for population rather than occupational epidemiology, so it will be valuable for only a small number of compounds (those to which exposure is ubiquitous and which pose relatively high population risks).

One crucial question is whether those statistical methods have demonstrated population thresholds for $PM_{2.5}$ or substantial departures from linearity. Another is whether the data would ever be rich enough to discriminate between a model with a threshold and a model without a threshold. Most studies that have used the methods (Schwartz and Zanobetti 2000; Daniels et al. 2000; Schwartz et al. 2002; Dominici et al. 2003; Samoli et al. 2005) have concluded that the functions are effectively linear throughout the range of observed concentrations, which, in the case of many time-series studies, approaches zero. Thus, in spite of the use of statistical models that could detect population thresholds, or at least low-dose nonlinearity, no thresholds appeared to be present in the range of observed concentrations. That finding has been attributed (Schwartz et al. 2002) to the fact that there is a wide distribution of individual thresholds and, in the case of cardiopulmonary mortality (a background disease process with which $PM_{2.5}$ exposures are associated), numerous genetic, environmental, disease-state, and behavioral risk factors each contribute to the distribution of the thresholds.

The extent of the distribution of individual thresholds was quantified by one study of PK and PD factors that influence heterogeneity in response to $PM_{2.5}$ (Hattis et al. 2001). The study assumed lognormality to describe the distribution for individual thresholds. The study concluded that the most susceptible (99.9th percentile) people would respond at doses only 0.2-0.7% of those needed to exhibit responses in people of median susceptibility. An extension of this analysis found results for subpopulations that were consistent with lognormal distributions for a very small number of cut points (Hattis 2008), suggesting the general population responses may be consistent with a mixture of lognormal distributions. Given that the analyses did not include all important aspects of coexposures and disease states that might influence vulnerability, the true heterogeneity could be greater. That provides good physiologic plausibility of low-dose linearity on a population basis, given ubiquitous exposures that imply that a substantial number of people will be found to be at least as sensitive as the 99.9th percentile individual.

Human heterogeneity in response has also been evaluated epidemiologically through the examination of effect modifiers to identify sensitive subpopulations. For example, multiple studies have found that the relative risk of cardiovascular end points (ranging from markers of systemic inflammation to hospitalization to death) was increased in people with diabetes, hypertension, or conduction disorders of the heart (Zanobetti et al. 2000; Dubowsky et al. 2006; Peel et al. 2007). In principle, such pooled evidence from multiple studies could allow a calculation of the risk of an effect of a defined dose in subpopulations with and without specific conditions. Instead of attempting to define risk-specific doses for a pooled population that includes a wide range of sensitivities, a stratified analysis could be performed of the range of thresholds possible in the population on the basis of what is known about unique and definable subgroups.

There are some aspects of $PM_{2.5}$ and other criteria pollutants that are not generalizable to other pollutants, but this case example illustrates the greater role that epidemiology could play in unified toxicity assessments. Opportunities to develop concentration-response functions for noncancer end points should be exploited by using statistical techniques to draw empirical inferences about the shape of the concentration-response function in the range of observed data, taking account of sensitive subpopulations. This case also serves as a reminder that EPA is already developing quantitative risk estimates for a few noncancer stressors that go beyond the threshold concept and has been doing so for some time.

Asthma Case Study

The $PM_{2.5}$ case provided an example of how "top-down" methods can be used to characterize the population distribution of vulnerability. "Bottom-up" approaches may also be informative, as described by this example. These approaches entail characterization of background processes of function loss, damage, disease, and concomitant exposures that will enable a description of the population distribution of vulnerability. That, in turn, can be used in assessing interindividual variability in toxicodynamic response at low doses and can inform the shape of the dose-response relationship at low doses. A case study of asthma is used to explore the concept. Here evidence from markers of disease susceptibility combined with analyses of genotypic differences in vulnerability and relatively high background asthma incidence are considered to evaluate the potential for asthmagenic chemicals to have linear-dose-response relationships at low doses.

Host markers of susceptibility to asthma have been developed and can be used to construct a vulnerability distribution. Asthma occurs in people who are hyperresponsive to allergens and irritants and are thus at the high end of the population distribution of airway responsiveness. The methacholine-challenge test is one of several probes used to screen populations for airway reactivity and used in the diagnosis of asthma. Methacholine is a cholinergic bronchoconstrictor in both normal and hyperreactive airways; there is a continuous distribution of airway reactivity as defined by the challenge dose required to decrease FEV_1 by a given percentage. FEV_1 is the volume of air that can be forced out of the lungs in 1 s after a person takes a deep breath. The PC_{20} is the provoking concentration of methacholine required to decrease FEV_1 by 20%. Among healthy, nonasthmatic people, this measure is distributed so that the majority have low reactivity (high PC_{20}) and a subset have high to very high reactivity. The PC_{20} of 8 mg/L has been used as a cut point to indicate airway hyperreactivity; a person with a PC_{20} below this value is considered to be hyperresponsive and is likely to be either asthmatic or vulnerable to becoming asthmatic. Those with reactive airways appear to be at increased risk for xenobiotic triggering of symptoms and the onset of clinically diagnosed asthma, as indicated in prospective studies that contrast "normoresponders" with asymptomatic "hyperresponders" (Laprise and Boulet 1997; Boutet et al. 2007). The hyperresponders tended to develop more asthmatic symptoms and have decreasing PC_{20}.

Boutet et al. (2007) evaluated the distribution of PC_{20} values in a population of 428 healthy vocational students in the province of Quebec, Canada. Figure 5-13 is constructed from the data presented in that study. Asymptomatic hyperresponsiveness (PC_{20} less than 8 mg/mL) was observed in 8.5% of the subjects. The increase in respiratory symptoms over a 3-year observation period differed dramatically in this population. Those most at risk had the highest baseline response to methacholine (PC_{20} less than 4 mg/mL); these high responders had a relative risk of symptoms of over 30 compared with baseline normal responders (PC_{20} over 32 mg/mL). The increase in symptoms in this population was apparently not related to workplace exposure and so may reflect a generalized trend toward the asthmatic phenotype in otherwise healthy people who are asymptomatic hyperresponders in the initial screening. This finding is reinforced by a similar earlier occupational study of animal workers and bakers (de Meer et al. 2003).

The findings indicate how an underlying disease factor, such as airway hyperresponsiveness, can influence the onset of new disease (in this case asthma) in the population. The more people are in the asymptomatic but vulnerable range, the more likely it is that new cases of disease will occur. Different populations may have different background distributions of predisposing risk factors, as shown in an analysis of PC_{20} data by Hattis (2008).

The background rate of airway hyperresponsiveness may be used to assess the number of people at risk for developing asthma symptoms in response to even low doses of a new insulting agent. If the background rate of hyperresponsiveness is low, the number of people near the threshold for symptoms may also be low, and the low-dose incremental effects of the toxicant may have a linear dose-response relationship but with a shallow slope. If many people are vulnerable, the slope at a low dose may be steeper, with a greater incremental effect increase per unit of exposure. Thus, variability in this precursor characteristic, airway hyperresponsiveness, may be a key input into a distributional analysis of the effects of ozone or other toxicants on asthma risk. It will be a challenge to toxicology and epidemiology to generate data that can inform understanding of the interaction of toxicants with predisposing disease factors in vulnerable populations. A simplistic approach to these relationships for asthma follows.

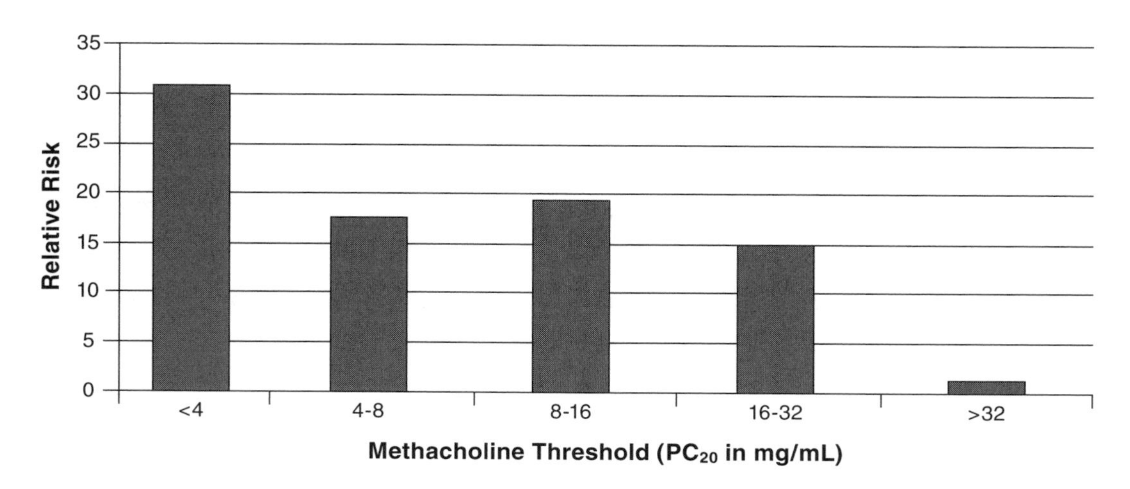

FIGURE 5-13 Baseline airway reactivity as vulnerability factor for allergen-induced respiratory effects expressed as relative risk. Source: Data from Boutet et al. 2007. Reprinted with permission; copyright 2007, *Thorax*.

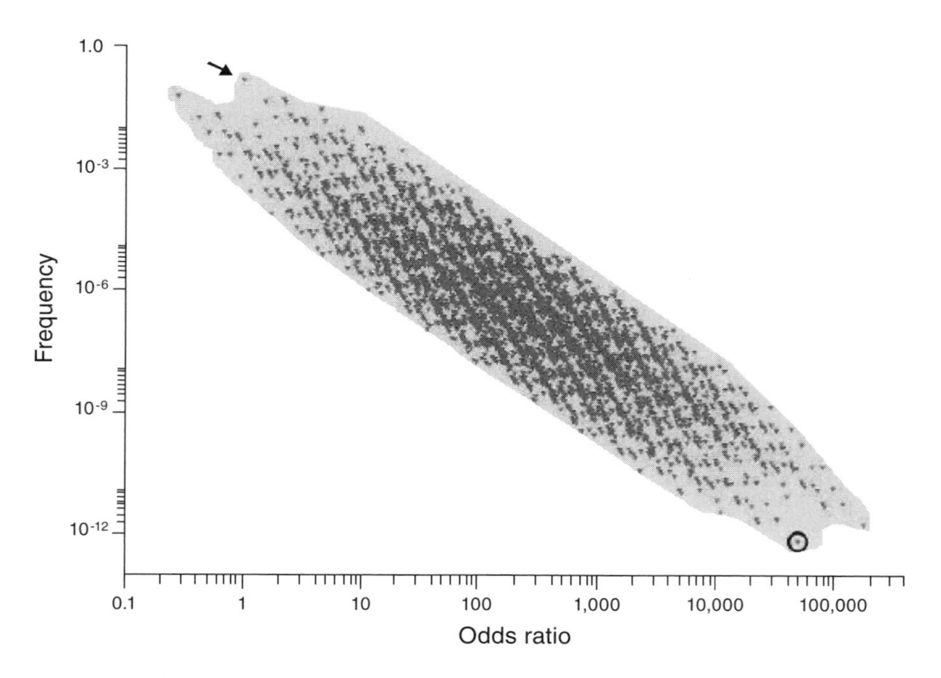

FIGURE 5-14 Effect of asthma-related gene polymorphisms on human vulnerability to asthma. Odds ratios and frequencies were calculated with assumption of 16 gene variants, each point representing a unique combination. Referent genotype (arrow) has odds ratio of 1, and profile composed of all variants (circle) is at other extreme. Source: Demchuk et al. 2007.

Progress has been made in identifying specific genetic factors that predispose to asthma. A recent publication breaks the factors into three broad categories: immune and inflammatory (12 genes), atopic (three genes), and metabolic (one gene) (Demchuk et al. 2007). By accounting for the population frequency of polymorphisms that affect gene expression or protein function and for the odds ratio associated with each polymorphism in terms of asthma risk, the analysis provided a population distribution of vulnerability to asthma, as shown in Figure 5-14. If some people had all the higher-risk polymorphisms (circle) and the sensitivity-enhancing effects acted multiplicatively when combined, these people would have a roughly 50,000-fold increase in risk of developing asthma compared with all wild-type people (arrow). Kramer et al. (2006) propose ways of identifying key candidate genes to better describe genetic susceptibility on PM induced asthma and how research might better support the regulatory standard-setting for PM. Modeling exercises can explore toxicant interaction with the polymorphic pathways to see how exposure in conjunction with host variability may combine to create a distribution of risk of asthma. In the absence of such an understanding, it would be reasonable to assume that chemicals that induce or exacerbate asthma do not have threshold dose-response relationships at the population level and that low-dose linearity prevails.

1,4-Dioxane in Animals Case Study

When epidemiologic data are lacking, diagnostic questions regarding vulnerability and background exposures may be difficult to answer. The background rate of toxicity in unexposed animals and the shape of the dose-response relationship may indicate whether background or endogenous processes will be important in evaluating the potential for low-dose linearity. Variability is expected to be much greater in the human population than in tester strains bred for use in the laboratory and exposed under controlled conditions, so it is important to reflect on potential human processes in reaching overall conclusions. However, animal studies can be more thorough in evaluating age-related and spontaneous toxicity in the control group than is typically possible in unexposed or reference human populations. Therefore, animal toxicity studies may provide important insights into the potential for low-dose linearity.

A case in point is 1,4-dioxane. This solvent produces histopathologic changes in the liver's Ito cells termed hepatic spongiosis—an inflammatory lesion of the sinusoidal and endothelial cells that can be progressive and is believed to be involved in the response to nitrosamines and other hepatocarcinogens in rodents (Karbe and Kerlin 2002; Bannasch 2003). This end point is sensitive to 1,4-dioxane exposure (Yamazaki et al. 1994) and is an example of a noncancer end point. However, evidence of its involvement as a precursor lesion in hepatocarcinogenesis could lead to its evaluation with a different analytic framework (for example, conceptual model 3). As shown below, control males have a high incidence (24%), whereas this lesion was not detected in the control and lowest-dose females. The sex-specific differences in background incidence of and sensitivity to liver disease mirror the pattern of hepatocarcinogenesis in rats and humans, with males more commonly affected than females (West et al. 2006).

As seen in Figure 5-15, the high background rate of the toxic end point in males is associated with a steeper dose-response curve at low dose in males than in females; this is consistent with the shape of the dose-response curve expected on the basis of the background rate of response.

The potential for background processes to affect the shape of the dose-response curve for specific toxicants as observed in animal studies should be considered in building PD variability distributions in humans and in evaluating the possibility of low-dose linearity. In the case of the hepatic effect caused by 1,4-dioxane, prefibrotic and precirrhotic findings

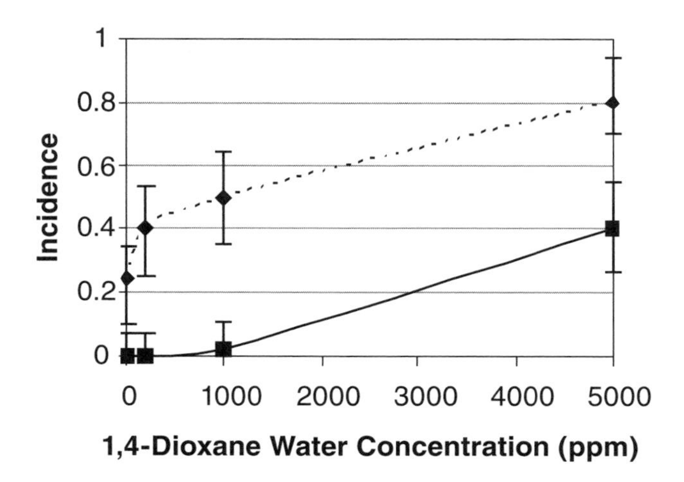

FIGURE 5-15 Dose-response relationship for liver spongiosis in 1,4-dioxane-exposed rats. Bars indicate the 95% confidence intervals. Source: Adapted from Yamazaki et al. 1994.

in the human population would be helpful in weighing the relevance of findings on animal vulnerability to that likely to occur in people. Diagnostic methods that can detect subtle liver damage in humans, such as ultrasonography and liver-function tests, may help in exploring background vulnerability to hepatotoxicants if developed further and applied to populations of healthy people (Hsiao et al. 2004; Maroni and Fanetti 2006). Existing underlying conditions and their causes could be considered in the context of potential mechanisms of 1,4-dioxane toxicity to evaluate whether the dose-response relationship should be treated as linear or nonlinear at low doses.

Default Modeling Approach for Conceptual Model 1: Linear Extrapolation for Phosgene

As described above, small chemical exposures in the presence of existing disease processes and other endogenous and exogenous exposures can have linear dose-response relationships at low doses. Thus, a simple methodologic default to address conceptual model 1 compounds is linear extrapolation from the POD, such as a benchmark dose, down to low doses. Greater information on MOA and chemical interactions with background disease processes and similarly acting chemicals may allow different low-dose extrapolations. For example, the slope of the line at the POD or another particular dose could be adjusted, as described below for conceptual model 3.

Linear low-dose extrapolation for a noncancer end point is illustrated with the case example of phosgene. This reactive respiratory toxicant damages the airways at high doses, and dose-response studies in rats exposed for 12 weeks report effects of inflammation and fibrosis of the bronchiolar region (Kodavanti et al. 1997; EPA 2005d). The BMD_{10} for this phosgene effect in rats is 170 $\mu g/m^3$ as a human equivalent concentration (HEC). The lower 95% confidence bound—the $BMDL_{10}$—is 30 $\mu g/m^3$. To this an adjustment is made because the study is subchronic rather chronic, and chronic exposure is of interest in calculating an alternative RfD.

In considering how this risk may be manifested in human populations, the background incidence of asthma—about 6% in children (CDC 2007)—is relevant. Asthmatics experience inflammation, fibrosis, and airway remodeling in response to environmental allergens and irritants and so constitute a large population potentially vulnerable to phosgene. In addition there are numerous medical conditions (for example, infection, environmental exposures, and pharmaceuticals) that lead to the lung inflammation and fibrosis that would potentially be worsened by phosgene exposure. Thus, there is a potential for background additivity that is consistent with conceptual model 1 and linear extrapolation to low dose. Further analysis of cell types and disease processes involved in phosgene toxicity and the other medical conditions may lead one to discover otherwise, but absent more definitive information indicating implausibility, background additivity would be assumed.

Box 5-2 shows that a linear extrapolation from the BMD derived by EPA would yield a risk-specific dose (median estimate) of 0.0085 $\mu g/m^3$ phosgene exposure. Theoretically, exposure at this dose is predicted to contribute to inflammation and fibrosis in 1 in 10^5 of exposed individuals. The phosgene RfC of 0.3 $\mu g/m^3$, set by EPA with a 100-fold cumulative uncertainty factor, corresponds to a theoretical risk that 1 in 3,000 (median estimate) individuals could be affected, on the basis of linear extrapolation. Implicit in the extrapolation are the assumptions that a 10-fold reduction in exposure will result in a 10-fold reduction in risk and that the $BMDL_{10}$ in terms of the HEC is the human 10% effect dose. This approach could be refined to explore the variability between individuals that is possible because of pharmacokinetics, the incidence and distribution of relevant respiratory health conditions, and many other factors, and to explore issues regarding species dose-effect concordance for

**BOX 5-2 Conceptual Model 1:
Default Linear Low-Dose Extrapolation for Phosgene**

1. Assume uncertainty in all parameters can be characterized by a lognormal distribution, with standard deviation represented by σ.
2. BMD_{10} (human equivalent concentration) = 170 $\mu g/m^3$, with 95%-tile lower bound 30 $\mu g/m^3$ variability in animal BMD, with a difference between lower 95% bound and median of 5.7-fold (because 5.7=170/30):
   $\sigma_{Animal\ BMD} = \log(5.7)/1.645 = 0.46$
   (Division by the 95% confidence bound is 1.645 standard deviations from the median in the standard normal distribution.)
3. The human equivalent concentration accounts for cross-species differerence in pharmacokinetics but not pharmacodynamics.
   Assume, as in Hattis et al. 2002, that $\sigma_{\log A \rightarrow H} = 0.42$
4. Median human POD:
   Adjust for subchronic to chronic study length, as in Hattis et al. 2002, by a factor of 2:
   170 $\mu g/m^3 \div 2 = 85\ \mu g/m^3$
   Assume the uncertainty ($\sigma_{\log SC \rightarrow C}{}^2$) in the adjustment, as in Hattis et al. 2002:
   $\sigma_{\log SC \rightarrow C} = \log[2.17] = 0.34$
5. Uncertainty in the human POD ($\sigma_{\log Human\ POD}$):
   $\sigma^2_{\log Human\ POD} = \sigma^2_{\log Animal\ BMD} + \sigma^2_{\log A \rightarrow H} + \sigma^2_{\log SC \rightarrow C}$
   $\sigma^2_{Human\ POD} = 0.46^2 + 0.42^2 + 0.34^2 = 0.71^2$
6. Lower 95% confidence bound on Human POD =
   (median human POD)/$10^{[(1.645)(\sigma \log Human\ POD)]} = 85/10^{[(1.645)(0.71)]} = 85/14.7 = 5.8\ \mu g/m^3$
7. Linear extrapolation to risk-specific dose - inflammation of 1 in 10^5 people would be affected:
   risk-specific dose = $10^{-5} \times (85/0.1) = 0.0085\ \mu g/m^3$, with lower bound 0.00058 $\mu g/m^3$
8. Estimate risk at different doses: for example, at 0.01 $\mu g/m^3$, three people in 10^5 (median estimate) would be affected.

phosgene. Here, as for conceptual model 3, an important issue is whether dose effectiveness is the same at high doses and low doses. Extrapolation methods for addressing that are discussed in the section below on the mathematical framework for conceptual model 3.

Conceptual Model 2: Low-Dose Nonlinear Dose-Response in Individuals and the Population, Low-Dose Response Independent of Background

The approach would be applied when there is sufficient evidence to reject the possibility of low-dose linearity on the basis of vulnerability and background assessments. As discussed above, the committee encourages the agency to conduct the necessary research and develop appropriate methods and practices for using probabilistic methods for low-dose nonlinear end points. To illustrate the approach, an example methodology using distributions for making calculations is laid out here and sample calculations are applied for a general case. The committee acknowledges that work is needed to further develop the underlying distributions and that methods are needed to support their use in a regulatory context.

Deriving a Reference Dose with Probabilistic Methods

Published methods and examples describing noncancer risk probabilistically (Gaylor et al. 1999; Evans et al. 2001; Hattis et al. 2002; Axelrad et al. 2005; Hattis and Lynch 2007;

Woodruff et al. 2007) illustrate a general approach or elements of it that can be used for this conceptual model. They can lead to an RfD based on a de minimis risk target, such as a specified fraction of the population exceeding a threshold, and the uncertainty in that estimate (for example, less than 1 in 100,000 people with some threshold response with a 95% confidence interval).

The general approach is to use distributions for the adjustments to the POD to derive a human-based POD and then to extrapolate from the human POD to lower doses on the basis of assumptions about how humans differ in susceptibility. Figure 5-16 shows how the adjustments from the animal to the human POD could be made. They are depicted here as distributions for the subchronic-to-chronic adjustment (if animal study was of less than chronic duration), database deficiencies, and animal-to-human adjustment, encompassing PK and PD across-species variability. As illustrated in Figure 5-16, the adjustment distributions can be convolved by using statistical or numerical approaches to form an overall adjustment and uncertainty distribution. Quantitatively accounting for the uncertainty in the adjustments enables a quantitative expression of the uncertainty in the overall adjustment. The adjustment distribution is applied to the animal POD to derive a distribution for the human POD. The extrapolation from the POD down the dose-response curve is driven by interhuman variability, broken out in Figure 5-16 into PK and PD elements. The application of adjustment and uncertainty distributions representing each of these elements effectively converts the animal POD (for example, the BMDL or the ED_{50}, the effective dose estimated to affect 50% of subjects) to a probabilistic dose-response relationship for the human population with confidence bounds based on the adjustment distributions.

It is possible in principle to derive the RfD on the basis of some upper percentile value selected from each of the distributions. That would yield a single estimate, similar to the current approach. The preferred method is to incorporate the full distributional information on each component factor by using probabilistic approaches, such as a Monte Carlo approach or a simple analytic approach (for example, when adjustments can be described by lognormal distributions). In that case, the RfD could be selected as a confidence point on the probability distribution for the fraction of the population with a defined risk. Alternatively, the population risk posed by a given dose could be described with a probability distribution reflecting the uncertainty in the estimate.

The approach relies on distributions for the adjustment factors. Researchers developing the method have defined distributions of each of the factors from empirical databases, as briefly summarized below. These distributions are provided to show how they might be derived, not as an endorsement of any specific distribution or their use by EPA. The distributions that lead to the adjustment of the animal POD to the human POD are described first, and then those used to extrapolate from the human POD to lower doses.

Distributions to Adjust Animal POD to Human POD

- *Subchronic-to-chronic factor.* Subchronic and chronic NOAELs from a database of 61 chemicals were compared and statistically analyzed (Weil and McCollister 1963; Nessel et al. 1995; Baird et al. 1996). A lognormal distribution was fitted to the data, which had a geometric mean of 2.01 (that is, the subchronic NOAEL was generally twice the chronic NOAEL) and a geometric standard deviation of 2.17 (Hattis et al. 2002). The standard 10-fold adjustment factor for subchronic-to-chronic extrapolation was about at the 98th

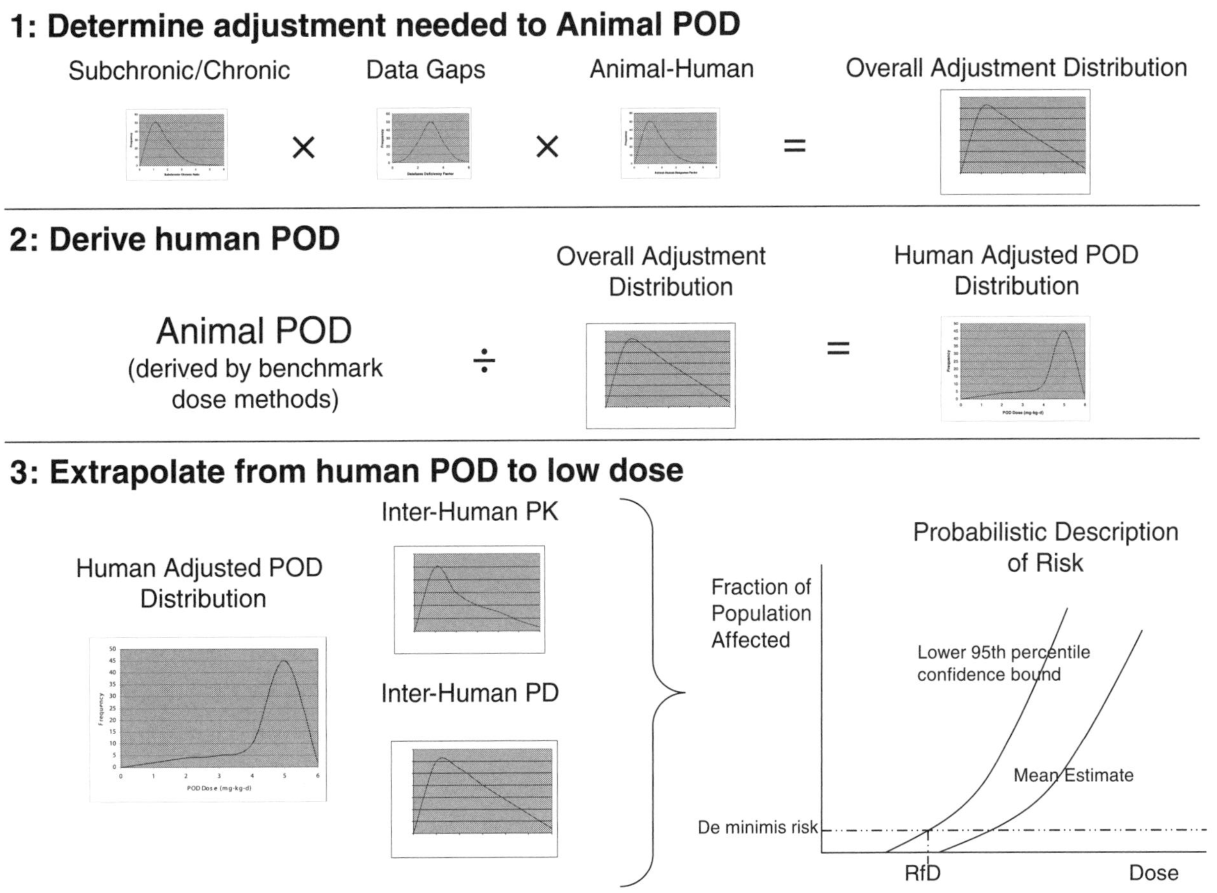

1: Determine adjustment needed to Animal POD

Subchronic/Chronic × Data Gaps × Animal-Human = Overall Adjustment Distribution

2: Derive human POD

Animal POD (derived by benchmark dose methods) ÷ Overall Adjustment Distribution = Human Adjusted POD Distribution

3: Extrapolate from human POD to low dose

Human Adjusted POD Distribution

Inter-Human PK

Inter-Human PD

Probabilistic Description of Risk

Fraction of Population Affected

Lower 95th percentile confidence bound

Mean Estimate

De minimis risk

RfD

Dose

FIGURE 5-16 Steps in derivation of risk estimates for low-dose nonlinear end points. In Step 1 a distribution to adjust the animal POD to a human POD is constructed. Step 2 adjusts animal POD by this cross-species distribution. Step 3 uses human variability distribution to extrapolate from POD to lower risk. It results in probabilistic statement reflecting the proportion of human population adversely affected by exposure and uncertainty bounds on that estimate. It also enables identification of the RfD, that is, lower-bound estimate of dose that results in de minimis risk (for example, 10^{-5} of population is affected).

percentile of this distribution (that is, 98th percentile ≈ 2.01 × 2.17 × 2.17 = 9.5) (Baird et al. 1996; Hattis et al. 2002[3]).

- *Database-deficiency factor*. A dataset for 35 pesticides with "complete" toxicity-testing profiles was analyzed to compare reproductive, developmental, and chronic NOAELs (Evans and Baird 1998). It was possible to develop distributions for missing reproductive, developmental, or chronic toxicity data in terms of how much the POD can change by the addition of the missing data (Hattis et al. 2002). The data source is limited in terms of the type of chemicals assessed (pesticides) and the end points analyzed, but it provides an example of a useful approach to developing a distribution for this factor.
- *Animal-to-human extrapolation*. Cross-species differences in acute and subacute toxicity of anticancer drugs have been generalized to draw conclusions about animal-human differences in noncancer and cancer toxicity (Freireich et al. 1966; Travis and White 1988; Watanabe et al. 1992; Hattis et al. 2002). Animal-human interspecies distributions have been inferred from rat-mouse comparisons of cancer potency (Crouch and Wilson 1979), although because of the nature of the underlying data the distributions are likely to underpredict actual species differences. The results for cancer chemotherapeutic agents may have limited applicability. First, the agents are mostly direct-acting, so species differences in PK may not be as great for environmental chemicals. Second, the results are for a narrow range of end points (lethality and tolerated dose), and may not be representative of species differences for the more variable critical end points for environmental toxicants. Third, results are for acute and subacute exposures, and may not adequately represent cross-species differences for chronic exposures and more subtle end points. Indeed, Rhomberg and Wolff (1998) have shown that cross-species scaling observed with single-dose-lethal toxicity differs from the subacute toxicity. These authors hypothesize that "dose-scaling patterns across differently sized species should be different for single-dose and repeated-dose regimes of exposure, at least for severe toxic effects." The number of animal species studied is also an important consideration in developing the cross-species extrapolation distribution (Hattis et al. 2003). Further exploration of the issues raised is needed in developing interspecies distributions for application in EPA assessments.
- *Example derivation of the human POD*. In the examples given above, lognormal distributions replace uncertainty factors, and each factor is independent of the other. The overall adjustment is simple to calculate and does not have to be done numerically, using for example, Monte Carlo treatment. To obtain the human POD, the animal POD is divided by the overall adjustment factor, which for the sake of discussion is called here "$F_{A \to H\ POD\ adjust}$."

$$\text{Human POD} = \text{Animal POD} \div F_{A \to H\ POD\ adjust}.$$

The overall adjustment is made up of three adjustments: for animal-to-human extrapolation, "$F_{A \to H}$"; for experiment duration from subchronic to chronic, "$F_{SC \to C}$"; and for data gaps, "F_{Gap}." Thus,

$$\text{Human POD} = (\text{Animal POD})/(F_{A \to H\ POD\ adjust}) = (\text{Animal POD})/(F_{A \to H} \times F_{SC \to C} \times F_{Gap}).$$

[3]A full version of this publication is available at http://www2.clarku.edu/faculty/dhattis. Updated results are published in Hattis and Lynch (2007).

If each factor is lognormally distributed, $F_{A \rightarrow H\ POD\ adjust}$ will be lognormally distributed; when a given adjustment is not needed its factor would be assigned a single value of 1.

The animal POD could be established as it is currently, or it could be described by the BMD distribution associated with its estimation. If the estimator of animal POD is lognormally distributed or is considered a constant, the human POD will be lognormally distributed. In this example, distribution of the animal POD is taken into account. Guidance on how a BMD should be defined for continuous outcomes would facilitate its current use as an animal POD (Gaylor et al. 1998; Sand et al. 2003) and for the probabilistic descriptions envisioned here.

The median value of the human POD distribution can be calculated by substituting the median values of the factors and the animal POD in the above equation.

$$\log\ (\text{Human POD}) = \log\ \text{Animal POD} - (\log F_{A \rightarrow H} + \log F_{SC \rightarrow C} + \log F_{Gap}).$$

For this case, each factor is assumed to be lognormally distributed,

$$\sigma^2_{\log \text{Human POD}} = \sigma^2_{\log \text{Animal POD}} + \sigma^2_{\log A \rightarrow H} + \sigma^2_{\log SC \rightarrow C} + \sigma^2_{\log Gap}.$$

The lower confidence bound on the human POD can be readily calculated. The human POD is the starting point for the extrapolation to lower doses based on information on human variability. A sample set of calculations is provided in Box 5-3 to illustrate how the above calculations can be made to derive a human POD.

Human Variability Distributions for Extrapolating from Human POD to Low Doses

- *Interindividual variability—PK Dimension.* Blood concentration information (AUC[4] and C_{max}[5]) were compiled for 471 data groups involving 37 drugs (Hattis et al. 2003) and summarized. A small number of data groups involved children under the age of 12 years. These PK data summaries that included young children (Ginsberg et al. 2002; Hattis et al. 2003) were then incorporated to yield a PK variability estimate for the overall population (Hattis and Lynch 2007). This work illustrates the feasibility of constructing PK variability distributions that are specific to particular age groups and clearance mechanisms. PK parameters have been derived from blood concentration data in children and adults and compiled according to type of agent, clearance pathway, or receptor (Ginsberg et al. 2002). Since these data come from a clinical setting in which the health of the studied subjects was impaired, and the characteristics of the treatment group may be similar, the data may not be representative of the general public. However, the researchers note the similarity of patterns of metabolizing-enzyme ontogeny in the databases and in vitro liver-bank specimens, suggesting that results from pharmaceutical studies may be generalizable.
- *Interindividual variability—PD Dimension.* From a database for 97 groups, Hattis et al. (2002) and Hattis and Lynch (2007) derived estimates of PD variability in (1) the chemical's reaching the target site after systemic absorption; (2) parameter change per delivered dose, the dose-response relationship at the active site (for example, beta-2-microglobulin spillage into urine in relation to urinary cadmium concentration); and (3) functional reserve,

[4]AUC is the area under the concentration-time curve that displays the complete time course of a chemical in a particular body compartment. AUC is sometimes used to represent the total dose in that compartment integrated over time.

[5]C_{max} is the maximum concentration of a chemical attained in a particular compartment after dosing.

BOX 5-3 Calculating a Risk-Specific Dose and Confidence Bound in Conceptual Model 2

I. Derivation of Human POD

$$\text{Human POD} = (\text{Animal POD})/F_{A\to H\ POD\ adjust} = (\text{Animal POD})/(F_{A\to H} \times F_{SC\to C} \times F_{Gap})$$
$$\log(\text{Human POD}) = (\log \text{Animal POD}) - (\log F_{A\to H} + \log F_{SC\to C} + \log F_{Gap})$$
$$\sigma^2_{\log F\ Human\ POD} = \sigma^2_{\log F\ Animal\ POD} + \sigma^2_{\log F\ A\to H} + \sigma^2_{\log F\ SC\to C}{}^2 + \sigma^2_{\log F\ Gap}$$

Assume:

Data gap is inconsequential:
$F_{Gap} = 1$, $\sigma_{\log F\ Gap} = 0$

Subchronic-to-chronic per Hattis et al. 2002:
50th percentile for $F_{SC\to C} = 2$, $\sigma_{\log F\ SC\to C} = \log[2.17] = 0.34$

Animal to human adjustment per Hattis et al. (2002) for sodium azide:
50th percentile for $F_{A\to H}$ 3.85, 95% upper bound 18.5, thus $\sigma_{\log A\to H} = \log(18.5/3.85)/1.645 = 0.42$ (Division by the 95% confidence bound is 1.645 standard deviations from the median in the standard normal distribution.)

Variability in animal POD:
lower 95% bound 2-fold difference from median; thus $\sigma_{Animal\ POD} = \log(2)/1.645 = 0.18$
$\Rightarrow$ Overall variability in human POD: $\sigma^2_{Human\ POD} = 0.34^2 + 0.18^2 + 0.42^2 = 0.32 = 0.57^2$

For animal POD (ED_{50}) of 1 mg/kg-d:
Human median POD (ED_{50}) = $1/(F_{A\to H}F_{SC\to C}F_{Gap}) = 1/(2 \times 3.85 \times 1) = 0.13$ mg/kg-d

Lower 95% confidence bound on human POD
= (median Human POD)/$10^{[(1.645)(\sigma \log Human\ POD)]} = 0.13/10^{[(1.645)(0.57)]} = 0.015$ mg/kg-d

II. Derivation of Risk-Specific Dose

Interindividual PK/PD variability (assume Hattis et al. 2002 distribution):
$\sigma_{\log H} = 0.476$ (This estimate also is uncertain, with geometric standard deviation of 1.45)
The 10^{-5} individual is 4.25 standard deviations from the estimated human ED_{50}:
$10^{[(4.25)(0.476)]} = 105$

Median human dose with 10^{-5} risk:
(Median POD)/105 = 0.13/105 = 0.0012 mg/kg-d
Lower 95% bound on human dose with 10^{-5} risk: 0.006 µg/kg-d (This is calculated using a Monte Carlo procedure. It takes into account $\sigma_{Human\ POD}$ and the uncertainty in $\sigma_{\log H}$.)

a factor inherent in many of the PD datasets but of which direct measurements in humans were not available. Hattis et al. (2002) took the first listed component as a component of PD rather than PK variability because it was related to reaching a specific organ, cell type, or subcellular constituent that is not typically addressed in physiologically-based pharmacokinetic models. The human interindividual variabilities derived for those components were combined to estimate the overall interhuman PD variability.

- *Overall distribution for human interindividual variability.* For the example here, overall human interindividual variability is described by a lognormal distribution with median of 1 and logarithmic (base 10) standard deviation σ_{logH}. Hattis et al. (2002) derived such a distribution for both PK and PD components from data for general systemic toxic effects on different agents, with a geometric standard deviation of 2.99 (σ_{logH} = 0.476; $10^{0.476}$ = 2.99), indicating the median and upper 98th percentile human differ in sensitivity by a factor of 9. Human variability in response is chemical dependent. For some chemicals the difference between the median and 98th percentile is greater than a factor of 9, for others it will be less. Hattis and Lynch (2007) also describe the uncertainty in the variability estimate. The estimate of 0.476 for σ_{logH} has its own geometric standard deviation of 1.45. Because these characterizations of variability are limited by the relatively small numbers upon which the estimates are based, this uncertainty estimate may have a downward bias.

Calculation of Risk-Specific Dose and Confidence Bound

A distribution of human variability can be applied to move from the human POD down the dose-response curve, as illustrated in the set of calculations in Box 5-3. These calculations illustrate a generic case with an animal median ED_{50} value of 1 mg/kg-day.

In Box 5-3, as done by Hattis et al. (2002) and (Evans et al. 2001), the ED_{50} was chosen as the POD. Because the ED_{50} is at the center of the animal dose-response curve, there is less uncertainty in its measurement, and it is not as heavily influenced by interanimal variability as a response at the tail of the distribution might be. In addition, in many animal experimental datasets, the ED_{50} is not likely to be as influenced by the dose-response model selected to analyze the data relative to other effect levels. But there are other factors, such as intra-individual variability and the extent that this may play a role in the dose-response relationship. Any implementation of this approach by EPA would have to develop a process for selecting the POD for risk extrapolation for nonlinear end points.

Interhuman PK and PD distributions would ideally be derived with chemical-specific data on the differences possible among human populations. However, this type of information is usually lacking. Therefore, generic distributions based on surrogate chemicals and end points will be needed. Specific distributions for related chemicals and end points of interest may be possible. The first tier of a default distribution may be one built on a broad array of structurally dissimilar chemicals tested in different types of systems (from in vitro to in vivo) for different end points. The Hattis et al. (2002) effort to collect and analyze mostly clinical human data is a good initial effort at characterizing human PD variability. However, an important consideration with regard to this and related exercises is whether they fully capture PD variability, given the limited array of data studied. Data on small numbers of people may be a useful beginning but provide little information on overall interhuman variability. Even when multiple studies are combined so that data on greater numbers of people are tabulated, they still might not capture the broad spectrum of PD variability caused by differences in age, genetics, diet, health status, medications, and exposure to other agents.

Greater relevance may be achieved by applying PD variability information on prototypical chemicals in the same class as the chemical of interest. When there is a much larger and substantial database on one particular toxicant in a structural series, there is the potential to apply the information to others in the series on the basis of relative-potency approaches, as described in Chapter 6. A similar analogy may also be useful for assessing interhuman PD variability if the toxicity end points of the prototype and of the chemical of interest match well. For example, human variability in the renal response to cadmium, as assessed on the basis of beta-2-microglobulin leakage, may be relevant to other heavy metals, such as mercury and uranium, that can also damage the kidney (Kobayashi et al. 2006). Another possibility is that the degree of interhuman variability can be gleaned from studies of environmental mixtures to which populations are exposed. Biomarkers of exposure—such as urinary 1-hydroxypyrene, a marker of exposure to polycyclic aromatic hydrocarbons (PAHs)—can be related to biomarkers of effective internal dose (such as bulky DNA adducts and urine mutagenicity) and effect (such as chromosomal damage in peripheral lymphocytes). Evaluations of such markers in coke-oven workers, bus drivers, and the general population ingesting charcoal-broiled meat or inhaling cigarette smoke provide a database from which interindividual variability in response to carcinogenic PAHs may be deduced (Santella et al. 1993; Kang et al. 1995; Autrup et al. 1999; Siwinska et al. 2004).

Thus, the data gap represented by interhuman PD variability presents a critical research need that can be approached by mining the existing epidemiology literature and by designing new studies in which biomarkers of exposure and effect are used to describe variability in sensitivity to health outcomes in similarly exposed people.

There are likely to be a number of cases in which the approach illustrated above can be used to derive an RfD. Sometimes, however, there will be a well-defined sensitive subgroup. The RfD for the pesticide alachlor is based on hemolytic anemia in dogs (EPA 1993); the background incidence of hemolytic anemia in humans is generally very low except in ethnic groups in which, because of inherited traits (such as glucose-6-phosphate dehydrogenase deficiency), the risk is higher (Sackey 1999). In cases like this one, an analysis focusing on describing risks to the sensitive subgroups would be needed (see Figure 5-12).

Conceptual Model 3: Low-Dose Linear Individual and Population Dose-Response Relationship

Here linear dose-response processes govern the dose-response relationship for individuals, as may occur for cancer and other complex toxic processes, and consequently the population dose response relationship is low-dose linear. This is unlike the previous two conceptual models, which described population dose-response distributions that arise when the dose-response relationship in an individual has a threshold. A possible approach to default analysis following this conceptual model is presented below. It emphasizes probabilistic descriptions of the uncertainty in the dose-response relationship and descriptions of variability among individuals exposed to the same dose.

Approach

This approach to dose-response analysis begins, as do the other examples above, with the derivation of the human POD distribution. When derived from animal data, the human POD is based on the animal POD and distributions of adjustment factor, such as for interspecies differences and study duration less than a lifetime. Here, the POD is taken from a model fitted at a dose in the lower end of the observable response range, and does not use an ED_{50}. Risk at lower dose than this POD for the median person is estimated by linear extrapolation, that is, risk is assumed to decrease linearly with dose below the POD. However, as illustrated in Chapter 4 (Table 4-1), people exposed to the same dose will differ in risk. Estimates of the spectrum of individual risks at a given dose can be based on a distribution that describes interhuman variability. The individual dose-response relationships allow the calculation of the population dose-response curve. This approach to dose-response assessment is illustrated in Figure 5-17 and through the case study for 4-aminobiphenyl.

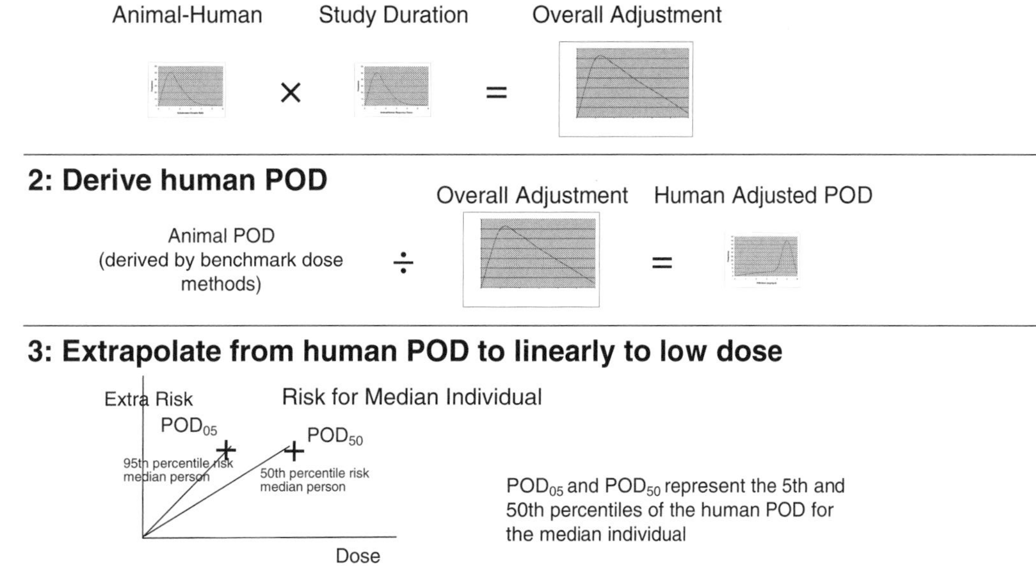

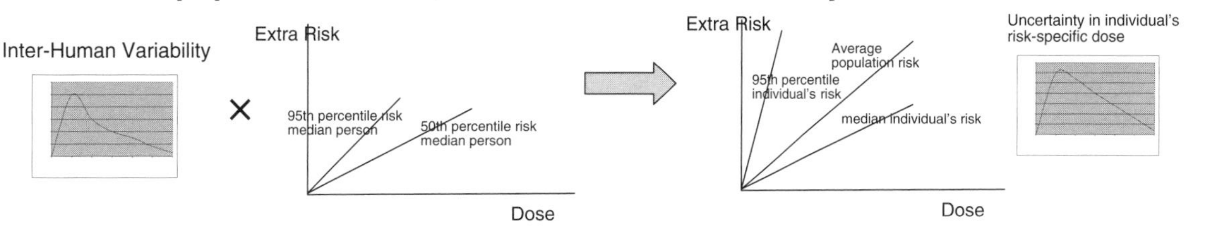

FIGURE 5-17 Steps to derive population and individual risk estimates, with uncertainty in estimates from animal data. Step 1 involves derivation of adjustment distribution to convert animal POD to human POD. Step 2 involves derivation of human POD from this distribution. Step 3 is linear extrapolation from POD to lower doses for median person. Lower bound on human POD is used to derive upper-bound risks for median person. Step 4 involves applying interindividual variability distribution to estimate average risk to population and risks to individuals with different degrees of sensitivity, with uncertainty in estimates.

Implications of the Approach

Functionally, the approach would change dose-response characterizations for low-dose linear carcinogens in two basic ways. First, there would be an explicit characterization of uncertainty in the human POD that accounted for uncertainty in the cross-species extrapolation and the statistical fit to the dose-response data. EPA could choose to report particular percentile values, such as the upper 95th percentile. EPA could describe the population excess cancer risk associated with dose D as the plausible upper bound of the excess risk, taking into account uncertainty in the population dose-response relationship and variability in the individual dose-response relationship. The excess-risk estimate for a person whose susceptibility puts him or her at the 95th or some other percentile of the population could also be separately reported.

Second, when the underlying variability distributions are right-skewed, as in the case of the lognormal distribution, the population risk estimate emerging from the analysis will be greater than the estimate for the median individual. The mean or "expected value" will exceed the median value by some amount that depends on the assumed shape of the distribution of interindividual variability in susceptibility.

Recommended Default for Interindividual Variability in Cancer Susceptibility

An assumption that the distribution is lognormal is reasonable, as is an assumption of a difference of a factor of 10-50 between median and upper 95th percentile people, as indicated by the series of examples provided in Chapter 4. It is clear that the difference is significantly greater than a factor of 1, the current implicit assumption in cancer risk assessment. In the absence of further research leading to more accurate distributional values or chemical-specific information, the committee recommends that EPA adopt a default distribution or fixed adjustment value for use in cancer risk assessment. A factor of 25 would be a reasonable default value to assume as a ratio between the median and upper 95th percentile persons' cancer sensitivity for the low-dose linear case, as would be a default lognormal distribution. A factor of twenty-five could be interpreted as a factor of 10 for pharmacokinetic variability, and a factor of 2.5 for pharmacodynamic variability. For some chemicals, as in the 4-aminobiphenyl case study below, variability due to interindividual PK differences can be greater. In a cancer process, with long latency and multiple determinants, PD variability could be considerably greater than the suggested default. PD differences would include the various degrees among people in DNA repair and misrepair, surveillance of mutated cells, and accumulation of additional mutations and other factors involved in progression to malignancy.

A common assumption for noncancer end points is an overall factor of 10 to account for interindividual variability—3.2 or 4 uncertainty factor for PK differences and 3.2 or 2.5 for PD differences (EPA 2002a; IPCS 2005). For genotoxic metabolically activated carcinogens, Hattis and Barlow (1996), considering activation, detoxification and DNA repair alone, found greater PK variability with individuals at the median and the 95th percentile differing by a factor of 10. The factor was a central estimate, some chemicals exhibited greater and others lesser PK variability. In the 4-aminobiphenyl case discussed below, additional physiologic factors such as storage in the bladder contributed to human variability in PK elements.

The suggested default of 25 will have the effect of increasing the population risk (average risk) relative to the median person's risk by a factor of 6.8: For a lognormal distribution, the mean to median ratio is equal to $\exp(\sigma^2/2)$. When the 95th percentile to median ratio is 25,

σ is 1.96 [=ln(25)/1.645], and the mean exceeds the median by a factor of 6.8. If the risk to the median human were estimated to be 10^{-6}, and a population of one-million persons were exposed, the expected number of cases of cancer would be 6.8 rather than 1.0.

Thus under this new default, the value for the median person would remain as provided by the current approach to cancer risk assessment; for a default of a factor of 25, the average would be higher by a factor of 6.8. It would be important for the cancer risk assessment to express interindividual variability by showing the median and average population risks, as well as the range of individual risks for risk-management consideration.

Case Study: 4-Aminobiphenyl

4-Aminobiphenyl is a known cause of human bladder cancer. It was once used as a dye intermediate and rubber antioxidant, but its use was curtailed after findings of bladder cancer in substantial numbers of workers. Current exposures are due mostly to cigarette-smoking, which increases bladder-cancer risk by 2-10 times. The compound binds to bladder DNA and is mutagenic in a variety of test systems, including human cell culture. It has the hallmarks of low-dose linearity and is implicated as a cause of bladder cancer in smokers exposed to relatively low doses and quite recently in female never-smokers in Los Angeles County exposed to environmental tobacco smoke (Jiang et al. 2007). The compound has been extensively studied and found to have marked interindividual differences in activation and detoxification, and higher risk has been observed in slow acetylators, who detoxify it less efficiently (Gu et al. 2005, Inatomi et al. 1999). It is presented to illustrate how human interindividual variability can be addressed in dose-response assessment when reasonably high-quality data are available.

Estimating Variability in Human Susceptibility to 4-Aminobiphenyl

Bois et al. (1995) modeled interindividual heterogeneity in human cancer risk using data on differences among humans in their PK and physiologic handling of 4-aminobiphenyl. Briefly, the compound is thought to be activated via *N*-hydroxylation by CYP1A2, although recently other enzymes have also been found to be involved (Tsuneoka et al. 2003; Nakajima et al. 2006). A major detoxification pathway is *N*-acetylation. To simulate interindividual variability in pharmacokinetics, parameters describing the absorption, distribution, activation, detoxification, and urinary excretion were varied according to human ranges found in the literature. Distributions of the formation of the proximate carcinogen and its binding to urinary-bladder DNA were simulated. The latter can be used to describe possible differences in susceptibility due to physiologic and PK factors and is shown in Figure 5-18.

The DNA-binding distribution accounts for human differences only up to the point of binding and does not address PD differences. The DNA-binding distribution therefore can be considered an undercharacterization of overall human variability. The upper and lower bounds for the PK-based distributions shown in Figure 5-18 differed from the geometric mean by factors of 16 and 26, respectively. The distribution of human interindividual variability would be greater than indicated by the PK-based distributions because of PD differences among people.

For the 4-aminobiphenyl case study an estimate of interindividual variability of a range of 50 (ratio of 95th percentile to median person) is assumed for the purposes of illustrating the incorporation of variability into cancer dose-response modeling. It reflects the factor of roughly 20-30 between median and upper 95th percentile individual sensitivity in pharmacokinetics and a modest factor for variability factors pertinent to PD differences in carci-

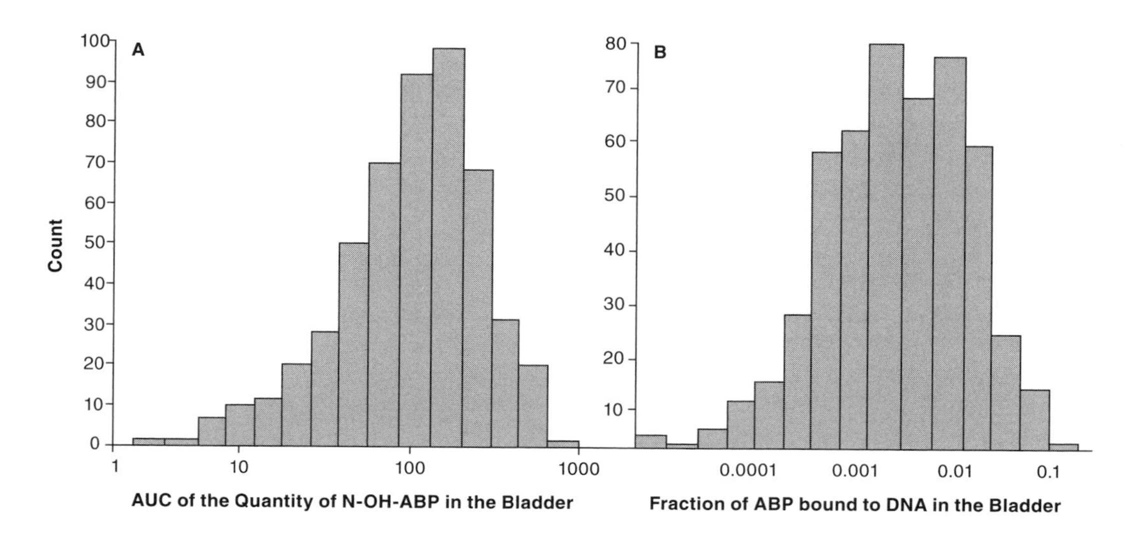

FIGURE 5-18 *A*, AUC for proximate carcinogen in bladder in units of nanograms-minutes simulated for 500 people. *B*, simulated fraction bound in bladder, presumed to indicate differences in susceptibility due to PK and physiologic parameters. Upper 95% confidence limit is factor of 16 above geometric mean of 0.0034 and factor of 26 above lower confidence limit. Source: Bois et al. 1995. Reprinted with permission; copyright 1995, *Risk Analysis*.

nogenesis. As noted above, this may be an underestimate, the range of 50, assumed in the calculations for this case study, corresponds to a geometric standard deviation of 10.8 and a standard deviation in natural log space of 2.38 ($\sigma_{\ln H}$), and in base 10 log space of 1.03.

Derivation of Median Human POD and Slope for 4-Aminobiphenyl

Despite the known causal association between human bladder-cancer risk and 4-aminobiphenyl, human exposure estimates in occupational studies may be insufficient for establishing reliable dose-response relationships, and the assessment may have to be based on animal data, as is done here. Fitting dose-response models to a sensitive site observed in the animals—liver tumors in female mice exposed by gavage—results in an ED_{10} of 0.1 mg/kg-d with a lower 95% confidence bound of 0.070 mg/kg-d. That corresponds roughly to a $\sigma_{\log\text{Animal POD}}$ of 0.09, assuming a lognormal uncertainty distribution. For cross-species extrapolation to adjust the animal POD to the human POD, doses are assumed to have equal effectiveness if the human dose is reduced consistently with three-fourths bodyweight scaling. As described above, data are available on acute and subacute toxicity in different species. Hattis et al. (2002) derived an uncertainty estimate of 0.416 for $\sigma_{\log A\rightarrow H}$ from those data, and also found that an additional small factor slightly increasing the uncertainty estimate was merited. However, for cancer end points, which result from more protracted and complex biologic processes, that value can be presumed to substantially understate the actual uncertainty. Nonetheless, it is adopted for the illustrative example here with the recognition that the estimate may be low. The median estimate for cross-adjustment scaling would be 7.3 [that is, $(70/0.025)^{(1-0.75)}$, assuming bodyweights of 70 kg and 0.025 kg for humans and mice, respectively]. Thus, the median human POD would be 0.014 mg/kg-d (0.1/7.3), and the slope of the dose-response curve at the POD would be 7.5 $(\text{mg/kg-d})^{-1}$ [$-\ln(0.9)/0.014$].

The confidence interval would take into account uncertainty resulting from fitting the dose-response model to the data, the cross-species extrapolation, and other factors. For the sake of illustration, the first two factors are accounted for here, and the resulting $\sigma_{humanPOD}$ is 0.43 $[(0.09^2 + 0.42^2)^{1/2}]$, reflecting a lower 95% bound on the median human POD of 0.003 mg/kg-d ($0.014/10^{1.645\sigma} = 0.014/5.1$) and an upper 95% bound of 38 (mg/kg-d)$^{-1}$ [(7.5)(5.1)].

Derivation of Individual and Population Dose-Response Relationships for 4-Aminobiphenyl

On the basis of the interindividual variability estimate noted above, the population and individual percentile dose-response relationships can be estimated with the uncertainty estimates for those functions. The slope of the population dose-response curve can be calculated from the risk, averaged among individuals, at a given dose. For a lognormally distributed variable, the derivation of the mean, μ, from the median involves a simple calculation [μ = (median)(exp$\{\sigma^2/2\}$), where σ is expressed in base e units and "exp$\{\sigma^2/2\}$" represents base e to the power $\{\sigma^2/2\}$]; for this case, with the human interindividual variability estimate, $\sigma_{lnH} = 2.38$, the mean potency is 126 (mg/kg-d)$^{-1}$ [(7.5)(exp$\{2.38^2/2\}$) = (7.5)(16.9)]. At low doses, population risk is calculated by multiplying the population potency by the dose. The uncertainty bound on this estimate is derived by considering the uncertainty in the adjustment factors and in the model fit at the POD.

At low doses, the risk for the 95th percentile person is given by multiplying the dose by 376 (mg/kg-d)$^{-1}$ [376 = (7.5)(50)], and the dose associated with a 10^{-5} risk for the 95th percentile sensitive person would be 3 µg/kg (that is, [14 µg/kg]/50). The uncertainty bounds around this dose estimate would be given by the human POD distribution, represented by $\sigma_{humanPOD}$. The lower confidence bound on the estimate for this person would be determined by σ as described above (for example, the 95% lower bound would be [(3 µg/kg-d)/$10^{1.645\sigma human\ POD}$ = 0.6 µg/kg-d]). This example does not capture all sources of uncertainty and is provided only to illustrate an approach.

Mathematical Framework for Conceptual Model 3

Human low-dose risk[6] from a given dose D of toxicant could be expressed as

$$Risk_H = Slope_H D = (Slope_{BMD} F_{H\text{-}A})D. \quad (1)$$

$Risk_H$ here is the incremental increase in risk above background, also called "extra risk." In current practice, $Slope_{BMD}$ is the slope[7] of the dose-response curve at the BMD. The cross-species factor, $F_{H\text{-}A}$, adjusts for differences in effect in humans compared with animals exposed to the same dose and is usually greater than 1. As discussed above, $F_{H\text{-}A}$ is typically expressed as two factors: one to account for human-animal differences in pharmacokinet-

[6]If a quantal linear-regression model is fitted and the risk over the dose-response range (π_D) can be given by $\pi_D = 1 - \exp(-\beta_0 - \beta_1 D)$. Extra risk (ER) can be defined as: $ER(D) = (\pi_D - \pi_0)/(1 - \pi_0)$. This model reduces to $ER(D) = 1 - \exp(-\beta_1 D)$. For a specified benchmark response (BMR), the BMD is defined in this model as $BMD = -\ln(1 - BMR)/\beta_1$. When the relationship between extra risk and dose is quadratic, $\pi_D = 1 - \exp(-\beta_0 - \beta_1 D - \beta_2 D^2)$.

[7]The $Slope_{BMD}$ could be defined as the slope of the line tangent to the ER(D) curve at D = BMD, that is, $Slope_{BMD} = ER(BMD) = d/dD[ER(D)]$ evaluated at D = BMD. For ER(D) defined in the context of a quantal linear model, this reduces to $Slope_{BMD} = ER'(BMD) = \beta_1 \exp(-\beta_1 BMD)$ evaluated at the estimated BMD. For simplicity and transparency, however, the following approximation can be used: $Slope_{BMD} = BMR/BMD$, which corresponds to the slope of the line connecting (BMD, BMR) and (0,0).

ics and the other to account for human-animal differences in pharmacodynamics ($F_{H\text{-}A} = F_{H\text{-}A\ PK}F_{H\text{-}A\ PD}$). It is a means of converting the animal slope (for example, in $[mg/kg\text{-}d]^{-1}$) to a human slope and can be thought of as going from the median animal to the median human. In cases in which cross-species differences in pharmacokinetics were used to derive the $Slope_{BMD}$, $F_{H\text{-}A}$ would be represented by $F_{H\text{-}A\ PD}$.

Each factor in Equation 1 may represent a model, a single number, or a distribution, depending on the nature of the data and the goal of the analysis. Variability in exposure could also be incorporated through some distribution of exposures (for example, as $D \sim G_D()$). Uncertainty in dose extrapolation or in the animal-human extrapolation could be addressed by $F_{H\text{-}A}$ as distributions (for example, $F_{H\text{-}A} \sim G_{H\text{-}A}()$). It is important to distinguish variability in risk among individuals—that is, the difference in risk from individual to individual—from uncertainty, which describes our lack of knowledge of the risk. The goal here is to enable such expressions as "The risk of effect does not exceed *x* for the *y*th percentile individual, stated with a confidence interval of *z1*- *z2*." The 4-aminobiphenyl case provided an example of how that might be done.

In some cases, as a default, it may be convenient and appropriate to describe uncertainty in $Risk_H$ mathematically with a lognormal distribution, for example, if the uncertainty in each factor in Equation 1 can be represented by a lognormal distribution. In this case, Equation 1 may be re-expressed as

$$\text{Log Risk} = \log \text{Slope}_{BMD} + \log F_{H\text{-}A} + \log D.$$

For this simplistic case,

$$\sigma^2 = \sigma_{logSLOPE}^2 + \sigma_{logF}^2 + \sigma_{logD}^2. \quad (2)$$

To the extent that σ_{logD} represents differences in exposure rather than uncertainty, it would not be incorporated as above but tracked separately to be combined with the human susceptibility distributions described below. The 4-aminobiphenyl case illustrates how variability in PK factors may lead to considerably greater risks in some people than others and how this might be taken into account quantitatively. Formally introducing human PD variability into mathematical descriptions is more challenging, and in the case example below a default distribution is assumed. The risk for the *y*th percentile person may be described by

$$\text{Risk}_{H\ yth} = \text{Slope}_{BMD}F_{H\text{-}A}DV_{H\ yth}, \quad (3)$$

where V_{Hyth} is the *y*th quantile of the distribution that describes the ratio of the *y*th percentile person to the median person. If the uncertainty in V_{Hyth} and the other elements of the uncertainty are described by a lognormal distribution, the overall uncertainty represented by σ^2 would be described by adding a term.

σ_{logV}^2, to the terms given in Equation 2.

Multiple Dose Dependent Modes of Action

The most recent EPA (2001c) dose-response assessment for chloroform, the drinking-water disinfection byproduct, assumed that sustained or repeated cytotoxicity followed by regenerative hyperplasia was probably the cause of kidney and liver cancer observed in rodent bioassays. A margin-of-exposure (MOE) approach was recommended for the evaluation of carcinogenic exposures to the compound. However, the EPASAB (2000, p. 12) noted there is "some possibility that genotoxicity could contribute to the dose-response at low doses" for the observed kidney tumors and called for the agency to address the general issue of mixed modes of action by "beginning to develop a reasonable means of estimating the most likely and upper bound estimate of potential contribution of a 'genotoxic' component to the carcinogenic activity."

Dose-response analysis of chemicals whose end points are associated with multiple MOAs is challenging. The EPA (2005b) *Guidelines for Carcinogen Risk Assessment* state (p. 3-22) that

> if there are multiple modes of action at a single tumor site, one linear and another nonlinear, then both approaches are used to decouple and consider the respective contributions of each mode of action in different dose ranges. For example, an agent can act predominantly through cytotoxicity at high doses and through mutagenicity at lower doses where cytotoxicity does not occur. Modeling to a low response level can be useful for estimating the response at doses where the high-dose mode of action would be less important.

Although that may have been the case for chloroform, the agency decided to take a low-dose nonlinear approach to characterize the risks associated with the chemical, and applied noncancer RfD methodology. In cases like that of chloroform, the slope at high doses would not give a good indication of the low-dose slope. For cases with low-dose linearity in an individual's response in which the high-dose response may be significantly influenced by a nonlinear MOA neither conceptual model 2 nor projection of low-dose risk from a high-dose BMD is satisfactory. In such cases an alternative default approach is suggested.

At low doses, the linear MOA can be expected to dominate. A modifying factor, M_S, could account for the change in slope. The adjustment factor would be based on mechanistic understanding. In this case risk (that is, "extra risk" as defined above) would be given by

$$\text{Risk}_H = [\text{Slope}_H]D = [\text{Slope}_{BMD} M_S F_{H\text{-}A}]D, \quad (4)$$

where the terms Slope_{BMD}, $F_{H\text{-}A}$, and D are as defined above for Equation 1, with Slope_{BMD} estimated as described above.

For cases like 4-aminobiphenyl, M_S would have a value of 1. For cases like chloroform, it would have a value less than 1 and would probably be the subject of controversy and debate. Nonetheless. this M_s provides a vehicle for addressing potential low-dose linearity in cases in which there is strong evidence that the slope observed at high doses overpredicts the low-dose slope.

M would serve the same purpose as the "dose and dose-rate effectiveness factor" adopted to adjust the slope of the dose-response curve observed at relatively high doses to predict radiation risk at low doses (ICRP 1991; EPA 1998; NRC 1998b; ICRP 2006; NRC 2006c; Wrixon 2008). Multiple mechanisms of toxicity may exist for a single agent, some of these mechanisms may have nonlinear-dose-response characteristics, or so-called dose-dependent transitions (Slikker et al. 2004). In considering values for M, any dose-dependent transitions would be considered in the context of background exposures and disease processes affecting these toxicity mechanisms. The selection of M would be a science policy call.

IMPLEMENTATION

The committee recognizes that the unified framework introduces additional needs for data and analyses into the risk-assessment process. The data and analyses may take time to develop, and development of an implementation strategy will be important. The committee notes that the framework can be implemented in the short term by establishing default distributions. For noncancer end points, the defaults will enable a probabilistic basis of establishing the RfD and characterizing noncancer risks; for cancer-risk characterization, they will enable incorporation of interhuman variability. Use of default distributions for adjustments in extrapolations, rather than default point-estimate uncertainty factors, provides an improved representation of variability and uncertainty and offers an opportunity for further refinements and incentives to gather and analyze existing information and to generate new data targeted to specific extrapolation needs. As experience accrues, guidelines will also be important to aid in the application of the defaults and to ensure consistency in the implementation of the framework. In the development of guidelines, the committee encourages attentiveness to issues regarding the use of defaults addressed in Chapter 6 and has concerns about the approach taken to ascertain a mutagenic MOA for genotoxic carcinogens (see discussion in the Mode-of-Action Assessment section above) in application of the guidelines to address early-life sensitivity to cancer (EPA 2005c). The committee has illustrated the ideas advocated in this chapter with conceptual models and example calculations. Assumptions and simplifications are used to make the examples tractable and clear, not to prescribe any particular approach or value.

Table 5-1 summarizes major aspects of the unified framework in terms of data needs, potential utility of defaults as interim placeholders for better-researched and better-defined distributions, and implementation. A number of other sources of uncertainty and variability that often arise in dose-response assessment are not peculiar to the proposed unified framework and so are not addressed in the table; some of these issues and their associated default approaches are described in Chapter 6.

An implementation plan can be devised to phase in the unified framework. Some considerations and suggestions for developing the plan are presented in Table 5-1. Default distributions can initially be based on datasets that can be augmented with adjustments or other distributional assumptions to account for inferences that generalize from small numbers of people, of chemical case studies, and of end points to large populations, numbers of chemicals, and numbers of effects. As more data are collected and variability is better understood, the uncertainty portion of the default distribution may decrease. Emerging technologies, such as toxicogenomics and high-throughput assays, will highlight pathways that are at the crossroads of disease causation and toxicant action and will assist in the incorporation of background additivity and variability components. The implementation plan should be associated with a research agenda that will, over time, enable refinement of distributional approaches to dose-response assessment. Finally, EPA guidance will clearly be needed in order to implement the unified framework, including conduct of the background exposure and vulnerability assessments, departure from the linear default, establishment of distributions for the analysis, model selection, and so on. The development and roll out of guidelines and policies will be an essential component of any implementation plan, as well as ample opportunity for stakeholder involvement, scientific peer review and mid-course correction to address false starts and mis-steps.

TABLE 5-1 Potential Approaches to Establish Defaults to Implement the Unified Framework for Dose-Response Assessment

Analytic Step	Data Need	Testing and Implementation Issues	Potential Approach for Establishing Defaults in the Near Term
Cross-species extrapolation	Relative sensitivity to toxicants, comparing rodent with human	Moving from default point estimate to distribution adds complexity and encounters data limitations; literature on acute and subacute effects and direct-acting drugs is used mostly in comparisons, and small numbers of people studied may not be representative of human population	Base default distribution on wide sample of drugs and toxicants for which there are data on rodents and humans (drug trials, clinical toxicologic and epidemiologic studies) for similar end points and on adjustments to address data gaps; look to specify distributions to particular classes of chemicals and comparisons of particular rodents (mouse vs rat vs hamster) with humans; consider using bodyweight scaling for PK portion of extrapolation if overall distribution covering PK and PD cannot be derived; develop default distribution to describe uncertainty in bodyweight scaling
Interindividual PK variability in humans	PK differences among life stages, disease states, genetic polymorphisms, drug interactions	PK datasets on susceptible groups (such as children) are difficult to obtain; default may have to be based primarily on drug literature, which is also limited	Derive default distribution of PK variability based on analogy with drug literature and, to extent possible, made spceific to particular enzyme pathways, types of receptors, and classes of chemicals; use PBPK Bayesian and Monte Carlo approaches to evaluate implications of variability in enzyme pathways for overall PK variability; consider adjustments to address small samples and other biases in derivation
Interindividual PD variability in humans	PD differences in population with respect to various types of end points, including cancer	Human PD response is likely to vary widely, especially in groups that are difficult to study (such as children, elderly); it is unclear how to consider and integrate clinical, precursor, and other upstream end points and how to separate PK from PD variability	Base default distribution on broad array of human responses, chemicals, and end points from drug testing and high-quality epidemiologic studies; use information on background exposures and vulnerabilities to develop default; develop distributions specific to chemical classes, end points (such as cancer, endocrine, and acute toxicity), and humans to extent possible; consider adjustments to address small samples and other biases in derivation

TABLE 5-1 Continued

Analytic Step	Data Need	Testing and Implementation Issues	Potential Approach for Establishing Defaults in the Near Term
Background exposures	Low-dose interaction studies for chemicals with similar MOA	Human population has numerous background exposures; MOAs are difficult to define; when they are defined, interaction with other chemicals can be difficult to predict at different doses and dose ratios and in different species, ages, and organs; mechanistic and interaction data are limited	Develop guidance to judge whether background exposures (and vulnerability) are sufficiently unimportant to reject linearity at low doses; when it is rejected, use probabilistic approach to develop RfD, using interindividual variability and other distributions
Background vulnerability[1]	Sensitive epidemiologic and mechanistic studies relating chemical exposures and disease processes; biomonitoring data	Human population has numerous degenerative and disease processes; it is difficult to sort relevance to particular MOA; data on chemical-disease interaction are insufficient	Establish guidance to judge whether background vulnerability, conditions, and exposures are sufficiently unimportant to reject linearity at low doses; when it is rejected, use probabilistic approach to develop RfD, using interindividual variability and other distributions
Low-dose extrapolation defaults	MOA information defining chemical effect at target and interaction with background processes	It is difficult to obtain low-dose data in relevant test systems; chemicals can have mixed MOAs; different models can fit high-dose data equally but differ at low dose	Continue assumption that carcinogens have low-dose linear response unless sufficient data support other approaches; develop guidance for noncancer low-dose response and linear extrapolation due to background additivity and vulnerability (conceptual model 1); formally adopt assumption that genotoxic chemicals (clastogens, mutagens) cause cancer via a mutagenic MOA
Low-dose linear slope factor—M adjustment	Dose-response data over wide dose ranges in human and animal studies and related mechanistic data	Data from epidemiologic and toxicologic studies are limited; there is need to know how to use biologic models in considering mechanistic data	Develop series of default M factors based on mechanistic considerations and human and animal observations to apply in different situations (such as saturation phenomena or high-dose cytotoxicity that influences carcinogenicity of chemicals with some genotoxic activity)

[a]Susceptibility to endogenous (for example, age, gender, genetics, pre-existing health deficits and disease) and exogenous factors (exposure to agents) and due to variability in exposure.

CONCLUSIONS AND RECOMMENDATIONS

Conclusions

This chapter reviews the current paradigm for characterizing the dose-response relationships of compounds for both cancer and noncancer end points and supports the following conclusions:

- Separation of cancer and noncancer outcomes in dose-response analysis is artificial because noncancer end points can occur without a threshold or low-dose nonlinearity on the population level and in some cases on the individual level. Similarly, the MOA for carcinogens varies and requires a flexible but consistent analytic framework. The separation not only is scientifically unjustified but leads to undesirable risk-management outcomes, including inadequate attention to noncancer end points, especially in benefit-cost analyses.
- The current formulation of the RfD is problematic because of its application as a determinant of risk vs no risk of regulatory importance, and it lacks a quantitative description of the risk at different doses. It hinders risk-risk and risk-benefit comparisons and risk-management decision-making and does not make the best possible use of available scientific evidence.
- Cancer risk assessment typically lacks a quantitative description of interindividual variability. That leads to an incomplete description of the range of risk possible in the population. Noncancer risk assessment addresses interindividual variability, but cancer risk assessment typically does not; this reflects the implicit default assumption that human cancer susceptibility does not vary (see Chapters 4 and 6). The argument that the linear dose-response extrapolation procedure covers the omission (EPA 2005b) is unsupported and presents a separate consideration that should not be confused with the need to describe risk differences among individuals in addition to high-dose–low-dose extrapolation. The approach adopted in the current carcinogen guidelines (EPA 2005b) that considers variability only when a sensitive subpopulation can be identified for a particular chemical is limited by a lack of chemical-specific data. It also ignores the appreciable scientific knowledge of human interindividual variability in sensitivity (see, for example, Table 4-1), which can form the basis of general assumptions regarding variability when chemical-specific data are absent. The supplemental guidance regarding children (EPA 2005c) is an important step in the right direction, but variability in the general population should also be addressed.
- Uncertainty factors are generally used to make adjustments whose accuracy is unknown. The uncertainty factors comprise elements of the adjustment for uncertainty and variability. The default factors should be replaced with distributions that separate the elements transparently. Default distributions that characterize PK and PD variability, cross-species dose adjustments, and adjustments for the lack of sensitive studies will be needed as starting points that can be improved as the research advances.
- The committee considers that the term *safety factor*, to characterize uncertainty factors in noncancer risk assessments, is inappropriate and misleading. The term *uncertainty factor* is also inappropriate as it does not reflect the variability and adjustment elements that the factor represents.
- The underlying scientific and risk-management considerations point to the need for unification of cancer and noncancer approaches in which chemicals are put into a common analytic framework regardless of type of outcome. There are core differences among end points, but in this analytic framework a dose corresponding to a specified increase in risk in the population could be derived for both cancer and noncancer end points, and this would

add transparency and quantitative insight to risk-management decisions. Among other changes, this would involve a redefinition of the RfD. The committee acknowledges that the risk estimates and risk specific RfDs derived from this methodology will often be uncertain. This would nonetheless be an improvement over the RfDs derived from the traditional BMD and uncertainty factor approach. The results are more transparent, presenting variability and uncertainty, and are more amenable to refinements as better data are obtained. Further, quantification of risk (along with the attendant uncertainty) not only at the RfD but along the dose continuum is an important advance for risk benefit analysis.

- The committee finds that a common analytic framework best reflects the underlying science. The main elements of this framework are shown in Figure 5-8 and include

  – Systematic assessments of the MOAs, vulnerable populations, and background exposures and disease processes that may affect a chemical's human dose-response relationships and human vulnerability. This includes an evaluation of the potential background exposures and processes (for example, damage and repair processes, disease, and aging) that interact with a chemical's MOAs and thus contribute to variability in and vulnerability to the toxicant response and that can result in a population dose-response relationship that is linear at low doses.

  – Selection of a conceptual model for individual and population dose-response relationships. The following three are described in the chapter:

    i. Low-dose nonlinear individual response, low-dose linear population response with background dependence.

    ii. Low-dose nonlinear individual and population response independent of background.

    iii. Low-dose linear individual and population response.

  – Selection of a conceptual model and dose-response method that best reflects MOA and background considerations and the form of risk characterization needed for risk management. Where feasible, methods that result in quantitative descriptions of risk and uncertainty should be selected.

- The key advantages of the framework are

  – Risk descriptors that are quantitative and probabilistic. The RfD would be redefined as a risk-specific dose (for example, the dose associated with a 1 in 100,000 risk of a particular end point), and the risk could be estimated at doses above and below the RfD. This would allow all end points to be more formally incorporated into risk-tradeoffs and benefit-cost analyses.

  – Characterization of variability and uncertainty for critical end points. This would address concerns about population heterogeneity in risk and inform value-of-information and other priority-setting analyses that require quantitative uncertainty estimates. The sources of variability and uncertainty and their quantitative contributions in the derivation of risk estimates would be more transparent. This would in turn enable the quantitative characterization of uncertainties in such benefits.

  – A means to quantitatively describe health benefits from changes in exposure. This would enable the direct comparisons of costs of these changes with the benefits accruing from them.

  – The basis for more flexibility in decision-making. The risk manager can use the risk specific RfD in the same manner the current RfD is used in regulatory decision-making. However, additional quantitative risk information can accompany the RfD, including risk and uncertainty estimates above and below the RfD. This will enable a more robust consideration of options and trade-offs in risk-risk and risk-benefit analyses.

- The key disadvantages of the framework are

  – The need for increased analysis to consider in detail the background factors that may add to the exposure in question and that may contribute to variability. This can increase the complexity of the analysis and pose a challenge for communicating the analysis and its results. Training will be needed for both risk assessors and risk managers. The agency has already included some elements of the framework in a few assessments (for example, EPA 2001b; EPA 2004), and explored other elements in case studies (for example, Axelrad et al. 2005; Woodruff et al. 2007). EPA laboratories also conduct research that is supportive of the characterizations envisioned by the committee. Thus, EPA has internal capacity for the development of these methods. Realizing full use will take further development and staff training. The risk assessment community external to the agency provides several examples that are cited above and is also a resource for developing further cases and expanding the methodology. The agency also has considerable expertise translating risk information and using it in decision-making. Approaches currently used in risk management may have to be adapted to make full use of the new information and risk managers may need to be trained on how to best use the new and different risk characterizations.

  – Because of the limitations of data on which some elements of the framework would be built, this necessarily entails development of defaults. Depending on the level of analysis, that would provide incentives for chemical-specific information on background exposures, interaction with baseline aging and disease processes, and interindividual variability. It comes at a time when toxicology and risk-assessment resources are already challenged by the expanding role of risk assessment in decision-making and the lack of basic toxicology information on many chemicals. However, it also comes at a time of rapid scientific and technologic innovation in the biologic sciences and testing that can be developed to support novel and improved approaches (NRC 2006d, 2007a,b).

- Establishing reasonable and scientifically supported default approaches (such as linear extrapolation to low dose for chemicals that are subject to background additivity) and default distributions (such as interindividual variability) to implement the framework will encourage research and a healthy discussion of the science that underpins risk assessment. The resulting default approaches are part of the anticipated advances in the use of defaults in risk assessment described in Chapter 6. The process of establishing the defaults will bring about a better understanding of how chemical-specific information should be used to inform toxicity assessment and low-dose extrapolation.

Recommendations

The committee has divided its recommendations on the unified framework into short- and long- term recommendations. If the short term recommendations are implemented, the committee envisions significant progress in the next 2-5 years. The time horizon for substantial progress for the long term recommendations is further out, 10-20 years.

Short-Term Recommendations

- The committee recommends the phase-in of the unified framework for dose-response assessment as new chemicals are assessed or old ones are reassessed for Integrated Risk Information System or program offices or incorporated in comparative or cost-benefit analyses. The initial test cases should be used as a proof of concept. The committee recommends a flexible approach in which different conceptual models can be applied in the unified framework, as illustrated by the three conceptual models presented in this chapter. This approach would involve

– Incorporation of probabilistic and distributional methods into dose-response analysis for agents believed to have low-dose nonlinear responses and the later redefinition of the RfD on the basis of the probabilistic description.

– Evaluation of each chemical in terms of MOA, background exposure and disease processes, and vulnerable populations. This would add a step to the dose-response analysis in which background exposures and vulnerabilities of the target population are analyzed and used to decide between analytic options based on conceptual models, according to the unified framework outlined in Figure 5-8.

– Incorporation of background additivity to account for

  ○ Additional sources of exposure to the same chemical or to similarly acting chemicals (including endogenous sources).

  ○ Chemical MOA interaction with relevant disease or aging processes that lead to a background vulnerability distribution.

– Development of defaults and guidance for assessing the MOA, background exposure and disease processes, and vulnerable populations, and selection of conceptual model. The committee recommends that cancer and noncancer responses be assumed to be linear as a default. An alternative analytic option (conceptual model 2) is available for cases in which it can be shown that background is unlikely to be an important contributor to risk, according to the recommended evaluation of MOAs and background.

– Formal introduction of human variability into cancer dose-response modeling and risk characterization. This will require chemical-specific distributions or the use of default variability distributions. The committee recommends that as the distributions are being developed, EPA use a default for interindividual variability that assumes a lognormal distribution and immediately begin to explicitly address human variability in cancer response estimates. A reasonable assumption would be that the 95% upper-bound person is about 10-50 times as sensitive as the median person.

- The committee recommends that EPA develop case studies to explore the use of the new unified framework. The goal of the case studies would not be simply to compare the results of the current approach and new framework. Rather, the case studies would be used to explore and gain experience with the framework in the MOA, vulnerability, and background assessments; using improved information on variability (for example, genetic polymorphisms, disease, and aging-related vulnerabilities) and coexposures in RfD derivation; incorporating variability into cancer risk analysis; and quantitative uncertainty characterizations of dose-response relationships.
- The committee recommends that EPA gather information from epidemiology, the pharmaceutical literature, and clinical toxicology and use it to develop default interhuman variability PK and PD distributions. Some possible approaches are outlined in Table 5-1.
- The committee recommends that the agency develop default-adjustment distributions that quantitatively characterize the adjustments and key uncertainties typical in dose-response assessment, including cross-species extrapolation in PK and PD and extrapolations among dose route, dosing intervals (for example, subchronic to chronic), and data gaps. Some possible approaches are outlined in Table 5-1. Maximum use of existing human datasets is encouraged. Studies with well-defined exposure information, such as biomarker measurements on individuals, could be examined to understand the heterogeneity in response. Such datasets could be used to build variability distributions that may be applicable to sets of chemicals (with similar structure, MOA, target sites, and effects) and increase understanding of interhuman PD variability.

- The agency should develop formal guidance for dose-response analysis under the unified framework. For example, guidance will be needed for the conduct of background vulnerability and exposure assessments, MOA evaluation, default dose-response modeling, nondefault chemical specific analyses.
- The committee recommends as default distributions are developed for the different adjustments used in dose-response assessment, they should be assigned accurate labels (such as *human variability distribution*). This should lessen the opportunity for transferring to the new default distributions the misunderstanding commonly associated with use of the term *uncertainty factor*.
- Over the next 5 years, the committee recommends that EPA further develop the issue of vulnerability by gathering data and developing a broad array of human-vulnerability information from the biomedical literature, focusing on diseases that are likely to interact with the MOAs of prevalent-exposure and high-priority chemicals (for example, pulmonary, cardiovascular, hepatic, and renal diseases and various cancers). This could involve working with clinicians, biochemists, epidemiologists, and other biomedical specialists to develop preclinical-disease biomarkers as upstream indicators of vulnerability to toxicant MOAs.

Long-Term Recommendations

- The committee recommends that EPA expand its research on the issues of vulnerability and susceptibility. The agency could conduct studies itself and coordinate with other agencies for more in-depth research on the determinants of vulnerability and the development of approaches for more accurate consideration of vulnerability in agency assessments. This could involve using epidemiologic studies to explore how the response to toxicants may be affected by pre-existing diseases and vulnerabilities in the population. Biomarkers of vulnerability and effect could be developed for applications as predictive screens in exposed populations. When analyzed with exposure biomarkers, they could be used to assess human exposure-response relationships and interindividual variability. Regional and national datasets, such as those from National Health and Nutrition Examination Surveys and environmental and public-health tracking, could be used to evaluate whether people with background vulnerability or background exposure are at increased risk of the effects of exposure to toxicants. This work could lead to vulnerability distributions for use in dose-response assessment. Pharmacogenetic and polymorphism probes could be incorporated into epidemiologic studies to explore key interindividual susceptibility factors and their frequency in the population. Animal models, such as genetically modified knockout mice, could be used to define the functional importance of particular genes and their polymorphisms in determining risk.
- The committee recommends computational research that applies systems-biology techniques to analyze how -omics end points might inform the development of distributions outlined in Table 5-1. For example, analyzing data from high-throughput screens with genomics end points may result in interpretable upstream indicators of disease vulnerability. The biochemical processes that lead to pathologic conditions or functional loss could be described by continuous parameters that may be suitable as disease biomarkers in the population. These approaches could also provide interpretable biochemical end points reflective of key steps in a toxicant's MOA.
- The committee recommends exploration into interactions of exposures to chemicals that have similar or different MOAs but affect the same toxicologic process. Such research should improve understanding of issues related to background additivity. The research would also affect approaches to mixtures and combined exposures and to the question of whether it

is more appropriate to assume effect additivity (now assumed in noncancer risk assessment), dose additivity, or some other characteristic in a given risk-assessment circumstance.

REFERENCES

Angelosanto, F.A., G.R. Blackburn, C.A. Schreiner, and C.R. Mackerer. 1996. Benzene induces a dose-responsive increase in the frequency of micronucleated cells in rat Zymbal glands. Environ. Health Perspect. 104(Suppl. 6):1331-1336.

Autrup, H., B. Daneshvar, L.O. Dragsted, M. Gamborg, A.M. Hansen, S. Loft, H. Okkels, F. Nielsen, P.S. Nielsen, E. Raffn, H. Wallin, and L.E. Knudsen. 1999. Biomarkers for exposure to ambient air pollution—comparison of carcinogen-DNA adduct levels with other exposure markers and markers for oxidative stress. Environ. Health Perspect. 107(3):233-238.

Axelrad, D.A., K. Baetcke, C. Dockins, C.W. Griffiths, R.N. Hill, P.A. Murphy, N. Owens, N.B. Simon, and L.K. Teuschler. 2005. Risk assessment for benefits analysis: Framework for analysis of a thyroid-disrupting chemical. J. Toxicol. Environ. Health A. 68(11-12):837-855.

Axelrad, D.A., D.C. Bellinger, L.M. Ryan, and T.J. Woodruff. 2007. Dose-response relationship of prenatal mercury exposure and IQ: An integrative analysis of epidemiologic data. Environ. Health Perspect. 115(4):609-615.

Baird, S.J.S., J.T. Cohen, J.D. Graham, A.I. Shiyakhter, and J.S. Evans. 1996. Non-cancer risk assessment: A probabilistic alternative to current practice. Hum. Ecol. Risk Asses. 2(1):79-102.

Bannasch, P. 2003. Comments on 'R. Karbe and R. L. Kerlin. (2002). Cystic Degeneration/Spongiosis Hepatis [Toxicol. Pathol. 30(2):216-227].' Toxicol. Pathol. 31(5):566-570.

Bois, F.Y., G. Krowech, and L. Zeise. 1995. Modeling human interindividual variability in metabolism and risk: The example of 4-aminobiphenyl. Risk Anal. 15(2):205-213.

Boutet, K, J.L. Malo, H. Ghezzo, and D. Gautrin. 2007. Airway hyperresponsiveness and risk of chest symptoms in an occupational model. Thorax 62(3):260-264.

Brouwer, A., M.P. Longnecker, L.S. Birnbaum, J. Cogliano, P. Kostyniak, J. Moore, S. Schantz, and G. Winneke. 1999. Characterization of potential endocrine-related health effects at low-dose levels of exposure to PCBs. Environ. Health Perspect. 107(Suppl. 4):639-649.

CDC (Centers for Disease Control and Prevention). 2007. Asthma: Asthma's Impact on Children and Adolescents. U.S. Department of Health and Human Services, Centers for Disease Control and Prevention, Washington, DC [online]. Available: http://www.cdc.gov/asthma/children.htm [accessed Sept. 14, 2007].

Crofton, K.M., E.S. Craft, J.M. Hedge, C. Gennings, J.E. Simmons, R.A. Carchman, W.H. Carter Jr., and M.J. DeVito. 2005. Thyroid-hormone-disrupting chemicals: Evidence for dose-dependent additivity or synergism. Environ. Health Perspect. 113(11):1549-1554.

Crouch, E., and R. Wilson. 1979. Interspecies comparison of carcinogenic potency. J. Toxicol. Environ. Health 5(6):1095-1118.

Crump, K.S., D.G. Hoel, C.H. Langley, and R. Peto. 1976. Fundamental carcinogenic processes and their implications for low dose risk assessment. Cancer Res. 36(9 Pt. 1):2973-2979.

Daniels, M.J., F. Dominici, J.M. Samet, and S.L. Zeger. 2000. Estimating particulate matter-mortality dose-response curves and threshold levels: An analysis of daily time-series for the 20 largest U.S. cities. Am. J. Epidemiol. 152(5):397-406.

de Meer, G., D.S. Postma, and D. Heederik. 2003. Bronchial responsiveness to adenosine-5'-monophosphate and methacholine as predictors for nasal symptoms due to newly introduced allergens: A follow-up study among laboratory animal workers and bakery apprentices. Clin. Exp. Allergy 33(6):789-794.

Demchuk, E., B. Yucesoy, V.J. Johnson, M. Andrew, A. Weston, D.R. Germolec, C.T. De Rosa, and M.I. Luster. 2007. A statistical model for assessing genetic susceptibility as a risk factor in multifactorial diseases: Lessons from occupational asthma. Environ. Health Perspect. 115(2):231-234.

Dominici, F., A. McDermott, S.L. Zeger, and J.M. Samet. 2003. National maps of the effects of particulate matter on mortality: Exploring geographical variation. Environ. Health Perspect. 111(1):39-43.

Dubowsky, S.D., H. Suh, J. Schwartz, B.A. Coull, and D.R. Gold. 2006. Diabetes, obesity, and hypertension may enhance associations between air pollution and markers of systemic inflammation. Environ. Health Perspect. 114(7):992-998.

EPA (U.S. Environmental Protection Agency). 1993. Alachlor (CASRN 15972-60-8). Integrated Risk Information System, U.S. Environmental Protection Agency, Washington, DC [online]. Available: http://www.epa.gov/iris/subst/0129.htm [accessed Sept. 12, 2007].

EPA (U.S. Environmental Protection Agency). 1998. Health Risks from Low-Level Environmental Exposure to Radionuclides. Federal Guidance Report No.13 -Part 1, Interim Version. EPA 402-R-97-014. Office of Radiation and Indoor Air, U.S. Environmental Protection Agency, Washington, DC. January 1998 [online]. Available: http://homer.ornl.gov/VLAB/fedguide13.pdf [accessed Aug. 7, 2008].

EPA (U.S. Environmental Protection Agency). 2000a. Benchmark Dose Technical Guidance Document. EPA/630/R-00/001. Risk Assessment Forum, U.S. Environmental Protection Agency, Washington, DC. October 2000 [online]. Available: http://www.epa.gov/ncea/pdfs/bmds/BMD-External_10_13_2000.pdf [accessed Jan. 4, 2008].

EPA (U.S. Environmental Protection Agency). 2000b. Risk Characterization Handbook. EPA 100-B-00-002. Office of Science Policy, Office of Research and Development, U.S. Environmental Protection Agency, Washington, DC. December 2000 [online]. Available: http://www.epa.gov/OSA/spc/pdfs/rchandbk.pdf [accessed Jan. 7, 2008].

EPA (U.S. Environmental Protection Agency). 2001a. Trichloroethylene Health Risk Assessment: Synthesis and Characterization. External Review Draft. EPA/600/P-01/002A. Office of Research and Development, Washington, DC. August 2001 [online]. Available: http://rais.ornl.gov/tox/TCEAUG2001.PDF [accessed Aug. 2, 2008].

EPA (U.S. Environmental Protection Agency). 2001b. Water Quality Criterion for the Protection of Human Health: Methylmercury, Chapter 4: Risk Assessment for Methylmercury. EPA-823-R-01-001. Office of Science and Technology, Office of Water, U.S. Environmental Protection Agency, Washington, DC. January 2001 [online]. Available: http://www.epa.gov/waterscience/criteria/methylmercury/pdf/merc45.pdf [accessed Aug. 6, 2008].

EPA (U.S. Environmental Protection Agency). 2001c. Chloroform (CASRN 67-66-3). Integrated Risk Information System, U.S. Environmental Protection Agency, Washington, DC [online]. Available: http://www.epa.gov/iris/subst/0025.htm [accessed Sept. 12, 2007].

EPA (U.S. Environmental Protection Agency). 2002a. A Review of the Reference Dose and Reference Concentration Processes. EPA/630/P-02/002F. Risk Assessment Forum, U.S. Environmental Protection Agency, Washington, DC. December 2002 [online]. Available: http://cfpub.epa.gov/ncea/cfm/recordisplay.cfm?deid=55365 [accessed Jan. 4, 2008].

EPA (U.S. Environmental Protection Agency). 2002b. Guidance on Cumulative Risk Assessment of Pesticide Chemicals That Have a Common Mechanism of Toxicity. Office of Pesticide Programs, U.S. Environmental Protection Agency, Washington, DC. January 14, 2002 [online]. Available: http://www.epa.gov/pesticides/trac/science/cumulative_guidance.pdf [accessed Jan. 7, 2008].

EPA (U.S. Environmental Protection Agency). 2004. Exposure and Human Health Reassessment of 2,3,7,8-Tetrachlorodibenzo-p-Dioxin (TCDD) and Related Compounds, NAS Review Draft. EPA/600/P-00/001Cb. National Center for Environmental Assessment, Office of Research and Development, U.S. Environmental Protection Agency, Washington, DC [online]. Available: http://cfpub.epa.gov/ncea/cfm/recordisplay.cfm?deid=87843 [accessed Aug. 6, 2008].

EPA (U.S. Environmental Protection Agency). 2005a. Approaches for the Application of Physiologically Based Pharmacokinetic (PBPK) Models and Supporting Data in Risk Assessment. External Review Draft. EPA/600/R-05/043A. National Center for Environmental Assessment, Office of Research and Development, U.S. Environmental Protection Agency, Washington, DC. June 2005 [online]. Available: http://cfpub.epa.gov/ncea/cfm/recordisplay.cfm?deid=135427 [accessed Jan. 22, 2008].

EPA (U.S. Environmental Protection Agency). 2005b. Guidelines for Carcinogen Risk Assessment. EPA/630/P-03/001F. Risk Assessment Forum, U.S. Environmental Protection Agency, Washington, DC. March 2005 [online]. Available: http://cfpub.epa.gov/ncea/cfm/recordisplay.cfm?deid=116283 [accessed Feb. 7, 2007].

EPA (U.S. Environmental Protection Agency). 2005c. Supplemental Guidance for Assessing Susceptibility for Early-Life Exposures to Carcinogens. EPA/630/R-03/003F. Risk Assessment Forum, U.S. Environmental Protection Agency, Washington, DC. March 2005 [ónline]. Available: http://cfpub.epa.gov/ncea/cfm/recordisplay.cfm?deid=160003 [accessed Jan. 4, 2008].

EPA (U.S. Environmental Protection Agency). 2005d. Toxicological Review of Phosgene (CAS 75-44-5) In Support of Summary Information on the Integrated Risk Information System (IRIS). EPA/635/R-06/001. U.S. Environmental Protection Agency, Washington, DC. December 2005 [online]. Available: http://www.epa.gov/iris/toxreviews/0487-tr.pdf [accessed Aug. 7, 2008].

EPA (U.S. Environmental Protection Agency). 2006a. Background on Risk Characterization. Technology Transfer Network, 1999 National-Scale Air Toxics Assessment, U.S. Environmental Protection Agency, Washington, DC [online]. Available: http://www.epa.gov/ttn/atw/nata1999/riskbg.html [accessed Sept. 10, 2007].

EPA (U.S. Environmental Protection Agency). 2006b. Harmonization in Interspecies Extrapolation: Use of BW¾ as Default Method in Derivation of the Oral RfD. External Review Draft. EPA/630/R-06/001. Risk Assessment Forum Technical Panel, U.S. Environmental Protection Agency, Washington, DC. February 2006 [online]. Available: http://cfpub.epa.gov/si/si_public_record_Report.cfm?dirEntryID=148525 [accessed Aug. 7, 2008].

EPA (U.S. Environmental Protection Agency). 2006c. Toxicological Review of Dibutyl Phthalate (Di-n-Phthalate) (CAS No. 84-74-2) In Support of Summary Information on the Integrated Risk Information System (IRIS). NCEA-S-1755. U.S. Environmental Protection Agency, Washington, DC [online]. Available: http://cfpub.epa.gov/ncea/cfm/recordisplay.cfm?deid=155707 [accessed Dec. 10, 2007].

EPA SAB (U.S. Environmental Protection Agency Science Advisory Board). 2000. Review of the EPA'S Draft Chloroform Risk Assessment. EPA-SAB-EC-00-009. Science Advisory Board, U.S. Environmental Protection Agency, Washington, DC. April 2000 [online]. Available: http://yosemite.epa.gov/sab/sabproduct.nsf/D0E41CF58569B1618525719B0064BC3A/$File/ec0009.pdf [accessed Jan. 22, 2008].

EPA SAB (U.S. Environmental Protection Agency Science Advisory Board). 2002. Workshop on the Benefits of Reductions in Exposure to Hazardous Air Pollutants: Developing Best Estimates of Dose-Response Functions. EPA-SAB-EC-WKSHP-02-001. Science Advisory Board, U.S. Environmental Protection Agency, Washington, DC. January 2002 [online]. Available: http://yosemite.epa.gov/sab/sabproduct.nsf/34355712EC011A358525719A005BF6F6/$File/ecwkshp02001+appa-g.pdf [accessed Jan. 22, 2008].

Evans, J.S., and S.J.S. Baird. 1998. Accounting for missing data in noncancer risk assessment. Hum. Ecol. Risk Assess. 4(2):291-317.

Evans, J.S., L.R. Rhomberg, P.L. Williams, A.M. Wilson, and S.J.S. Baird. 2001. Reproductive and developmental risks from ethylene oxide: A probabilistic characterization of possible regulatory thresholds. Risk Anal. 21(4):697-717.

Freireich, E.J., E.A. Gehan, D.P. Rall, L.H. Schmidt, and H.E. Skipper. 1966. Quantitative comparison of toxicity of anti-cancer agents in mouse, rat, hamster, dog, monkey, and man. Cancer Chemother. Rep. 50(4):219-244.

Gaylor, D., L. Ryan, D. Krewski, and Y. Zhu. 1998. Procedures for calculating benchmark doses for health risk assessment. Regul. Toxicol. Pharmacol. 28(2):150-164.

Gaylor, D.W., R.L. Kodell, J.J. Chen, and D. Krewski. 1999. A unified approach to risk assessment for cancer and noncancer endpoints based on benchmark doses and uncertainty/safety factors. Regul. Toxicol. Pharmacol. 29(2 Pt 1):151-157.

Ginsberg, G., D. Hattis, B. Sonawane, A. Russ, P. Banati, M. Kozlak, S. Smolenski, and R. Goble. 2002. Evaluation of child/adult pharmacokinetic differences from a database derived from the therapeutic drug literature. Toxicol. Sci. 66(2):185-200.

Gray, L.E., J. Ostby, J. Furr, C.J. Wolf, C. Lambright, L. Parks, D.N. Veeramachaneni, V. Wilson, M. Price, A. Hotchkiss, E. Orlando, and L. Guillette. 2001. Effects of environmental antiandrogens on reproductive development in experimental animals. Hum. Reprod. Update 7(3):248-264.

Gu, J., D. Liang, Y. Wang, C. Lu, X. Wu. 2005. Effects of N-acetyl transferase 1 and 2 polymorphisms on bladder cancer risk in Caucasians. Mutat Res. 581(1-2):97-104.

Hass, U., M. Scholze, S. Christiansen, M. Dalgaard, A.M. Vinggaard, M. Axelstad, S.B. Metzdorff, and A. Kortenkamp. 2007. Combined exposure to anti-androgens exacerbates disruption of sexual differentiation in the rat. Environ. Health Perspect. 115(Suupl.1):122-128.

Hattis, D. 2008. Distributional analyses for children's inhalation risk assessments. J. Toxicol. Environ. Health Part A 71(3):218-226.

Hattis, D., and K. Barlow. 1996. Human interindividual variability in cancer risks -technical and management challenges. Hum. Ecol. Risk Assess. 2(1):194-220.

Hattis, D., and M.K. Lynch. 2007. Empirically observed distributions of pharmacokinetic and pharmacodynamic variability in humans: Implications for the derivation of single point component uncertainty factors providing equivalent protection as existing RfDs. Pp. 69-93 in Toxicokinetics and Risk Assessment, J.C. Lipscomb, and E.V. Ohanian, eds. New York: Informa Healthcare.

Hattis, D., A. Russ, R. Goble, P. Banati, and M. Chu. 2001. Human interindividual variability in susceptibility to airborne particles. Risk Anal. 21(4):585-599.

Hattis, D., S. Baird, and R. Goble. 2002. A straw man proposal for a quantitative definition of the RfD. Drug Chem. Toxicol. 25(4):403-436.

Hattis, D., G. Ginsberg, B. Sonawane, S. Smolenski, A. Russ, M. Kozlak, and R. Goble. 2003. Differences in pharmacokinetics between children and adults. II. Children's variability in drug elimination half-lives and in some parameters needed for physiologically-based pharmacokinetic modeling. Risk Anal. 23(1):117-142.

Hsiao, T.J., J.D. Wang, P.M. Yang, P.C. Yang, and T.J. Cheng. 2004. Liver fibrosis in asymptomatic polyvinyl chloride workers. J. Occup. Environ. Med. 46(9):962-966.

IARC (International Agency for Research on Cancer). 2000. Di(2-ethylhexyl)phthalate. Pp. 41-148 in Some Industrial Chemicals. IARC Monographs on the Evaluation of Carcinogenic Risks to Humans Vol. 77. Lyon: IARC.

IARC (International Agency for Research on Cancer). 2006. Preamble. IARC Monographs on the Evaluation of Carcinogenic Risks to Humans. International Agency for Research on Cancer, World Health Organization, Lyon [online]. Available: http://monographs.iarc.fr/ENG/Preamble/CurrentPreamble.pdf [accessed Aug. 6, 2008].

ICRP (International Commission on Radiological Protection). 1991. 1990 Recommendations of the International Commission on Radiological Protection. ICRP Publication 60. Annals of the ICPR 21(1/3). New York: Pergamon.

ICRP (International Commission on Radiological Protection). 2006. Low Dose Extrapolation of Radiation-Related Cancer Risk. ICRP Publication 99. Annals of the ICPR 35(4). Oxford: Elsevier.

Inatomi, H., T. Katoh, T. Kawamoto, and T. Matsumoto. 1999. NAT2 gene polymorphism as a possible marker for susceptibility to bladder cancer in Japanese. Int. J. Urol. 6(9):446-454.

IPCS (International Programme for Chemical Safety). 2005. Guidance for the use of data in development of chemical-specific adjustment factors for interspecies differences and human variability. Pp. 25-48 in Chemical-Specific Adjustment Factors for Interspecies Differences and Human Variability: Guidance Document for Use of Data in Dose/Concentration-Response Assessment. Harmonization Project Document No. 2. International Programme for Chemical Safety, World Health Organization, Geneva [online]. Available: http://whqlibdoc.who.int/publications/2005/9241546786_eng.pdf [accessed Aug. 6, 2008].

Ito, Y., O. Yamanoshita, N. Asaeda, Y. Tagawa, C.H. Lee, T. Aoyama, G. Ichihara, K. Furuhashi, M. Kamijima, F.J. Gonzalez, and T. Nakajima. 2007. Di(2-ethylhexyl)phthalate induces hepatic tumorigenesis through a peroxisome proliferator-activated receptor alpha-independent pathway. J. Occup. Health. 49(3):172-182.

Jeong, Y.C., N.J. Walker, D.E. Burgin, G. Kissling, M. Gupta, L. Kupper, L.S. Birnbaum, and J.A. Swenberg. 2008. Accumulation of M(1)dG DNA adducts after chronic exposure to PCBs, but not from acute exposure to polychlorinated aromatic hydrocarbons. Free Radic Biol. Med. 45(5):585-591.

Jiang, X, J.M. Yuan, P.L. Skipper, S.R. Tannenbaum, and M.C. Yu. 2007. Environmental tobacco smoke and bladder cancer risk in never smokers of Los Angeles county. Cancer Res. 67(15):7540-7545.

Kang, D.H., N. Rothman, M.C. Poirier, A. Greenberg, C.H. Hsu, B.S. Schwartz, M.E. Baser, J.D. Groopman, A. Weston, and P.T. Strickland. 1995. Interindividual differences in the concentration of 1-hydroxypyrene-glucuronide in urine and polycyclic aromatic hydrocarbon-DNA adducts in peripheral white blood cells after charbroiled beef consumption. Carcinogenesis 16(5):1079-1085.

Karbe, R., and R.L. Kerlin. 2002. Cystic degeneration/Spongiosis hepatis in rats. Toxicol. Pathol. 30(2):216-227.

Klaunig, J.E., M.A. Babich, K.P. Baetcke, J.C. Cook, J.C. Corton, R.M. David, J.G. DeLuca, D.Y. Lai, R.H. McKee, J.M. Peters, R.A. Roberts, and P.A. Fenner-Crisp. 2003. PPARalpha agonist-induced rodent tumors: Modes of action and human relevance. Crit. Rev. Toxicol. 33(6):655-780.

Kobayashi, E., Y. Suwazono, M. Uetani, T. Inaba, M. Oishi, T. Kido, M. Nishijo, H. Nakagawa, and K. Nogawa. 2006. Estimation of benchmark dose as the threshold levels of urinary cadmium, based on excretion of total protein, beta2-microglobulin, and N-acetyl-beta-D-glucosaminidase in cadmium nonpolluted regions in Japan. Environ. Res. 101(3):401-406.

Kodavanti, U.P., D.L. Costa, S.N. Giri, B. Starcher, and G.E. Hatch. 1997. Pulmonary structural and extracellular matrix alterations in Fischer F344 rats following subchronic phosgene exposure. Fundam. Appl. Toxicol. 37(1):54-63.

Kramer, C.B., A.C. Cullen, and E.M. Faustman. 2006. Policy implications of genetic information on regulation under the Clean Air Act: The case of particulate matter and asthmatics. Environ. Health Perspect. 114(3):313-319.

Laprise, C., and L.P. Boulet. 1997. Asymptomatic airway hyperresponsiveness: A three-year follow-up. Am. J. Respir. Crit. Care Med. 156(2 Pt. 1):403-409.

Lutz, W.K. 1990. Dose-reponse relationship and low dose extrapolation in chemical carcinogenesis. Carcinogenesis 11(8):1243-1247.

Lutz, W.K. 2001. Susceptibility differences in chemical carcinogenesis linearize the dose-response relationship: Threshold doses can be defined only for individuals. Mutat. Res. 482(1-2):71-76.

Maroni, M., and A.C. Fanetti. 2006. Liver function assessment in workers exposed to vinyl chloride. Int. Arch. Occup. Environ. Health 79(1):57-65.

Metzdorff, S.B., M. Dalgaard, S. Christiansen, M. Axelstad, U. Hass, M.K. Kiersgaard, M. Scholze, A. Kortenkamp, and A.M. Vinggaard. 2007. Dysgenesis and histological changes of genitals and perturbations of gene expression in male rats after in utero exposure to antiandrogen mixtures. Toxicol. Sci. 98(1):87-98.

Nakajima, M., M. Itoh, H. Sakai, T. Fukami, M. Katoh, H. Yamazaki, F.F. Kadlubar, S. Imaoka, Y. Funae, and T. Yokoi. 2006. CYP2A13 expressed in human bladder metabolically activates 4-aminobiphenyl. Int. J. Cancer 119(11):2520-2526.

Nessel, C.S., S.C. Lewis, K.L. Stauber, and J.L. Adgate. 1995. Subchronic to chronic exposure extrapolation: Toxicologic evidence for a reduced uncertainty factor. Hum. Ecol. Risk Assess. 1(5):516-526.

NRC (National Research Council). 1993. Pesticides in the Diets of Infants and Children. Washington, DC: National Academy Press.
NRC (National Research Council). 1994. Science and Judgment in Risk Assessment. Washington, DC: National Academy Press.
NRC (National Research Council). 1998a. Assessment of Exposure-Response Functions for Rocket-Emission Toxicants. Washington, DC: National Academy Press.
NRC (National Research Council). 1998b. Health Effects of Exposure to Low Levels of Ionizing Radiations: Time for Reassessment? Washington, DC: National Academy Press.
NRC (National Research Council). 2000. Copper in Drinking Water. Washington, DC: National Academy Press.
NRC (National Research Council). 2006a. Assessing the Human Risks of Trichloroethylene. Washington, DC: The National Academies Press.
NRC (National Research Council). 2006b. Health Risks for Dioxin and Related Compounds. Washington, DC: The National Academies Press.
NRC (National Research Council). 2006c. Health Risks from Exposures to Low Levels of Ionizing Radiation: BEIR VII. Washington, DC: The National Academies Press.
NRC (National Research Council). 2006d. Toxicity Testing for Assessment of Environmental Agents: Interim Report. Washington, DC: The National Academies Press.
NRC (National Research Council). 2007a. Toxicity Testing in the Twenty-First Century: A Vision and a Strategy. Washington, DC: The National Academies Press.
NRC (National Research Council). 2007b. Applications of Toxicogenomic Technologies to Predictive Toxicology and Risk Assessment. Washington, DC: The National Academies Press.
Peel, J.L, K.B. Metzger, M. Klein, W.D. Flanders, J.A. Mulholland, and P.E. Tolbert. 2007. Ambient air pollution and cardiovascular emergency department visits in potentially sensitive groups. Am. J. Epidemiol. 165(6):625-633.
Rhomberg, L.R., and S.K. Wolff. 1998. Empirical scaling of single oral lethal doses across mammalian species based on a large database. Risk Anal. 18(6):741-753.
Sackey, K. 1999. Hemolytic anemia: Part 1. Pediatr. Rev. 20(5):152-158.
Samoli, E., A. Analitis, G. Touloumi, J. Schwartz, H.R. Anderson, J. Sunyer, L. Bisanti, D. Zmirou, J.M. Vonk, J. Pekkanen, P. Goodman, A. Paldy, C. Schindler, and K. Katsouyanni. 2005. Estimating the exposure–response relationships between particulate matter and mortality within the APHEA multicity project. Environ. Health Perspect. 113(1):88-95.
Sand, S.J., D. von Rosen, and A.F. Filipsson. 2003. Benchmark dose calculations in risk assessment using continuous dose-response information: The influence of variance and the determinants of a cut-off value. Risk Anal. 23(5):1059-1068.
Santella, R.M., K. Hemminki, D.L. Tang, M. Paik, R. Ottman, T.L Young, K. Savela, L. Vodickova, C. Dickey, R. Whyatt, and F.P. Perera. 1993. Polycyclic aromatic hydrocarbon-DNA adducts in white blood cells and urinary 1-hydroxypyrene in foundry workers. Cancer Epidemiol. Biomarkers Prev. 2(1):59-62.
Schwartz, J., and A. Zanobetti. 2000. Using meta-smoothing to estimate dose-response trends across multiple studies, with application to air pollution and daily death. Epidemiology 11(6):666-672.
Schwartz, J., F. Laden, and A. Zanobetti. 2002. The concentration–response relation between $PM_{2.5}$ and daily deaths. Environ. Health Perspect. 110(10):1025-1029.
Schwartz, J., B. Coull, F. Laden, and L. Ryan. 2008. The effect of dose and timing of dose on the association between airborne particles and survival. Environ. Health Perspect. 116(1):64-69.
Siwinska, E., D. Mielzynska, and L. Kapka. 2004. Association between urinary 1-hydroxypyrene and genotoxic effects in coke oven workers. Occup. Environ. Med. 61(3):e10.
Slikker, W. Jr, M.E. Andersen, M.S. Bogdanffy, J.S. Bus, S.D. Cohen, R.B. Conolly, R.M. David, N.G. Doerrer, D.C. Dorman, D.W. Gaylor, D. Hattis, J.M. Rogers, R. Setzer, J.A. Swenberg, and K. Wallace. 2004. Dose-dependent transitions in mechanisms of toxicity. Toxicol. Appl. Pharmacol. 201(3):203-225.
Subramaniam, R.P., P. White, and V.J. Cogliano. 2006. Comparison of cancer slope factors using different statistical approaches. Risk Anal. 26(3):825-830.
Travis, C.C., and R.K. White. 1988. Interspecific scaling of toxicity data. Risk Anal. 8(1):119-125.
Tsuneoka, Y., T.P. Dalton, M.L. Miller, C.D. Clay, H.G. Shertzer, G. Talaska, M. Medvedovic, and D.W. Nebert. 2003. 4-aminobiphenyl–induced liver and urinary bladder DNA adduct formation in Cyp1a2(–/–) and Cyp1a2(+/+) mice. J. Natl. Cancer Inst. 95(16):1227-1237.
Watanabe, K., F.Y. Bois, and L. Zeise. 1992. Interspecies extrapolation: A reexamination of acute toxicity data. Risk Anal. 12(2):301-310.
Weil, C.S., and D.D. McCollister. 1963. Safety evaluation of Chemicals: Relationship between short- and long-term feeding studies in designing an effective toxicity test. Agr. Food Chem. 11(6):486-491.

West, J., H. Wood, R.F. Logan, M. Quinn, and G.P. Aithal. 2006. Trends in the incidence of primary liver and biliary tract cancers in England and Wales 1971-2001. Br. J. Cancer 94(11):1751-1758.

Wolf, C.J., G.A. LeBlanc, and L.E. Gray Jr. 2004. Interactive effects of vinclozolin and testosterone propionate on pregnancy and sexual differentiation of the male and female SD rat. Toxicol. Sci. 78(1):135-143.

Woodruff, T.J., E.M. Wells, E.W. Holt, D.E. Burgin, and D.A. Axelrad. 2007. Estimating risk from ambient concentrations of acrolein across the United States. Environ. Health Perspect. 115(3):410-415.

Wrixon, A.D. 2008. New ICRP recommendations. J. Radiol. Prot. 28(2):161-168.

Yamazaki, K, H. Ohno, M. Asakura, A. Narumi, H. Ohbayashi, H. Fujita, M. Ohnishi, T. Katagiri, H. Senoh, K. Yamanouchi, E. Nakayama, S. Yamamoto, T. Noguchi, K. Nagano M. Enomoto, and H. Sakabe. 1994. Two-year toxicological and carcinogenesis studies of 1, 4-dioxane in F344 rats and BDF1 mice: Drinking studies. Pp. 193-198 in Proceedings of the Second Asia-Pacific Symposium on Environmental and Occupational Health, 22-24 July, 1993, Kobe, Japan, K. Sumino, and S. Sato, eds. Kobe: International Center for Medical Research Kobe, University School of Medicine.

Zanobetti, A., J. Schwartz, and D. Gold. 2000. Are there sensitive subgroups for the effects of airborne particles? Environ. Health Perspect. 108(9):841-845.

6

Selection and Use of Defaults

As described in Chapter 2, the authors of the National Research Council report *Risk Assessment in the Federal Government: Managing the Process* (NRC 1983), known as the Red Book, recommended that federal agencies develop uniform inference guidelines for risk assessment. The guidelines were to be developed to justify and select, from among available options, the assumptions to be used for agency risk assessments. The Red Book committee recognized that distinguishing the available options on purely scientific grounds would not be possible and that an element of what the committee referred to as risk-assessment policy—often referred to later as science policy (NRC 1994)[1]—was needed to select the options for general use. The need for agencies to specify the options for general use was seen by the committee as necessary to avoid manipulation of risk-assessment outcomes and to ensure a high degree of consistency in the risk-assessment process.

The specific inference options that now appear in EPA's risk-assessment guidelines, and that permeate risk assessments performed under those guidelines, have come to be called default options, or more simply defaults. The Red Book committee defined a default option as the inference option "chosen on the basis of risk assessment policy that appears to be the best choice in the absence of data to the contrary." As the authors of *Science and Judgment in Risk Assessment* (NRC 1994) observed, many of the key inference options selected as defaults by EPA are based on relatively strong scientific foundations, although none can be demonstrated to be "correct" for every toxic substance. Because generally applicable defaults are necessary, the ultimate choice of defaults involves an element of policy. Since 1983, EPA has updated its set of defaults and has made strides in providing more detailed explanations for the choice of defaults that emphasize their theoretical and evidentiary foundations and the policy and administrative considerations that may have influenced the choices (EPA 2004a).

[1]The Red Book committee did not use the phrase *risk-assessment policy* in the usual sense in which *science policy* is used but far more narrowly to describe the policy elements of risk assessments. The committee distinguished between the policy considerations in risk assessment and those pertaining to risk management.

The Red Book emphasized both the need for generically applicable defaults and the need for flexibility in their application. Thus, the Red Book and *Science and Judgment* pointed out that scientific data could shed light, in the case of specific substances, on one or more of the information gaps in a risk assessment for which a generally applicable default had been applied. The substance-specific data might reveal that a given default might be inapplicable because it is inconsistent with the data. The substance-specific data might not show that the default had been ill chosen in the general sense but could show its inapplicability in the specific circumstance. Thus, there arose the notion of substance-specific *departures from defaults* based on substance-specific data. Much discourse and debate have attended the question of how many data, and of what type, are necessary to justify such departures, and the committee addresses the matter in this chapter. EPA recently altered its view on the question of "departures from defaults," and this chapter begins by examining this view in relation to its central theme.

CURRENT ENVIRONMENTAL PROTECTION AGENCY POLICY ON DEFAULTS

The committee recognizes that defaults are among the most controversial aspects of risk assessments. Because the committee considers that defaults will always be a necessary part of the risk-assessment process, the committee examined EPA's current policy on defaults, beginning with an eye toward understanding its applications, its strengths and weaknesses, and how the current system of defaults might be improved.

EPA began articulating a shift toward its current policy on defaults in the *Risk Characterization Handbook* (EPA 2000a) when it stated,

> For some common and important data gaps, Agency or program-specific risk assessment guidance provides default assumptions or values. Risk assessors should carefully consider all available data before deciding to rely on default assumptions. If defaults are used, the risk assessment should reference the Agency guidance that explains the default assumptions or values (p. 41).

EPA's staff paper titled *Risk Assessment Principles and Practices* (EPA 2004a) reflected a further shift in the agency's practices on defaults:

> EPA's current practice is to examine all relevant and available data first when performing a risk assessment. When the chemical- and/or site-specific data are unavailable (that is, when there are data gaps) or insufficient to estimate parameters or resolve paradigms, EPA uses a default assumption in order to continue with the risk assessment. Under this practice EPA invokes defaults only after the data are determined to be not usable at that point in the assessment—this is a different approach from choosing defaults first and then using data to depart from them (p. 51).

EPA's revised cancer guidelines (EPA 2005a) emphasize that the policy is consistent with EPA's mission and make clear that the general policy applies to cancer risk assessments:

> As an increasing understanding of carcinogenesis is becoming available, these cancer guidelines adopt a view of default options that is consistent with EPA's mission to protect human health while adhering to the tenets of sound science. Rather than viewing default options as the starting point from which departures may be justified by new scientific information, these cancer guidelines view a critical analysis of all of the available information that is relevant to assessing the carcinogenic risk as the starting point from which a default option may be invoked if needed to address uncertainty or the absence of critical information (p. 1-7).

Those statements may reflect the agency's current perspective on the primacy of scientific data and analysis in its risk assessments; the agency commits to examining all relevant and available data before selecting defaults. The committee struggled with what the current policy means in terms of both literal interpretation and application to the risk-assessment process. The lack of clarity has the potential to lead to multiple interpretations. It raised questions regarding the implications of the policy for risk decision-making. It is difficult to argue with a more robust examination of available science, which the committee strongly supports; however, the committee expressed concern that without clear guidelines on the extent to which science should be evaluated, the open-ended approach could lead to delays and undermine the credibility of defaults and the ultimate decision process. The committee notes that the risk-characterization handbook (EPA 2000a) provides some statements regarding the need to identify key data gaps and avoid delays in the risk-assessment process in the planning and scoping phase, but it is concerned that such statements may not be adequate to address complications resulting from the current policy:

> Another discussion during the planning and scoping process concerns the identification of key data gaps and thoughts about how to fill the information needs. For example, can you fill the information needs in the near-term using existing data, in the mid-term by conducting tests with currently available test methods to provide data on the agents(s) of interest, and over the long-term to develop better, more realistic understandings of exposure and effects, and to construct more realistic test methods to evaluate agents of concern? In keeping with [transparency, clarity, consistency, and reasonableness] TCCR, care must be taken not to set the risk assessment up for failure by delaying environmental decisions until more research is done (p. 29).

The policy may be appealing at first glance: it creates a two-phase process that obligates the agency to give full attention to all available and relevant scientific information and in the absence of some needed information to use defaults rather than allow uncertainties to force an end to an assessment and to related regulatory decision-making. On closer examination, the current policy carries a number of disadvantages.

Concerns with EPA's Current Policy on Defaults

Depending on implementation, the position in the current policy as articulated in the 2004 staff paper (EPA 2004a) and 2005 cancer guidelines (EPA 2005a) could represent a radical departure from previous policies. Rather than starting with a default that represents a culmination of a thorough examination of "all the relevant and available scientific information," this policy has the potential to promote with each assessment a full ad hoc examination of data and the spectrum of inferences they may support without being selective or contrasting them with the default to reflect on their plausibility. There are then no real defaults, and every inference is subject to ready replacement. By definition, a full evaluation of the evidence identifies the best available assumption, whether it is based on chemical-specific information or more general information. Thus, EPA takes on, even more than before, the burden of establishing that existing science does not warrant use of an inference different from the default. There is also the commitment "to examine all relevant and available data" first. Pushed to the extreme for some chemicals, that can mean retrieving, cataloging, and demonstrating full consideration of thousands of references, many of little utility but nonetheless "relevant." It also could lead to the reopening of the basis of some of the generic defaults on an ad hoc basis, as discussed below. Those possibilities create further vulnerability to challenge and delay that could affect environmental protection and public health. From a practical management perspective, the mandate to consider "all relevant and available data" may be unworkable for an overburdened and underresourced EPA (EPA SAB 2006, 2007) that is struggling to keep up with demands for analysis of hazard and dose-response information (Gilman 2006; Mills 2006). It may also have profound ripple effects on regulatory and risk-management efforts by other agencies at both the federal and state levels. And there is a lack of clarity as to what the policy means in cases in which the database supports a different inference from the default and does not merely replace a default with data.[2]

What Is Needed for an Effective Default Policy?

Both the current and previous EPA policies on defaults raise a crucial question: How should the agency determine that the available data are or are not "usable," that is, that

[2]One member of the committee concluded that the new EPA policy is *not* unclear, but instead represents a definitive and troubling shift away from a decades-old system that appropriately valued sound scientific information and avoided the paralysis of having to re-examine generic information with every new risk assessment. During its deliberations, the member heard two things clearly from EPA that make the intent of its above language unambiguous: (1) that EPA regards "data" and inferences as two concepts that can be compared to each other, and that the former should trump the latter (the member heard, for example, that the new policy is intended to repudiate the historical use of "risk assessment without data—just defaults"); and (2) that the goal of the policy shift is to "reduce reliance on defaults" (EPA SAB 2004a; EPA 2007d).

This member of the committee questioned both of these premises. First, the member concluded that there are two problems with the notion of pitting "data" against defaults. The logical problem, in this member's opinion, was that the actual choice EPA faces is a choice *among* models (inferences, assumptions), which are not themselves "data" but which are ways of making sense of data. For example, reams of data may exist on some biochemical reaction that *might* suggest that a particular rodent tumor was caused via a mechanism that does not operate in humans. EPA's task, however, is whether or not to make the assumption that the rodent tumors are relevant, in the absence of a well-posed *theory* to the contrary, one that is supported by data. Without the alternative assumption being articulated, EPA has nothing coherent to do with the data. The more important practical problem with EPA's new formulation, in this member's opinion, is that a policy of "retreating to the default" if the chemical- or site-specific data are "not usable" ignores the vast quantities of data (interpretable via inferences with a sound theoretical basis) that *already support* most of the defaults EPA has chosen over the past 30 years. In order for a decision to not "invoke" a default to be made fairly, data supporting the inference that a rodent tumor response was irrelevant would have to be weighed against the *data* supporting the default inference that such responses are generally relevant (see, for example, Allen et al. 1988), data supporting a possible nonlinearity in cancer dose-response would have to be weighed against the data supporting linearity as a general rule (Crawford and Wilson 1996), data on pharmacokinetic parameters would have to be weighed against the data and theory supporting allometric interspecies scaling (see, for example, Clewell et al. 2002), and so on. In other words, having no chemical-specific data other than bioassay data does not imply there is a "data gap," as EPA now claims—it may well mean that vast amounts of data support a time-tested inference on how to interpret this bioassay, and that no data to the contrary exist because no plausible inference to the contrary exists in this case. In short, this committee member sees most of the common risk assessment defaults *not* as "inferences retreated to because of the absence of information," but rather as "inferences generally endorsed *on account of the information*."

Therefore, this committee member concluded that EPA's stated goal of "reducing reliance on defaults" per se is problematic; it begs the question of why a scientific-regulatory agency would ever want to reduce its reliance on those inferences that are supported by the most substantial theory and evidence. Worse yet, the committee member concluded, it seems to prejudice the comparison between default and alternative models before it starts—if EPA accomplishes part of its mission by ruling against a default model, the "critical analysis of all available information" may be preordained by a distaste for the conclusion that the default is in fact proper.

This committee member certainly endorses the idea of reducing EPA's reliance on *those defaults* that are found to be outmoded, erroneous, or correct in the general case but not in a specific case—but identifying those inferior assumptions is exactly what a system of departures from defaults, as recommended in the Red Book, in *Science and Judgment*, and in this report, is designed to do. EPA should modify its language to make clear that across-the-board skepticism about defaults is not scientifically appropriate. Thus, the committee member concludes that recommendations in this chapter apply whether or not EPA believes it has "evolved beyond defaults." A system that evaluates every inference for every risk assessment still needs ground rules, of the kind recommended in this chapter, to show interested parties *how* EPA will decide what data are "usable" or which inference is proper. This committee member urges EPA to delineate what evidence will determine how it makes these judgments, and how that evidence will be interpreted and questioned—and EPA's current policy sidesteps these tasks.

they do or do not support an inference alternative to the default? The question underscores the need for guidance to implement a default policy and evaluate its effect on risk decisions and efforts to protect the environment and public health. The committee did not conduct a detailed evaluation, but a cursory examination of some recent assessments shows detailed presentations and analyses of the available data bearing on each assessment, explicit determinations that identified data that do not support an inference alternative to such defaults as low-dose linearity and the cross-species scaling of risk, but thus far not the wholesale reconsideration of generic defaults. No matter how one interprets EPA's current policy on defaults, an effective policy requires criteria to guide risk assessors on factors that would render data "not usable" or supportable of inference alternatives to a default, and therefore requiring that a default be invoked.

Therefore it remains the case that

- Defaults need to be maintained for the steps in risk assessment that require inferences beyond those that can be clearly drawn from the available data or to otherwise fill common data gaps.
- Criteria should be available for judging whether, in specific cases, data are adequate for direct use or to support an inference in place of a default.

The "data" that may be usable in place of a default will depend on the role of the particular default in question. For example, some defaults regarding exposure may be readily inferred from observations and in this sense are "measurable," but many defaults for biologic end points will continue to be based on science and policy judgments. The latter type of defaults is the focus of this report.

Readily observable and measurable defaults, such as the amount of air breathed each day or the number of liters of water consumed, may be chosen to make assessments manageable or consistent with one another but not to support inferences beyond the available data or what can be readily observed, and they are therefore generally less difficult to justify. Decisions about replacing them with distributions (for variability analysis) or specific values based on survey data tend to be less controversial.

In contrast, the defaults involving science and policy judgments, such as the relevance of a rodent cancer finding in predicting low-dose-human risk, are used to draw inferences "beyond the data," that is, beyond what may be directly observable through scientific study. The next section gives examples of important defaults of that kind related to the hazard-identification and dose-response assessment steps. Inferences are needed when underlying biologic knowledge is uncertain or absent. Indeed, fundamental lack of understanding of key biologic phenomena can remain after many years of research. In some cases, however, research "data"—typically on pharmacokinetic (PK) behavior and modes of toxic action—support an inference *different* from that implicit in the default. Determining whether such "data" are adequate to support a different inference is often difficult and controversial. Much of the emphasis of this chapter is on the defaults chosen as "inferences" in the presence of considerable uncertainty, not on those chosen to represent observed parameters or to fill gaps in data on readily observable phenomena.

In the discussions in this chapter, simply for ease of presentation, the committee uses the term *departures* in offering its views regarding the use of inferences based on substance-specific data rather than defaults. *Departures* in the sense used in this report is related to the decision in specific cases as to whether data are adequate to support an inference different from the default and to make it unnecessary to adopt the default. Recognizing the challenge of interpreting EPA's policy, the committee, to be consistent with its charge, offers its discussions and recommendations in the context of current EPA policy.

THE ENVIRONMENTAL PROTECTION AGENCY'S SYSTEM OF DEFAULTS

Explicit Defaults

The system of inferences used in EPA risk assessments is contained in the agency's reports, staff papers, procedural manuals and guidance documents. These materials provide some advice and information on interpreting the strengths and limitations of various types of scientific datasets and on data synthesis, including whether a body of data supports a default or alternative inference, and risk assessment methods. Guidance is given on assessment of risks of cancer (EPA 2005a), neurotoxicity (EPA 1998a), developmental toxicity (EPA 1991a), and reproductive toxicity (EPA 1996); on Monte Carlo analysis (EPA 1997); on assessment of chemical mixtures (EPA 1986, 2000b); on reference-dose (RfD) and reference-concentration (RfC) processes (EPA 1994, 2002a,b); and on how to judge data on whether, for example, male rat kidney tumors (EPA 1991b) or rodent thyroid tumors (EPA 1998b) are relevant to humans (see, for example, Box 2-1 and Table D-1). The toxicity guidance documents also identify some defaults commonly used in assessments covered by the guidance. Tables 6-1 and 6-2 list some of the important defaults for carcinogen and noncarcinogen risk assessments.

Missing Defaults

In addition to explicitly recognized defaults, EPA relies on a series of implicit or "missing" defaults[3]—assumptions that may sometimes exert great influence on risk characterization. For a risk assessment to be completed, *every* "inference gap" must have been "bridged" with some assumption, whether explicitly stated or not. Assumptions analogous to missing defaults are made in every field. For example, it is common to treat a pair of variables as independent when no information exists about any relationship between them. That assumption may well be reasonable, but it imposes a powerful condition on the analysis: that the correlation coefficient between the variables is exactly 0.0 rather than any other value between -1 and 1.

Use of missing defaults has become so ingrained in EPA risk-assessment practice that it is as though EPA has chosen the same assumptions explicitly. The committee recommends that EPA systematically examine the risk-assessment process and identify key instances of the bridging of an inference gap with a missing default, examine its basis, and consider alternatives if such a default is not sufficiently justified.

This committee is concerned particularly about two missing defaults. First, agents that have not been examined sufficiently in epidemiologic or toxicologic studies are insufficiently included in or even excluded from risk assessments. Typically, there is no description of the risks potentially posed by these agents in the risk characterization, so their presence often carries no weight in decision-making. With few notable exceptions (for example, dioxin-like compounds), they are treated as though they pose no risk that should be subject to regulation in EPA's air, drinking-water, and hazardous-waste site programs. Also with very few

---

[3]*Science and Judgment in Risk Assessment* (NRC 1994) coined the term *missing default* to describe the use of de facto assumptions by EPA without explicit explanation. These *de facto* assumptions may also be thought of as "implicit defaults."

TABLE 6-1 Examples of Explicit EPA Default Carcinogen Risk-Assessment Assumptions

Issue	EPA Default Approach
Extrapolation across human populations	"When cancer effects in exposed humans are attributed to exposure to an agent, the default option is that the resulting data are predictive of cancer in any other exposed human population." (EPA 2005a, p. A-2)
	"When cancer effects are not found in an exposed human population, this information by itself is not generally sufficient to conclude that the agent poses no carcinogenic hazard to this or other populations of potentially exposed humans, including susceptible subpopulations or lifestages." (EPA 2005a, p. A-2)
Extrapolation of results from animals to humans	"Positive effects in animal cancer studies indicate that the agent under study can have carcinogenic potential in humans." (EPA 2005a, p. A-3)
	"When cancer effects are not found in well-conducted animal cancer studies in two or more appropriate species and other information does not support the carcinogenic potential of the agent, these data provide a basis for concluding that the agent is not likely to possess human carcinogenic potential, in the absence of human data to the contrary." (EPA 2005a, p A-4)
Extrapolation of metabolic pathways across species, age groups, and sexes	"There is a similarity of the basic pathways of metabolism and the occurrence of metabolites in tissues in regard to the species-to-species extrapolation of cancer hazard and risk" (EPA 2005a, p. A-6).
Extrapolation of toxicokinetics across species, age groups, and sexes	"As a default for oral exposure, a human equivalent dose for adults is estimated from data on another species by an adjustment of animal applied oral dose by a scaling factor based on body weight to the 3/4 power. The same factor is used for children because it is slightly more protective than using children's body weight." (EPA 2005a, p. A-7)
Shape of dose-response relationship	"When the weight of evidence evaluation of all available data are insufficient to establish the mode of action for a tumor site and when scientifically plausible based on the available data, linear extrapolation is used as a default approach, because linear extrapolation generally is considered to be a health-protective approach. Nonlinear approaches generally should not be used in cases where the mode of action has not been ascertained. Where alternative approaches with significant biological support are available for the same tumor response and no scientific consensus favors a single approach, an assessment may present results based on more than one approach." (EPA 2005a, p. 3-21)

exceptions, EPA treats all adults as equally susceptible to carcinogens that act via a linear mode of action (MOA) (see Chapter 5 and, for a recent example, EPA 2007a). Table 6-3 lists those and several other apparently missing EPA defaults.

Both explicit and missing defaults used by EPA are a cornerstone of the agency's approach to facilitating human health risk assessment in the face of inherent scientific limitations that may prevent verification of any particular causal model. Understanding of the complications introduced by EPA's policy and practice regarding defaults is central to evaluating EPA's management of uncertainty.

TABLE 6-2 Examples of Explicit EPA Default Noncarcinogen Risk-Assessment Assumptions

Issue	EPA Default Approach
Relevant human health end point and extrapolation from animals to humans	"The effect used for determining the NOAEL, LOAEL,[a] or benchmark dose in deriving the RfD or RfC is the most sensitive adverse reproductive end point (that is, the critical effect) from the most appropriate or, in the absence of such information, the most sensitive mammalian species." (EPA 1996, p. 77)
Adjustment to account for differences between humans and animal test species	Factor of 1, 3, or 10. (EPA 2002a, p. 2-12)
Heterogeneity among humans	Factor of 1, 3, or 10. (EPA 2002a, p. 2-12)
Shape of dose-response relationship	"In quantitative dose-response assessment, a nonlinear dose-response is assumed for noncancer health effects unless mode of action or pharmacodynamic information indicates otherwise." (EPA 1996, p. 75)
Human risk estimate	Division of the point of departure (for example, NOAEL, LOAEL, or benchmark dose) by the appropriate uncertainty factors to take into account, for example, the magnitude of the LOAEL compared with the NOAEL, interspecies differences, or heterogeneity among members of the human population produces "an estimate (with uncertainty spanning perhaps an order of magnitude) of a daily exposure to the human population (including sensitive subgroups) that is likely to be without an appreciable risk of deleterious effects during a lifetime." (EPA 1998a, p. 57)

[a]NOAEL = no-observed-adverse-effect level, LOAEL = lowest-observed-adverse-effect level.

COMPLICATIONS INTRODUCED BY USE OF DEFAULTS

The National Research Council (NRC 1994) noted that although EPA had justified the selection of some of its defaults, many had received incomplete scrutiny by the agency. In the agency's *Guidelines for Carcinogen Risk Assessment* (EPA 2005a), it elucidated more fully the bases of many of its defaults. Selection of defaults by EPA has been controversial, and the controversies were described in *Science and Judgment in Risk Assessment* (NRC 1994, Chapter 6 and Appendices N-1 and N-2). Because choice of defaults involves a blend of science and risk-assessment policy, controversy is inevitable. Some have argued that EPA has selected defaults at each opportunity that are needlessly "conservative" and result in large overestimates of human risk (OMB 1990; Breyer 1992; Perhac 1996). Others have argued—given the large scientific uncertainties surrounding risk assessment, human variability in both exposure to and response to toxic substances, and various missing defaults with "nonconservative" biases—that risk overestimation might not be common in EPA's practices and that risk underestimation may occur (Finkel 1997; EPA SAB 1997, 1999). EPA (2004a, p. 20) states that the sum of conservative risk estimates for a chemical mixture overstates risk to a relatively modest extent (a factor of 2-5). In general, estimates based on animal extrapolations have been found to be generally concordant with those based on epidemiologic studies (Allen et al. 1988; Kaldor et al. 1988; Zeise 1994), and in several cases human

TABLE 6-3 Examples of "Missing" Defaults in EPA "Default" Dose-Response Assessments

- *For low-dose linear agents, all humans are equally susceptible during the same life stage* (when estimates are based on animal bioassay data) (EPA 2005a). The agency assumes that the linear extrapolation procedure accounts for human variation (explained in Chapter 5), but does not formally account for human variation in predicting risk. For low-dose nonlinear agents, an RfD is derived with an uncertainty factor for interhuman variability of 1-10 (EPA 2004a, p. 44; EPA 2005a, p. 3-24).

- *Tumor incidence from conventional chronic rodent studies is treated as representative of the effect of lifetime human exposures after species dose equivalence adjustments* (EPA 2005a). For chemicals established as operating by a mutagenic mode of action, that holds after adjustment for early-life sensitivity (EPA 2005b). This assumes (1) that humans and rodents have the same "biologic clock," that is, that rodents and humans exposed for a lifetime to the same (species-corrected) dose will have the same cancer risk, and (2) that a chronic rodent bioassay, which doses only in adulthood and misses late old age (EPA 2002a, p. 41), is representative of a lifetime of rodent exposure.

- *Agents have no in utero carcinogenic activity.* Although the agency notes that in utero activity is a concern, default approaches do not take carcinogenic activity from in utero exposure into account, and risks from in utero exposure are not calculated (EPA 2005b; EPA 2006a, p. 29).

- *For known or likely carcinogens not established as mutagens, there is no difference in susceptibility at different ages* (EPA 2005b).

- *Nonlinear carcinogens and noncarcinogens act independently of background exposures and host susceptibility* (see Chapter 5 for full discussion).

- *Chemicals that lack both adequate epidemiologic and animal bioassay data are treated as though they pose no risk of cancer worthy of regulatory attention*, with few exceptions. They are typically classified as having "inadequate information to assess carcinogenic potential" (EPA 2005a, Section 2.5); consequently, no cancer dose-response assessment is performed (EPA 2005a, p. 3-2). Integrated Risk Information System and provisional peer-reviewed toxicity values are then based on noncancer end points, and cancer risk estimates are not presented.

data have indicated that animal-based estimates were not conservative for the population as a whole (see discussion in Chapter 4).

In any event, the committee observes that any set of defaults will impose value judgments on balancing potential errors of overestimation and underestimation of risk even if the judgments dictate that the balance be exactly indifferent between the two. Thus, the issue is not whether to accept a value-laden system of model choice but which value judgments EPA's assessments will reflect. Some members of the *Science and Judgment in Risk Assessment* committee endorsed the view that risk-assessment policy should seek a "plausible conservatism"[4] in the choice of default options rather than seeking to impose the alternative value judgment that models should strive to balance errors of underestimation and overestimation exactly (Finkel 1994); others took the view that relative scientific plausibility alone should govern the choice of defaults and the motivation for departing from them (McClellan and North 1994). EPA (2004a, pp. 11-12) acknowledged the debate:

> EPA seeks to adequately protect public and environmental health by *ensuring that risk is not likely to be underestimated.* However, because there are many views on what "adequate" protection is, some may consider the risk assessment that supports a particular protection

[4]This use of *conservatism* is intended to describe the situation in which the assumptions and defaults used in risk assessment are likely to overstate the true but unknowable risk. It is derived from the public-health dictum that when science is uncertain, judgments based on it should err on the side of public-health protection.

> level to be "too conservative" (that is, it overestimates risk), while others may feel it is "not conservative enough" (that is, it underestimates risk). . . .
>
> Even with an optimal cost-benefit solution, in a heterogeneous society, some members of the population will bear a disproportionate fraction of the costs while others will enjoy a disproportionate fraction of the benefits (Pacala et al. 2003). Thus, inevitably, different segments of our society will view EPA's approach to public health and environmental protection with different perspectives.

In addition to the debate over how "conservative" default assumptions should be, there is tension between their use and the complete characterization of uncertainty. For example, it is possible to imagine eliminating defaults and instead using ranges of plausible assumptions in their place. Doing so, however, could produce such a broad range of risk estimates, with no clear way to distinguish their relative scientific merits, that the result could be useless for the purpose of choosing among various risk-management options for decision-making (see Chapter 8). As explained above, using defaults ameliorates that problem but at the cost of reporting only a portion of the complete range of risk estimates that is consistent with available scientific knowledge. In some cases, use of defaults overstates the central tendency of the complete range; in other cases, it underestimates the central tendency. As discussed below, that pitfall is important because of the ubiquitous nature of tradeoffs that surround most risk-management decisions.

How EPA has responded to suggestions to improve its system of defaults reveals three related issues. First, the agency has not published clear, general guidance on what level of evidence is needed to justify use of chemical-specific evidence and not use a default, although EPA has provided some specific guidance for a small number of particular defaults (see below).

Second, as part of its current practice of using defaults, EPA often does not quantify the portion of the total uncertainty characterized in the resulting risk estimate or RfD that is due to the presence of competing plausible causal models. EPA in its various guidance documents and reviews has provided a scientific justification for many of its defaults (for example, EPA 1991a, 2002b, 2004a, 2005a,b). In some cases, it has demonstrated that the defaults are plausible, but not the extent to which a default may produce an estimate of the risk or RfD different from that produced by a plausible alternative model. Tables 6-1 and 6-2 list explicit defaults used by EPA. A notable example is the use of the linear no-threshold dose-response relationship for extrapolation of cancer risk below the point of departure when there is no evidence of an MOA that would introduce nonlinearity. That assumption is based on both mechanistic hypotheses and empirical evidence. "Low-dose nonlinear" carcinogens and chemicals without established carcinogenic properties are assumed to follow threshold-like dose-response relationships[5] even when, as in the case of chloroform, it is acknowledged that multiple modes of action, including genotoxicity, cannot be ruled out (EPA SAB 2000, p. 1; EPA 2001, p. 42). The nonlinear effects are also presumed to act independently of background processes although for many mechanisms (such as receptor-mediated ones) there can be endogenous and exogenous agents that contribute to the same disease process present in the population that the toxicant under study contributes to (see Chapter 5).

EPA risk-assessment guidance acknowledges that defaults are uncertain (EPA 2002a, 2005a). In practice, the agency addresses the uncertainty by discussing it qualitatively. EPA

[5]The agency's most recent cancer and noncancer guidelines do not strictly assume biologic thresholds, because of "the difficulty of empirically distinguishing a true threshold from a dose-response curve that is nonlinear at low doses"; instead, it refers to the dose-response relationships as low-dose nonlinear (EPA 2005a).

has recently been criticized, however, for not describing the range of risk estimates associated with alternative assumptions quantitatively (NRC 2006a), and it has been encouraged in various forums to begin to develop the methodology and data to describe the uncertainty in dose-response modeling quantitatively (EPA SAB 2004b; NRC 2007a).

Third, EPA has not established a clear set of standards to apply when evidence of an alternative assumption is sufficiently robust not to invoke a default. EPA (2005a, p. 1-9) states that "with a multitude of types of data, analyses, and risk assessments, as well as the diversity of needs of decision makers, it is neither possible nor desirable to specify step-by-step criteria for decisions to invoke a default option." The committee agrees that it is neither possible nor desirable to reduce the evaluation of defaults to a checklist. However, failure to establish clear guidelines detailing the issues that must be addressed to depart from a default and the type of evidence that would be compelling can have a number of adverse consequences. The lack of clear standards may reduce the incentive for further research (Finkel 2003). With no guidance on criteria for using an alternative assumption, it is difficult for an interested party to understand the type of scientific information that might be required by the agency, and a lack of clear standards can make the process of deciding whether new research data (instead of a default) are usable appear to be arbitrary. The committee considers that clear evidence standards for deciding to retain or depart from defaults can make the process more transparent, consistent, and fair for all stakeholders involved and enhance their trust in the process. Examples from EPA (discussed below) demonstrate that it is possible to specify criteria for departure from defaults.

Risk estimates developed with defaults focus on a portion of the scientifically plausible risk-estimate range. However, because some defaults may lead to the overstatement of the risk posed by a chemical and others to an understatement of risk, EPA needs to be mindful of the influence of defaults on risk estimates when the estimates will influence risk-management decisions. Intervention options often involve tradeoffs, and the tradeoffs being considered (such as replacement of one chemical with another in a production process) might result in risk estimates whose health protectiveness depends on the defaults used in estimation. An example is the tradeoff between the risks resulting from exposure to mercury and PCBs in fish and the nutritional benefit of fish consumption (Cohen et al. 2005).

When chemical risks are being compared, the agency can minimize the differential effects of defaults by ensuring that they are applied consistently. When chemical risks are being compared with other considerations whose estimated effects are not influenced by defaults, EPA should emphasize the quantitative characterization of the contribution of the defaults to uncertainty (as discussed below).

ENHANCEMENTS OF THE ENVIRONMENTAL PROTECTION AGENCY'S DEFAULT APPROACH

This section describes the committee's recommendations for improving how defaults are chosen, used, and modified. These recommendations include continued and expanded use of the best, most current science to choose, justify, and, when appropriate, revise EPA's default assumptions; development of a clear standard to determine when evidence supporting an alternative assumption is robust enough that the default need not be invoked and development of various sets of scientific criteria for identifying when an alternative has met that standard; making explicit the existing assumptions or developing new defaults to address the missing defaults, such as treatment of chemicals with limited information as though they pose risks that do not require regulatory action; and quantifying the risk estimates emerging from more than one model (assumption) when EPA has determined that an alternative model is sufficiently well developed and validated to be presented alongside the risks resulting from use of the default.

Best Use of Current Science to Define Defaults

The defaults selected for EPA's risk assessments and described in the agency's guidelines should be periodically reviewed to determine their consistency with evolving science. The advance of scientific knowledge relevant to the selection of defaults is typically associated with studies of specific agents that provide insights into the applicability of alternative models to those agents (and perhaps also to related agents). As knowledge accumulates, it may point to the need for revision of one or more defaults for entire classes of related agents or even for all agents. Because general scientific understanding is continually evolving, it is essential that EPA remain committed to evaluating the bases of its defaults. Chapter 5 provides an example of how EPA might evaluate and revise its default dose-response assessment assumptions in order to take into account the growing understanding of how dose-response assessment depends on interindividual variability and background exposures to a particular chemical and to chemicals that have similar MOAs.

Guidelines describing defaults should include a detailed description of the underlying science to justify the plausibility of the default for a wide array of circumstances. For example, the assumed relevance of rodent carcinogenicity testing to human risk might be justified by the high degree of common genetics across mammalian species and by empirical evidence that rodents are useful models of human disease processes. The documentation should also include the known and suspected limitations of the default's applicability in any specific case. In the example above, limitations might include known differences in organ sensitivity and enzyme pathways between rodents and humans. The documentation should systematically establish grounds for departing from the default.

None of the possible inference options that is evaluated for its scientific strengths can be shown with high certainty to be generically applicable, but a default must be chosen from among them. As the Red Book pointed out, an element of "risk-assessment policy" will need to be invoked for the selection of defaults. EPA should use available science to the maximum extent and clearly specify the basis of its final selection of defaults. The same process should be used when new defaults are being considered to replace existing ones.

Clear Standards for Departures from Defaults

In keeping with the Red Book's recommendations concerning the need for flexibility in the application of EPA's inference guidelines, EPA has accepted alternatives to defaults in several specific cases. For example, the last decade saw major advances in the development of physiologically based pharmacokinetic (PBPK) models, and the agency has found these models useful to replace defaults in cross-route and cross-species extrapolation. In the agency's toxicologic review of 1,1,1-trichloroethane (EPA 2007a), for example, it evaluated 14 PBPK models that had been published in peer-reviewed journals, selected those it judged to be best supported, and then used model results to assess animal-to-human differences in the pharmacokinetic behavior of 1,1,1-trichloroethane. The typical default uncertainty factor (UF) of 10, used to extrapolate animal findings to humans, is assumed by default to be made

BOX 6-1 Boron: Use of Data-Derived Uncertainty Factors

EPA has been struggling with characterization of uncertainty in risk assessments for decades. In most cases involving noncancer health effects, default uncertainty factors are used to account for conversion of subchronic to chronic exposure data, the adequacy of the database, extrapolation from the lowest-observed-adverse-effect level to a no-observed-adverse-effect level, interspecies extrapolation, and human variability. Inadequacies in the database often compel the agency to rely on default assumptions to compensate for gaps in data. In the case of the boron risk assessment, data were available, so EPA could apply a "data-derived approach" to develop uncertainty factors. This approach "uses available toxicokinetic and toxicodynamic data in the determination of uncertainty factors, rather than relying on the standard default values" (Zhao et al. 1999). The boron case illustrates issues surrounding the development and use of data-derived uncertainty factors by the agency.

Without endorsing the specifics, the committee notes that in the boron risk assessment the availability of data lowered the uncertainty factor by roughly one-third, from 100 to 66. Chemical-specific pharmacokinetic and physiologic data were used to derive the factors (DeWoskin et al. 2007). Specifically, data on renal clearance from studies of pregnant rats and pregnant humans were used in determining data-driven interspecies pharmacokinetic adjustments, and glomular-filtration variability in pregnant women was used to develop the nondefault values for intraspecies pharmacokinetic adjustments.

The data-derived approach used in the risk assessment was largely supported by the three external reviewers of the risk assessment (see EPA 2004b, p. 110):

> All three reviewers agreed that the new pharmacokinetic data on clearance of boron in rats and humans should be used for derivation of an uncertainty factor instead of a default factor. Comments included statements that EPA should always attempt to use real data instead of default factors and a statement that this use of clearance data is a significant step forward in the general EPA methodology for deriving uncertainty.

The use of data-driven uncertainty factors was not without controversy, as reported in a 2004 *Risk Policy Report*: "environmentalists are concerned EPA is eroding its long-standing practice of using established safety factors when faced with scientific uncertainties. 'Our major concern is that this represents a major move by EPA away from the concept of defaults, and towards a concept of default if we think that it's required, and if there are data to support a default'," a scientist with the Natural Resources Defense Council says. "EPA may use a 'scrap of evidence' to support the idea that one chemical is like another, reducing the need for important safety factors, the source says" (Risk Policy Report 2004, p. 3).

up of two factors of about 3: one for PK differences and the other for pharmacodynamic (PD) differences.[6] In the draft 1,1,1-trichloroethane assessment, the agency used PBPK model results instead of the default UF of 3; but in the absence of information on PD differences, it retained the default UF of 3. This example reflects increased agency recognition of the value of reliable scientific information to reduce model uncertainties in risk assessment.

In another recent example (see Box 6-1), EPA used chemical-specific PK and physiologic data to derive two UFs (for extrapolating from animal to humans and for human variability) in establishing the RfD for boron.

Those examples show that EPA has departed from default assumptions in specific cases; however, the committee believes that EPA and the research community would benefit from the development of clear standards and criteria for such departures.

Developing clear standards and criteria for departing from defaults requires a system

[6]The assumption that PK and PD are similar in their contribution to interindividual heterogeneity is likely to be incorrect. Hattis and Lynch (2007) argued that PD factors are likely to be more important.

that has two components: a single "evidentiary standard" governing how EPA considers alternative assumptions in relation to the default and the specific scientific criteria that EPA will use to gauge whether an alternative model has met the evidentiary standard.

Evidentiary Standard

Because of the effort that EPA has invested in selecting its current defaults and the consistency that defaults confer on the risk-assessment process, the use of an alternative to the default in specific cases faces a substantial hurdle and should be supported by specific theory and evidence. The committee recommends that EPA adopt an alternative assumption in place of a default when it determines that the alternative is "clearly superior,"[7] that is, that its plausibility clearly exceeds the plausibility of the default.

Specific Criteria to Judge Alternatives

The scientific questions that should be addressed to assess whether an alternative to a default is clearly superior will depend on the particular inference gap that is to be bridged. The committee recommends that EPA establish issue-specific criteria for bridging inference gaps. Important issues that require development of criteria include the use of PBPK models vs allometry to scale doses across species, the relevance of animal tumors to humans, and PD differences between animals and humans. Many of those issues are relevant to the unification of cancer and noncancer dose-response modeling described in Chapter 5.

EPA in specific cases has developed criteria for departing from defaults. Three examples are presented below. The committee notes that these cases are presented as starting points for the development of criteria for departing from defaults; and their use does not imply that the committee agrees with their rationale in every detail.

Low-dose extrapolation for thyroid follicular tumors in rodents. In 1998, EPA developed guidance for when and how to depart from the default assumption that a substance that causes thyroid follicular tumors in rodents will have a linear dose-response relationship in humans (EPA 1998b). That guidance states clearly that EPA will consider a margin-of-exposure, rather than a linear approach, when it can be demonstrated that a particular rodent carcinogen is not mutagenic, that it acts to disrupt the thyroid-pituitary axis, and that no MOA other than antithyroid activity can account for the observed rodent tumor formation. EPA then presents eight criteria for determining whether the substance disrupts the thyroid-pituitary axis and states that the first five must be satisfied (the remaining three are "desirable").

Relevance to humans of animal α2μ-globulin carcinogens. In the case of criteria for setting aside the relevance of renal tumors that occurred after exposure to agents that act through the α2μ-globulin MOA, EPA developed clear criteria for departure from the default assumption that animal tumors are relevant to human risk. EPA (1991b) specified two conditions that must be satisfied to replace that default. First, for the agent in question, α2μ-globulin must be shown to be involved in tumor development. For this condition, EPA requires three findings (p. 86): "(1) Increased number and size of hyaline droplets in

[7]In legal parlance, a "beyond a reasonable doubt" standard would be "clearly superior." The term *clearly superior* should not be interpreted quantitatively, but the committee notes that statistical P values can also be used as an analogy. For example, rejecting the null in favor of the alternative only when $P < 0.05$ could be viewed as insisting that the alternative hypothesis is "clearly superior" to the "default null."

renal proximal tubule cells of treated male rats," "(2) Accumulating protein in the hyaline droplets is α2μ-g[lobulin]," and "(3) Additional aspects of the pathological sequence of lesions associated with α2μ-g[lobulin] nephropathy are present." If the first condition is satisfied, EPA states that the extent to which α2μ-globulin is responsible for renal tumors must be established. Establishing that it is largely responsible for the observed renal tumors is grounds for setting aside the default assumption of their relevance to humans. EPA states (p. 86) that this step "requires a substantial database, and not just a limited set of information confined to the male rat. For example, cancer bioassay data are needed from the mouse and the female rat to be able to demonstrate that the renal tumors are male-rat specific." EPA lists the type of data that are helpful, for example, data showing that the chemical in question does not cause renal tumors in the NBR rat (which does not produce substantial quantities of α2μ-globulin), evidence that the substance's binding to α2μ-globulin is reversible, sustained cell division of the P2 renal tubule segment that is typical of the α2μ-globulin renal-cancer mode of action, structure-activity relationship data similar to those on other known α2μ-globulin MOA substances, evidence of an absence of genotoxicity, and the presence of positive renal-carcinogenicity findings only in male rats and negative findings in mice and female rats (EPA 1991b).

Applicability of the safety factor[8] *of 10 under the Food Quality Protection Act.* EPA's treatment of the safety factor of 10 to protect infants and children when setting pesticide exposure limits is an example of how the agency could establish a process to determine regularly whether data are sufficient to depart from what is, in effect, a default. The 1996 Food Quality Protection Act (FQPA) mandates the use of a safety factor of 10 unless EPA has sufficient evidence to determine that a different value is more appropriate [§ 408 (b)(2)(c)]. The EPA Office of Pesticide Programs (EPA 2002b) has developed a systematic weight-of-evidence approach that addresses a series of considerations, including prenatal and postnatal toxicity, the nature of the dose-response relationship, PK, and MOA. On the basis of the framework, EPA had found it unnecessary to apply the safety factor of 10 in 48 of 59 cases (reviewed in NRC 2006b).

Committee's Evaluation

Those examples provide a starting point for the agency's development of a standardized approach to departures from defaults. An improvement based on these examples would be greater specificity regarding the type of evidence that is sufficient to justify a departure.

Consider, for example, EPA's guidance for chemicals that cause follicular tumors. Section 2.2.4 of EPA 1998b (p. 21) requires that "enough information on a chemical should be given to be able to identify the sites that contribute the major effect on thyroid-pituitary function," but EPA does not indicate what quantity and quality of information are "enough" for a researcher to make such a determination. In addition, the key statement that "where thyroid-pituitary homeostasis is maintained, the steps leading to tumor formation are not expected to develop, and the chances of tumor development are negligible" refers throughout the document to humans in general and does not address interindividual variability in homeostasis.

EPA has presented guidance (EPA 2002b) for departing from the use of a safety factor of 10 as provided for in the FQPA. The guidance includes a list of issues to consider and the type of evidence to evaluate. Some of the guidelines provide sufficient specificity as to

[8]In Chapter 5, the committee takes exception to the term *safety factor*, but it uses it here to avoid confusion with EPA terminology.

evaluation of departures. For example, a finding of effects in humans or in more than one species militates against departure, as does a finding that the young do not recover as quickly from the adverse effects of a chemical as do adults. In contrast, some of the guidelines lack specificity. In particular, an MOA supporting the human relevance of effects observed in animals militates against departure from the default; this guideline would be more useful if it spelled out specific MOA findings that support the relevance to humans.

The committee recommends that EPA review those and other cases in which it has used substance-specific data and not invoked defaults and that it catalog the principles characterizing those departures. The principles can be used in developing more general guidance for deciding when data clearly support an inference that can be used in place of a default.

Crafting Defaults That Replace (or Make Explicit) *Missing* Assumptions: The Case of Chemicals with Inadequate Toxicity Data

EPA should work toward developing explicit defaults to use in place of missing defaults. To the extent possible, the new, explicit defaults should characterize the uncertainty associated with their use. Although there appear to be a number of missing defaults, this section focuses on the "untested-chemical assumption" and outlines an approach for characterizing the toxicity of untested or inadequately tested chemicals.[9] The approach attempts to strike a balance between gathering enough information to reduce uncertainty sufficiently to make the resulting estimate useful and making the approach applicable for characterizing a large number of chemicals.

In the absence of data to derive a quantitative, chemical-specific estimate of toxicity, EPA treats such chemicals as though they pose risks that do not require regulatory action in its air, drinking-water, and hazardous-waste programs. In the case of carcinogens, EPA assigns no potency factor to a chemical and thus implicitly treats it as though it poses no cancer risk, for example, chemicals whose evidence meets the standard of "inadequate information to assess carcinogenic potential" in the carcinogen guidelines (EPA 2005a, p. 1-12). For noncancer end points, EPA practice limits the product of the uncertainty factors applied to no more than 3,000. When a larger value would be required to address the uncertainty (for example, when "there is uncertainty in more than four areas of extrapolation" [EPA 2002a, p. xvii]), EPA does not derive an RfD or RfC. The vast majority of chemicals now produced lack a cancer slope factor, RfD, RfC, or a combination of these.

The effective assumption that many chemicals pose no risk that should be subject to regulation can compromise decision making in a variety of contexts, as it is not possible to meaningfully evaluate net health risks and benefits associated with the substitution of one chemical for another in a production process or interpret risk estimates where there can be a large number of untested chemicals (for example, a Superfund site) that have not been examined sufficiently in epidemiologic or toxicologic studies.

To develop a distribution of dose-response relationship estimates for chemicals on which agent-specific information is lacking, a tiered series of default distributions could be constructed. The approach is based on the notion that for virtually all chemicals it is possible to say something about the uncertainty distribution regarding dose-response relationships. The process begins by selecting a set of cancer and noncancer end points and applying the full distribution of chemical potencies (including a data-driven probability of zero potency) to

[9]Chapter 5 addresses other missing defaults including that in the absence of chemical-specific data, EPA treats all members of the human population as though they are de facto equally susceptible to carcinogens that act via a linear MOA.

the unknown chemical in question. That initial distribution can then be narrowed by using the various types and levels of intermediate toxicity information.

At the simplest level, information on chemical structure can be used to bin chemicals in much the way that EPA uses chemical structures and physicochemical properties to perform quantitative structure activity relationship (QSAR) analyses for premanufacturing notices and for developing distributions of toxicity parameter values derived from data on representative data-rich chemicals (The Toxic Substances Control Act [TSCA] Section 5 New Chemicals Program [EPA 2007b]). At the next level, the distributions can be further refined by including toxicologic tests and other model or experimental data to create chemical categories. That has been done to fill in data gaps in the U.S. and Organisation for Economic Co-operation and Development high-production-volume chemical programs (OECD 2007). Chemical categories in those programs have been created to help to estimate actual values for the programs' short-term toxicity tests, but the underlying concepts could be applied to the development of distributions of cancer potencies or dose-response parameters for other chronic-toxicity end points. In the future, the results of intermediate mechanistic tests, in the context of growing understanding of toxicity networks and pathways, are likely to assist in selecting end points and estimating potency distributions. There are descriptions of how to make use of the observed correlation between carcinogenic potency and short-term toxicity values, such as the maximum tolerated dose (Crouch et al. 1982; Gold et al. 1984; Bernstein et al. 1985) and acute LD_{50} (Zeise et al. 1984, 1986; Crouch et al. 1987). The approach can be updated and expanded to include other data on toxicity from structure-activity and short-term tests. EPA is building databases that could facilitate such development (EPA 2007c; Dix et al. 2007); the National Research Council (NRC 2007b) advocates eventually relying on high and medium throughput assays for risk assessment. Finally, the most sophisticated level can involve development of toxic-potency distributions for chemicals whose structures are clearly similar to those of well-studied substances, such as polycyclic aromatic hydrocarbons and dioxin-like compounds, in a manner like current extrapolation methods (for example, see Boström et al. 2002; EPA 2003; van den Berg et al. 2006). In that way, the agency can take advantage of the wealth of intermediate toxicity data being generated in multiple settings at a stage when their precise implications for traditional dose-response estimation are not fully understood. EPA over the long term can develop probability distributions based on results of the intermediate assays, and the potency distribution for a chemical can become narrower as more data become available.

Those approaches have a number of limitations. For now, they would be based on results with chemicals that have already been tested in long-term bioassays. If selection for long-term bioassay testing is already associated with indications of toxicity, generalization of the results to untested chemicals could lead to an overestimation of the toxicity of the untested chemicals. The creation of potency distributions for unknown chemicals will have to include a database estimation of the probability of zero potency to reduce the possibility of systematic overestimation. Characterization of the uncertainty surrounding the potency estimates will be necessary, but it should be facilitated by the probabilistic nature of the approach. The lack of sufficient data to estimate potency distributions for a wide variety of end points poses a serious challenge. Creation of such a database may be feasible now for cancer and a small number of noncancer end points but not for many of the end points of great concern, such as developmental neurotoxicity, immune toxicity, and reproductive toxicity. Full implementation of such a system will require about 10-20 years of data and method development. The committee urges EPA to begin to develop the methods for such a system by using existing data and the wealth of intermediate toxicity data being generated now by U.S. and international chemical priority-setting programs (EC 1993, 1994, 1998, 2003; 65 Fed. Reg. 81686[2000]; NRC 2006b).

When necessary, EPA can prioritize efforts to establish missing default information based on the potential impact of this information on the estimated benefits of regulatory action. This impact is most likely to be substantial for chemicals that have exposure levels that could change substantially in response to regulation (for example, chemicals that might be substituted for other chemicals that undergo more stringent control), and for chemicals whose physical and chemical properties increase the likelihood of their relative toxicity.

PERFORMING MULTIPLE RISK CHARACTERIZATIONS FOR ALTERNATIVE MODELS

The current management of defaults resembles an all-or-none approach in that EPA often quantifies the dose-response relationship for one set of assumptions—either the default or whatever alternative to the default the agency adopts. Model uncertainty is discussed qualitatively; EPA discusses the scientific merits of competing assumptions.

In the long term, the committee envisions research leading to improved descriptions of model uncertainty (see Chapter 4). In the near term, sensitivity analysis could be performed when risk estimates for alternative hypotheses that are sufficiently supported by evidence are reported. This approach would require development of a framework with criteria for judging when such an analysis should be performed. The goal is not to present the multitude of possible risk estimates exhaustively but to present a small number of exemplar, plausible cases to provide the risk manager a context for understanding additional uncertainty contributed by considering assumptions other than the default. The committee acknowledges the difficulty of assigning probabilities to alternative estimates in the face of a lack of scientific understanding related to the defaults and acknowledges that much work is needed to move toward a more probabilistic approach to model uncertainty (see Chapter 4).

The standard for reporting alternative risk estimates should be less stringent than the "clearly superior" standard recommended for use of alternatives in place of the default. The committee finds that alternative risk estimates should be reported if they are "comparably" plausible relative to the risk estimate based on the default. The standard of comparability should not be interpreted to mean that the alternative must be at least as plausible as the default; this makes sense given that the alternative risk estimates provide information on the implications of tradeoffs associated with the interventions or options to address a given risk and that a risk manager might be interested in possible outcomes even if they are less than 50% probable. The comparability standard, however, does rule out risk estimates that are possibly valid but that are based on assumptions that are substantially less plausible than the default. The purposes are to help to ensure that the set of risk estimates to be considered by the risk manager remains manageable and to prevent distraction by risk estimates that are unlikely to be valid. In the final analysis, making the term *comparable* operational will depend on EPA's deciding how large a probability it is willing to accept that its risk assessment omitted the true risk. EPA should consider developing guidance that explicitly directs risk assessors to present a broader array of risk estimates in "high stakes" risk assessment situations, that is, situations where there are potentially important countervailing risks or economic costs associated with mitigation of a target risk. The guidance should take into account the analytic cost of developing more extensive information, including the potential additional delay (see discussion of value of information in Chapter 3).

As in the case of the "clearly superior" standard to replace the default, the agency should establish guidance for evaluation of plausibility and should issue specific criteria for the demonstration that an alternative is "comparably plausible." EPA should exclude from consideration alternative risk estimates that fail to satisfy the "reasonably" plausible criteria, because they can distract attention from the possibilities that have a reasonable level of scientific support. Specifically, the committee discourages EPA from the regular (pro forma) reporting that the risk posed by an evaluated chemical "may be as small as zero" unless there is scientific evidence that raises this possibility to the requisite level of plausibility. Under the proposed approach, the risk assessor would describe, to the extent possible, the relative scientific merits of alternative assumptions and the factors that make the assumptions as "comparably plausible" relative to the default (and the factors that cause it to fall short of a "clearly superior" standard). Such a characterization would identify the risk estimate associated with the default assumptions and identify that estimate as the appropriate basis of risk management. Nonetheless, the risk assessment would also report a small number of other plausible exemplar assessments to convey the uncertainty associated with the preferred risk estimate. That recommendation is consistent with the National Research Council recommendation (NRC 2006a) that encouraged EPA to report risk estimates corresponding to alternative assumptions in its risk assessments.

The level of detail in and scientific support for the alternative risk estimates should be tailored to be appropriate for the type of questions that the risk assessment is addressing (see Chapter 3). If potential tradeoffs associated with intervention options under evaluation are modest, less detail is needed to discriminate among the intervention options. For example, while maintaining designation of the risk calculated with the default assumptions as the primary estimate, it may be sufficient to provide a range of risk estimates without detailed information about the relative plausibility of alternative values within the range; the information can then be used in screening assessments to identify options whose desirability can be established robustly in the face of uncertainty. Because it is not always possible to know what options will be evaluated, simple characterizations of uncertainty can serve as a starting point for later assessments of alternative options. In all cases, refinement of the uncertainty characterization can proceed in an iterative fashion as needed to address either more serious tradeoffs or the evaluation of options and tradeoffs that were not initially contemplated. The key point is that the options to be evaluated drive the level of detail needed in the assessment (see Chapter 3).

Advantages of Multiple Risk Characterizations

Presenting a full risk characterization for models other than the default confers several benefits on the risk-assessment process. Retaining alternative risk estimates in the final risk-assessment results gives the risk manager wider latitude to understand the tradeoffs among the risk-management options. However, it is important that any evaluation of the range of risk-assessment outcomes take into account EPA's mandate to protect public health and the environment. The committee recommends that EPA quantify the implications of using an alternative assumption when it elects to depart from a default assumption. In particular, EPA should describe how use of a default and the selected alternative influences the risk estimate for the risk-management options under consideration. For example, if a risk assessment that departs from default assumptions identifies chemical A as the lowest-risk chemical to use in a production process rather than chemical B, it should also describe which chemical would pose the lower risk if the default assumption were used.

It is important for EPA to emphasize that only one assumption deserves primary consideration for risk characterization and risk management. If alternative assumptions are presented as "comparably plausible," the default must be highlighted and given deference.

The proposed approach more completely characterizes the uncertainty in the resulting risk estimate. As explained in Chapter 3, identifying the most appropriate course of action may depend on the degree of uncertainty associated with a risk estimate. Under the framework (Chapter 8), when there are multiple control options and multiple causal models, highlighting the model uncertainty can facilitate finding the optimal choices. Clear standards for departure from defaults can provide incentives for third parties to produce research in that they will know what data need to be produced that could influence the risk-assessment process. Finally, the approach facilitates the setting of priorities among research needs as a necessary component of value-of-information analysis (see Chapter 3).

CONCLUSIONS AND RECOMMENDATIONS

EPA's current policy on defaults calls for evaluating all relevant and available data first and considers defaults only when it is determined that data are not available or unusable. It is not known to what extent that is practiced, in contrast with judging the adequacy of available data to depart from a default. Whatever the case, defaults need to be maintained for the steps in risk assessment that require inferences or to fill common data gaps. Criteria are needed for judging whether, in specific cases, data are adequate to support a different inference from the default (or whether data are sufficient to justify departure from a default). The committee urges EPA to delineate what evidence will determine how it makes these judgments, and how that evidence will be interpreted and questioned. Providing a credible and consistent approach to defaults is essential to have a risk-assessment process to support regulatory decision-making.

The committee provides the following recommendations to strengthen the use of defaults in EPA:

- EPA should continue and expand use of the best, most current science to support or revise its default assumptions. The committee is reluctant to specify a schedule for revising these default assumptions. Factors EPA should take into consideration in setting priorities for such revisions include (1) the extent to which the current default is inconsistent with available science; (2) the extent to which a revised default would alter risk estimates; and (3) the public health (or ecologic) importance of risk estimates that would be influenced by a revision to the default.
- EPA should work toward the development of explicitly stated defaults to take the place of implicit or missing defaults. Key priorities should be development of default approaches to support risk estimation for chemicals lacking chemical-specific information to characterize individual susceptibility to cancer (see Chapter 5) and to develop a dose-response relationship. With respect to chemicals that have inadequate data to develop a dose-response relationship, information is currently available to make progress on cancer and a limited number of noncancer end points. EPA should also begin developing methods that take advantage of information already available in the U.S. or by international prioritization programs with a goal of creating a comprehensive system over the next 10 to 20 years. When necessary, EPA can prioritize efforts to target chemicals for which this information is most likely to influence the estimated benefits of regulatory action.
- In the next 2-5 years, EPA should develop clear criteria for the level of evidence needed to justify use of alternative assumptions in place of defaults. The committee recommends that departure should occur only when the evidence of the plausibility of the alternative is clearly superior to the evidence of the value of the default. In addition to a general standard for the level of evidence needed for use of alternative assumptions, EPA should

describe specific criteria that must be addressed for use of alternatives to each particular default.

- When none of the alternative risk estimates achieves a level of plausibility sufficient to justify use in place of a default, EPA should characterize the impact of the uncertainty associated with use of the default assumptions. To the extent feasible, the characterization should be quantitative. In the next 2-5 years, EPA should develop criteria for the listing of the alternative values, limiting attention to assumptions whose plausibility is at least comparable with that of the plausibility of the default. The goal is not to present the multitude of possible risk estimates exhaustively but to present a small number of exemplar, plausible cases to provide a context for understanding the uncertainty in the assessment. The committee acknowledges the difficulty of assigning probabilities to alternative estimates in the face of a lack of scientific understanding related to the defaults and acknowledges that much work is needed to move toward a more probabilistic approach to model uncertainty.
- When EPA elects to depart from a default assumption, it should quantify the implications of using an alternative assumption, including describing how use of the default and the selected alternative influences the risk estimate for risk-management options under consideration.
- EPA needs to more clearly elucidate a policy on defaults and provide guidance on its implementation and on evaluation of its impact on risk decisions and on efforts to protect the environment and public health.

REFERENCES

Allen, B.C., K.S. Crump, and A.M. Shipp. 1988. Correlations between carcinogenic potency of chemicals in animals and humans. Risk. Anal. 8(4):531-544.

Bernstein, L., L.S. Gold, B.N. Ames, M.C. Pike, and D.G. Hoel. 1985. Some tautologous aspects of the comparison of carcinogenic potency in rats and mice. Fundam. Appl. Toxicol. 5(1):79-86.

Boström, C.E., P. Gerde, A. Hanberg, B. Jernström, C. Johansson, T. Kyrklund, A. Rannug, M. Törnqvist, K. Victorin, and R. Westerholm. 2002. Cancer risk assessment, indicators, and guidelines for polycyclic aromatic hydrocarbons in the ambient air. Environ. Health Perspect. 110(Suppl. 3):451-488.

Breyer, S. 1992. Breaking the Vicious Circle: Toward Effective Risk Regulation. Cambridge, MA: Harvard University Press.

Clewell, H.J. III, M.E. Andersen, and H.A. Barton. 2002. A consistent approach for the application of pharmacokinetic modeling in cancer and noncancer risk assessment. Environ. Health Perspect. 110(1):85-93.

Cohen, J., D. Bellinger, W. Connor, P. Kris-Etherton, R. Lawrence, D. Savitz, B. Shaywitz, S. Teutsch, and G. Gray. 2005. A quantitative risk-benefit analysis of changes in population fish consumption. Am. J. Prev. Med. 29(4):325-334.

Crawford, M., and R. Wilson. 1996. Low-dose linearity: The rule or the exception? Hum. Ecol. Risk Assess. 2(2):305-330.

Crouch, E.A.C., J. Feller, M.B. Fiering, E. Hakanoglu, R. Wilson, and L. Zeise. 1982. Health and Environmental Effects Document: Non-Regulatory and Cost Effective Control of Carcinogenic Hazard. Prepared for the Department of Energy, Health and Assessment Division, Office of Energy Research, by Energy and Environmental Policy Center, Harvard University, Cambridge, MA. September 1982.

Crouch, E., R. Wilson, and L. Zeise. 1987. Tautology or not tautology? Toxicol. Environ. Health 20(1-2):1-10.

DeWoskin, R.S., J.C. Lipscomb, C. Thompson, W.A. Chiu, P. Schlosser, C. Smallwood, J. Swartout, L. Teuschler, and A. Marcus. 2007. Pharmacokinetic/physiologically based pharmacokinetic models in integrated risk information system assessments. Pp. 301-348 in Toxicokinetics and Risk Assessment, J.C. Lipscomb and E.V. Ohanian, eds. New York: Informa Healthcare.

Dix, D.J., K.A. Houck, M.T. Martin, A.M. Richard, R.W. Setzer, and R.J. Kavlock. 2007. The ToxCast program for prioritizing toxicity testing of environmental chemicals. Toxicol. Sci. 95(1):5-12.

EC (European Commission). 1993. Commission Directive 93/67/EEC of 20 July 1993, Laying down the Principles for the Assessment of Risks to Man and the Environment of Substances Notified in Accordance with Council Directive 67/548/EEC. Official Journal of the European Communities L227:9-18.

EC (European Commission). 1994. Commission Regulation (EC) No. 1488/94 of 28 June 1994, Laying down the Principles for the Assessment of Risks to Man and the Environment of Existing Substances in Accordance with Council Regulation (EEC) No793/93. Official Journal of the European Communities L161:3-11 [online]. Available: http://www.unitar.org/cwm/publications/cbl/ghs/Documents_2ed/C_Regional_Documents/85_EU_Regulation148894EC.pdf [accessed Jan. 25, 2008].

EC (European Commission). 1998. Directive 98/8/EC of the European Parliament and of the Council of 16 February 1998 Concerning the Placing of Biocidal Products on the Market. Official Journal of the European Communities L123/1-L123/63 [online]. Available: http://ecb.jrc.it/legislation/1998L0008EC.pdf [accessed Jan. 28, 2008].

EC (European Commission). 2003. Technical Guidance Document in Support of Commission Directive 93/67/EEC on Risk Assessment for New Notified Substances and Commission Regulation (EC) 1488/94 on Risk Assessment for Existing Substances, and Directive 98/8/EC of the European Parliament and the Council Concerning the Placing of Biocidal Products on the Market, 2nd Ed. European Chemicals Bureau, Joint Research Centre, Ispra, Italy [online]. Available: http://ecb.jrc.it/home.php?CONTENU=/DOCUMENTS/TECHNICAL_GUIDANCE_DOCUMENT/EDITION_2/ [accessed Jan. 28, 2008].

EPA (U.S. Environmental Protection Agency). 1986. Guidelines for the Health Risk Assessment of Chemical Mixtures. EPA/630/R-98/002. Office of Research and Development, U.S. Environmental Protection Agency, Washington, DC. September 1986 [online]. Available: http://www.epa.gov/ncea/raf/pdfs/chem_mix/chemmix_1986.pdf [accessed Jan. 24, 2008].

EPA (U.S. Environmental Protection Agency). 1991a. Guidelines for Developmental Toxicity Risk Assessment. EPA/600/FR-91/001. Risk Assessment Forum, U.S. Environmental Protection Agency, Washington, DC. December 1991 [online]. Available: http://www.epa.gov/NCEA/raf/pdfs/devtox.pdf [accessed Jan. 10, 2008].

EPA (U.S. Environmental Protection Agency). 1991b. Alpha-2μ-Globulin: Association with Chemically-Induced Renal Toxicity and Neoplasia in the Male Rat. EPA/625/3-91/019F. Prepared for Risk Assessment Forum, U.S. Environmental Protection Agency, Washington, DC. February 1991.

EPA (U.S. Environmental Protection Agency). 1994. Methods for Derivation of Inhalation Reference Concentrations and Application of Inhalation Dosimetry. EPA/600/8-90/066F. Environmental Criteria and Assessment Office, Office of Health and Environmental Assessment, Office of Research and Development, U.S. Environmental Protection Agency, Research Triangle Park, NC. October 1994 [online]. Available: http://cfpub.epa.gov/ncea/cfm/recordisplay.cfm?deid=71993 [accessed Jan. 24, 2008].

EPA (U.S. Environmental Protection Agency). 1996. Guidelines for Reproductive Toxicity Risk Assessment. EPA/630/R-96/009. Risk Assessment Forum, U.S. Environmental Protection Agency, Washington, DC. October 1996 [online]. Available: http://www.epa.gov/ncea/raf/pdfs/repro51.pdf [accessed Jan. 10, 2008].

EPA (U.S. Environmental Protection Agency). 1997. Guiding Principles for Monte Carlo Analysis. EPA/630/R-97/001. Risk Assessment Forum, U.S. Environmental Protection Agency, Washington, DC. March 1997 [online]. Available: http://www.epa.gov/ncea/raf/montecar.pdf [accessed Jan. 7, 2008].

EPA (U.S. Environmental Protection Agency). 1998a. Guidelines for Neurotoxicity Risk Assessment. EPA/630/R-95/001F. Risk Assessment Forum, U.S. Environmental Protection Agency, Washington, DC. April 1998 [online]. Available: http://www.epa.gov/NCEA/raf/pdfs/neurotox.pdf [accessed Jan. 24, 2008].

EPA (U.S. Environmental Protection Agency). 1998b. Assessment of Thyroid Follicular Cell Tumors. EPA/630/R-97-002. Risk Assessment Forum, U.S. Environmental Protection Agency, Washington, DC. March 1998 [online]. Available: http://www.epa.gov/ncea/pdfs/thyroid.pdf [accessed Jan. 25, 2008].

EPA (U.S. Environmental Protection Agency). 2000a. Risk Characterization Handbook. EPA-100-B-00-002. Office of Science Policy, Office of Research and Development, U.S. Environmental Protection Agency, Washington, DC. December 2000 [online]. Available: http://www.epa.gov/OSA/spc/pdfs/rchandbk.pdf [accessed Feb. 6, 2008].

EPA (U.S. Environmental Protection Agency). 2000b. Supplementary Guidance for Conducting Health Risk Assessment of Chemical Mixtures. EPA/630/R-00/002. Risk Assessment Forum, U.S. Environmental Protection Agency, Washington, DC. August 2000 [online]. Available: http://www.epa.gov/ncea/raf/pdfs/chem_mix/chem_mix_08_2001.pdf [accessed Jan. 7, 2008].

EPA (U.S. Environmental Protection Agency). 2001. Toxicological Review of Chloroform (CAS No. 67-66-3) In Support of Summary Information on the Integrated Risk Information System (IRIS). EPA/635/R-01/001. U.S. Environmental Protection Agency, Washington, DC. October 2001 [online]. Available: http://www.epa.gov/iris/toxreviews/0025-tr.pdf [accessed Jan. 25, 2008].

EPA (U.S. Environmental Protection Agency). 2002a. A Review of the Reference Dose and Reference Concentration Processes. Final report. EPA/630/P-02/002F. Risk Assessment Forum, U.S. Environmental Protection Agency, Washington, DC. December 2002 [online]. Available: http://www.epa.gov/iris/RFD_FINAL%5B1%5D.pdf [accessed Jan. 14, 2008].

EPA (U.S. Environmental Protection Agency). 2002b. Determination of the Appropriate FQPA Safety Factor(s) in Tolerance Assessment. Office of Pesticide Programs, U.S. Environmental Protection Agency, Washington, DC. February 28, 2002 [online]. Available: http://www.epa.gov/oppfead1/trac/science/determ.pdf [accessed Jan. 25, 2008].

EPA (U.S. Environmental Protection Agency). 2003. Exposure and Human Health Reassessment of 2,3,7,8-Tetrachlorodibenzo-*p*-Dioxin (TCDD) and Related Compounds. NAS Review Draft. National Center for Environmental Assessment, Office of Research and Development, U.S. Environmental Protection Agency, Washington, DC. December 2003 [online]. Available: http://www.epa.gov/NCEA/pdfs/dioxin/nas-review/ [accessed Jan. 9, 2008].

EPA (U.S. Environmental Protection Agency). 2004a. Risk Assessment Principles and Practices: Staff Paper. EPA/100/B-04/001. Office of the Science Advisor, U.S. Environmental Protection Agency, Washington, DC. March 2004 [online]. Available: http://www.epa.gov/osa/pdfs/ratf-final.pdf [accessed Jan. 9, 2008].

EPA (U.S. Environmental Protection Agency). 2004b. Toxicological Review of Boron and Compounds (CAS No. 7440-42-8) In Support of Summary Information on the Integrated Risk Information System (IRIS). EPA 635/04/052. U.S. Environmental Protection Agency, Washington, DC. June 2004 [online]. Available: http://www.epa.gov/iris/toxreviews/0410-tr.pdf [accessed Jan. 25, 2008].

EPA (U.S. Environmental Protection Agency). 2005a. Guidelines for Carcinogen Risk Assessment. EPA/630/P-03/001F. Risk Assessment Forum, U.S. Environmental Protection Agency, Washington, DC. March 2005 [online]. Available: http://cfpub.epa.gov/ncea/cfm/recordisplay.cfm?deid=116283 [accessed Feb. 7, 2007].

EPA (U.S. Environmental Protection Agency). 2005b. Supplemental Guidance for Assessing Susceptibility for Early-Life Exposures to Carcinogens. EPA/630/R-03/003F. Risk Assessment Forum, U.S. Environmental Protection Agency, Washington, DC. March 2005 [online]. Available: http://cfpub.epa.gov/ncea/cfm/recordisplay.cfm?deid=160003 [accessed Jan. 4, 2008].

EPA (U.S. Environmental Protection Agency). 2006. Modifying EPA Radiation Risk Models Based on BEIR VII. Draft White Paper. Office of Radiation and Indoor Air, U.S. Environmental Protection Agency. August 1, 2006 [online]. Available: http://www.epa.gov/rpdweb00/docs/assessment/white-paper8106.pdf [accessed Jan. 25, 2008].

EPA (U.S. Environmental Protection Agency). 2007a. Toxicological Review of 1,1,1-Trichloroethane (CAS No. 71-55-6) In Support of Summary Information on the Integrated Risk Information System (IRIS). EPA/635/R-03/013. U.S. Environmental Protection Agency, Washington, DC. August 2007 [online]. Available: http://www.epa.gov/IRIS/toxreviews/0197-tr.pdf [accessed Jan. 25, 2008].

EPA (U.S. Environmental Protection Agency). 2007b. Chemical Categories Report. New Chemicals Program, Office of Pollution Prevention and Toxics, U.S. Environmental Protection Agency [online]. Available: http://www.epa.gov/opptintr/newchems/pubs/chemcat.htm [accessed Jan. 25, 2008].

EPA (U.S. Environmental Protection Agency). 2007c. Distributed Structure-Searchable Toxicity (DSSTox) Database Network. Computational Toxicology Program, U.S. Environmental Protection Agency [online]. Available: http://www.epa.gov/comptox/dsstox/ [accessed Jan. 25, 2008].

EPA (U.S. Environmental Protection Agency). 2007d. Human Health Research Program: Research Progress to Benefit Public Health. EPA/600/F-07/001. Office of Research and Development, U.S. Environmental Protection Agency, Washington, DC. April 2007 [online]. Available: http://www.epa.gov/hhrp/files/g29888-gpi-gpo-epa-brochure.pdf [accessed Oct. 21, 2008]

EPA SAB (U.S. Environmental Protection Agency, Science Advisory Board). 1997. An SAB Report: Guidelines for Cancer Risk Assessment. Review of the Office of Research and Development's Draft Guidelines for Cancer Risk Assessment. EPA-SAB-EHC-97-010. Science Advisory Board, U.S. Environmental Protection Agency, Washington, DC. September 1997 [online]. Available: http://yosemite.epa.gov/sab/sabproduct.nsf/6A6D30CFB1812384852571930066278B/$File/ehc9710.pdf [accessed Jan. 25, 2008].

EPA SAB (U.S. Environmental Protection Agency, Science Advisory Board). 1999. Review of Revised Sections of the Proposed Guidelines for Carcinogen Risk Assessment. Review of the Draft Revised Cancer Risk Assessment Guidelines. EPA-SAB-EC-99-015. Science Advisory Board, U.S. Environmental Protection Agency, Washington, DC. July 1999 [online]. Available: http://yosemite.epa.gov/sab/sabproduct.nsf/857F46C5C8B4BE4985257193004CF904/$File/ec15.pdf [accessed Jan. 25, 2008].

EPA SAB (U.S. Environmental Protection Agency, Science Advisory Board). 2000. Review of EPA's Draft Chloroform Risk Assessment. EPA-SAB-EC-00-009. Science Advisory Board, U.S. Environmental Protection Agency, Washington, DC. April 2000 [online]. Available: http://yosemite.epa.gov/sab/sabproduct.nsf/D0E41CF58569B1618525719B0064BC3A/$File/ec0009.pdf [accessed Jan. 25, 2008].

EPA SAB (U.S. Environmental Protection Agency, Science Advisory Board). 2004a. Commentary on EPA's Initiatives to Improve Human Health Risk Assessment. Letter from Rebecca Parkin, Chair of the SAB Integrated Human Exposure, and William Glaze, Chair of the Science Advisory Board, to Michael O. Levitt, Administrator, U.S. Environmental Protection Agency, Washington, DC. EPA-SAB-COM-05-001. October 24, 2004 [online]. Available: http://yosemite.epa.gov/sab/sabproduct.nsf/36a1ca3f683ae57a85256ce9006a32d0/733E51AAE52223F18525718D00587997/$File/sab_com_05_001.pdf [accessed Oct. 21, 2008].

EPA SAB (U.S. Environmental Protection Agency, Science Advisory Board). 2004b. EPA's Multimedia, Multpathway, and Multireceptor Risk Assessment (3MRA) Modeling System. EPA-SAB-05-003. Science Advisory Board, U.S. Environmental Protection Agency, Washington, DC [online]. Available: http://yosemite.epa.gov/sab/sabproduct.nsf/99390EFBFC255AE885256FFE00579745/$File/SAB-05-003_unsigned.pdf [accessed Jan. 25, 2008].

EPA SAB (U.S. Environmental Protection Agency, Science Advisory Board). 2006. Science and Research Budgets for the U.S. Environmental Protection Agency for Fiscal Year 2007. EPA-SAB-ADV-06-003. Science Advisory Board, Office of the Administrator, U.S. Environmental Protection Agency, Washington, DC. March 30, 2006 [online]. Available: http://www.epa.gov/science1/pdf/sab-adv-06-003.pdf [accessed Dec. 5, 2007].

EPA SAB (U.S. Environmental Protection Agency, Science Advisory Board). 2007. Comments on EPA's Strategic Research Directions and Research Budget for FY 2008. EPA-SAB-07-004. Science Advisory Board, Office of the Administrator, U.S. Environmental Protection Agency, Washington, DC. March 13, 2007 [online]. Available: http://www.epa.gov/science1/pdf/sab-07-004.pdf [accessed Dec. 5, 2007].

Finkel, A.M. 1994. The case for "plausible conservatism" in choosing and altering defaults. Appendix N-1 in Science and Judgment in Risk Assessment. Washington, DC: National Academy Press.

Finkel, A.M. 1997. Disconnect brain and repeat after me: "Risk Assessments is too conservative." Ann. N.Y. Acad. Sci. 837:397-417.

Finkel, A.M. 2003. Too much of the "Red Book" is still (!) ahead of its time. Hum. Ecol. Risk Assess. 9(5):1253-1271.

Gilman, P. 2006. Response to "IRIS from the Inside." Risk Anal. 26(6):1413.

Gold, L.S., C.B. Sawyer, R. Magaw, G.M. Backman, M. de Veciana, R. Levinson, N.K. Hooper, W.R. Havender, L. Bernstein, R. Peto, M.C. Pike, and B.N. Ames. 1984. A carcinogenic potency database of the standardized results of animal bioassays. Environ. Health Perspect. 58:9-319.

Hattis, D., and M.K. Lynch. 2007. Empirically observed distributions of pharmacokinetic and pharmacodynamic variability in humans: Implications for the derivation of single point component uncertainty factors providing equivalent protection as existing RfDs. Pp. 69-93 in Toxicokinetics and Risk Assessment, J.C. Lipscomb, and E.V. Ohanian, eds. New York: Informa Healthcare.

Kaldor, J.M., N.E. Day, and K Hemminki. 1988. Quantifying the carcinogenicity of antineoplastic drugs. Eur. J. Cancer Clin. Oncol. 24(4):703-711.

McClellan, R.O., and D.W. North. 1994. Making full use of scientific information in risk assessment. Appendix N-2 in Science and Judgment in Risk Assessment. Washington, DC: National Academy Press.

Mills, A. 2006. IRIS from the Inside. Risk Anal. 26(6):1409-1410.

NRC (National Research Council). 1983. Risk Assessment in the Federal Government: Managing the Process. Washington, DC: National Academy Press.

NRC (National Research Council). 1994. Science and Judgment in Risk Assessment. Washington, DC: National Academy Press.

NRC (National Research Council). 2006a. Health Risks from Dioxin and Related Compounds: Evaluation of the EPA Reassessment. Washington, DC: The National Academies Press.

NRC (National Research Council). 2006b. Toxicity Testing for Assessment of Environmental Agents: Interim Report. Washington, DC: The National Academies Press.

NRC (National Research Council). 2007a. Quantitative Approaches to Characterizing Uncertainty in Human Cancer Risk Assessment Based on Bioassay Results. Second Workshop of the Standing Committee on Risk Analysis Issues and Reviews, June 5, 2007, Washington, DC [online]. Available: http://dels.nas.edu/best/risk_analysis/workshops.shtml [accessed Nov. 27, 2007].

NRC (National Research Council). 2007b. Toxicity Testing in the Twenty-first Century: A Vision and a Strategy. Washington, DC: The National Academies Press.

OECD (Organisation for Economic Co-operation and Development). 2007. Guidance on Grouping Chemicals. Series on Testing and Assessment No. 80. ENV/JM/MONO(2007)28. Environment Directorate, Joint Meeting of the Chemicals Committee and the Working Party on Chemicals, Pesticides and Biotechnology, Organisation for Economic Co-operation and Development. September 28, 2007 [online]. Available: http://appli1.oecd.org/olis/2007doc.nsf/linkto/env-jm-mono(2007)28 [accessed Jan. 25, 2008].

OMB (Office of Management and Budget). 1990. Current Regulatory Issues in Risk Assessment and Risk Management in Regulatory Program of the United States, April 1, 1990-March 31, 1991. Office of Management and Budget, Washington, DC.

Pacala, S.W., E. Bulte, J.A. List, and S.A. Levin. 2003. False alarm over environmental false alarms. Science 301(5637):1187-1188.

Perhac, R.M. 1996. Does Risk Aversion Make a Case for Conservatism? Risk Health Saf. Environ. 7:297.

Risk Policy Report. 2004. EPA Boron Review Reflects Revised Process to Boost Scientific Certainty. Inside EPA's Risk Policy Report 11(8):3.

van den Berg, M., L.S. Birnbaum, M. Denison, M. De Vito, W. Farland, M. Feeley, H. Fiedler, H. Hakansson, A. Hanberg, L. Haws, M. Rose, S. Safe, D. Schrenk, C. Tohyama, A. Tritscher, J. Tuomisto, M. Tysklind, N. Walker, and R.E. Peterson. 2006. The 2005 World Health Organization reevaluation of human and mammalian toxic equivalency factors for dioxins and dioxin-like compounds. Toxicol. Sci. 93(2):223-241.

Zeise, L. 1994. Assessment of carcinogenic risks in the workplace. Pp. 113-122 in Chemical Risk Assessment and Occupational Health: Current Applications, Limitations and Future Prospects, C.M. Smith, D.C. Christiani, and K.T. Kelsey, eds. Westport, CT: Auburn House.

Zeise, L., R. Wilson, and E.A.C. Crouch. 1984. Use of acute toxicity to estimate carcinogenic risk. Risk Anal. 4(3):187-199.

Zeise, L., E.A.C. Crouch, and R. Wilson. 1986. A possible relationship between toxicity and carcinogenicity. J. Am. Coll. Toxicol. 5(2):137-151.

Zhao, Q., J. Unrine, and M. Dourson. 1999. Replacing the default values of 10 with data-derived values: A comparison of two different data-derived uncertainty factors for boron. Hum. Ecol. Risk Asses. 5(5):973-983.

7

Implementing Cumulative Risk Assessment

INTRODUCTION AND DEFINITIONS

In the previous chapters, the committee proposed modifications of multiple risk-assessment steps to provide better insight into the health risks associated with exposure to individual chemicals, including characterization of uncertainty and variability. That reflects the focus of many risk-assessment applications in the Environmental Protection Agency (EPA) and elsewhere, which are often centered on evaluating risks associated with individual chemicals in the context of regulatory requirements or isolated actions, such as the issuance of an air permit for an industrial facility.

However, there is increasing concern among stakeholder groups (especially communities affected by environmental exposure) that such a narrow focus does not accurately capture the risks associated with exposure, given simultaneous exposure to multiple chemical and nonchemical stressors and other factors that could influence vulnerability. More generally, a primary aim of risk assessment should be to inform decision-makers about the public-health implications of various strategies for reducing environmental exposure, and omission of the above factors may not provide the information needed to discriminate among competing options accurately. Without additional modifications, risk assessment might become irrelevant in many decision contexts, and its application might exacerbate the credibility and communication gaps between risk assessors and stakeholders.

In part to address those complex issues, EPA has developed the Framework for Cumulative Risk Assessment (EPA 2003a). *Cumulative risk* is formally defined as the combination of risks posed by aggregate exposure to multiple agents or stressors in which *aggregate exposure* is exposure by all routes and pathways and from all sources of each given agent or stressor. Chemical, biologic, radiologic, physical, and psychologic stressors are all acknowledged as affecting human health and are potentially addressed in the multiple-stressor, multiple-effects assessments (Callahan and Sexton 2007). *Cumulative risk assessment* is therefore defined as analysis, characterization, and possible quantification of the combined risks to health or the environment posed by multiple agents or stressors (EPA 2003a).

As noted recently (Callahan and Sexton 2007), there are four key differences between EPA's cumulative risk-assessment paradigm and traditional human health risk assessments:

- Cumulative risk assessment is not necessarily quantitative.
- Cumulative risk assessment by definition evaluates the combined effects of multiple stressors rather than focusing on single compounds.
- Cumulative risk assessment focuses on population-based assessments rather than source-based assessments.
- Cumulative risk assessment extends beyond chemicals to include psychosocial, physical, and other factors.

In addition, an explicit component of the cumulative risk-assessment paradigm defined by EPA involves an initial planning, scoping, and problem-formulation phase (EPA 2003a), which the committee previously proposed as an important component of any risk assessment in Chapter 3. That involves bringing risk managers, risk assessors, and various stakeholders together early in the process to determine the major factors to be considered, the decision-making context, the timeline and related depth of analysis, and so forth. Planning and scoping ensure that the right questions are asked in the context of the assessment and that the appropriate suite of stressors is considered (NRC 1996).

The committee acknowledges the conceptual framework and broadened definitions of cumulative risk assessment as constituting a move toward making risk assessments more relevant to decision-making and to the concerns of affected communities. Many components of cumulative risk assessment (such as planning and scoping or explicit consideration of vulnerability) should be considered as standard features of any risk assessment in principle. In practice, however, EPA assessments conducted today can fall short of what is possible and what is supported by the agency's framework, and this chapter is directed at improvements in agency practice.

The chapter considers in detail some of the specific reasons why cumulative risk assessment might be needed, because the risk-management needs will inform necessary revisions of the analytic framework. First, even if the regulatory decision of interest were related to strategies to address a single chemical with a single route of exposure, consideration of other compounds and other factors may be necessary to inform the decision. Ignoring numerous agents or stressors that affect the same toxic process as the chemical of interest and omitting background processes could lead to risk assessments that, for example, assume population thresholds in circumstances when such thresholds may not exist. That issue has been largely addressed in Chapter 5 in relation to the need to evaluate background exposure and vulnerability factors to determine the likelihood that these factors could "linearize" an otherwise nonlinear mode of action (MOA). We do not treat this issue in further detail in this chapter other than to note that it is a crucial component of cumulative risk assessment and that it leads to potentially important exposure-assessment and epidemiologic and toxicologic data requirements.

Second, as alluded to above, the types of questions that are increasingly being asked of EPA require the tools and concepts of cumulative risk assessment. Communities concerned about environmental toxicants often wish to know whether environmental factors can explain observed or hypothesized disease trends or whether specific facilities are associated with important health burdens (and whether specific interventions could reduce those burdens). The relevance of standard risk-assessment methods in settings with vulnerable populations and multiple coexposures is being challenged by stakeholders, especially those with concerns about environmental justice (Israel 1995; Kuehn 1996). Addressing those issues requires an ability to evaluate multiple agents or stressors simultaneously—to consider exposures not in isolation but in the context of other community exposures and risk factors. In addition, many of the decisions faced by EPA and other stakeholders involve tradeoffs and complex interactions among multiple risk factors, and any analytic tool must be able to address these factors reasonably.

Although we propose in this chapter some modifications of the framework and practice of cumulative risk assessment to help EPA and other stakeholders to determine high-risk populations and discriminate among competing options, we recognize that the topic of cumulative risk assessment raises important questions about the bounds between risk assessment and other lines of evidence that may inform risk-related decisions. As the number and types of stressors and end points under consideration increase, decisions must be made about which dimensions should be considered as components of risk assessment as defined and used by EPA and others and which dimensions should be considered as ancillary information that can inform risk-management decisions but not considered as a components of risk assessment itself. That is in part a semantic distinction, but defining the bounds will be important in articulating recommendations for improving risk-analysis methods in EPA. Similarly, decisions must be made about the levels of complexity and quantification necessary for a given cumulative risk assessment in light of the decision context. This chapter emphasizes methods that can allow for the quantification of human health effects associated with exposure to chemical and nonchemical stressors, but we note that cumulative risk assessment can involve qualitative analyses and is not necessarily quantitative (EPA 2003a; Callahan and Sexton 2007), given that such analyses may be sufficient at times to discriminate among competing risk-management options.

Another boundary issue involves the contexts in which cumulative risk assessment would be able to yield useful information. Some of the questions that communities or other stakeholders are concerned about cannot and should not be answered by risk assessment even if refined techniques addressing cumulative risks are used. For example, questions like "What are the sources of environmental contaminants in our community that may be causing the most health problems?" or "What intervention strategies that we can adopt would most improve community health?" can be answered in principle with risk-assessment methods, but questions like "Should yet one more polluting facility be sited in our community?" or "Should there be mitigation because this low-income population lives much closer to sources of environmental contaminants than high-income populations?" are broader questions than can be answered by cumulative risk assessment alone. Clarifying the types of questions that cumulative risk assessment can and cannot answer but can support will be important in refining the cumulative risk-assessment tools and considering complementary analyses to aid in decision-making.

In this chapter, we briefly discuss some key settings in which cumulative risk assessment has been developed and applied in EPA, focusing on the problem context, the analytic methods used, and refinements that may be warranted. We consider proposed approaches derived from such fields as ecologic risk assessment and social epidemiology to construct cumulative risk models in the light of numerous stressors or end points, while maintaining focus on decisions relevant to EPA. We conclude by providing some specific guidance about how the committee believes that cumulative risk assessment needs to be developed further, including the use of clear and consistent terminology; methods to incorporate interactions between chemical and nonchemical stressors; the use of biomonitoring, epidemiologic, and surveillance data; the need to develop simpler analytic tools to support more wide-ranging analyses; and the related need to engage stakeholders throughout the cumulative risk-assessment process.

HISTORY OF CUMULATIVE RISK ASSESSMENT

The formal cumulative risk-assessment framework at EPA was developed recently, but relevant activity has occurred for decades. This historical overview is not meant to be exhaustive but rather aims to illustrate some of the different ways in which cumulative-risk issues have been addressed at different times in different offices in EPA.

One of the early applications of cumulative risk assessment in EPA was in the context of the Superfund program. Given the focus on specific hazardous-waste sites rather than single compounds, risk assessments need to capture the health effects of simultaneous exposures. EPA issued guidance documents focused on methods for addressing chemical mixtures (EPA 1986), which were relatively undetailed but established the general approach of first looking for evidence of health effects of the mixture of concern, then considering effects of a similar mixture if no such information were available, then addressing pairwise interactions if data were available, and finally presuming additivity if none of the prior information was available. The 1986 guidelines also distinguished between dose additivity (appropriate if the compounds of interest had the same MOA and the same health effects) and response additivity (which presumes independent MOAs). Data were available on some complex mixtures, such as diesel emissions and polychlorinated biphenyls, or mixtures similar to them; but in the majority of cases, dose additivity when the same MOA could be assumed was the default. Analyses of chemical mixtures constitute only one component of cumulative risk assessment, and the Superfund risk assessments did not extend beyond this realm, but the early assessments helped to establish the rationale and framework for consideration of multiple stressors.

Similarly, the 1996 amendments to the Safe Drinking Water Act required consideration of chemical mixtures in drinking water by explicitly stating that EPA shall conduct studies that "develop new approaches to the study of complex mixtures . . . especially to determine the prospects for synergistic or antagonistic interactions that may affect the shape of the dose-response relationship of the individual chemicals or microbes" (Pub. L. No. 104-182, 104th Cong. [1996]). These approaches have been most commonly applied to disinfection byproducts (DBPs): characterization of multiple routes of exposure to multiple DBPs with the same MOA, physiologically based pharmacokinetic models for each individual DBP, and risk characterization that used relative potency factors to aggregate across constituents (Teuschler et al. 2004). Although aggregate exposure assessments have been thoroughly constructed and the combination of dose addition for chemicals with similar MOAs and response addition for mixtures with different MOAs helped to expand the scope of the assessments, the scope of cumulative risk assessment did not consider nonchemical stressors, and insight about synergistic or antagonistic effects remained minimal. Uncertainty quantification was also minimal, and variability was characterized for some components of the risk assessment (such as heterogeneity in food and water consumption) but not others (such as vulnerability).

An important recent example of cumulative risk assessment was related to the Food Quality Protection Act (FQPA), which explicitly required EPA to assess aggregate exposures to pesticides across multiple exposure routes and to consider the cumulative effects of exposures to pesticides with the same MOAs (Pub. L. No. 104-170, 104 Cong. [1996]). Key work completed to date has included a cumulative risk assessment of organophosphorus (OP) pesticides (EPA 2006a). Given the fact that the OP pesticides have a common MOA (inhibition of cholinesterase activity), a cumulative assessment of all pesticides in the family was used. Components of the analysis that deviated from single-chemical risk assessment included consideration of coexposures through various exposure pathways (that is, in the case of a given food item, which pesticides are likely to be found together), consideration of aggregate exposures across multiple pathways, and calculation of relative potency factors to allow cumulative noncancer hazard indexes to be calculated. That work produced among the most detailed and comprehensive cumulative risk assessments conducted to date. However, no evidence was available to determine potential deviations from dose additivity, to incorporate pharmacokinetics explicitly into the dose-response assessment, or to consider interactions with nonchemical stressors or vulnerability other than mandated safety factors of 10 for infants and children. In addition, uncertainty quantification was not extensive, and the focus on margin-of-exposure calculations for individual routes of exposure makes it difficult to quantify the magnitude of harm at various exposure levels (as discussed in Chapter 5). As a general point, most publications in the peer-reviewed literature related to cumulative risk assessment have focused on pesticide health risks both because of the structure of the FQPA and because of availability of data on pesticides.

A final example of cumulative risk assessment in EPA is the National-Scale Air Toxics Assessment, an attempt to estimate the cancer and noncancer health effects of joint exposure to air toxics across the United States. The most recent assessment (EPA 2006b) considered 177 air toxics, used atmospheric-dispersion models to estimate concentrations on the basis of a national emissions inventory, linked the concentrations to population exposure, and estimated health risks. Cancer risks were calculated individually for each compound, given inhalation unit risks from EPA's Integrated Risk Information System database and other resources; synergistic and antagonistic effects were not considered. Noncancer effects were determined by estimating reference concentrations (RfCs) and adding the hazard quotients of individual compounds that had similar adverse health effects (not necessarily similar MOAs). Thus, the analysis clearly captured multiple agents or stressors, but, like the previous applications, did not introduce evidence beyond simple additivity, did not consider nonchemical stressors or vulnerability, and did not provide extensive insight about uncertainties. The study is also an example of the importance of characterizing exposures to multiple compounds in the current and modified noncancer risk-assessment frameworks: acrolein concentrations exceeded the RfC for a majority of the U.S. population, and this implies that other respiratory irritants (in spite of being below their individual RfCs) were considered to contribute to population health risks.

Thus, in part because of the risk-management questions and regulatory issues historically facing EPA, cumulative risk assessments to date have largely focused on aggregate exposure assessment and have generally not considered nonchemical stressors. However, in segments of EPA and the stakeholder community interested in environmental justice, discussions about cumulative risk assessment have focused on different dimensions of the methodology and extended beyond aggregate chemical-exposure issues. For example, a 2004 National Environmental Justice Advisory Council (NEJAC) report provided guidance about the short-term and long-term actions that EPA should take to implement the concepts in its Framework for Cumulative Risk Assessment with a focus on environmental justice (NEJAC 2004; Hynes and Lopez 2007). Among the important insights in the report were

- The need to distinguish between cumulative risks and cumulative impacts; although the report does not formally define these terms, both are mentioned explicitly throughout.
- The importance of considering nonchemical stressors in the context of a community assessment.
- The significance of vulnerability as a critical component of cumulative risk assessment, including differential sensitivity and susceptibility, differential exposure, differential preparedness to respond to an environmental insult, and differential ability to recover from the effects of an insult or stressor.

- The significance of community-based participatory research to implement cumulative risk assessment, both for capacity-building and to incorporate local data and knowledge into the analysis.
- The need to avoid analytic complexity that seriously delays decision-making and, in parallel, the value of efficient screening and priority-setting tools that can be used by all stakeholders and the necessity of qualitative information in domains where quantitative assessment is not likely in the near term.

The NEJAC report emphasized risks to communities, so some of the components (such as community-based participatory research) may not be applicable to national-scale or other broad-based cumulative risk assessments. Although cumulative risk assessment and community-based risk assessment have many features in common, they are not identical. Other components emphasized in the NEJAC report (such as explicit consideration of vulnerability and having a level of analytic complexity appropriate for the decision context) can be generalized beyond cumulative risk assessment to all forms of risk assessment, as stated in earlier chapters (such as Chapters 3 and 5). Regardless, the NEJAC report emphasized that multiple stakeholders perceive that the potential of cumulative risk assessment as articulated by EPA has not yet been met, primarily because many of the dimensions beyond aggregate chemical exposure assessment have not been formally incorporated.

Related to those issues are recent efforts at EPA to develop tools and techniques for community-based risk assessment, including assessment in the Community Action for a Renewed Environment program (EPA 2008a). Resources and simplified approaches for risk-based priority-setting are made available to communities (EPA 2004), but the approaches do not yet consider key dimensions of cumulative risk, such as nonchemical stressors, vulnerability, or multiple routes of exposure.

A final setting outside EPA in which the general concepts of cumulative risk assessment have been applied is the assessment of the global burden of disease related to environmental and other risk factors. It may not be directly relevant to EPA, given the primary focus on multifactorial global risk rankings (including many nonenvironmental stressors), but it provides some additional lessons related to the analytic challenges and potential information value of assessments that consider an array of diverse risk factors. As articulated by Ezzati et al. (2003), these global burden of disease analyses estimate the *population attributable fractions* associated with various risk factors, defined as the proportional reductions in population disease or mortality that would occur if exposure to a given risk factor were reduced to an alternative exposure scenario. The risk factors in question are as varied as diet, physical activity, smoking, and environmental and occupational exposures. Given the number of factors considered and the desire to develop indicators applicable to numerous countries (Ezzati et al. 2003), the methods used in connection with any individual risk factor were relatively simple. For example, the burden of disease associated with urban air pollution was estimated on the basis of particulate-matter concentrations, and the concentration-response function from a cohort mortality study in the United States was applied to all countries included in the analysis. The analytic methods took account of potential interactions between risk factors and distinguished between situations in which the direct effects of a risk factor are mediated through intermediate factors, in which effect modification occurs, and in which effects may be independent but exposures may be correlated. The analyses demonstrated approaches in which relatively simplified exposure and dose-response assessment could be applied to yield insight about relative contributions to disease patterns and approaches by which interactions among risk factors could be considered. However, it is important to note the considerable opportunities for mischaracterization of factors when attributable-risk methods are used

(Cox 1984, 1987; Greenland and Robins 1988; Greenland 1999; Greenland and Robins 2000), and these issues may grow in significance when the marginal benefits of control strategies are considered.

In conclusion, cumulative risk assessment has been applied in EPA and elsewhere in an increasing number of contexts over the past two decades, and, given the recent development of the Framework for Cumulative Risk Assessment and growing interest in numerous arms of EPA, the applications are expected to grow. The studies have generally been thorough in modeling distributions of aggregate exposures (albeit with limited characterization of uncertainty), and the approach to evaluate cumulative risk posed by multiple chemicals with similar MOAs has been developed reasonably as well (although with generally modest treatment of synergistic and antagonistic effects). However, cumulative risk assessments have generally not yet reached the potential implied by the stated definition; there has been less than optimal formal consideration of nonchemical stressors, aspects of vulnerability, background processes, and other factors that could be of interest to stakeholders concerned about effects of cumulative exposures. Stakeholder involvement has not been as comprehensive as guidelines would indicate would be optimal in most of the above applications, and the tools have not yet been developed to allow communities to engage in even simplified cumulative risk assessment (screening methods are generally restricted to single media and standard risk-assessment practice). Cumulative risk assessment has also been used to determine the risks posed by baseline exposures rather than the benefits of various risk-management strategies, and this use has implications for the methods developed and their interpretations.

Some of the omissions can be attributed to the fact that formal consideration of numerous simultaneous chemical, physical, and psychosocial exposures with evaluation of background disease processes and other dimensions of vulnerability could quickly become analytically intractable if the standard risk-assessment paradigm is followed, both because of the computational burden and because of the likelihood that important exposure and dose-response data will be missing. That points toward the need for simplification of risk-assessment tools in the spirit of iterative risk assessment, and it emphasizes that cumulative human health risk assessment could learn a great deal from such fields as ecologic risk assessment and social epidemiology, which have had to grapple with similar issues related to evaluation of the effects of numerous stressors on defined populations or geographic areas. The expanded scope of cumulative risk assessment that would be theoretically desired includes many elements outside EPA's standard practice, expertise, and regulatory functions, so there is clearly a need to define carefully how nonchemical stressors and aspects of vulnerability should most appropriately be considered. The following sections present approaches that can be used to expand the scope of cumulative risk assessment while keeping in mind the need for timeliness and EPA's regulatory mandates, in part by developing screening tools and by orienting analyses around well-defined risk-management objectives.

APPROACHES TO CUMULATIVE RISK ASSESSMENT

From the definitions and examples above, it is clear that cumulative risk assessment has a broad scope and an extremely ambitious mandate. In fact, it is difficult to imagine any risk assessment in which it would not be important to understand the effects of coexposures to agents or stressors that have similar MOAs (as articulated in Chapter 5) or to identify characteristics of the affected populations that could contribute to vulnerability to a given exposure. That is salient in a context of risk management, in which numerous chemical and nonchemical stressors could be simultaneously affected. The critical challenge from the perspective of the risk assessor is to devise an analytic scope and a level of complexity that are appropriate to the context in which cumulative risk assessment is used. Following some of the approaches outlined below could allow EPA to incorporate the aforementioned dimensions of cumulative risk assessment.

A few general approaches have been proposed in the literature; the most appropriate approach clearly is driven by the problem and decision context. Using approaches from ecologic risk assessment, Menzie et al. (2007) develop one type of application, an effects-based assessment. In this case, epidemiologic analyses or general surveillance data provide an indication that a defined population may be at increased risk, and the objective of the analysis is to determine which stressors influence the observed effects. An effects-based assessment is retrospective, so it does not fit neatly into a risk-management framework in which various control options are being weighed; but there are contexts in which strategies would be developed around specific end points, and many of the methods could be generalized to other approaches (including stressor-based assessments, as described below).

Menzie et al. recommend that risk assessors begin with a conceptual model that considers the subset of stressors that are plausibly associated with the health outcomes or other effects of interest. That step would dovetail with the proposed MOA assessment steps proposed in Chapter 5, including MOA evaluation, background and vulnerability assessment, and selection of a conceptual model, but beginning with the health outcome rather than the individual chemical. The next step proposed by Menzie et al. would be a screening assessment to determine a manageable number of factors that are most likely to contribute substantially to the observed effects; this is based in part on simple comparisons with reference values or discussions with stakeholders, and it may be a crucial element of the planning and scoping for the analysis. Stressors are then evaluated individually, then in combination without consideration of interactions, and finally with consideration of interactions and a reliance in part on standard epidemiologic techniques. Although many characteristics are shared by this approach and epidemiologic assessment, this is not identical with proposing that a formal site-specific epidemiologic investigation be conducted. In many community circumstances, epidemiologic investigations will not have adequate statistical power to link defined environmental exposures with observed health outcomes. However, epidemiologic concepts could be useful in framing the analysis and providing insight into the subset of stressors that merit more careful consideration, and knowledge could be leveraged from previously conducted epidemiologic studies. The primary value of this approach is that it emphasizes the need for characterization of coexposures and background processes that could influence the health outcomes of interest and the need to conduct initial screening assessments to construct an analytically tractable model.

A more common approach to risk management would be a stressor-based assessment, in which the cumulative risk assessment is initiated not by questions about the stressors that may explain observed or hypothesized health effects but by questions about the effects that may be associated (generally in a prospective assessment) with a defined set of stressors. A stressor-based assessment would often arise in a source-oriented analysis, in which stakeholders wish to assess the effects of a source (or the benefits of control strategies that address the source) but want to take account of the full array of chemical and nonchemical stressors that have similar health effects. The framework proposed (Menzie et al. 2007) begins with a conceptual model and involves a screening assessment followed by consideration of individual stressors followed by interactions among stressors, but a stressor-based assessment begins with the stressors and identification of the populations and end points that would be influenced by them. The MOA assessment steps outlined above would be central to this process, in that they would help to characterize the end points of interest, the related stressors, and factors that could influence variability in response to the stressors.

An important modification in the approach to cumulative risk assessment that could potentially alleviate some of the analytic challenges would involve an orientation around evaluation of risk-management options rather than characterization of problems (see Chapter 8 for a more extensive discussion of this proposed framework). The approaches presented above and most previous case examples would help to determine which stressors are of greatest concern with respect to a defined outcome in a defined subpopulation or what the burden of disease is in the context of simultaneous exposure to a number of stressors. However, cumulative risk assessment would be most valuable to both communities and decision-makers when it can provide information about the health implications of alternative control options. For example, a community may be choosing among alternatives for drinking-water disinfection, and it would be important to consider the effects of the changes in concentrations of all disinfection byproducts jointly, to consider simultaneous exposure to a number of waterborne pathogens, to consider all routes of exposure to key compounds of interest, and to identify vulnerable populations. Many of the analytic tools would be similar, but in a decision context different factors may be correlated or affected on the margin from those when baseline conditions are considered, and the stressors that are important to include may also differ. In other words, it is important to include a stressor only to the extent that it will influence the estimated benefits of a control strategy either in its estimation or in its interpretation. In principle, focusing on stressors relevant to risk-management strategies will help to ensure that analyses are aligned with EPA's mandated focus on chemical or biologic stressors while acknowledging the influence of nonchemical stressors. A modified version of the stressor-based paradigm from Menzie et al. oriented around discriminating among risk-management options is presented in Table 7-1.

Following that approach would have multiple fringe benefits. For example, evaluating background exposures and vulnerability factors will not only allow cumulative risk assessment after the committee's proposed revisions to the cancer and noncancer dose-response assessment paradigm (Chapter 5) but will also provide information that can be used in environmental-justice analyses focused on inequality in outcomes and help to bring risk assessment and environmental justice into a single analytic framework (Levy et al. 2006; Morello-Frosch and Jesdale 2006). The geospatial components of the exposure and vulnerability assessment could be mapped to communicate key information to stakeholders, who would be engaged throughout the analytic process in a community risk setting. Most important, as alluded to above, the approach would potentially result in a need to model only a subset of stressors formally; the remainder would contribute to a general understanding about background processes but would otherwise not need to be quantitatively characterized to determine the benefits of risk-management options.

In spite of the benefits, there clearly are limitations of both the bottom-up stressor-based and top-down effects-based approaches. In cumulative risk assessment, the scope and complexity of the problem can quickly exceed the capacity of stressor-based analyses, although the approach outlined above can help to maintain focus on the key stressors. Given the analytic challenges, there is a temptation to think that effects-based analyses would be more practical even though risk-management decisions are often stressor-based. However, the size and subtlety of the effects are generally beyond the reach of standard epidemiologic tools. The relative influence of stressor-based vs effects-based analyses clearly will depend on the problem framework, including the decision context and the geographic scale of the analysis.

In addition, although the proposed approaches provide guidance on how a complex system can be systematically evaluated to develop an analytically tractable cumulative risk assessment, data limitations may make quantitative analyses impractical for some cumula-

TABLE 7-1 Modified Version of Stressor-Based Cumulative-Risk-Assessment Approach from Menzie et al. (2007) Oriented Around Discriminating among Risk-Management Options

Step 1:

- Develop a conceptual model for the stressors of primary interest for the analysis (stressors that would be significantly influenced by any of the risk-management options under study). The model includes an MOA assessment, an assessment of background exposures to chemical and nonchemical stressors that may affect the same health outcome, and a vulnerability assessment that takes into account underlying disease processes in the population to which the chemicals in question may be adding.
- Identify the receptors and end points affected by these stressors.
- Review the conceptual model and stressors, receptors, and end points of interest with stakeholders in initial planning and scoping.

Step 2:

- Use epidemiologic and toxicologic evidence and screening-level benefit calculations to provide an initial evaluation of which stressors should be included in the cumulative risk assessment. Gather stakeholder feedback and review and re-evaluate planning and scoping for the analysis.
- Focus the assessment only on stressors that contribute to end points of interest for risk-management options (for example, stressors that contribute significantly to monetized benefits in benefit-cost analyses or stressors that influence an identified high-risk subpopulation) and are either differentially affected by different control strategies or influence the benefits of stressors that are differentially affected.

Step 3:

- Evaluate the benefits of different risk-management options with appropriate characterization of uncertainty, including quantification of the effects of individual stressors and bounding calculations of any possible interaction effects.

Step 4:

- If Step 3 is sufficient to discriminate among risk-management options given other economic, social, and political factors, conclude the analysis; otherwise, sequentially refine the analysis as needed, taking into account potential interactions among stressors.

tive risk assessments. In ecologic risk assessment, a rank-oriented approach has been used in a relative-risk model (RRM) to account for the fact that addressing cumulative effects of multiple chemical and nonchemical stressors may not otherwise be viable. The RRM was developed to evaluate simultaneously and comparatively the risk posed by multiple, dissimilar stressors to multiple receptors in heterogeneous environments on landscape scales. It was first developed in 1997 for an ecologic risk assessment of chemical stressors at Port Valdez, AK (Landis and Wiegers 1997) and later applied successfully to other risk assessments of ecosystems on various scales and with other stressors and end points (Landis et al. 2000; Obery and Landis 2002). One of its specific strengths is an ability to incorporate stakeholders' values readily in evaluating risks in multiple geographic areas with multiple stressors, habitats, and receptors. Although originally designed for ecologic concerns, risk to humans can be readily accommodated in its flexible framework.

Similarly, in the realm of social epidemiology, the complexities of simultaneous exposures to numerous physical and social environmental factors have been addressed in some applications with cumulative risk models based on summing dichotomous classifications (for example, 1 if more than one standard deviation above the mean for a given risk factor, otherwise 0) for numerous risk factors of interest. Those indicators are acknowledged as not capturing the relative weights of the various factors, but they avoid the need for numerous multiplicative interaction models and have been shown to be more predictive of health end points than single-risk-factor models (Evans 2003). When data are sufficient, more refined approaches based on relative risks rather than simply distributions of exposures may be useful.

A disadvantage of the approaches is their focus on ranking and scoring systems where weights do not necessarily correspond with relative risks, which can be difficult to interpret in situations where different risk-management strategies lead to different combinations of risk factor reductions without one strategy leading to greater reductions for all risk factors. Practices that move away from quantitative risk characterizations within a core component of risk assessment should be considered and implemented judiciously because the applicability and interpretability of the resulting assessments in a decision context can be severely limited. At a minimum, ranking approaches should be evaluated for their sensitivity to key input assumptions, and in settings where quantitative information is available, these approaches could be helpful in initial assessments for organizing information and determining whether a solution can be easily chosen or more complex analysis is needed to distinguish among options.

KEY CONCERNS AND PROPOSED MODIFICATIONS

The EPA cumulative risk-assessment paradigm recognizes an important issue and provides a useful conceptual framework, but substantial logistical barriers remain, and some core issues are largely unaddressed by the current framework. For example, as articulated by EPA (2003a), that about 20,000 pesticide products are on the market and 80,000 existing chemicals are on the Toxic Substances Control Act inventory makes it impractical to try to account for all relevant synergisms and antagonisms. More broadly, cumulative risk assessment requires extensive information beyond chemical toxicity and MOAs, including aggregate exposure data and information on population characteristics and nonchemical stressors. Therefore, EPA concludes in its Framework for Cumulative Risk Assessment that "identification of critical information and research needs may be the primary result of many cumulative risk assessment endeavors" (EPA 2003a, p. xii).

That statement may be correct, and it does reflect one important aim of risk assessment (to provide insight about key uncertainties that should be addressed to discriminate among risk-management options), but it implies that cumulative risk assessment would be largely uninformative for near-term decision-making, and this is a matter of concern, given the salience of the questions asked by cumulative risk assessment from the perspective of many stakeholders. The committee feels that the conclusion understates the value of less complex but more wide-ranging risk assessments and ignores the fact that an analysis focused on specific mitigation measures in a community will potentially have a more narrow scope than an attempt to characterize relative contributors to the burden of disease (as described in Table 7-1). That is, although there may be numerous theoretical combinations of exposures, only a subset will be relevant in choosing among various intervention options for a well-defined problem.

We propose below a series of short-term and long-term efforts, focusing on measures that could enhance the utility of cumulative risk assessment in the context of environmental decision-making.

Clarification of Terminology

Although the definition of cumulative risk assessment as articulated by EPA is comprehensive and well crafted, the fact that a cumulative risk assessment as defined (including nonchemical stressors and vulnerability) has never been done in the agency raises questions about whether the definition is practical in the near term without some modifications of current practice. The Framework for Cumulative Risk Assessment was published relatively recently, but research and regulatory action related to cumulative risks have been conducted for decades without much advancement beyond chemical stressors in a small number of contexts. In addition, the ways in which cumulative risk assessment is being considered vary greatly among offices in EPA and among different stakeholder groups, and this indicates the need for greater clarity in its aims and scope.

We propose that EPA explicitly define and maintain a conceptual distinction among cumulative risk assessment, cumulative impact assessment, and community-based risk assessment, which overlap but are conflated in many discussions. The terms have been defined (CEQ 1997) and recently discussed (NEJAC 2004), but a clear and consistent delineation of EPA's interpretation of the boundaries and degree of overlap would help to reduce confusion about the intended scope of any given assessment.

The committee proposes that cumulative risk assessment be defined as evaluating an array of stressors (chemical and nonchemical) to characterize—quantitatively to the extent possible—human health or ecologic effects, taking account of such factors as vulnerability and background exposures. Cumulative impact assessment would consider a wider array of end points, including effects on historical resources, quality of life, community structure, and cultural practices (CEQ 1997), some of which may not lend themselves to quantification following the *Risk Assessment in the Federal Government: Managing the Process* (NRC 1983; the Red Book) paradigm and are beyond the scope of the present report. Community-based risk assessment would follow the practices and principles of community-based participatory research (CBPR), involving active engagement of the community throughout the entire assessment process (Israel et al. 1998).

Although those are conceptually distinct definitions, there will be overlaps in practice. For example, it will often be desirable to use CBPR approaches in cumulative risk assessments, although in principle a community-based risk assessment might not address cumulative risks, and a cumulative risk assessment (such as the pesticide analyses under the FQPA) may not always follow CBPR approaches. Similarly, cumulative impact assessments would generally include the outputs of cumulative risk assessment and other considerations; but, depending on the nature of the decision, the quantitative cumulative risk component may have more or less significance in a cumulative impact assessment.

The definition of cumulative risk assessment above is meant to be functionally identical with that of cumulative risk assessment in the Framework for Cumulative Risk Assessment (EPA 2003a) and that of cumulative impact assessment[1] by the California Environmental Protection Agency (CalEPA 2005). This difference in nomenclature further emphasizes the need for clear definitions. In addition, although it is preferable to have quantitative information as the primary health risk-assessment output, it will often be useful to provide qualitative information about potential health effects when risks cannot be fully quantified and to have terminology that distinguishes the full discussion of possible health effects from the myriad other effects that may be considered in a cumulative impact assessment and that may be important for a decision at hand.

We further propose that EPA apply the term *cumulative risk assessment* only to an

[1]As defined by the California EPA, cumulative impact means exposures, public health, or environmental effects from the combined emissions and discharges in a geographic area, including environmental pollution from all sources, whether single or multi-media, routinely, accidentally, or otherwise released. Impacts will take into account sensitive populations and socio-economic factors, where applicable and to the extent data are available (CalEPA 2005).

analysis that considers in some capacity all the components mentioned in EPA's definition of cumulative risk assessment. An analysis that does not consider nonchemical stressors, that considers only a subset of routes and pathways of exposure, or that does not consider vulnerability should not be termed a cumulative risk assessment. This does not imply that all cumulative risk assessments will formally quantify all of these dimensions - if an initial screening assessment or qualitative examination demonstrates that it is not necessary to consider nonchemical stressors, vulnerability, or specified routes of exposure given a defined decision context, they need not be included in the final assessment for it to be deemed a cumulative risk assessment. That may appear to be a largely semantic distinction, but it emphasizes the primary aims and objectives of cumulative risk assessment and would encourage EPA and other investigators to develop methods to address the aforementioned elements when relevant to a regulatory decision. The committee recognizes that these modified definitions may run counter to the language in the FQPA and elsewhere; this may make redefinition impractical in the near term, but this inconsistency within the agency reinforces the need for greater clarity. Following these modified definitions, many of the previous assessments by EPA and others would be more appropriately termed mixture risk assessments, inasmuch as they consider aggregate exposures to multiple chemicals in the same family but do not consider other components mentioned above. To be clear, this does not imply that such assessments were not well done or informative for policy decisions, as analyses of the effects of chemical mixtures can have great utility, but simply that they do not answer the same questions asked by cumulative risk assessment.

More generally, EPA should emphasize that even cumulative impact assessment cannot by itself bridge the gap between community concerns about environmental risks and decisions made by EPA and other stakeholders. Some communities are concerned principally about the cumulative burden of environmental exposures or the local burden of disease, but others may be more concerned about unfairness in siting processes, ensuring that low-socioeconomic-status (low-SES) communities are at the table with other stakeholders articulating their concerns, and so forth. Some of those concerns can be addressed through cumulative impact assessment, but not all of them. EPA should recognize that cumulative impact assessment has the potential to greatly inform concerns related to outcomes but cannot by itself address concerns about process (although, as articulated later, stakeholder involvement is a crucial component of cumulative risk assessment and cumulative impact assessment, which could help to address some process concerns). The clarification about the decision contexts in which cumulative impact assessment will and will not be useful should provide more realistic expectations on the part of all stakeholders.

Integrating Nonchemical Stressors

In spite of the fact that cumulative risk assessment by definition considers psychosocial, physical, and other factors, no cumulative risk assessments by EPA have formally incorporated nonchemical stressors. That may be in large part because data have been inadequate and because many nonchemical stressors are beyond EPA's regulatory mandate, but the omission means that cumulative risk assessment has a much narrower scope than originally expected or desired by many stakeholders. Moreover, as illustrated in the global analyses of burden of disease described above, data are available on the effects of a number of dietary, physical, and psychosocial risk factors, and extensive exposure data are available on many of these stressors. In addition, ecologic risk assessments commonly apply methods that simultaneously consider numerous chemical and nonchemical stressors in a single assessment in spite of the complexity of the system and the limitations of data availability. In this section, we give examples of some data sources that EPA could use to incorporate nonchemical stressors and use a case example to demonstrate the utility of a cumulative risk assessment that includes nonchemical stressors.

An initial recommendation is that EPA develop databases and default approaches that would allow the incorporation of key nonchemical stressors in the absence of population-specific data. From an exposure perspective, a parallel effort would be the *Exposure Factors Handbook* (EPA 1997), which synthesizes extensive data from disparate sources to allow default estimates of activity patterns and intake rates for defined subpopulations. EPA should work to synthesize and develop datasets related to exposures to nonchemical stressors that influence similar health end points as key chemical stressors to allow these factors to be readily incorporated into cumulative risk assessments in settings where population-specific assessment is infeasible or impractical. Emphasis should be on characterization of distributions for key subpopulations and on evaluation of correlations between factors to allow more realistic assessments. For some factors (for example, smoking, diet, and alcohol consumption), extensive data are already available from other sources, such as the National Health and Nutrition Examination Survey (NHANES), but would need to be compiled and processed in a format suitable for cumulative risk assessment. For example, cumulative risk assessments may require information about correlations between exposures to chemical and nonchemical stressors (cross-sectional or longitudinal), which may not generally be calculated and compiled for other purposes. Factors such as temperature and humidity (which may interact with air pollution effects) and various infectious agents would similarly have readily-available data sets which may require additional analysis to be incorporated into cumulative risk assessments. In general, EPA should collaborate with other agencies and organizations with more expertise in nonchemical stressors to build these databases.

For other factors (such as psychosocial stress), additional methodologic research and data-collection efforts would potentially be needed. With individual stressors for which exposures could be quantified, EPA should compile relevant data related to socioeconomic status (SES), which may serve as a proxy for numerous individual risk factors (O'Neill et al. 2003) and may be a more direct measure of vulnerability than could reasonably be assembled by looking at all relevant individual risk factors. The key is to understand correlations between SES and exposure-related activities and later the degree to which SES acts as an effect modifier for given chemical stressors and health outcomes. Efforts such as these may be beyond the expertise and purview of EPA in the near term, and knowledge of other agencies (such as CDC) and stakeholders should be leveraged.

Incorporating nonchemical stressors also requires information on modes of action among disparate types of exposures. EPA not only should focus on pharmacokinetic and pharmacodynamic models and approaches typically used in MOA determinations for chemicals (following the modified approach proposed in Chapter 5) but should make use of epidemiologic evidence on effect modification when it is available. For example, there may be epidemiologic studies that demonstrate differential relative risks by SES (for example, risk of death related to particulate matter) or interactions between smoking status and chemical exposures (for example, to radon). The importance of epidemiologic evidence can be seen by considering socioeconomic factors and stressors, which could not be incorporated by using evidence only from animal bioassays. Although direct epidemiologic evidence may not be available on a specific chemical of interest, insight from similar compounds may provide useful default assumptions about interactions between chemical and nonchemical stressors. The potential importance of epidemiology in cumulative risk assessment is discussed in more detail later in this chapter.

To illustrate how a cumulative risk assessment could in principle capture both chemical and nonchemical stressors while maintaining focus on the subset of stressors influenced by risk-management strategies under study, we provide an illustrative example. Suppose that the risk-management decisions in question were related to various strategies to reduce the public-health effects of airport emissions on the surrounding communities. Some of the strategies (such as changes in fuel composition or control technologies) would influence only air-pollution exposures and related health risks, and others (such as changes in flight paths or runway use) could also influence noise exposures and related psychosocial stress. The committee recognizes that such an example does not neatly correspond with EPA's regulatory mandates and would cross the jurisdictional boundaries of multiple agencies; this example is simply meant to illustrate the steps that would need to be taken within a cumulative risk assessment.

Following the paradigm proposed in Table 7-1, the first step in a stressor-based assessment oriented around risk-management options would involve building a conceptual model to provide insight into the various stressors of concern and their linkages with the health outcomes of interest. Given a focus on evaluating the benefits of proposed risk-management strategies rather than burden-of-disease assessments, the stressors of interest should include either the ones that would be influenced differentially by potential risk-management strategies or the ones that would not be influenced by the strategies but would have a quantitative influence on risk estimates.

In this case, it clearly would be important to include psychosocial stress as a key nonchemical stressor in at least two dimensions. First, it would be important to know whether the effect of air-pollution exposure reductions depended at all on the level of psychosocial stress (related to noise and other causes). That would be important even for the interventions that did not influence psychosocial stress, provided that the level of psychosocial stress influenced the effects of changes in air pollution (that is, by contributing to background processes or acting as an effect modifier). Second, it would be important to develop the quantitative relationship between interventions and levels of psychosocial stress, as a potential cobenefit of risk-management strategies targeted at air pollutant emissions. If the effect of air pollution were independent of the level of psychosocial stress and the interventions did not have any differential influence on psychosocial stress, it would not be an important stressor to consider in this decision context even if it were an important contributor to the general burden of disease.

Given that structure, the approach in Table 7-1 involves a MOA assessment and consideration of background exposures that may affect the same health outcome. A comprehensive evaluation for this case is beyond the scope of the present report, but one example could involve cardiovascular disease as a significant end point of concern and hypertension as the mechanistic link between the various stressors and this end point. Previous studies (Evans et al. 1998) have demonstrated that airport noise and the associated stress can increase blood pressure (and epinephrine, norepinephrine, and cortisol). Air pollution has similarly been associated with blood pressure (Künzli and Tager 2005), and this indicates that both exposures would be important to model, given either an underlying model linking hypertension and cardiovascular end points or a quantal cutpoint for hypertension itself. It would also be necessary to be able to model the relationship between the risk-management strategies and exposures to both air pollution and noise. Following the conceptual model makes this relatively straightforward: methods are readily available to model the influence of airport activities on noise (which could be presumed to be a surrogate for airport-related psychosocial stress), and the aforementioned studies can link noise with such key health-relevant end points as blood pressure. Thus, a physiologically based conceptual model can readily incorporate nonchemical stressors into the cumulative risk assessment.

That example was simplified and did not formally go through all the steps in Table 7-1; for example, characterizing the baseline distribution of blood pressures in the population would be necessary, as would characterizing the distribution of other underlying disease processes to which the stressors could contribute. Other issues are potentially raised by the above approach, such as a focus on only pathways that are well understood and quantifiable, as well as the complexity of a real-world case that would potentially involve multiple federal agencies. However, in spite of those concerns, this simple example demonstrates the general feasibility of the approach and highlights that a focus on specific risk-management strategies would greatly narrow the scope of the analysis. Often, more epidemiologic evidence is available on nonchemical stressors than on chemical stressors, so inclusion of nonchemical stressors may be plausible in many contexts.

The inclusion of nonchemical stressors as outlined above can lead to more informative assessments and correspondingly better decisions if used appropriately but can run the risk of contributing to less informative assessments if used in the wrong way. Information on the varied risk factors should not be used solely for risk comparisons that are uninformative from the perspective of the decisions faced by EPA. For example, if the inputs for cumulative risk assessment are used not to determine the impacts of alternative risk-management strategies but to determine contributors to disease burdens in a community, analyses may find that cigarette-smoking confers a greater disease burden than outdoor exposures to air toxics. Even setting aside the risk-communication limitations of such a comparison (given the different nature of the risks), the comparison is largely uninformative from the perspective of EPA, industry, or other agency decision-making. In other words, it is difficult to imagine a context in which EPA must decide whether to require industrial facilities to install pollution-control devices or to lobby other agencies to increase funding for smoking-cessation efforts. The problem would be avoided by the framework proposed in detail in Chapter 8, in that a focus on options to achieve a defined objective (that is, a functional-unit definition) would make these sorts of burden-of-disease comparisons less relevant. The simple fact that stressors other than chemicals may contribute a substantial portion of the burden of disease in a community does not by itself imply that reduction of chemical exposures would not have net benefits that would exceed the costs, and an emphasis on this comparative dimension of the analysis will only widen the gulf between risk assessors and community stakeholders. This is not to say that there is no rationale for risk communication efforts that attempt to contextualize risk assessment outputs by comparison with other risk factors, but simply to emphasize that such comparisons should not be the primary intent of cumulative risk assessment.

In addition, especially with nonchemical stressors, such as psychosocial stress, analytic boundaries need to be carefully established. If the existence of industrial facilities or other environmental problems serves as a social stressor, control strategies could reduce both chemical exposures and psychosocial stress (provided that the affected community perceived the reduction as important and substantive). The Agency for Toxic Substances and Disease Registry (ATSDR) recently (Tucker 2002) emphasized the psychosocial ramifications of living near hazardous-waste sites and the potential need to consider psychosocial factors in remediation decisions, although these factors are rarely formally quantified or characterized. That raises the broader question of whether stress related to an environmental exposure should "count" as part of quantifying the benefits of an intervention. Counting those benefits would in principle provide a more accurate estimate of benefits, but one could imagine a situation in an extreme case in which a community is greatly concerned about a chemical in its drinking water that has no direct effects on health but in which an intervention measure could result in health benefits through the reduction of psychosocial stress. That would be somewhat more important than a placebo effect, but it would be awkward to estimate health benefits associated with controlling a benign chemical. Such extreme cases should be avoided by well-formulated problem scoping and risk-management option development, but the example highlights the importance of stakeholder involvement at multiple stages in the assessment process.

Finally, even with the triage indicated in Table 7-1, addition of all relevant chemical and nonchemical stressors runs the risk of making the assessment analytically intractable and impossible to complete in a limited amount of time and of jeopardizing timely decision-making. In addition to limiting the number of stressors under consideration, there is a need for relatively simple risk-assessment methods that can be applied to address the stressors in a timely fashion; this is discussed in more detail later.

In summary, approaches to incorporate nonchemical stressors into cumulative risk assessment are feasible in the near term although there are many situations in which site-specific data needs may not be met. We recommend that EPA start to address nonchemical stressors in settings in which sufficient epidemiologic or pharmacokinetic and pharmacodynamic data are available to understand interactions with chemical stressors, following the tiered strategy articulated by Menzie et al. (2007) and reoriented in Table 7-1 to focus on discriminating among risk-management options. Databases and default approaches should be developed regarding exposure patterns and plausible interactions with chemical stressors. In the long term, we recommend that EPA and other agencies invest in research related to interactions between chemical and nonchemical stressors, including epidemiologic investigations and pharmacokinetic and pharmacodynamic or other study types as relevant. The direction of the research should be informed by pending risk-management decisions in which the agency identifies critical data gaps that impede decision-making in specific contexts rather than broadly considering all the combinations of chemical and nonchemical stressors that could potentially be investigated.

Role of Biomonitoring

As summarized recently (Ryan et al. 2007), biomonitoring has a potentially important role in cumulative risk assessment, with significant roles to be played by biomarkers of exposure, susceptibility, and effect. For example, if multiple stressors are thought to influence acetylcholinesterase inhibition (that is, in the case of OP pesticides), simultaneous collection of compound-specific biomarkers, nonspecific biomarkers of the OP family, and biomarkers of effect can provide insight into the joint effects of these exposures. Collection of biologic samples can allow characterization of simultaneous exposure to multiple stressors, which may be difficult to determine accurately by modeling exposures to each of the compounds individually.

Ryan et al. (2007) view the primary capabilities of biomonitoring in the framework of cumulative risk assessment as the ability to disaggregate disease burden into specific risk factors and the ability to infer contributions of different sources and pathways. The former approach provides one route for effects-based or burden-of-disease assessments, and the latter approach can in principle inform stressor-based and later cumulative risk assessments focused on interventions.

A potential limitation of biomonitoring data is the difficulty of linking biomarkers to contributions from individual sources of emissions. Even if the distribution of biomarkers of exposure or effect is well characterized for a defined subpopulation, including an understanding of routes of exposure and contributing source categories, it is difficult to model how changes in emissions from a small number of identified sources would influence the distribution. Biomarkers may therefore be suitable for developing mechanistic understanding and contributing to effects-based cumulative risk assessment but may be of limited use to stressor-based cumulative risk assessment directly in a risk-management context, especially in situations with relatively small marginal changes in exposures. Research efforts related to reverse dosimetry (Sohn et al. 2004; Tan et al. 2006) indicate a possible approach to reconstructing exposures from dose data, but such methods are not sensitive enough to determine marginal changes in emissions from individual facilities and therefore may not be suitable for discriminating among risk-management options for more narrowly-defined or community-scale control strategies. In this context, biomonitoring may be most useful as a validation check against modeled doses or as an input to epidemiologic investigations.

Regardless, the existence of the Centers for Disease Control and Prevention (CDC) large-scale biomarker databases, the *Third National Report on Human Exposure to Environmental Chemicals* (CDC 2005), indicates that data on the distribution of doses among representative samples of the U.S. population are increasingly available. The full set of data available through the NHANES could also provide a means of characterizing correlations between biomarkers for chemical and nonchemical stressors, demographic predictors of magnitudes of those stressors, and other relationships that could form the basis of a cumulative risk assessment. Thus, although it seems unlikely, because of both cost and limited interpretability, that biomarkers could be used directly to quantify the benefits of control strategies leading to marginal changes in exposures, biomarker studies can provide enhanced mechanistic understanding of the relationships among chemical and nonchemical stressors, and insight about highly-exposed populations or source category contributions that can allow for the development of targeted control strategies.

Role of Epidemiology and Surveillance Data

The cumulative risk-assessment paradigm, given its focus on communities or defined populations and consideration of such nonchemical stressors as SES and access to health care, lends itself to being informed by epidemiology. In fact, many of the key interactions among chemical and nonchemical stressors, given numerous simultaneous coexposures, would be impossible to capture in toxicologic studies. The call for more "realistic" risk assessment in community settings is in part a call for better epidemiology that can characterize the effects of varied coexposures in the presence of background processes and differences in vulnerability. This raises the question of whether sufficient epidemiologic information is available, or could be developed, to enable EPA to generate cumulative risk assessments that include physical, chemical, biologic, and social factors with a sufficient degree of scientific plausibility. This section briefly provides examples of advances in epidemiologic methods that show promise for improving the information base needed for the advancement of cumulative risk assessment, and in parallel it describes the role that surveillance data and systems could play in facilitating the transition from single chemical risk assessment to cumulative risk assessment.

At the outset, limitations of epidemiology in the context of cumulative risk assessment must be acknowledged. Because of relatively low ambient exposures, multiple concurrent exposures, weak statistical power, exposure misclassification, and other issues, it is often difficult for epidemiology to capture main effects, let alone interaction effects, of environmental exposures. In spite of those limitations, there is growing epidemiologic evidence of interactions between environmental stressors and place-based and individual-based psychosocial stressors, driven in part by the spatial and demographic concordance between physical and chemical environmental exposures and socioeconomic stressors (IOM 1999; O'Neill et al. 2003; Clougherty et al. 2007). The evidence adds to historical examples of well-documented interactions between environmental and nonenvironmental risk factors in humans, such as synergistic effects between radon or asbestos and cigarette-smoking. In addition, by definition, reliance on epidemiology reduces the ability to be preventive and to evaluate the risk of new stressors to which humans have not yet been exposed. Epidemiology is best suited to cumulative risk assessments directed at remediation of existing problems, which would be expected to be the majority of applications, given the inherent focus on populations at risk.

Two growing categories of inquiry in epidemiology may help to bolster the evidence base and inform cumulative risk assessment. Problems of characterizing exposure and outcomes in observational epidemiology have generated increasing attention to molecular epidemiology, which involves incorporating biologic events at the physiologic, cellular, and molecular levels into epidemiologic studies. Aside from enhancing the biologic understanding of epidemiologic findings, the biomarkers used in molecular epidemiology can be used in some circumstances to reconstruct exposure (albeit with some of the limitations listed above). The combination of better exposure assessment and better understanding of disease pathways helps to reduce their misclassification in epidemiologic studies. That provides better statistical power and biologic insight that can improve characterization of potential synergies among risk factors and factors that contribute to vulnerability, including age, sex, inherited genetic variation, nutrition, and pre-existing health impairments. Such studies, although it may be difficult to apply them directly to quantitative population risk assessment, may have a greater likelihood of detecting subtle effects in relatively small populations and demonstrating the biologic plausibility of synergistic relationships.

A somewhat different direction of epidemiologic inquiry potentially informative for cumulative risk assessment involves the emerging field of social epidemiology, which has shed light on the relations between social factors and disease in populations (Kaufman and Cooper 1999). There is little room for disagreement about the importance of "social factors" as predictors of health risks; the consistent documentation of these patterns in a wide variety of outcomes is an important achievement of health and medical science. Of significance for cumulative risk assessment is the recent work of social epidemiologists who are examining the biologic underpinnings of social factors and considering interactions with environmental exposures (Berkman and Glass 2000). Aside from elucidating those interactions, social epidemiology may provide methodologic lessons for cumulative risk assessment in general; as mentioned above, methods have been developed to characterize cumulative risks (Evans 2003), and studies addressing allostatic load (the long-term effect of the various physiologic responses to stress) have both considered the effects of numerous stressors and developed measures of allostatic load that integrate multiple outcomes (McEwen 1998).

To benefit from developments in molecular and social epidemiology and related sciences and technology with the potential to reduce exposure-measurement error (that is, environmental sensors, biologic sensors, and geographic information systems), there will need to be greater interactions between epidemiologic research and risk assessment, as opposed to treating risk assessment simply as an end user of epidemiologic output. Epidemiologic studies conducted with cumulative risk assessment in mind may use different exposure-assessment and analytic strategies from those used by epidemiologic studies conducted for other purposes. For example, an epidemiologic analysis done for its own sake will tend to focus on disentangling the contributions of individual risk factors in the presence of potential confounding, whereas an epidemiologic analysis done for cumulative risk assessment might characterize the risks of defined "bundles" of exposures without further decomposition.

The interaction between epidemiology and cumulative risk assessment can be enhanced as risk assessments identify key uncertainties related to interactions among chemical and nonchemical stressors, shaping the research agenda and stimulating demand for more relevant (to risk assessment) epidemiologic research. In general, as mentioned above, EPA and other agencies should pursue a long-term research agenda related to enhanced epidemiologic insight into interactions among chemical and nonchemical stressors and in the short term should work to develop internal capacity in a variety of epidemiologic disciplines to foster the development of new methods and knowledge.

Although epidemiologic approaches may improve understanding of the effects of exposure to multiple stressors, for effects-based assessments, surveillance data may be needed both to identify the at-risk populations and to characterize patterns of disease and background exposures. Surveillance for various diseases is well established in the public-health system, including monitoring networks and registries that collect data in several ways. For example, nearly all states have some form of infectious-disease and chronic-disease reporting laws that require hospitals, physicians, or schools to report cases that are considered to be of public-health importance to the state or to CDC. Such information is available at various levels of spatial resolution, influenced in part by confidentiality considerations and by the nature and prevalence of the disease in question. In addition, federal agencies, such as CDC, maintain active or passive surveillance on a wide variety of diseases and health-status measures for populations in various geographic areas. A relatively new component of public-health surveillance involves biosurveillance, the early detection of abnormal disease patterns and nontraditional early disease indicators, such as pharmaceutical sales, school and work absences, and cases of animal disease.

Another form of surveillance system is the toxic-substance registry. As mandated by Superfund legislation, the ATSDR established a National Exposure Registry (ATSDR 2008) with the goal of assessing and evaluating relationships between adverse health effects and exposure to hazardous waste, particularly between chronic health effects and long-term, low-level chemical exposure. For example, NER's trichloroethylene subregistry has been used to demonstrate increased rates of hearing impairment and other conditions associated with historical exposure to trichloroethylene.

Those surveillance systems have substantial utility in some contexts but have been limited in multiple respects in the context of environmental risk factors. In particular, little information has been routinely and systematically collected on many health outcomes potentially linked to environmental pollutants, such as birth defects, developmental disorders, childhood leukemia, and lupus. More generally, many chronic diseases (such as diabetes and asthma) have not been given sufficient attention. In addition, given numerous data streams, it has been difficult to relate members of populations included in one health-information system to members in another system.

For those reasons, CDC in 2001 began the development of a health-tracking network to monitor the prevalence of chronic conditions of potential interest for human health risk assessment. Known as the Environmental Public Health Tracking (EPHT) Program, its purpose is to provide information from a nationwide network of integrated health and environmental data to be used as the basis of risk assessment and risk management. An important distinction between EPHT and traditional surveillance is the emphasis on data integration across health, human-exposure, and hazard-information systems, which will enhance efforts of risk assessors to evaluate the spatial and temporal relations between environmental factors and health outcomes. If the EPHT surveillance systems were linked with registries from private health-care organizations, more comprehensive disease-prevalence estimates could be readily obtained.

Of particular interest to the cumulative risk-assessment process is the potential of EPHT to identify susceptible populations and to provide an important foundation for environmental epidemiology addressing chemical and nonchemical stressors. Developing the relations between environmental and health outcomes will require individual-level data not routinely collected by any surveillance system, so there will be the need for both targeted research and methods for data linkage with the EPHT Program. In general, the goals of EPHT are ambitious and resources are limited, in particular for data-linkage efforts that are expensive in both time and money (Kyle et al. 2006). Investing more resources in EPHT could be a useful mechanism to develop the information base necessary for cumulative risk assessment or community-based risk assessment.

Need for Simpler Analytic Tools

Given the breadth of exposure pathways and types of stressors considered in cumulative risk assessment, there is a danger that it could become analytically intractable and therefore uninformative for making decisions in a timely fashion. Application of more advanced methods for dose-response assessment as proposed in Chapters 4 and 5 would appear to make this issue even more problematic. The problem is more acute in community-based risk assessments, in which the sheer number of communities and environmental risks that could potentially be evaluated could quickly outstrip the available resources for conducting such analyses and in which the CBPR emphasis implies that analytic tools should be able to be understood and implemented by community stakeholders. It should be clear that not all decisions will need to be informed by the most advanced analytic methods (see Chapters 3 and 8), just as not all risk-management decisions will necessarily involve quantifying all theoretical dimensions of cumulative risk assessment.

To enhance the utility of cumulative risk assessment, there will need to be increased reliance on relatively simple methods to determine whether more refined methods are required or information is adequate to inform policy decisions. Developing simpler tools seems to contradict the complexity of cumulative risks, but methods can be developed that capture the breadth of chemical and nonchemical stressors with less computational burden, at least for initial screening calculations. There will also need to be techniques to develop indicators or ranking approaches that could categorize the benefits of different strategies ordinally as has been done in ecologic risk assessment; for example, Thomas (2005) has shown that the RRM, a rank-based method, can be used to analyze alternative decisions involving multiple stressors and receptors on various spatial scales. The critical issue is to ensure that any simplified methods used in the context of cumulative risk assessment retain the key attributes of quantitative risk assessment, that is, consideration of both exposure and toxicity, notions of probability rather than just possibility, and information about the severity of health effects. It will be difficult to interpret outputs that do not retain those features, especially in the contexts of tradeoffs or comparisons with control costs.

While development of simpler approaches will not be straightforward, fields such as ecologic risk assessment and life cycle analysis have successfully developed and utilized tools to address similar concerns, and these methods will be relevant to cumulative risk assessment. One example focused on exposure assessment comes from the field of intake-fraction estimation (Bennett et al. 2002a). An intake fraction is the population exposure per unit of emission from a defined source or source category. Intake fractions are generally derived from dispersion modeling or from the combination of monitoring data and emissions-inventory assessment, in either case linked with population patterns. They therefore use detailed information about exposures but summarize this information as single unitless measures directly interpretable for risk assessment; in cases in which the dose-response function is linear in the range of exposures of interest or is well defined and nonlinear, intake fractions can be used directly to estimate population health risks. Intake fractions vary with the compound, source, and setting, but values can be extrapolated to unstudied settings given known characteristics of the setting (such as population density). Intake fractions have been adopted by the life-cycle analysis community for incorporating population-exposure concepts in settings where more complex modeling is implausible and where the alternative is priority-setting with no consideration of exposure (Bennett et al. 2002b; Evans et al. 2002). As another example of simplified methods for exposure assessment in the context of screening-level risk assessment, the *Community Air Screening How-To Manual* (EPA 2004) includes look-up tables for concentration effects, given stack characteristics and distance from a source.

Although those approaches address only exposure assessment, they provide useful lessons about how simpler methods can be applied to yield reasonable and timely insight without sacrificing the critical components of quantitative risk assessment. The concept of using a limited number of more extensive analyses to determine approximate relationships for an unstudied setting can be extended to exposures to nonchemical stressors or interactions among compounds. This can provide effective defaults in the absence of more detailed site-specific data. The committee therefore recommends that EPA develop guidelines and methods for less analytically complex cumulative risk assessments to be used for screening assessments. The guidelines should give insight into approaches for choosing the appropriate level of analytic complexity and into recommended methods for simplified assessments, including both exposure assessment and dose-response assessment. The selection of the appropriate analytic model would be a component of the planning and scoping and problem-formulation steps and would be driven by the risk-management decisions at hand and the priorities of the various stakeholders. In other words, drawing on the example above, simplifying exposure assessment by using intake fractions is valuable only if total population benefits without distributional considerations were the measure of interest to risk managers. The simplified tools would need to be tailored to the decision context and the outputs of interest.

The databases, methods and other modeling resources developed by EPA for less analytically complex cumulative risk assessments would have an important ancillary benefit. Local community participation could be greatly enhanced if analytic tools were easier to understand or, ideally, could be used by community groups and other stakeholders to determine the benefits of control strategies in a cumulative risk context quickly but reasonably. That is clearly difficult given the numerous decision contexts and types of models required, but examples could be drawn from the life-cycle analysis community, in which generally applicable software packages and on-line resources have been developed that can be used by people who lack expertise in the specific scientific disciplines that underlie life-cycle impact assessment. The general issue of the need for and approaches to enhancing stakeholder involvement in cumulative risk assessment is discussed in more detail in the next section.

Need for Stakeholder Involvement

The issue of increased stakeholder involvement in the risk-assessment process has been discussed at length in previous National Research Council reports and EPA guidance documents. The committee agrees with many of the core principles articulated in those reports, such as the mutual and recursive analytic-deliberative process articulated in *Understanding Risk Informing Decisions in a Democratic Society* (NRC 1996) and the need for stakeholder participation throughout the risk-assessment process, including participation in planning and scoping and in problem formulation (EPA 2003b). A key insight from the previous reports is that stakeholder involvement should go well beyond risk communication or risk characterization and should include substantive involvement in the assessment process (often following

CBPR principles) and explicit attempts to build capacity to ensure that all stakeholders have an equal opportunity to participate substantively in collaborative problem-solving (NEJAC 2004). That is not simply a means of improving public relations and acceptability of risk-assessment outputs but a means of enhancing the technical quality of the analysis and ensuring that risk-management strategies are reasonable and well developed.

The cumulative risk-assessment framework further emphasizes the value of bringing stakeholders together at the outset, devising clear and explicit project planning and scoping, and focusing on a specific decision problem to guide the analysis. However, the added complexity of cumulative risk assessment creates some substantial barriers: if there is to be substantive stakeholder involvement, all parties must have access to and in-depth understanding of relevant databases, models, and information resources. It is not realistic to hope that all stakeholders will become expert risk assessors, but the use of simpler analytic tools, as proposed above, may provide some of the necessary resources for community members and other stakeholders to understand and participate in the analytic portions of an assessment.

In addition to models for cumulative risk assessment, information resources would need to be developed to allow stakeholders to be sufficiently informed to participate in the process. EPA has developed a substantial array of public resources and databases, but none provides adequate information to allow stakeholders to understand the intricacies of cumulative risks in specific communities or subpopulations. For example, EPA has made available such public resources as Envirofacts (EPA 2007a), EnviroMapper (EPA 2006c), and TRI Explorer (EPA 2007b), which provide extensive information about the locations of key emission sites for any given ZIP code, information about environmental-justice assessments, and links to related concentration data. However, none of the available resources provides the information or tools needed for stakeholders to understand their cumulative risks associated with chemical and nonchemical stressors or, more important, the potential benefits associated with specific control strategies. Models of the benefits of control strategies may be beyond the scope of on-line resources, but well-developed and publicly available databases could provide both the foundation for cumulative risk models and the information for communities to use in understanding their exposures and background disease patterns. Linking environmental databases described above with surveillance-system data in a framework of geographic information systems would be a good starting point for such efforts, using high spatial resolution to provide maximal insight into community-scale risks.

EPA has numerous programs and guidance documents related to stakeholder involvement (EPA 2008b), whose formal evaluation is beyond the scope of this chapter. The committee recommends that EPA adhere to its guidance when conducting cumulative risk assessments, including planning and budgeting for public and other stakeholder involvement, working to identify interested parties, providing financial or technical assistance and resources to facilitate involvement, providing information and outreach materials, engaging in other activities to build community capacity to participate in the process, involving the public in the decision process at a stage where substantive input can be made, and formally evaluating the process to ensure that adequate stakeholder participation (in depth and breadth) has been incorporated (EPA 2003b).

RECOMMENDATIONS

The committee recommends the following short-term and long-term actions to enhance the utility of cumulative risk assessment for discriminating among risk-management options:

- EPA should maintain the core definitional components of cumulative risk assessment from its 2003 framework document—including planning, scoping, and problem-formulation phases; explicit consideration of vulnerability; and the use of screening tools and other methods to ensure analytic complexity appropriate for the decision context. The analytic structure of ecologic risk assessment should continue to serve as an important guide for human health cumulative risk assessment, given the conceptual similarities.
- EPA should use a revised framework for risk-based decision making (see Chapter 8), focused on discriminating among risk-management options, to narrow the scope of cumulative risk assessments to those stressors that would be influenced by risk-management options or would modify the risks of other stressors influenced by risk-management options. This would allow for the inclusion of nonchemical stressors within a decision framework relevant to EPA. For stressor-based assessments, EPA should follow a tiered assessment strategy that parallels the mode-of-action and background-process determination to ascertain the subset of stressors that would substantially influence the benefits of proposed risk-management strategies.
- EPA should explicitly define and maintain conceptual distinctions among cumulative risk assessment, cumulative impact assessment, and community-based risk assessment to avoid confusion about the scope of work expected of a given assessment. These definitions should be consistently used and applied across the agency.
- In the near term, EPA should develop databases and default approaches to allow the incorporation of key nonchemical stressors in cumulative risk assessments in the absence of population-specific data, considering exposure patterns, contributions to relevant background processes, and interactions with chemical stressors. EPA should use existing nationally representative biomarker and surveillance databases and databases related to nonchemical stressors to help to construct the approaches, leveraging insight from social epidemiology and ecologic risk assessment.
- In the long term, EPA should invest in research programs and develop internal capacity related to interactions between chemical and nonchemical stressors, including epidemiologic investigations with sufficient power to evaluate interactions and physiologically based pharmacokinetic and other study types as relevant. Given the need for substantial epidemiologic research conducted in a form and direction suitable for cumulative risk assessment, EPA should build internal capacity in various epidemiologic disciplines and ensure close collaboration between epidemiologists and risk assessors. EPA should also develop partnerships with other federal agencies with expertise related to nonchemical stressors, and should work with these agencies on large-scale cumulative risk assessments that cross jurisdictional boundaries.
- In the process of refining cumulative risk assessment, EPA should focus on development of guidelines and methods for simplified analytic tools that could allow screening-level cumulative risk assessment and could provide tools for communities and other stakeholders to use in conducting assessments. These tools can be used as the foundation of an enhanced stakeholder-participation process that builds on current guidance but expands it by providing cumulative risk models that can be applied and interpreted by nonpractitioners. EPA should work to ensure that cumulative risk assessments both guide future information and research needs and inform near-term decisions, recognizing that decisions must be made with incomplete information.

REFERENCES

ATSDR (Agency for Toxic Substances and Disease Registry). 2008. National Exposure Registry [online]. Available: http://www.atsdr.cdc.gov/ner/index.html [accessed Aug. 12, 2008].

Bennett, D.H., T.E. McKone, J.S. Evans, W.W. Nazaroff, M.D. Margni, O. Jolliet, and K.R. Smith. 2002a. Defining intake fraction. Environ. Sci. Technol. 36(9):207A-211A.

Bennett, D.H., M.D. Margni, T.E. McKone, and O. Jolliet. 2002b. Intake fraction for multimedia pollutants: A tool for life cycle analysis and comparative risk assessment. Risk Anal. 22(5):905-918.

Berkman, L.F., and T.A. Glass. 2000. Social integration, social networks, social support and health. Pp. 137-173 in Social Epidemiology, L.F. Berkman, and I. Kawachi, eds. New York, NY: Oxford University Press.

CalEPA (California Environmental Protection Agency). 2005. Cal/EPA EJ Action Plan Pilot Projects Addressing Cumulative Impacts and Precautionary Approach. California Environmental Protection Agency. March 25, 2005 [online]. Available: http://www.calepa.ca.gov/envjustice/ActionPlan/ [accessed Jan. 28, 2008].

Callahan, M.A., and K. Sexton. 2007. If cumulative risk assessment is the answer, what is the question? Environ. Health Perspect. 115(5):799-806.

CDC (Centers for Disease Control and Prevention). 2005. Third National Report on Human Exposure to Environmental Chemicals. U.S. Department of Health and Human Services, Centers for Disease Control and Prevention, Atlanta, GA [online]. Available: http://www.jhsph.edu/ephtcenter/Third%20Report.pdf [accessed Jan. 24, 2008].

CEQ (Council on Environmental Quality). 1997. Considering Cumulative Effects under the National Policy Act. Council on Environmental Quality, Executive Office of the President, Washington, DC. January 1997 [online]. Available: http://www.nepa.gov/nepa/ccenepa/ccenepa.htm [accessed Jan. 28, 2008].

Clougherty, J.E., J.I. Levy, L.D. Kubzansky, P.B. Ryan, S.F. Suglia, M.J. Canner, and R.J. Wright. 2007. Synergistic effects of traffic-related air pollution and exposure to violence on urban asthma etiology. Environ. Health Perspect. 115(8):1140-1146.

Cox, L.A., Jr. 1984. Probability of causation and the attributable proportion of risk. Risk Anal. 4(3):221-230.

Cox, L.A., Jr. 1987. Statistical issues in the estimation of assigned shares for carcinogenesis liability. Risk Anal. 7(1):71-80.

EPA (U.S. Environmental Protection Agency). 1986. Guidelines for Health Risk Assessment of Chemical Mixtures. EPA/630/R-98/002. Risk Assessment Forum, U.S. Environmental Protection Agency, Washington, DC. September 1986 [online]. Available: http://www.epa.gov/ncea/raf/pdfs/chem_mix/chemmix_1986.pdf [accessed Jan. 7, 2008].

EPA (U.S. Environmental Protection Agency). 1997. Exposure Factors Handbook. EPA/600/P-95/002F. National Center for Environmental Assessment, Office of Research and Development, U.S. Environmental Protection Agency, Washington, DC [online]. Available: http://www.epa.gov/ncea/efh/ [accessed Jan. 28, 2008].

EPA (U.S. Environmental Protection Agency). 2003a. Framework for Cumulative Risk Assessment. EPA/600/P-02/001F. National Center for Environmental Assessment, Risk Assessment Forum, U.S. Environmental Protection Agency, Washington, DC [online]. Available: http://cfpub.epa.gov/ncea/cfm/recordisplay.cfm?deid=54944 [accessed Jan. 4, 2008].

EPA (U.S. Environmental Protection Agency). 2003b. Public Involment Policy of the U.S. Environmental Protection Agency. EPA 233-B-03-002. Office of Policy, Economics and Innovation, U.S. Environmental Protection Agency. May 2003 [online]. Available: http://www.epa.gov/stakeholders/pdf/policy2003.pdf [accessed Jan. 28, 2008].

EPA (U.S. Environmental Protection Agency). 2004. Community Air Screening How-To Manual, A Step-by-Step Guide to Using Risk-Based Screening to Identify Priorities for Improving Outdoor Air Quality. EPA 744-B-04-001. U.S. Environmental Protection Agency, Washington, DC [online]. Available: http://www.epa.gov/oppt/cahp/pubs/community_air_screening_how-to_manual.pdf [accessed Jan. 28, 2008].

EPA (U.S. Environmental Protection Agency). 2006a. Organophosphorus Cumulative Risk Assessment-2006 Update. Office of Pesticide Programs, U.S. Environmental Protection Agency, Washington, DC. August 2006 [online]. Available: http://www.epa.gov/pesticides/cumulative/2006-op/index.htm [accessed Jan. 28, 2008].

EPA (U.S. Environmental Protection Agency). 2006b. 1999 National-Scale Air Toxics Assessment. Technology Transfer Network, U.S. Environmental Protection Agency [online]. Available: http://www.epa.gov/ttn/atw/nata1999/ [accessed Jan. 28, 2008].

EPA (U.S. Environmental Protection Agency). 2006c. EnviroMapper. U.S. Environmental Protection Agency [online]. Available: http://www.epa.gov/enviro/html/em/ accessed Jan. 29, 2008].

EPA (U.S. Environmental Protection Agency). 2007a. Envirofacts Data Warehouse. U.S. Environmental Protection Agency [online]. Available: http://www.epa.gov/enviro/ [accessed Jan. 29, 2008].

EPA (U.S. Environmental Protection Agency). 2007b. TRI Explorer. Toxics Release Inventory, U.S. Environmental Protection Agency [online]. Available: http://www.epa.gov/triexplorer/ [accessed Jan. 29, 2008].

EPA (U.S. Environmental Protection Agency). 2008a. Community Action for a Renewed Environment (CARE). Office of Air and Radiation, U.S. Environmental Protection Agency [online]. Available: http://www.epa.gov/air/care/index.htm [accessed Jan. 29, 2008].

EPA (U.S. Environmental Protection Agency). 2008b. Tools for Public Involvement. U.S. Environmental Protection Agency [online]. Available: http://www.epa.gov/stakeholders/involvework.htm [accessed Jan. 29, 2008].

Evans, G.W. 2003. A multimethodological analysis of cumulative risk and allostatic load among rural children. Dev. Psychol. 39(5):924-933.

Evans, G.W., M. Bullinger, and S. Hygge. 1998. Chronic noise exposure and physiological response: A prospective study of children living under environmental stress. Psychol. Sci. 9(1):75-77.

Evans, J.S., S.K. Wolff, K. Phonboon, J.I. Levy, and K.R. Smith. 2002. Exposure efficiency: An idea whose time has come? Chemosphere 49(9):1075-1091.

Ezzati, M., S.V. Hoorn, A. Rodgers, A.D. Lopez, C.D. Mathers, and C.J. Murray. 2003. Estimates of global and regional potential health gains from reducing multiple major risk factors. Lancet 362(9380):271-280.

Greenland, S. 1999. Relation of probability of causation to relative risk and doubling dose: A methodologic error that has become a social problem. Am. J. Public Health 89(8):1166-1169.

Greenland, S., and J.M. Robins. 1988. Conceptual problems in the definition and interpretation of attributable fractions. Am. J. Epidemiol. 128(6):1185-1197.

Greenland, S., and J.M. Robins. 2000. Epidemiology, justice and the probability of causation. Jurimetrics 40(3):321-340.

Hynes, H.P., and R. Lopez. 2007. Cumulative risk and a call for action in environmental justice communities. J. Health Disparities Res. Pract. 1(2):29-57.

IOM (Institute of Medicine). 1999. Toward Environmental Justice: Research, Education and Health Policy Needs. Washington, DC: National Academy Press.

Israel, B.A., A.J. Schulz, E.A. Parker, and A.B. Becker. 1998. Review of community-based research: Assessing partnership approaches to improving public health. Annu. Rev. Public Health 19:173-202.

Israel, B.D. 1995. An environmental justice critique of risk assessment. New York U. Environ. Law J. 3(2):469-522.

Kaufman, J.S., and R.S. Cooper. 1999. Seeking causal explanations in social epidemiology. Am. J. Epidemiol. 150 (2):113-120.

Kuehn, R.R. 1996. The environmental justice implications of quantitative risk assessment. U. Illinois Law Rev. 1996(1):103-172.

Künzli, N., and I.B. Tager. 2005. Air pollution: From lung to heart. Swiss Med. Wkly. 135(47-48):697-702.

Kyle, A.D., J.R. Balmes, P.A. Buffler, and P.R. Lee. 2006. Integrating research, surveillance, and practice in environmental public health tracking. Environ. Health Perspect. 114(7):980-984.

Landis, W.G., and J.A. Wiegers. 1997. Design considerations and a suggested approach for regional and comparative ecological risk assessment. Hum. Ecol. Risk Assess. 3(3):287-297.

Landis, W.G., M. Luxon, and L.R. Bodensteiner. 2000. Design of a relative risk model regional-scale risk assessment with conformational sampling for the Willamette and McKenzie Rivers, Oregon. Pp. 67-88 in Environmental Toxicology and Risk Assessment: Recent Achievements in Environmental Fate and Transport, Vol. 9., F.T. Prince, K.V. Brix, and N.K. Lane, eds. STP1381. West Conshohocken, PA: American Society for Testing and Materials.

Levy, J.I., S.M. Chemerynski, and J.L. Tuchmann. 2006. Incorporating concepts of inequality and inequity into health benefits analysis. Int. J. Equity Health 5:2.

McEwen, B.S. 1998. Protective and damaging effects of stress mediators. N. Engl. J. Med. 338(3):171-179.

Menzie, C.A., M.M. MacDonell, and M. Mumtaz. 2007. A phased approach for assessing combined effects from multiple stressors. Environ. Health Perspect. 115(5):807-816.

Morello-Frosch, R., and B.M. Jesdale. 2006. Separate and unequal: Residential segregation and estimated cancer risks associated with ambient air toxics in U.S. metropolitan areas. Environ. Health Perspect. 114(3):386-393.

NEJAC (National Environmental Justice Advisory Council). 2004. Ensuring Risk Reduction in Communities with Multiple Stressors: Environmental Justice and Cumulative Risks/Impacts. National Environmental Justice Advisory Council, Cumulative Risks/Impacts Working Group. December 2004 [online]. Available: http://www.epa.gov/enforcement/resources/publications/ej/nejac/nejac-cum-risk-rpt-122104.pdf [accessed Jan. 29, 2008].

NRC (National Research Council). 1983. Risk Assessment in the Federal Government: Managing the Process. Washington, DC: National Academy Press.

NRC (National Research Council). 1996. Understanding Risk: Informing Decisions in a Democratic Society. Washington, DC: National Academy Press.

Obery, A.M., and W.G. Landis. 2002. A regional multiple stressors assessment of the Codorus Creek watershed applying the relative risk model. Hum. Ecol. Risk Assess. 8(2):405-428.

O'Neill, M.S., M. Jerrett, I. Kawachi, J.I. Levy, A.J. Cohen, N. Gouveia, P. Wilkinson, T. Fletcher, L. Cifuentes, and J. Schwartz. 2003. Health, wealth, and air pollution: Advancing theory and methods. Environ. Health Perspect. 111(16):1861-1870.

Ryan, P.B., T.A. Burke, E.A. Cohen Hubal, J.J. Cura, and T.E. McKone. 2007. Using biomarkers to inform cumulative risk assessment. Environ. Health Perspect. 115(5):833-840.

Sohn, M.D., T.E. McKone, and J.N. Blancato. 2004. Reconstructing population exposures from dose biomarkers: Inhalation of trichloroethylene (TCE) as a case study. J. Expo. Anal. Environ. Epidemiol. 14(3):204-213.

Tan, Y.M., K.H. Liao, R.B. Conolly, B.C. Blount, A.M. Mason, and H.J. Clewell. 2006. Use of a physiologically based pharmacokinetic model to identify exposures consistent with human biomonitoring data for chloroform. J. Toxicol. Environ. Health A 69(18):1727-1756.

Teuschler, L.K., G.E. Rice, C.R. Wilkes, J.C. Lipscomb, and F.W. Power. 2004. A feasibility study of cumulative risk assessment methods for drinking water disinfection by-product mixtures. J. Toxicol. Environ. Health A 67(8-10):755-777.

Thomas, J.F. 2005. Codorus Creek: Use of the relative risk model ecological risk assessment as a predictive model for decision-making. Pp. 143-158 in Regional Scale Ecological Risk Assessment: Using the Relative Risk Model, W.G. Landis, ed. Boca Raton, FL: CRC Press.

Tucker, P. 2002. Report of the Expert Panel Workshop on the Psychological Responses to Hazardous Substances. Office of the Director, Division of Health Education and Promotion, Agency for Toxic Substances and Disease Registry, Atlanta, GA [online]. Available: http://www.atsdr.cdc.gov/HEC/PRHS/psych5ed.pdf [accessed Oct. 19, 2007].

8

Improving the Utility of Risk Assessment

The committee's primary charge was to propose ways to improve risk assessment in the Environmental Protection Agency (EPA). As described in Chapter 1, we decided to focus on two broad criteria for improvement. The first criterion for improvement involves the technical content of risk assessment, which has been addressed in Chapters 4-7. The second concerned opportunities for making risk assessments more useful for informing risk-management decisions. Risk assessment in EPA is not an end in itself but a means to develop policies that make the best use of resources to protect the health of the public and of ecosystems. In Chapter 3, the committee demonstrated the importance of increased attention to risk-assessment planning and to ensuring that the levels and complexity of risk assessment (their "design") are consistent with the goals of decision-making. Increased attention to planning and scoping and to problem formulation, referred to in EPA guidance for ecologic risk assessment (EPA 1998) and cumulative risk assessment (EPA 2003), was shown in Chapters 3 and 7 to provide opportunities for increasing the relevance, and hence the utility, of the products of risk assessment.

Environmental problems arise in many forms, and new ones are always emerging. Some are large in scope, involving multiple sources of potential harm and many pathways from their sources to the creation of exposures of large human and ecologic populations. At the other extreme, a problem may involve a single source of harm and a single pathway of exposure, perhaps of relatively small populations (of production workers, for example). In some cases, a problem concerns the entire life cycle of a product or line of products; in others, it may concern approvability of a new pesticide by EPA or of a new food ingredient by the Food and Drug Administration, both driven by highly specific legislative requirements. Concerns raised by a community regarding emissions from nearby sources are increasingly common, as are concerns about the safety of various products moving in international commerce. All those problems have in common their origins in the environment and their potential to threaten human health or ecosystems; many involve not only chemicals but biologic, radiologic, and physical agents, and their potential interactions. The scope of environmental problems is increasingly enlarged to include the search for methods of resource use and product manufacture that are likely to be more sustainable—a criterion that includes health and environmental factors but others as well. Moreover, decisions in EPA often require consideration of difficult questions of costs, benefits, and risk-risk tradeoffs. Much of the discussion of Chapter 7, for example, revealed the difficulties encountered in current approaches as attempts are made to apply them to complex problems of cumulative and communitywide risks.

As the complexities of the problems and of needed decisions faced by EPA increase, so do the challenges to risk assessment to provide evaluations of clear relevance to the questions posed. That means, of course, that the questions posed to risk assessors must be both relevant to the problems and decisions faced and sufficiently comprehensive to ensure that the best available options for managing risks are given due consideration. This chapter provides guidance on the development and application of questions, methods, and decision processes to enhance the utility of risk assessment; although many elements of the guidance are applicable in the near term, our emphasis is on the longer-term future.

BEYOND THE RED BOOK

The model described in *Risk Assessment in the Federal Government: Managing the Process* (NRC 1983), referred to as the Red Book, was discussed in Chapters 1 and 2; in this model, risk assessment occupies a place between research and risk management. Risk assessment is seen as a framework[1] within which complex and often inconsistent, and always incomplete, research information is interpreted and put into usable form for risk managers. The Red Book committee was concerned principally with defining risk assessment and identifying the steps necessary to complete an assessment. It was also concerned with ensuring that risk characterization (the fourth and final step) is faithful to the underlying science and its uncertainties. Finally, and perhaps most important, the committee was concerned with protecting risk assessments from the inappropriate intrusions of policy-makers and other stakeholders, and from that concern came recommendations for the *conceptual* separation of assessment and management and for the development of risk-assessment guidelines and the elucidation and selection of "inference options" (defaults; see Chapters 2 and 6). Those and other recommendations of the Red Book have served for 25 years as sources of clarity and guidance for regulatory and public-health officials throughout the world and for stakeholders of many types.

The present committee supports retention and advancement of the major recommendations of the Red Book as they pertain to definitions, the content of risk assessment, the need for guidelines and defaults, and the conceptual separation of assessment from management. Many of our recommendations advance those aspects of the recommendations in the Red Book (and the National Research Council's 1994 report *Science and Judgment in Risk Assessment*).

To the extent that risk assessment is perceived as becoming less relevant to many important decisions or as contributing to protracted scientific debate and regulatory gridlock, that perception may result from interpretations of the Red Book that take the conceptual distinctions and separations as representing the committee's guide to a preferred decision-making process. In fact, the Red Book's concern with "process" focused heavily on protecting the integrity of risk assessment, and the committee offered little discussion of how all the necessary elements of decision-making should be arranged to achieve good decisions. That

[1]The term *framework* as used here refers to the entire decision process, of which risk assessment is one element. Risk assessment has its own framework, as described in Chapters 1 and 2 and the Red Book.

committee did not discuss the process whereby risk assessment might achieve maximum relevance, how it might be tailored in scientific depth to match the decision-making context, or how various stakeholders might influence the question of specifically what risk assessment should focus on in specific decision contexts. Those were not central issues for the Red Book committee. They clearly are issues for today in the evolution of risk assessment.

A DECISION-MAKING FRAMEWORK THAT MAXIMIZES THE UTILITY OF RISK ASSESSMENT

To ensure that risk assessments are maximally useful for risk-management decisions, the questions that risk assessments need to address must be raised before risk assessment is conducted and may need to be different from the questions that risk assessors have traditionally been tasked with answering. The more complex and multifaceted the problem to be dealt with, the more important the need to operate in that fashion. As noted in the previous section, the Red Book framework was not oriented to identifying the optimal process for complex decision-making but rather to ensuring the conceptual separation of risk assessment and risk management. A framework for risk-based decision-making (Figure 8-1, "the framework") is proposed here to provide the guidance that was missing from the Red Book. Its principal purpose, in the context of the present report, is to ensure that risk assessment is maximally useful for decision-making; as noted, this would fulfill the second of our two criteria for improving risk assessment. The framework is also intended to ensure that the methodologic changes recommended in Chapters 4-7 are put to the best use, given the repeated emphasis on analytic efforts that are appropriate to decision-making in scope and content. We offer some background on the framework in this section and then describe it more fully in the next section.

Perhaps the easiest way to explain the basic difference between the framework and the traditional assessment-management relationship is to look first at the beginnings and ends of each process. We start with an assumption that in either model no analysis would be done and no decision would be needed unless some "signal" of potential harm had come to EPA's attention. The signal can arrive in many forms, but it would generally involve a set of environmental conditions that appear to pose a threat to human or environmental health. The traditional process receives that signal and begins immediately with the question, What are the probability and consequence of one or more adverse health (or ecologic) effects posed by the signal? The framework (in Figure 8-1), in contrast, receives the signal and asks, What *options* are there to reduce the *hazards or exposures* that have been identified, and how can risk assessments be used to evaluate the merits of the various options?

Beginning the inquiry with the latter type of question immediately focuses attention on the *options* for dealing with a potential problem—the risk-management options. The options are often thought of as possible *interventions*—actions designed both to provide adequate public-health and environmental protection and to satisfy the criterion of well-supported decision-making. We note that, in most cases, "no intervention required" is one of the options to be considered explicitly.

In the framework, the questions to be posed for risk assessment arise from early consideration of the types of assessments needed to judge the relative merits of the options considered. By examining both the options and the types of assessments available, one may expand the scope of the options considered to embrace other possible interventions. Risk management involves choosing among the options after the appropriate assessments have been undertaken and evaluated. Assessments of relevant risk-management factors other than risk—such as costs, technical feasibility, and other possible benefits—also require early planning.

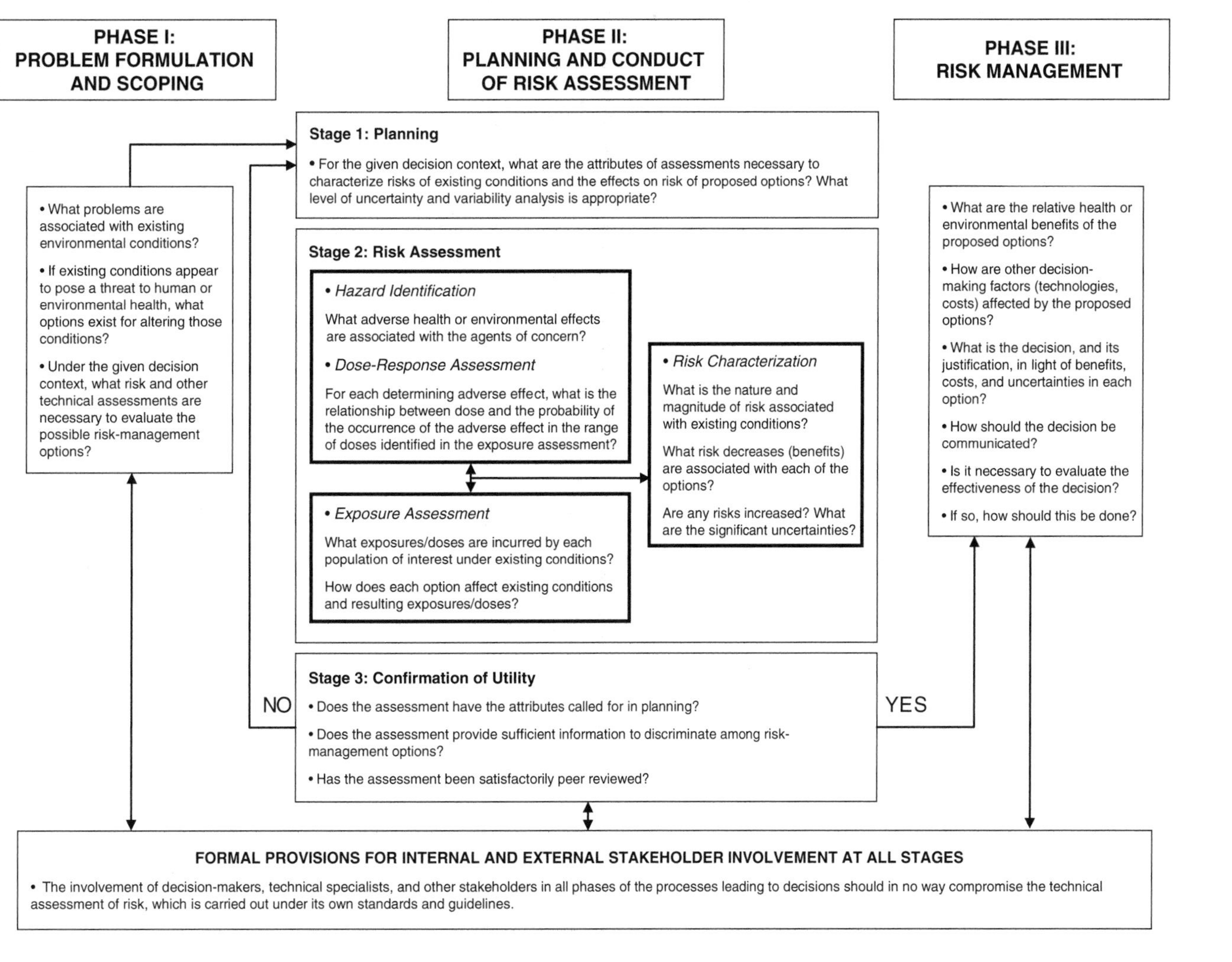

FIGURE 8-1 A framework for risk-based decision-making that maximizes the utility of risk assessment.

Risk assessment, in the framework of Figure 8-1, would typically be asked to examine risks associated with the "no intervention" option in addition to examining risk reductions (and possible increases) associated with each of the proposed interventions. Questions arising from consideration of options need to be well formulated (including a sufficient precision and breadth of issues) to ensure that important risk issues are not inadvertently overlooked; this requires that the array of options not be unnecessarily restricted.

As emphasized in Chapter 3 and elsewhere, without early and careful consideration of the decision-context, risk assessors cannot identify the types of assessments and the required level of their scientific depth necessary to support decisions (or, indeed, whether risk assessment is even the appropriate decision support tool, as shown in Figure 3-1). Without such a well-defined context, assessments will often lack well-defined stopping points and may yield ancillary analyses (for example, highly detailed quantitative uncertainty analyses) that are not essential for the decision at hand, prolonging the decision process unnecessarily (Chapter 4). By focusing on early and careful problem formulation and on the options for managing the problem, implementation of the framework can do much to improve the utility of risk assessment. Indeed, without such a framework, risk assessments may be addressing the wrong questions and yielding results that fail to address the needs of risk managers.

The framework is based on a re-examination of one of the misinterpretations of the Red Book—that assessors should be shielded from the specific decision-making issues that their analyses are intended to support. Instead, it asserts that risk assessment is of little usefulness, and can even waste resources, if it is not oriented to help discriminate among risk-management options that have to be informed by risk (and often nonrisk) considerations. More important, the framework should ensure that decisions themselves will be improved if risk-assessment information is presented to demonstrate how it affects the worth of competing choices, not for how it sheds light on an isolated substance or "problem." To be clear, the framework maintains the conceptual distinction between risk assessment and risk management articulated in the Red Book, and it remains intent on not allowing the manipulation of risk-assessment calculations to support predetermined policy choices. The conduct of risk assessments used to evaluate the risk-management options are in no way to be influenced by the preferences of risk managers.

The proposed decision-making framework resembles the well-known decision-analytic process that has been used in diverse fields for many decades (Raiffa 1968; Weinstein et al. 1980; Lave and Omenn 1986; Lave et al. 1988; Clemen 1991), in which the utility of various concrete policy options is evaluated according to the benefits that each provides. Similarly, the need to ensure that the full range of policy options is considered for the analysis has been emphasized by others including Finkel (2003); Hattis and Goble (2003); and Ashford and Caldart (2008). The committee also recognizes that numerous previous reports and guidance documents, and EPA practice in some settings, have anticipated this framework to some extent. For example, *Science and Judgment in Risk Assessment* (NRC 1994) emphasized that "risk assessment is a tool, not an end in itself," and recommended that resources be focused on obtaining information that "helps risk managers to choose the best possible course of action among the available options." The 1996 National Research Council report *Understanding Risk: Informing Decisions in a Democratic Society* (NRC 1996) emphasized that "risk characterization should be a decision driven activity, to inform choices in solving problems." The latter report also called for attention to problem formulation, with an explicit options-selection step, and representation of interested and affected parties from the earliest stages of the process. The framework also builds on but goes beyond the recommendations of the 1997 Presidential/Congressional Commission on Risk Assessment and Risk Management report (PCCRARM 1997) that called for a six-stage risk-management framework: formulate the problem in broad context, analyze the risks, define the options, make sound decisions, take actions to implement the decisions, and perform an evaluation of the effectiveness of the actions taken. Yet another National Research Council report, *Estimating the Public Health Benefits of Proposed Air Pollution Regulations* (NRC 2002), focused on evaluating the benefits of air pollution regulations and emphasized that EPA should evaluate multiple regulatory options in any benefit-cost analysis to make best use of the insights available through quantitative risk assessment. However, none of those recommendations to think more systematically about risk-management options moves consideration of options to the beginning of the assessment process in EPA, which is the key procedural change that we recommend. As articulated in more detail below, the present committee views the framework as a step beyond previous proposals and current practice—one that can possibly meet multiple objectives:

- Systematically identify problems and options that risk assessors should evaluate at the earliest stages of decision-making.
- Expand the range of effects assessed beyond individual end points (for example, cancer, respiratory problems, and individual species) to include broader questions of health status and ecosystem protection.
- Create opportunities to integrate regulatory policy with other decision-making options and strategies that expand environmental protection (for example, economic incentives, public-private partnerships, energy and other resource efficiencies, material substitution, public awareness, and product-stewardship programs).
- Serve the needs of a greatly expanded number of decision-makers (for example, government agencies, private companies, consumers, and various stakeholder organizations) whose individual and institutional roles in environmental decision-making continue to expand.
- Increase understanding of the strengths and limitations of risk assessment by decision-makers at all levels.

We expand on some of those objectives in later sections. First, we present the framework and discuss its key elements.

THE FRAMEWORK: AN OVERVIEW

Three broad phases of the framework are evident in Figure 8-1: enhanced problem formulation and scoping, planning and conduct of risk assessment, and risk management. Risk assessment and other technical and cost assessments necessary to evaluate risk-management options are carried out in the assessment phase of the process, although Figure 8-1 focuses on the planning and conduct of risk assessment given the charge of this committee. It is critical that those assessments be undertaken with assurance of their scientific integrity; technical guidelines are necessary to achieve this end, as are procedures to ensure they are followed. At the same time, it is important to recognize that risk assessments and other technical assessments are not undertaken simply because research data are available and assessments are *possible*; they are undertaken, in the proposed framework, only when the reasons for them are understood and the necessary level of their technical detail has been clarified.

The utility of assessments will be enhanced if they are undertaken within the framework. The framework will have particular importance, given the potential complexity of our proposed unified approach for dose-response assessment (Chapter 5) or methods for cumulative risk assessment of chemical and nonchemical stressors (Chapter 7), in that it emphasizes that these methodologic advances should not occur in a vacuum and are most valuable if they are clearly linked to and can inform risk-management decisions.

We emphasize that our promotion of the framework is focused on improving the utility of risk assessment to support better decision-making. As noted earlier, the framework is intended to provide guidance that was not provided by the Red Book.

Elements of the Framework: A Process Map

In this section, we outline the content of each of the elements of the framework. Each element involves a set of discrete activities, which are briefly suggested in Figure 8-1 and more fully described in Boxes 8-1 through 8-5. Some of the institutional issues associated with implementation of the framework are described in Chapter 9.

BOX 8-1 Key Definitions Used in the Framework for Risk-Based Decision-Making

PROBLEM: Any *environmental condition* (a method of product manufacture, residence near a manufacturing facility, exposure to a consumer product, occupational exposure to a pesticide, exposure of fish to manufacturing effluents, a transboundary or global environmental challenge, and so on) that is suspected to pose a threat to human or ecosystem health. It is assumed that early screening-level risk assessments may sometimes be used to identify problems or to eliminate concerns.

RISK-MANAGEMENT OPTION: Any *intervention* (a change of manufacturing process, imposition of an environmental standard, the development of warnings, use of economic incentives, voluntary initiatives, and so on) that may alter the environmental condition, reduce the suspected threat, and perhaps provide ancillary benefits. Any given problem may have several possible risk-management options. In most cases, “no intervention” will be one of the options.

LIFE-CYCLE ANALYSIS: A formal process for evaluating and managing problems associated with each stage of a product’s manufacture, distribution, uses, and disposal. It includes problems as defined above and can include evaluations of such issues as resource use and sustainability.

POPULATION: Any group of general or occupational populations or populations of nonhuman organisms.

AGENT: Any chemical (including pharmaceuticals and nutrients), biologic, radiologic, or other physical entity.

MEDIA: Air, water, food, soils, or substances having direct contact with the body.

RISK SCENARIO: A combination of agents, media, and populations in which risks to human or ecosystem health can arise.

BENEFITS: The changes (positive or negative) in health and environmental attributes that are associated with an intervention. Typically, a risk assessment will estimate the number of cases of disease, injury, or death associated with a problem—which is equivalent to the benefits of eliminating the problem. Any intervention that reduces risk without eliminating it will have benefits estimated by the difference between the status quo and the risks remaining after the intervention.

STAKEHOLDER: Any individual or organization that may be affected by the identified problem (defined above). Stakeholders may include community groups, environmental organizations, academics, industry, consumers, and government agencies.

BOX 8-2 Phase I of the Framework for Risk-Based Decision-Making (Problem Formulation and Scoping)

Identification of Risk-Management Options and Required Assessments

a. What is the problem to be investigated, and what is its source?
b. What are the possible opportunities for managing risks associated with the problem? Has a full array of possible options been considered, including legislative requirements?
c. What types of risk assessments and other technical and cost assessments are necessary to evaluate existing conditions, and how do the various risk-management options alter the conditions?
d. What impacts other than health and ecosystem threats will be considered?
e. How can the assessments be used to support decisions?
f. What is the required timeframe for completion of assessments?
g. What resources are needed to undertake the assessments?

Scope of the Framework and Definitions of Key Terms

The framework is intended for broad applicability, as can be discerned from the definitions (see Box 8-1) of terms used to describe activities in the elements of the framework.

Phase I. PROBLEM FORMULATION AND SCOPING

Two types of activities are associated with Phase I of the risk-based decision-making framework (Figure 8-1): problem formulation and the simultaneous (and recursive) identification of risk-management options and identification of the types of technical analyses, including risk assessments, that will be necessary to evaluate and discriminate among the options. The expected contents of Phase I are outlined in Box 8-2, as a series of questions to be pursued.[2]

Agency decisions related to premarket product approvals (for example, for new pesticides) depend on long-established requirements for toxicology and exposure data, and there are also well-established guidelines for risk assessments and criteria for premarket approvability. Those well-established requirements can be said to constitute Phase I planning for this type of decision-making, and the committee sees no need to alter the existing arrangement; but we do note that the proposed framework of Figure 8-1 accommodates this specific category of regulatory decision-making.

Phase II. PLANNING AND CONDUCT OF ASSESSMENT

Risk assessments designed to evaluate the risk-management options set out in Phase I are undertaken during Phase II. Phase II consists of three stages; planning, assessment, and confirmation of the utility of the assessment (see Box 8-3).

The first stage of Phase II involves the development of a careful set of plans for the necessary risk assessments. Risk assessments should not be conducted unless it is clear that they are designed to answer specific questions, and that the level of technical detail and

[2]The committee acknowledges that there may be cases following completion of appropriate problem formulation and scoping in which it is determined that risk assessment is not needed.

BOX 8-3 Phase II of the Framework for Risk-Based Decision-Making (Planning and Conduct of Risk Assessment)

Stage 1: Planning for Risk Assessment

a. What are the goals of the required risk assessments?
b. What specific risk scenarios (agents, media, and populations, including possible consideration of background exposures and cumulative risks) are to be investigated?
c. What scenarios are associated with existing conditions and with conditions after application of each of the possible risk-management options, and how should they be evaluated?
d. What is the required level of risk quantification and uncertainty/variability analysis?
e. Will life-cycle impacts be considered?
f. Are there critical data gaps that prevent completion of the required assessment? If so, what should be done?
g. How are the risk assessments informed by the other technical analyses of options (technical feasibility, costs, and so on)? How will communication with other analysts be ensured?
h. What processes should be in place to ensure that the risk assessments are carried out efficiently and with assurance of their relevance to the decision-making strategy, including time requirements?
i. What procedures are in place to ensure that risk assessments are conducted in accordance with applicable guidelines?
j. What are the necessary levels and timing of peer review?

Stage 2: Risk Assessment

a. *Hazard Identification:*
   - What adverse health or environmental effects are associated with each of the agents of potential interest?
   - What is the weight of scientific evidence supporting the classification of each effect?
   - What adverse effects are the likely risk determinants?
b. *Exposure Assessment:*
   - For the agents under study, what exposures and resulting doses are incurred by each relevant population under existing conditions?
   - What do the technical analyses (Box 8-4) reveal about how existing conditions and resulting exposures/doses would be altered by each proposed risk-management option?
c. *Dose-Response Assessment:*
   - For each determining adverse effect, what is the relationship between dose and the probability of the occurrence of the adverse effect in the dose region identified in the exposure assessment?
d. *Risk Characterization:*
   - For each population, what is the nature and magnitude of risk associated with existing conditions?
   - How are risks altered by each risk-management option (both decreases and increases)?
   - What is the distribution of individual risks in the population and subpopulations of concern, and what is the distribution of benefits under each option?
   - Considering the weight-of-evidence classification of hazards, the dose-response assessment, and the exposure assessment, what degree of scientific confidence is associated with the risk characterization?
   - What are the important uncertainties, and how are they likely to affect the risk results?

Stage 3: Confirmation of the Utility of the Risk Assessment

- Does the assessment have the attributes called for in planning?
- Does the assessment provide sufficient information to discriminate among risk-management options?
- Has the assessment been satisfactorily peer-reviewed, and have all peer-reviewer comments been explicitly addressed?

BOX 8-4 Other Technical Analyses Necessary for the Framework for Risk-Based Decision-Making

- How does each of the proposed risk-management options alter existing conditions, and with what degree of certainty?
- Are there important impacts other than those directly affecting existing conditions (as revealed, for example, by life-cycle analysis)?
- What costs are associated with no intervention to alter existing conditions and with each of the proposed risk-management options?
- What are the uncertainties in the cost assessments and the variabilities in the distribution of costs?
- Do the assessments conform to the requirements set forth in the planning phase?

uncertainty and variability analysis is appropriate to the decision context. Such attention to planning should ensure the most efficient use of resources and the relevance of the risk assessment to decision-makers. The typical questions addressed during the risk assessment planning process are set out in Box 8-3, Stage 1 (Planning).

Other technical analyses are typically required to evaluate how specific interventions will alter existing conditions; the information developed through such technical analyses (see Box 8-4) must be communicated to risk assessors, so that the effects of these interventions on risk can be evaluated.

Once the planning has been completed, risk assessments are conducted (Phase II, Stage 2). Risk assessments are conducted under agency guidelines. The guidelines should include defaults and explicit criteria for departures from defaults with other elements recommended in the present report, including those related to uncertainty assessment, unification of cancer and noncancer dose-response methods, and cumulative or community-based risk assessment (Chapters 4-7).

Once risk assessments have been completed, the framework calls for an evaluation of the *utility* of what has been produced (Stage 3 of Phase II). Thus, an evaluation of whether the assessments have the attributes called for in planning, and of whether they allow discrimination among the risk-management options, is necessary to determine whether they are useful for decision-making. If the assessments are not determined to be adequate given the problem formulation and risk-management options, the framework calls for a return to the planning stage. If they are adequate, Phase III of the framework is entered.

Phase III. RISK MANAGEMENT

In Phase III of the framework, the relative health or environmental benefits of the proposed risk-management options are evaluated, as are other factors relevant to decisions. Legislative requirements are also critical to the decision process.

The purpose of Phase III is to reach decisions, fully informed by the risk assessments. A justification for the decision, with full elucidation of the roles played by the risk information, and other pertinent factors, should be offered. A discussion of how uncertainties in all of the information used to develop decisions influenced those decisions is essential. Some of the questions that are central to risk management are set out in Box 8-5.

BOX 8-5 Elements of Phase III of the Framework for Risk-Based Decision-Making (Risk Management)

Analysis of Risk-Management Options

- What are the relevant health or environmental benefits of the proposed risk-management options? How are other decision-making factors (technologies, costs) affected by the proposed options?
- Is it indicated, with a sufficient degree of certainty given the preference of risk managers, that any of the options are preferred to a "no intervention" strategy?
- What criteria are used to assess the relative merits of the proposed options (for example, does the risk manager consider population benefits, reductions below a predefined *de miminis* level, or equity considerations)?

Risk-Management Decisions

- What is the preferred risk-management decision?
- Is the proposed decision scientifically, economically, and legally justified?
- How will it be implemented?
- How will it be communicated?
- Is it necessary to evaluate the effectiveness of the decision? How should this be done?

Stakeholder Involvement

A critical feature of the framework is related to stakeholder involvement. A continuing theme in earlier National Research Council and other expert reports on risk assessment, and loudly echoed in opinions offered to the present committee by many commenters, concerns the consistent failure to involve stakeholders adequately throughout the decision process. Without such involvement, the committee sees no way to ensure that the decision process will be satisfactory; indeed, without such involvement, it is inevitably deficient.

Figure 8-1 emphasizes that point through the box on the bottom, which spans all three phases. In addition, the two-headed arrows are meant to represent the fact that adequate communication among analysts and stakeholders, which is necessarily two-way, is critical to ensure efficiency and relevance of the analyses undertaken to support decisions. Adequate stakeholder involvement and communication among those involved in the policy and technical evaluations are difficult to achieve, but they are necessary for success. It is time that formal processes be established to ensure implementation of effective stakeholder participation in all stages of risk assessment.

For any given problem that requires EPA action, there are certain to be a number of affected parties seeking to influence the agency's course. Some stakeholders may wish to ensure that particular problems come to the attention of the agency and that their formulations be adequate. Others will hope that the agency consider various possible management options, sometimes including options that have not traditionally been part of regulatory thinking. Still others will have proposals that they believe will improve the scientific strength of agency risk assessments. And, of course, many parties will seek to influence ultimate decisions.

For cases in which agency actions will lead to regulations, formal procedural requirements are in place to allow members of the public to offer comments on proposed regulations. That type of stakeholder involvement in agency activities is obviously important, but it is insufficient in that it applies only to formal rule-making and typically comes only at the end of the process of decision-making. The present committee, like several that have come before it (Chapter 2), recommends that EPA make formal a process for gathering stakeholder views in each of the three broad phases of decision-making depicted in Figure 8-1; conflicts of interest will need to be considered in this process. It is critical that time limits for stakeholder involvement be well defined so that decision-making schedules can be met. In addition, effective stakeholder participation must consider incentives to allow for balanced participation including impacted communities and less advantaged stakeholders.

ADDITIONAL IMPROVEMENTS OFFERED BY THE FRAMEWORK

Operating under the framework can lead to improvements in the technical aspects of analysis (including economics and other nonrisk components) and can help to improve the basic research supporting risk assessment by allowing formal or informal value-of-information considerations (Chapter 3). But the major advances that the framework can bring about involve improving the quality of risk-based decision-making by raising the expectations for what risk assessments can provide. The framework could address the frustration among some that the current system channels substantial energy toward dissecting and comparing problems rather than advancing decisions that deal with problems. Other important advantages of the framework include the following:

1. *It augments and complements related trends in risk-assessment practice.* As described in Chapters 3 and 7, there is a need to design risk assessments to better inform the technical aspects of risk assessment and the ultimate decision context. EPA's *Framework for Cumulative Risk Assessment* (EPA 2003) and *Guidelines for Ecological Risk Assessment* (EPA 1998) endorse this approach and emphasize that it would be impossible to determine the appropriate scope or level of resolution of an assessment in the absence of the risk-management context. The framework takes the planning stage one step further by embedding the development of risk-management options as a formal step *before* the planning of the assessment, thereby encouraging the development of risk assessments that adequately capture important tradeoffs and cross-media exposures. In addition, the methodologic developments proposed in Chapter 5 and elsewhere are meant in part to provide greater insight for risk managers regarding the health-risk implications of specific management decisions, feeding directly into the proposed framework. A related trend involves the growth of life-cycle assessment, which includes many aspects of risk assessment but also evaluates a broader array of issues related to energy use, water consumption, and other characteristics of technologies, industrial processes, and products that determine their propensity to consume natural resources or to generate pollution. The term *life cycle* refers to the need to include all stages of a business process—raw-material extraction, manufacturing, distribution, use, and disposal, including all intervening transportation steps—to provide a balanced and objective assessment of alternatives. A critical component in the planning of a life-cycle assessment is the "functional-unit determination," in which various alternatives are compared on the basis of their ability to achieve a desired end point (for example, generation of a kilowatt-hour of electricity). The approach emphasizes the need to understand the objectives of the process or product under study, broaden the scope, and bring novel approaches and risk-management options to the forefront, including considering pollution prevention efforts. The framework builds on those important trends and emphasizes that risk assessments should be designed to provide risk managers with the necessary information to discriminate among risk-man-

agement options and that life-cycle and functional-unit thinking (if not analysis itself) will facilitate the development of a wide array of options.

2. *It makes it easier to discern "locally optimal" decisions.* The framework helps to identify locally optimal decisions (for example, choices among strategies to reduce risks posed by a given compound) by making it more difficult to make the fundamental mathematical error of averaging the predictions of incompatible models together. If, for example, there is a default estimate (including parameter uncertainty, perhaps, but small with respect to the model uncertainty) that predicts a risk X for a particular substance and a credible alternative model (with expert weight $1 - p$ assigned to it) that posits that the risk is zero, there is a temptation to declare that the "best estimate" of the risk is pX. In the traditional paradigm, if the risk assessment reports that the "best estimate" is pX, a decision-maker might be inclined to regulate as though the baseline risk is exactly pX. Following the framework in Figure 8-1 would bring the options to the fore and emphasize to all stakeholders that key uncertainties might imply that different options would be chosen, depending on key risk-assessment assumptions. In this setting, the risk characterization would more likely take the form of the statement "there is a probability p that the risk is X, in which case option B is preferred, and a probability $1 - p$ that the risk is zero, in which case option C is preferred." Thus, operating with the framework can sometimes help to avoid confusing "expected-value decision-making" (a coherent although ethically controversial approach) with "decision-making by expected value" (an incorrect and precarious approach—see Box 4-5). Careful consideration of uncertainty is not precluded by the conventional framework, but the framework in Figure 8-1 helps to determine the degree to which key uncertainties influence decisions among risk-management options and orients the risk assessor and other stakeholders around such questions about uncertainty.

3. *It makes it easier to identify and move toward "globally optimal" decisions.* More broadly, the framework opens the prospect of moving beyond a choice among strategies to deal with a single substance to the development, evaluation, and selection of alternative strategies to fulfill the function with minimum net risk. As implied by the *functional-unit* definition above, this involves expanding the lens of current environmental decision-making from primarily a single-issue and incremental-risk focus to address issues of comparative and cumulative risk, benefits and costs, life-cycle risks, technologic innovation and public values. We believe that questions about the risks posed by industrial processes can often be answered better by considering risk-risk tradeoffs and evaluating risk-management options than by studying risks in isolation from the feasible means of control. Although the expanded scope may exceed the bounds of EPA decision-making (either in a practical sense, given current regulations, or in a theoretical sense, given the agency's jurisdiction), functional-unit thinking will help to avoid considering only local optima that represent the peaks within a valley, will encourage the development of agencywide initiatives and strategies, and will encourage EPA to cooperate with other federal agencies (and vice versa) to work on more sweeping interventions that increase efficiency and minimize untoward risk-risk tradeoffs. In short, the framework would allow EPA to compare options with appropriate use of knowledge about uncertainty and would allow it to broaden (within reason) the set of options under consideration.

4. *It can provide the opportunity for improved public participation.* The framework can broaden the focus of inquiry from studying the risk—which may be dominated by highly technical discussions of potency, fate and transport, mode of action, and so on—to developing and evaluating alternative interventions, which should be a more accessible and interesting arena for affected stakeholders to participate in. Stakeholders (such as local communities) may also bring particular knowledge about the benefits, costs, and implementation of risk-

management options to a discussion. The process would recognize the roles, relationships, and capabilities of government and nongovernment decision-makers and would ensure that risk assessments serve their needs. The committee recognizes that effective implementation of the framework in many cases will not be possible without the involvement of other governmental agencies and other organizations.

5. *It would make economics and risk-risk tradeoffs more central in the analysis.* Although many regulatory, legislative, and logistical constraints complicate the simultaneous consideration of costs of control and benefits, the framework would, where applicable and feasible, encourage the use of similar methods between disciplines (such as the explicit incorporation of uncertainty and variability and the development of default assumptions and criteria for departure in economic analyses) and would spur collaboration between risk assessors and regulatory economists. As articulated above, the framework would also make consideration of potential risk-risk tradeoffs central in the assessment, inasmuch as the initial planning and scoping steps and the development of risk-management options under study would lead to an explicit discussion of the array of exposures that could be influenced by each option.

In Appendix F, the committee presents three case examples to demonstrate how the usefulness of risk assessments might be enhanced by implementation of the framework for risk-based decision-making.[3]

POTENTIAL CONCERNS RAISED BY THE FRAMEWORK

The framework has many desirable attributes that can allow risk assessment to be maximally informative for decision-making, but various concerns could be raised about it. Some of the concerns are misconceptions, and others are legitimate issues that would need to be addressed. We discuss various critiques and consider their potential implications below.

Concern 1: There are many contexts in which EPA is constrained to a narrow set of options by the structure of regulations or in which it is unclear at the outset whether a problem is of sufficient magnitude to require an intervention or whether a potential intervention exists, so the framework may waste effort in producing needless evaluations.

This concern has some legitimacy, but the framework does not preclude risk assessment solely to determine the potential magnitude of a problem or to compare the impacts of options within a severely constrained solution set. As to the former, the framework is intended to keep one eye continually on problems and one on interventions, and choosing between one and the other is a false dichotomy. The committee believes that the current use of risk assessment has disproportionately emphasized dissecting risks rather than implementing possible interventions, but the pendulum does not need to (and should not) swing past a middle ground. As to the latter, in situations where the regulatory requirements preclude consideration of a wide array of risk-management options, EPA could both formally evaluate the options that can be considered and use the framework to determine the extent to which current constraints preclude a better risk-management strategy. At a minimum, the

[3]The three case examples in Appendix F address electricity generation, decision support for drinking-water systems, and control of methylene chloride exposure in the workplace and general environment. These are stylized examples intended to illustrate how application of the framework for risk-based decision-making might lead to a process and outcome different from those of conventional application of risk assessment.

framework would emphasize the need for EPA to consider risk tradeoffs and alternative strategies explicitly when devising risk-management options.

Concern 2: The framework may exacerbate the problem of "paralysis by analysis," both because the analytic burdens will increase with the need to evaluate numerous options and because risk assessments may show that uncertainties are too great to permit discrimination among various options.

The committee proposed earlier that the framework will help risk assessments to come to closure by focusing on the information needed to discriminate among risk-management options rather than focusing on the information needed to "get the number right." However, it could be argued that the need to quantify benefits among multiple potential risk-management options, including tradeoffs and multimedia considerations, will greatly expand the analytic requirements of a given assessment, especially given that the uncertainties in a simpler assessment may prove too large for discrimination among options. That is an important concern, but many of the more analytically complex components (for example, cumulative risk assessment and multimedia exposure) would be needed for any risk assessment with a similar scope, regardless of what risk-management options are under consideration, and the marginal time to evaluate multiple risk-management options should be relatively small once a model has been constructed to evaluate the benefit of one option appropriately. In addition, if the uncertainties are too large for discrimination among options on a risk basis, it would imply simply that other considerations are central in the risk-management decisions or that further research is required.

Concern 3: The framework will not lead to better decisions and public-health protection, because the process does not provide for equal footing for competing interest groups.

Although the committee proposes that the framework will enhance public participation and will reduce asymmetries among stakeholders by focusing on early development of risk-management options, there would continue to be asymmetries in the ability of different stakeholders to get options "on the table," given issues of political power and imbalance in available information. More generally, the framework could potentially be manipulated if the set of options evaluated were constrained inappropriately. In addition, the importance of risk assessment is not reduced in the framework, so the technical imbalance would remain. The concern is relevant, but it is not introduced by the framework, but rather is endemic to processes that bring together government, communities, and industries to debate decisions that will have serious economic and public-health effects. The framework could improve on the current practice provided there is substantive stakeholder involvement throughout the process, if stakeholder groups have sufficient technical expertise (which can be developed over time through efforts by EPA and others), and if EPA formally addresses all suggested options in writing (either by evaluating them quantitatively or by discussing qualitatively how they are strictly dominated by other options and therefore do not need to be considered). The potential for manipulation is not created by the framework and in fact would be reduced by it: risk managers can now implicitly reduce the option set by asking risk assessors to evaluate the benefits of a preselected control scenario, and a public process to explicitly construct a wide-ranging set of options seems preferable. As a component of the development and implementation of the framework, EPA should propose guidelines for the options-development step of Phase I, focusing explicitly on stakeholder participation and formal processes for transparent selection of risk-management options to study.

Concern 4: The framework breaks down the firewall between risk assessment and risk management, creating a potential for manipulation.

That the framework allows assessors to see the choices facing the decision-maker does not imply that they would be involved in risk management, nor does it imply that decision-makers would have license or opportunity to impose their will on the analysis. The framework empowers risk *assessment* to drive the engine that determines which options perform best in the presence of uncertainty, variability, and public preferences, but it does not empower risk *assessors* to impose their preferences on the analysis. It will remain important in the framework to have clear risk-assessment guidelines (see, for example, Chapters 3, 5, and 7) that can be used to conduct the assessments needed to evaluate options.

Increasing the interaction between risk assessors and risk managers requires that there be further protection against the possibility that identified or preferred policy options will bias the evaluation of risks or, even more problematically, that risk managers will influence the content of the risk assessment to support preferred risk-management options. Ensuring the integrity of evaluations along the continuum of the risk-assessment–risk-management discussion fundamentally rests on maintaining an effective system of governance in EPA and other organizations applying risk assessment. The governance process should have the following elements:

- *Clarity and accountability of roles and responsibilities.* The extent to which risk assessors and risk managers understand their roles and are evaluated on the basis of their fulfilling their responsibilities will assist in mitigating concerns about potential compromise of scientific or policy-related assessments.
- *Greater transparency of the process.* Making information about the assumptions used and judgments reached in risk-assessment and policy deliberations more widely available is itself an important safeguard against abuse.
- *Documentation of the process.* There needs to be appropriate documentation of the rules and milestones of the process and of the relevant information base at all important stages of risk-assessor–risk-manager deliberations.
- *Oversight and periodic review.* EPA should submit selected decisions each year for independent review to ensure the integrity of the risk-assessment–risk-management process. Independent reviewers should issue a public report on their findings.

As mentioned above, the problems can occur with the current (conceptual or institutional) "firewall" between assessment and management. A risk manager who keeps analysts in the dark about the choices can still order them to "make the risk look smaller (bigger)." Safeguards against any form of manipulation of the risk-assessment process, whether related to the framework or not, must be in place; it seems to the committee that a process that emphasizes evaluation of risk-management options will by definition involve broader participation, which implies more "sunshine" and less opportunity for the type of manipulation that the Red Book committee was justifiably concerned with.

CONCLUSIONS AND RECOMMENDATIONS

Some features of the framework may be evident in EPA programs, but its full implementation will require a substantial transition period. The committee believes that the long-term utility of risk assessment as a decision-support tool requires that EPA operate in the proposed framework (or a very similar one) and so urges the agency to begin the transition. It is perhaps useful to conceive of the transition process as involving a period of experimentation and development of carefully selected "demonstration projects" to illustrate the application of the framework. Selection of a few important environmental problems to which the framework would be applied in full (with *formal* and *time-limited* stakeholder involvement at all stages) would constitute a learning period for agency assessors, managers, and stakeholders. Lessons from such demonstration projects could be recorded and used to improve the framework and its application. The committee believes strongly that gradual adoption of the framework will do much to improve the analytic power and utility of risk assessment and will reveal this power and utility to a much wider audience; its credibility and general acceptability will thereby be enhanced.

In summary, we recommend the following:

- The technical framework for risk assessment presented in the Red Book should remain intact but should be embedded in a broader framework in which risk assessment is used principally to help to discriminate among risk-management options.
- The framework for risk-based decision-making (Figure 8-1) should have as its core elements a problem-formulation and scoping phase in which the available risk-management options are identified, a planning and assessment phase in which risk-assessment tools are used to determine risks under existing conditions and with proposed options, and a management phase in which risk and nonrisk information is integrated to inform choices among options.
- EPA should develop multiple guidance documents relevant to the framework, including a more expansive development of the framework itself (with explicit steps to determine the appropriate scope of the risk assessment), formal provisions for stakeholder involvement at all stages, and methods for options development that ensure that a wide array of options will be formally evaluated.
- EPA should phase in the use of the framework with a series of demonstration projects that apply the framework and that determine the degree to which the approach meets the needs of the agency risk managers, and how risk-management conclusions differ as a result of the revised orientation.

REFERENCES

Ashford, N.A. and C.C. Caldart. 2008. Environmental Law, Policy, and Economics: Reclaiming the Environmental Agenda. Cambridge: MIT Press.

Clemen, R.T. 1991. Making Hard Decisions: An Introduction to Decision Analysis. Boston: PWS-Kent Pub. Co.

EPA (U.S. Environmental Protection Agency). 1998. Guidelines for Ecological Risk Assessment. EPA/630/R-95/002F. Risk Assessment Forum, U.S. Environmental Protection Agency, Washington, DC. April 1998 [online]. Available: http://oaspub.epa.gov/eims/eimscomm.getfile?p_download_id=36512 [accessed Feb. 9, 2007].

EPA (U.S. Environmental Protection Agency). 2003. Framework for Cumulative Risk Assessment. EPA/600/P-02/001F. National Center for Environmental Assessment, Risk Assessment Forum, U.S. Environmental Protection Agency, Washington, DC [online]. Available: http://cfpub.epa.gov/ncea/cfm/recordisplay.cfm?deid=54944 [accessed Jan. 4, 2008].

Finkel A.M. 2003. Too much of the [National Research Council's] 'Red Book' is still ahead of its time. Hum. Ecol. Risk Assess. 9(5):1253-1271.

Hattis, D., and R. Goble. 2003. The Red Book, risk assessment, and policy analysis: The road not taken. Hum. Ecol. Risk Assess. 9(5):1297-1306.

Lave, L.B., and G.S. Omenn. 1986. Cost-effectiveness of short-term tests for carcinogenicity. Nature 324(6092):29-34.

Lave, L.B., F.K. Ennever, H.S. Rosenkranz, and G.S. Omenn. 1988. Information value of rodent bioassay. Nature 336(6200):631-633.

NRC (National Research Council). 1983. Risk Assessment in the Federal Government: Managing the Process. Washington, DC: National Academy Press.

NRC (National Research Council). 1994. Science and Judgment in Risk Assessment. Washington, DC: National Academy Press.

NRC (National Research Council). 1996. Understanding Risk: Informing Decisions in a Democratic Society. Washington, DC: National Academy Press.

NRC (National Research Council). 2002. Estimating the Public Health Benefits of Proposed Air Pollution Regulations. Washington, DC: The National Academies Press.

PCCRARM (Presidential/Congressional Commission on Risk Assessment and Risk Management). 1997. Framework for Environmental Health Risk Management - Final Report, Vol. 1. [online]. Available: http://www.riskworld.com/nreports/1997/risk-rpt/pdf/EPAJAN.PDF [accessed Jan. 4, 2008].

Raiffa, H. 1968. Decision Analysis: Introductory Lectures on Choices under Uncertainty. New York: Random House.

Weinstein, M.C., H.V. Fineberg, A.S. Elstein, H.S. Frazier, D. Neuhauser, R.R. Neutra, and B.J. McNeil. 1980. Clinical Decision Analysis. Philadelphia: W.B. Saunders.

9

Toward Improved Risk-Based Decision-Making

The Framework for Risk-Based Decision-Making is designed to improve risk assessment by enhancing the value of risk assessment to policy-makers, expanding stakeholder participation, and more fully informing the public, Congress, and the courts about the basis of Environmental Protection Agency (EPA) decisions. That will require building on EPA's decision-making practices to expand consideration of options and developing a long-term strategy for renewal. To shape such a strategy, this chapter identifies three categories of prerequisites of successful transition to the framework:

- *Adopting transition rules.* The most successful experiences and practices that govern current risk assessment and risk-management decision-making in EPA and other institutions offer models for introducing agency leaders and staff to new issues and processes and for integrating new principles and practices into the framework outlined in Chapter 8.
- *Managing institutional processes.* Management issues include consideration of legal impediments to implementing the framework, changes in organizational structure, and strengthening institutional capacity, for example, skills, training and other forms of knowledge-building, and resources.
- *Providing leadership and management.* The transition will require support, including guidance and resources, from the EPA leadership community, the executive and legislative branches of government, and key stakeholders.

Those and related implementation recommendations signify the committee's recognition that assembling, evaluating, and interpreting information called for in the framework introduce major changes in EPA's various risk-assessment and decision-making processes. Some aspects of the framework (for example, new approaches to communication and participation) may not require major new investment in the short term; however, for an institution as large and diverse as EPA, the availability and allocation of resources—funding, time, and personnel—are central aspects of sustaining any institutional arrangements for agencywide change of the magnitude outlined in Chapter 8.[1] As in all enterprises, funding is a rate-limiting and quality-determining step.

TRANSITION TO THE FRAMEWORK FOR RISK-BASED DECISION-MAKING

Improving the utility of risk assessment to include upfront problem formulation and scoping and planning with an expanded array of options requires several practical steps to ensure that risk assessors and risk managers have a clear understanding of their roles and responsibilities and have sufficient guidance to administer them effectively. As a beginning, EPA should examine the key functions and attributes of its decision-making processes in relation to those recommended in this report. Although many activities are comparable (for example, hazard assessment and dose-response assessment), others, such as life-cycle assessment, will be new in many agency programs and will need to be integrated into the process of assessing risks and the options for managing them.

Historically, even though EPA risk assessment is generally linked to decision-making, guidance arising out of National Research Council risk-assessment reports has been directed mainly to improving agency risk assessments with little attention to future decision-making. The framework focuses attention on improving the utility of risk assessments to better inform decision-making. To implement the framework, the agency will need *innovative and instructive guidance that informs its scientists, economists, lawyers, regulatory staff, senior managers, and policy makers* of their roles and, most important, fosters interaction among them. Principles, examples, and practices drawn from "success stories" in which EPA and other entities have used processes similar to those proposed for the framework offer starting points for such guidance. Selected risk-based decision-making scenarios that provide realistic illustrations of how the framework can work can be especially instructive.

The framework promotes greater attention to and use of risk-related information from such fields as economics, psychology, and sociology—disciplines not usually involved to a great extent in EPA assessments. While those fields may not be central in the risk assessment itself, the framework integrates a variety of information in constructing risk-management decisions. Increased emphasis on those fields in the framework requires extending the kind of robust peer-review practices historically required by statute or policy for risk assessment to cost and benefit analyses, community impact assessments, life-cycle analyses, and related information.[2] The objective would be to give decision-makers, stakeholders, and the public confidence in, and understanding of, the insights and limitations of evaluations. Improved peer review of analyses will also add an important dimension of transparency.

INSTITUTIONAL PROCESSES

The framework presents opportunities for EPA to review and realign some institutional processes to foster consistent approaches to using risk assessment and other analyses (in-

[1]This committee comment is prompted by recent congressional testimony on the impact of budget cuts on EPA's capacity to meet the demands of risk assessment *as currently practiced* (Renner 2007). The budget cuts generate serious concern about the agency's capacity to undertake the advanced analyses recommended in this report and to implement a new, more data-intensive framework without concerted attention to funding and staffing as part of governmentwide and EPA strategic planning and annual budget processes.

[2]As in traditional risk assessment, peer reviewers would be experts in the discipline under review—sociologists for societal impacts, economists for economic impacts, and so on. However, especially valuable would be the addition of peer reviewers, expert in multiple disciplines, that can evaluate the risk and benefit-cost analyses that inform different decision options.

cluding technical and economic) to better inform risk-management decisions across EPA's various programs. Several processes warrant consideration.

Statutory Authority

The committee believes that it has achieved its goal of recommending substantial improvements that can be accomplished by refining and refocusing institutional processes within existing statutory authority. Committee recommendations for expanding risk-assessment activities to give more emphasis to, for example, cumulative risk, quantitative uncertainty and variability analysis, and harmonizing analyses for cancer and noncancer end points call for state-of-the-science improvements that easily fall within the agency's existing authority: for more than 20 years, EPA has regularly incorporated state-of-the-science improvements of this kind to develop and amend general risk-assessment guidelines and conduct individual assessments.

The committee's more far-reaching recommendations—such as broad-based discussion of risk-management options early in the process, extensive stakeholder participation throughout the process, and consideration of life-cycle approaches in a broader array of agency programs—can be viewed as common-sense extensions throughout the agency as a whole of practices that are now limited to selected programs or are unevenly and incompletely implemented. For example, EPA's *Guidelines for Ecological Risk Assessment* contemplates the kind of options-informed risk-assessment planning envisioned by the framework (EPA 1998, p. 10):

> Risk assessors and risk managers *both* consider the potential value of conducting a risk assessment to address identified problems. Their discussion explores what is known about the degree of risk, *what management options are available to mitigate or prevent it*, and the value of conducting a risk assessment compared with other ways of learning about and addressing environmental concerns [emphasis added].

Focused attention on integrated agencywide implementation of that and other existing guidance related to cumulative risk assessment, criteria for departing for defaults, and life-cycle analysis would lead to some of the improvements contemplated by the framework without new legislative initiatives.

Structural Change

In keeping with EPA's media-based organizational structure, agency decision-making processes are compartmentalized in line with media- and statute-specific environmental problems, legal requirements, case law, and programmatic history. This approach parallels EPA statutes but takes little cognizance of current understanding of the multimedia, cumulative-risk characteristics of environmental pollution and the need for multidisciplinary, cross-program, and cross-agency analyses of scientific issues and regulatory options. The committee's major recommendation that EPA move to a consistent and transparent process that ensures the right questions are being asked of the assessment will therefore require new approaches to coordination, communication, and framing of environmental-protection options.

To adapt its current decision-making process to the framework, EPA should establish an options-development team composed of Senior Executive Service environmental professionals from the major regulatory programs, the Office of Environmental Information, the Office of General Counsel, the Office of Research and Development, and other relevant offices. The team's primary responsibilities would include identifying prospective decisions (or categories of decisions) for which risk assessments will be needed and providing risk assessors with contextual information on the problem under review and the regulatory or other options then[3] under consideration. To provide guidance for EPA risk assessors and managers and information for stakeholders and the public, essential team functions would include

- Developing criteria for defining and selecting high-priority risk assessments for continuing attention by the team.
- Defining a suite of preliminary decision-making options that identify critical factors and suggest bounds for individual risk assessments.
- Providing an explicit statement of the problem that the agency is attempting to solve.
- Ensuring consideration of risk tradeoffs.
- Maintaining a system for tracking accountability in the preparation of individual risk assessments and the options-development process's contribution to and impact on the use of each assessment in decision-making.

The options-informed process recommended in this report recognizes both regulatory and nonregulatory options and gives EPA the flexibility to define options narrowly or broadly, depending on the nature and extent of the problem to be solved. The nature and scope of the options can be expected to vary from one problem to the next.

Skills, Training, and Knowledge-Building

Many risk assessments involve a complex, data-intensive, and multidisciplinary analyses. The data come from studies on highly inbred laboratory animals and from genetically diverse human populations, and basic monitoring data come from environmental media and sophisticated analyses of biochemical mechanisms, cancer pathology, and exposure pathways. Such analyses demand a multidisciplinary and scientifically sophisticated workforce, experienced not only in the underlying disciplines but in special aspects of the risk-assessment process.

Quantitative uncertainty analysis and cumulative risk assessment, for example, may well require expertise not now available in EPA or the larger scientific community in the numbers and experience levels needed to implement recommendations in this report. As a result, implementing many committee recommendations will require new expertise, and EPA may need to expand its programs to draw on expertise in other federal agencies and private entities. In all cases, training will be necessary on a continuing basis to ensure that staff are conversant with advances in disciplines that contribute to risk assessment and decision-making.

Training of managers and decision-makers on risk-assessment issues is essential for the assessor-manager discussion at the core of problem formulation and scoping, planning, and subsequent decision-making. Those senior participants in the process can participate fully and knowledgeably only if they are conversant with risk-assessment issues and methods. Such training is also essential for communication between senior agency officials, stakeholders, and other members of the public. It is equally important for technical staff to be trained to understand and appreciate the nontechnical factors that shape some risk-management and decision-making issues.

[3]As discussed in Chapter 3, the iterative nature of the overall process calls for continuing evaluation of options as a risk assessment proceeds. The initial set of options can therefore be expected to evolve through revision, deletion, and addition.

LEADERSHIP AND MANAGEMENT

Because the development of the framework has agencywide application, it is critical for the EPA top-leadership to participate in the development and implementation of the framework. The leadership and participation by the EPA administrator and assistant administrators, Congress, other arms of the executive branch (for example, the Office of Science and Technology Policy, the White House, and the Office of Management and Budget), and major stakeholders, including other federal agencies, will be essential for improvements in EPA's decision-making processes.

In this context, leadership attention to several management objectives will be critical:

- Developing explicit policies that commit EPA to implementing an options-informed process for risk assessment and risk management.
- Funding to implement these policies, including budgets adequate for preparing guidance and other documents, for training to prepare EPA personnel to undertake implementation activities, and for developing an expanded knowledge base and institutional capacity for more timely results.
- Adopting a common set of evaluation factors—applicable to all programs—for assessing the outcomes of policy decisions and the efficacy of the framework.

Other activities can advance the agency's implementation program. Ideally, the program would include a system of workshops for managers and staff to create a learning culture that emphasizes acquiring new knowledge, professional development, and decision-making practices and tools aimed at effective problem-solving. In this regard, a serious commitment to a consistent process for implementing the framework would include evaluating senior managers, in part, on the pace and success of applying new principles and practices in individual programs. Committed leadership would also pursue opportunities for partnerships and cooperative relationships with stakeholder organizations to expand the universe of options for problem-solving beyond traditional regulation.

In summary, informed and, in some cases, ground-breaking governance are intended to improve EPA risk-assessment processes, focus the assessment on the relevant questions, discourage political interference or pre-determined policy biases, and promote senior-level oversight of the timeliness, relevance, and impact of decision-making. The present report presents a major opportunity for EPA to re-examine its decision-making processes, innovate reforms, and expedite change that takes account of 21st century scientific developments, the faster pace of the global marketplace, and the needs of contemporary policy-making.

CONCLUSIONS AND RECOMMENDATIONS

The committee was given a broad charge to develop scientific and technical recommendations for improving risk-analysis approaches used by EPA. In its evaluation, the committee focused on the scientific underpinnings of risk assessment and its role in decision-making.

Risk assessment is at a crossroads, and the credibility of this essential tool is being challenged by stakeholders who have the potential to gain or lose from the outcome of an assessment. Although there appears to be an expanding need for risk-based decisions, the science underlying risk assessment and the decision contexts in which risk assessments are being used are increasingly complex, and the value and relevance of risk assessment are being questioned. The context of risk decisions has evolved since the development of the framework in the 1983 National Research Council report *Risk Assessment in the Federal Government:*

Managing the Process (NRC 1983), known as the Red Book, and challenges now often include broad consideration of multiple health and ecologic effects, costs and benefits, and risk-risk tradeoffs. The growing complexity of the process is compounded by the ever-changing nature of the science underlying many of the assumptions concerning measurement of adverse effects, exposures, dose and response, and uncertainty in the characterization of risks. As the science has advanced, so has the need to consider the social impacts of risk decisions to ensure that risk assessment is relevant to stakeholder concerns.

The following conclusions and recommendations aim to provide guidance to improve the scientific and technical basis of risk estimates, to address the characterization of variability and uncertainty, and ultimately to broaden the focus of risk analysis toward the development of improved public-health and environmental decisions. Implementation of the committee's recommendations will help to ensure that risk assessments are consistent with current and evolving scientific understanding and relevant to the various risk-management missions of EPA.

Design of Risk Assessment

The process of planning risk assessment and ensuring that its level and complexity are consistent with the needs to inform decision-making can be thought of as the "design" of risk assessment. The committee encourages EPA to focus greater attention on design in the formative stages of risk assessment, specifically on planning and scoping and problem formulation, as articulated in EPA guidance for ecologic and cumulative risk assessment (EPA 1998, 2003). Good design involves bringing risk managers, risk assessors, and various stakeholders together early in the process to determine the major factors to be considered, the decision-making context, and the timeline and depth needed and to ensure that the right questions are being asked in the context of the assessment.

Increased emphasis on planning and scoping and on problem formulation has been shown to lead to risk assessments that are more useful and better accepted by decision-makers (EPA 2002, 2003, 2004); however, incorporation of these stages in risk assessment has been inconsistent, as noted by their absence from various EPA guidance documents (EPA 2005a, b). An important element of planning and scoping is definition of a clear set of options for consideration in decision-making where appropriate. This should be reinforced by the up-front involvement of decision-makers, stakeholders, and risk assessors, who together can evaluate whether the design of the assessment will address the identified problems.

Recommendation: Increased attention to the design of risk assessment in its formative stages is needed. The committee recommends that planning and scoping and problem formulation, as articulated in EPA guidance documents (EPA 1998, 2003), should be formalized and implemented in EPA risk assessments.

Uncertainty and Variability

Addressing uncertainty and variability is critical for the risk-assessment process. Uncertainty stems from lack of knowledge, so it can be characterized and managed but not eliminated. Uncertainty can be reduced by the use of more or better data. Variability is an inherent characteristic of a population, inasmuch as people vary substantially in their exposures and their susceptibility to potentially harmful effects of the exposures. Variability cannot be reduced, but it can be better characterized with improved information.

There have been substantial differences among EPA's approaches to and guidance for addressing uncertainty in exposure and dose-response assessment. EPA does not have a consistent approach to determine the level of sophistication or the extent of uncertainty analysis needed to address a particular problem. The level of detail for characterizing uncertainty is appropriate only to the extent that it is needed to inform specific risk-management decisions appropriately. It is important to address the required extent and nature of uncertainty analysis in the planning and scoping phases of a risk assessment. Inconsistencies in the treatment of uncertainty among components of a risk assessment can make the communication of overall uncertainty difficult and sometimes misleading.

Variability in human susceptibility has not received sufficient or consistent attention in many EPA health risk assessments although there are encouraging exceptions, such as those for lead, ozone, and sulfur oxides. For example, although EPA's 2005 *Guidelines for Carcinogen Risk Assessment* (EPA 2005a) acknowledges that susceptibility can depend on one's stage in life, this requires greater attention in practice, particularly for specific population groups that may have greater susceptibility because of their age, ethnicity, or socioeconomic status. The committee encourages EPA to move toward the long-term goal of quantifying population variability more explicitly in exposure assessment and dose-response relationships. An example of progress that moves towards this goal is EPA's draft risk assessment of trichloroethylene (EPA 2001; NRC 2006), which considers how differences in metabolism, disease, and other factors contribute to human variability in response to exposures.

Recommendation: EPA should encourage risk assessments to characterize and communicate uncertainty and variability in all key computational steps of risk assessment—for example, exposure assessment and dose-response assessment. Uncertainty and variability analysis should be planned and managed to reflect the needs for comparative evaluation of the risk-management options. In the short term EPA, should adopt a "tiered" approach for selecting the level of detail to be used in the uncertainty and variability assessments, and this should be made explicit in the planning stage. To facilitate the characterization and interpretation of uncertainty and variability in risk assessments, EPA should develop guidance to determine the appropriate level of detail needed in uncertainty and variability analyses to support decision-making and should provide clear definitions and methods for identifying and addressing different sources of uncertainty and variability.

Selection and Use of Defaults

Uncertainty is inherent in all stages of risk assessment, and EPA typically relies on assumptions when chemical-specific data are not available. The 1983 Red Book recommended the development of guidelines to justify and select from among the available inference options, the assumptions—now called defaults—to be used in agency risk assessments to ensure consistency and avoid manipulations in the risk-assessment process. The committee acknowledges EPA's efforts to examine scientific data related to defaults (EPA 1992, 2004, 2005a), but recognizes that changes are needed to improve the agency's use of them. Much of the scientific controversy and delay in completion of some risk assessments has stemmed from the long debates regarding the adequacy of the data to support a default or an alternative approach. The committee concludes that established defaults need to be maintained for the steps in risk assessment that require inferences and that clear criteria should be available for judging whether, in specific cases, data are adequate for direct use or to support an inference in place of a default. EPA, for the most part, has not yet published clear, general guidance on what level of evidence is needed to justify use of agent-specific data and not resort to a default. There are also a number of defaults (missing or implicit defaults) that are engrained in EPA risk-assessment practice but are absent from its risk-assessment guidelines. For example, chemicals that have not been examined sufficiently in epidemiologic or toxicologic studies are often insufficiently considered in or are even excluded from risk assessments; because no description of their risks is included in the risk characterization, they carry no weight in decision-making. That occurs in Superfund-site and other risk assessments, in which a relatively short list of chemicals on which there are epidemiologic and toxicologic data tends to drive the exposure and risk assessments.

Recommendation: EPA should continue and expand use of the best, most current science to support and revise default assumptions. EPA should work toward the development of explicitly stated defaults to take the place of implicit defaults. EPA should develop clear, general standards for the level of evidence needed to justify the use of alternative assumptions in place of defaults. In addition, EPA should describe specific criteria that need to be addressed for the use of alternatives to each particular default assumption. When EPA elects to depart from a default assumption, it should quantify the implications of using an alternative assumption, including how use of the default and the selected alternative influences the risk estimate for risk-management options under consideration. EPA needs to more clearly elucidate a policy on defaults and provide guidance on its implementation and on evaluation of its impact on risk decisions and on efforts to protect the environment and public health.

A Unified Approach to Dose-Response Assessment

A challenge to risk assessment is to evaluate risks in ways that are consistent among chemicals, that account adequately for variability and uncertainty, and that provide information that is timely, efficient, and maximally useful for risk characterization and risk management. Historically, dose-response assessments at EPA have been conducted differently for cancer and noncancer effects, and the methods have been criticized for not providing the most useful results. Consequently, noncancer effects have been underemphasized, especially in benefit-cost analyses. A consistent approach to risk assessment for cancer and noncancer effects is scientifically feasible and needs to be implemented.

For cancer, it has generally been assumed that there is no dose threshold of effect, and dose-response assessments have focused on quantifying risk at low doses and estimating a population risk for a given magnitude of exposure. For noncancer effects, a dose threshold (low-dose nonlinearity) has been assumed, below which effects are not expected to occur or are extremely unlikely in an exposed population; that dose is a reference dose (RfD) or a reference concentration (RfC)—it is thought "likely to be without an appreciable risk of deleterious effects" (EPA 2002).

EPA's treatment of noncancer and low-dose nonlinear cancer end points is a major step by the agency in an overall strategy to harmonize cancer and noncancer approaches to dose-response assessment; however, the committee finds scientific and operational limitations in the current approaches. Noncancer effects do not necessarily have a threshold, or low-dose nonlinearity, and the mode of action of carcinogens varies. Background exposures and underlying disease processes contribute to population background risk and can lead to linearity at the population doses of concern. Because the RfD and RfC do not quantify risk for different magnitudes of exposure but rather provide a bright line between possible harm and safety, their use in risk-risk and risk-benefit comparisons and in risk-management decision-making is limited. Cancer risk assessments usually do not account for differences among humans in cancer susceptibility other than possible differences in early-life susceptibility.

Scientific and risk-management considerations both support unification of cancer and noncancer dose-response assessment approaches. The committee therefore recommends a consistent, unified approach for dose-response modeling that includes formal, systematic assessment of background disease processes and exposures, possible vulnerable populations, and modes of action that may affect a chemical's dose-response relationship in humans. That approach redefines the RfD or RfC as a risk-specific dose that provides information on the percentage of the population that can be expected to be above or below a defined acceptable risk with a specific degree of confidence. The risk-specific dose will allow risk managers to weigh alternative risk options with respect to that percentage of the population. It will also permit a quantitative estimate of benefits for different risk-management options. For example, a risk manager could consider various population risks associated with exposures resulting from different control strategies for a pollution source and the benefits associated with each strategy. The committee acknowledges the widespread applications and public-health utility of the RfD; the redefined RfD can still be used as the RfD has been to aid risk-management decisions.

Characteristics of the committee's recommended unified dose-response approach include use of a spectrum of data from human, animal, mechanistic, and other relevant studies; a probabilistic characterization of risk; explicit consideration of human heterogeneity (including age, sex, and health status) for both cancer and noncancer end points; characterization (through distributions to the extent possible) of the most important uncertainties for cancer and noncancer end points; evaluation of background exposure and susceptibility; use of probabilistic distributions instead of uncertainty factors when possible; and characterization of sensitive populations.

The new unified approach will require implementation and development as new chemicals are assessed or old chemicals are reassessed, including the development of test cases to demonstrate proof of concept.

Recommendation: The committee recommends that EPA implement a phased-in approach to consider chemicals under a unified dose-response assessment framework that includes a systematic evaluation of background exposures and disease processes, possible vulnerable populations, and modes of action that may affect human dose-response relationships. The RfD and RfC should be redefined to take into account the probability of harm. In developing test cases, the committee recommends a flexible approach in which different conceptual models can be applied in the unified framework.

Cumulative Risk Assessment

EPA is increasingly asked to address broader public-health and environmental-health questions involving multiple exposures, complex mixtures, and vulnerability of exposed populations—issues that stakeholder groups (such as communities affected by environmental exposures) often consider to be inadequately captured by current risk assessments. There is a need for cumulative risk assessments as defined by EPA (EPA 2003)—assessments that include combined risks posed by aggregate exposure to multiple agents or stressors; aggregate exposure includes all routes, pathways, and sources of exposure to a given agent or stressor. Chemical, biologic, radiologic, physical, and psychologic stressors are considered in this definition (Callahan and Sexton 2007).

The committee applauds the agency's move toward the broader definition in making risk assessment more informative and relevant to decisions and stakeholders. However, in practice, EPA risk assessments often fall short of what is possible and is supported by agency guidelines in this regard. Although cumulative risk assessment has been used in various contexts, there has been little consideration of nonchemical stressors, vulnerability, and background risk factors. Because of the complexity of considering so many factors simultaneously, there is a need for simplified risk-assessment tools (such as databases, software packages, and other modeling resources) that would allow screening-level risk assessment and could allow communities and stakeholders to conduct assessments and thus increase stakeholder participation. Cumulative human health risk assessment should draw greater insights from ecologic risk assessment and social epidemiology, which have had to grapple with similar issues. A recent National Research Council report on phthalates addresses issues related to the framework within which dose-response assessment can be conducted in the context of simultaneous exposures to multiple stressors (NRC 2008).

Recommendation: EPA should draw on other approaches, including those from ecologic risk assessment and social epidemiology, to incorporate interactions between chemical and nonchemical stressors in assessments; increase the role of biomonitoring, epidemiologic, and surveillance data in cumulative risk assessments; and develop guidelines and methods for simpler analytical tools to support cumulative risk assessment and to provide for greater involvement of stakeholders. In the short-term, EPA should develop databases and default approaches to allow for incorporation of key nonchemical stressors in cumulative risk assessments in the absence of population-specific data, considering exposure patterns, contributions to relevant background processes, and interactions with chemical stressors. In the long-term, EPA should invest in research programs related to interactions between chemical and nonchemical stressors, including epidemiologic investigations and physiologically based pharmacokinetic modeling.

Improving the Utility of Risk Assessment

Given the complexities of the current problems and potential decisions faced by EPA, the committee grappled with designing a more coherent, consistent, and transparent process that would provide risk assessments that are relevant to the problems and decisions at hand and that would be sufficiently comprehensive to ensure that the best available options for managing risks were considered. To that end, the committee proposes a framework for risk-based decision-making (see Figure 9-1). The framework consists of three phases: I, enhanced problem formulation and scoping, in which the available risk-management options are identified; II, planning and assessment, in which risk-assessment tools are used to determine risks under existing conditions and under potential risk-management options; and III, risk management, in which risk and nonrisk information is integrated to inform choices among options.

The framework has at its core the risk-assessment paradigm (stage 2 of phase II) established in the Red Book (NRC 1983). However, the framework differs from the Red Book paradigm, primarily in its initial and final steps. The framework begins with a "signal" of potential harm (for example, a positive bioassay or epidemiologic study, a suspicious disease cluster, or findings of industrial contamination). Under the traditional paradigm, the question has been, What are the probability and consequence of an adverse health (or ecologic) effect posed by the signal? In contrast, the recommended framework asks, implicitly, What *options* are there to reduce the *hazards* or *exposures* that have been identified, and how can risk assessment be used to evaluate the merits of the various options? The latter question

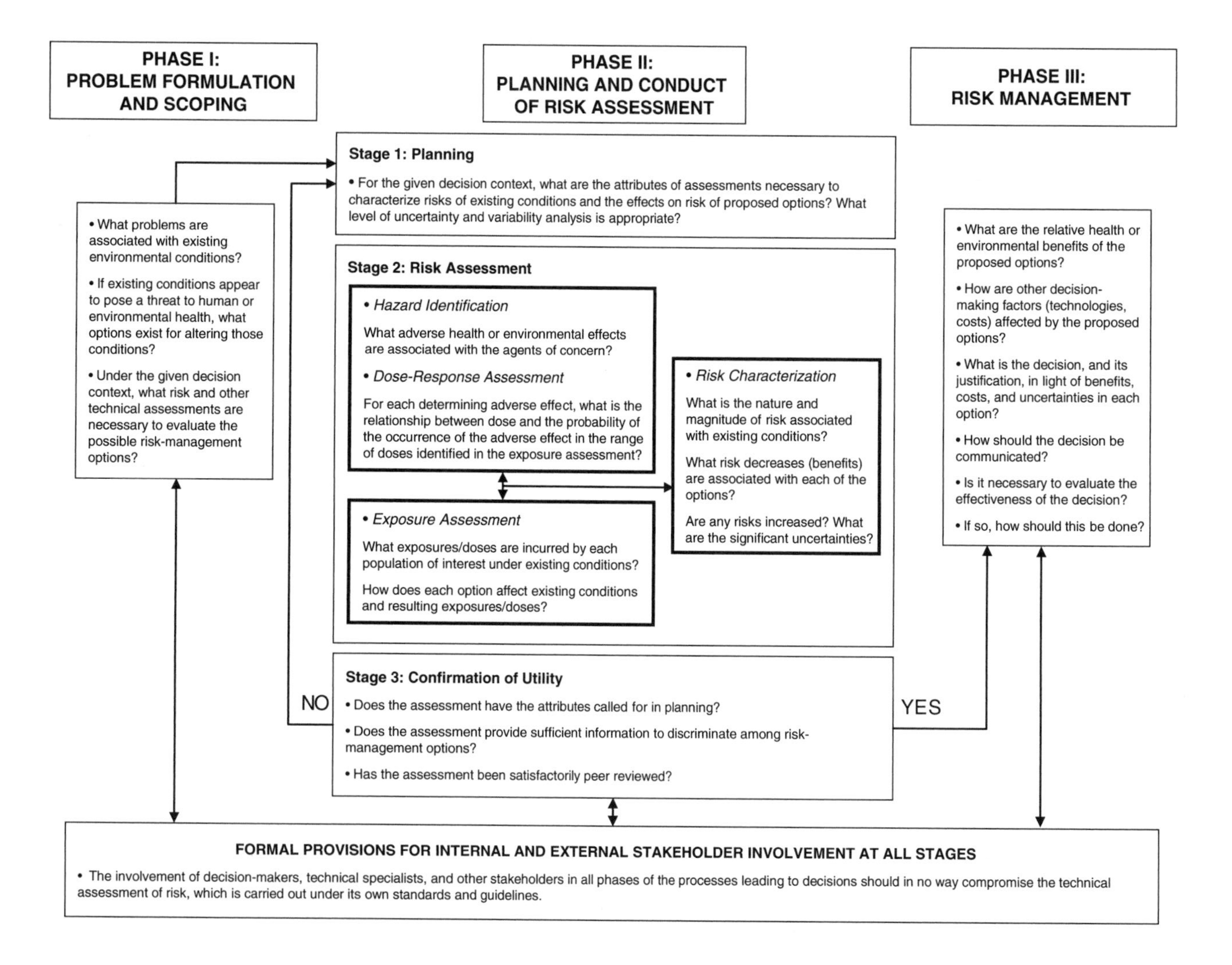

FIGURE 9-1 A framework for risk-based decision-making that maximizes the utility of risk assessment.

focuses on the risk-management options (or interventions) designed to provide adequate public-health and environmental protection and to ensure well-supported decision-making. Under this framework, the questions posed arise from early and careful planning of the types of assessments (including risks, costs, and technical feasibility) and the required level of scientific depth that are needed to evaluate the relative merits of the options being considered.[4] Risk management involves choosing among the options after the appropriate assessments have been undertaken and evaluated.

The framework begins with enhanced problem formulation and scoping (phase I), in which risk-management options and the types of technical analyses, including risk assessments, needed to evaluate and discriminate among the options are identified. Phase II consists of three stages: planning, risk assessment, and confirmation of utility. Planning (stage 1) is done to ensure that the level and complexity of risk assessment (including uncertainty and variability analysis) are consistent with the goals of decision-making. After risk assessment (stage 2), stage 3 evaluates whether the assessment was appropriate and whether it allows discrimination among the risk-management options. If the assessment is not determined to be adequate, the framework calls for a return to planning (phase II, stage 1). Otherwise, phase III (risk management) is undertaken: the relative health or environmental benefits of the proposed risk-management options are evaluated for the purpose of reaching a decision.

The framework systematically identifies problems and options that risk assessors should evaluate at the earliest stages of decision-making. It expands the array of impacts assessed beyond individual effects (for example, cancer, respiratory problems, and individual species) to include broader questions of health status and ecosystem protection. It provides a formal process for stakeholder involvement throughout all stages but has time constraints to ensure that decisions are made. It increases understanding of the strengths and limitations of risk assessment by decision-makers at all levels, for example, by making uncertainties and choices more transparent.

The committee is mindful of concerns about political interference in the process, and the framework maintains the conceptual distinction between risk assessment and risk management articulated in the Red Book. It is imperative that risk assessments used to evaluate risk-management options not be inappropriately influenced by the preferences of risk managers.

With a focus on early and careful planning and problem formulation and on the options for managing the problem, implementation of the framework can improve the utility of risk assessment for decision-making. Although some aspects of the framework are achievable in the short term, its full implementation will require a substantial transition period. EPA should phase in the framework with a series of demonstration projects that apply it and that determine the degree to which it meets the needs of the agency risk managers, how risk-management conclusions differ as a result of its application, and the effectiveness of measures to ensure that risk managers and policy-makers do not inappropriately influence the scientific conduct of risk assessments.

Recommendation: To make risk assessments most useful for risk-management decisions, the committee recommends that EPA adopt a *framework for risk-based decision-making* (see Figure 9-1) that embeds the Red Book risk-assessment paradigm into a process with initial problem formulation and scoping, upfront identification of risk-management options, and use of risk assessment to discriminate among these options.

---

[4]The committee notes that not all decisions require or are amenable to risk assessment and that in most cases one of the options explicitly considered is "no intervention."

Stakeholder Involvement

Many stakeholders believe that the current process for developing and applying risk assessments lacks credibility and transparency. That may be partly because of failure to involve stakeholders adequately as active participants at appropriate points in the risk-assessment and decision-making process rather than as passive recipients of the results. Previous National Research Council and other risk-assessment reports (NRC 1996; PCCRARM 1997) and comments received by the committee (Callahan 2007; Kyle 2007) echo such concerns.

The committee agrees that greater stakeholder involvement is necessary to ensure that the process is transparent and that risk-based decision-making proceeds effectively, efficiently, and credibly. Stakeholder involvement needs to be an integral part of the risk-based decision-making framework, beginning with problem formulation and scoping.

Although EPA has numerous programs and guidance documents related to stakeholder involvement, it is important that it adhere to its own guidance, particularly in the context of cumulative risk assessment, in which communities often have not been adequately involved.

Recommendation: EPA should establish a formal process for stakeholder involvement in the framework for risk-based decision-making with time limits to ensure that decision-making schedules are met and with incentives to allow for balanced participation of stakeholders, including impacted communities and less advantaged stakeholders.

Capacity-Building

Improving risk-assessment practice and implementing the framework for risk-based decision-making will require a long-term plan and commitment to build the requisite capacity of information, skills, training, and other resources necessary to improve public-health and environmental decision-making. The committee's recommendations call for considerable modification of EPA risk-assessment efforts (for example, implementation of the risk-based decision-making framework, emphasis on problem formulation and scoping as a discrete stage in risk assessment, and greater stakeholder participation) and of technical aspects of risk assessment (for example, unification of cancer and noncancer dose-response assessments, attention to quantitative uncertainty analysis, and development of methods for cumulative risk assessment). The recommendations are tantamount to "change-the-culture" transformations in risk assessment and decision-making in the agency.

EPA's current institutional structure and resources may pose a challenge to implementation of the recommendations, and moving forward with them will require a commitment to leadership, cross-program coordination and communication, and training to ensure the requisite expertise. That will be possible only if leaders are determined to reverse the downward trend in budgeting, staffing, and training and to making high-quality, risk-based decision-making an agencywide goal.

Recommendation: EPA should initiate a senior-level strategic re-examination of its risk-related structures and processes to ensure that it has the institutional capacity to implement the committee's recommendations for improving the conduct and utility of risk assessment for meeting the 21st century environmental challenges. EPA should develop a capacity building plan that includes budget estimates required for implementing the committee's recommendations, including transitioning to and effectively implementing the framework for risk-based decision-making.

REFERENCES

Callahan, M.A. 2007. Improving Risk Assessment: A Regional Perspective. Presentation at the Third Meeting of Improving Risk Analysis Approaches Used by EPA, February 26, 2007, Washington, DC.

Callahan, M.A., and K. Sexton. 2007. If cumulative risk assessment is the answer, what is the question? Environ. Health Perspect. 115(5):799-806.

EPA (U.S. Environmental Protection Agency). 1992. Guidelines for Exposure Assessment. EPA600Z-92/001. Risk Assessment Forum, U.S. Environmental Protection Agency, Washington, DC [online]. Available: http://cfpub.epa.gov/ncea/raf/recordisplay.cfm?deid=15263 [accessed Jan. 14, 2008].

EPA (U.S. Environmental Protection Agency). 1998. Guidelines for Ecological Risk Assessment. EPA/630/R-95/002F. Risk Assessment Forum, U.S. Environmental Protection Agency, Washington, DC. April 1998 [online]. Available: http://oaspub.epa.gov/eims/eimscomm.getfile?p_download_id=36512 [accessed Feb. 9, 2007].

EPA (U.S. Environmental Protection Agency). 2001. Trichloroethylene Health Risk Assessment: Synthesis and Characterization. External Review Draft. EPA/600/P-01/002A. Office of Research and Development, Washington, DC. August 2001 [online]. Available: http://rais.ornl.gov/tox/TCEAUG2001.PDF [accessed Aug. 2, 2008].

EPA (U.S. Environmental Protection Agency). 2002. A Review of the Reference Dose and Reference Concentration Processes. EPA/630/P-02/002F. Risk Assessment Forum, U.S. Environmental Protection Agency, Washington, DC. December 2002 [online]. Available: http://cfpub.epa.gov/ncea/cfm/recordisplay.cfm?deid=55365 [accessed Jan. 4, 2008].

EPA (U.S. Environmental Protection Agency). 2003. Framework for Cumulative Risk Assessment. EPA/600/P-02/001F. National Center for Environmental Assessment, Risk Assessment Forum, U.S. Environmental Protection Agency, Washington, DC [online]. Available: http://cfpub.epa.gov/ncea/cfm/recordisplay.cfm?deid=54944 [accessed Jan. 4, 2008].

EPA (U.S. Environmental Protection Agency). 2004. Risk Assessment Principles and Practices: Staff Paper. EPA/100/B-04/001. Office of the Science Advisor, U.S. Environmental Protection Agency, Washington, DC. March 2004 [online]. Available: http://www.epa.gov/osa/pdfs/ratf-final.pdf [accessed Jan. 9, 2008].

EPA (U.S. Environmental Protection Agency). 2005a. Guidelines for Carcinogen Risk Assessment. EPA/630/P-03/001F. Risk Assessment Forum, U.S. Environmental Protection Agency, Washington, DC. March 2005 [online]. Available: http://cfpub.epa.gov/ncea/cfm/recordisplay.cfm?deid=116283 [accessed Feb. 7, 2007].

EPA (U.S. Environmental Protection Agency). 2005b. Supplemental Guidance for Assessing Susceptibility for Early-Life Exposures to Carcinogens. EPA/630/R-03/003F. Risk Assessment Forum, U.S. Environmental Protection Agency, Washington, DC. March 2005 [online]. Available: http://cfpub.epa.gov/ncea/cfm/recordisplay.cfm?deid=160003 [accessed Jan. 4, 2008].

Kyle, A. 2007. Community Needs for Assessment of Environmental Problems. Presentation at the Fourth Meeting of Improving Risk Analysis Approaches Used by EPA, April 17, 2007, Washington, DC.

NRC (National Research Council). 1983. Risk Assessment in the Federal Government: Managing the Process. Washington, DC: National Academy Press.

NRC (National Research Council). 1996. Understanding Risk: Informing Decisions in a Democratic Society. Washington, DC: National Academy Press.

NRC (National Research Council). 2006. Assessing the Human Risks of Trichloroethylene. Washington, DC: The National Academies Press.

NRC (National Research Council). 2008. Phthalates and Cumulative Risk Assessment: The Tasks Ahead. Washington, DC: The National Academies Press.

PCCRARM (Presidential/Congressional Commission on Risk Assessment and Risk Management). 1997. Framework for Environmental Health Risk Management - Final Report, Vol. 1. [online]. Available: http://www.riskworld.com/nreports/1997/risk-rpt/pdf/EPAJAN.PDF [accessed Jan. 4, 2008].

Renner, R. 2007. Budget cut increasingly damaging to EPA. Environ. Sci. Technol. News, May 9, 2007 [online]. Available: http://pubs.acs.org/subscribe/journals/esthag-w/2007/may/policy/rr_EPA.html [accessed Aug. 12, 2008].

Appendixes

Appendix A

Biographic Information on the Committee on Improving Risk Analysis Approaches Used by the Environmental Protection Agency

Thomas A. Burke *(Chair)* is associate dean for public health practice and professor of health policy and management at the Johns Hopkins University Bloomberg School of Public Health. He holds joint appointments in the Department of Environmental Health Sciences and the School of Medicine Department of Oncology. Dr. Burke is also director of the Johns Hopkins Risk Sciences and Public Policy Institute. His research interests include environmental epidemiology and surveillance, evaluation of population exposures to environmental pollutants, assessment and communication of environmental risks, and application of epidemiology and health risk assessment to public policy. Before joining the university, Dr. Burke was deputy commissioner of health for New Jersey and director of science and research for the New Jersey Department of Environmental Protection. In New Jersey, he directed initiatives that influenced the development of national programs, such as Superfund, the Safe Drinking Water Act, and the Toxics Release Inventory. Dr. Burke is a member of the U.S. EPA Science Advisory Board. He was the inaugural chair of the Advisory Board to the director of the Centers for Disease Control and Prevention National Center for Environmental Health and served two terms on the National Research Council Board on Environmental Studies and Toxicology. He has served on several National Research Council committees; he was chair of the Committee on Human Biomonitoring for Environmental Toxicants and the Committee on Toxicants and Pathogens in Biosolids Applied to Land and a member of the Committee on the Toxicological Effects of Methyl Mercury. In 2003, he was designated a lifetime national associate of the National Academies. He received his PhD in epidemiology from the University of Pennsylvania.

A. John Bailer is distinguished professor in the Department of Mathematics and Statistics, an affiliate member of the Department of Zoology, an affiliate member of the Department of Sociology and Gerontology, and a research fellow in the Scripps Gerontology Center at Miami University in Oxford, OH. His research interests include the design and analysis of environmental and occupational health studies and quantitative risk estimation. Dr. Bailer is a fellow of the American Statistical Association (ASA), a fellow of the Society for Risk

Analysis, and a recipient of the ASA Statistics and the Environment Distinguished Achievement Medal. He serves on the National Research Council Committee on Spacecraft Exposure Guidelines and has served on other National Research Council committees, including the Committee to Review the OMB Risk Assessment Bulletin and the Committee on Toxicologic Assessment of Low-Level Exposures to Chemical Warfare Agents. He also has served as a member of the Report on Carcinogens Subcommittee and the Technical Reports Review Subcommittee of the Board of Scientific Counselors of the National Toxicology Program. He received his PhD in biostatistics from the University of North Carolina at Chapel Hill.

John M. Balbus is the chief health scientist at Environmental Defense and adjunct professor of environmental health sciences at Johns Hopkins University. His expertise is in epidemiology, toxicology, and risk science. He spent 7 years at George Washington University, where he was the founding director of the Center for Risk Science and Public Health and served as acting chair of the Department of Environmental and Occupational Health; he was also an associate professor of medicine at the university. Dr. Balbus has served as a member of the Environmental Protection Agency (EPA) Children's Health Protection Advisory Committee, as a core panel member of EPA's Voluntary Children's Chemical Exposure Program, and on EPA review committees for air-toxics research, computational toxicology, and climate-change research. He serves on the National Research Council's Board on Environmental Studies and Toxicology. Dr. Balbus received his MD from the University of Pennsylvania and his BA from Harvard University.

Joshua T. Cohen is a research associate professor at Tufts Medical Center in the Institute for Clinical Care Research and Health Policy Studies. Dr. Cohen's research focuses on the application of decision analytic techniques to public-health risk-management problems with an emphasis on the characterization and analysis of uncertainty. He was the lead author on a study comparing the risks and benefits associated with changes in population fish-consumption patterns, an analysis of the risks and benefits associated with cellular-phone use during driving, and a study comparing the costs and health impacts of advanced diesel and compressed natural-gas urban-transit buses. He also has played a key role in a risk assessment of bovine spongiform encephalopathy ("mad cow disease") in the United States. Dr. Cohen served on the National Research Council Committee on EPA's Exposure and Human Health Reassessment of TCDD and Related Compounds and was a member of the Environmental Protection Agency Clean Air Science Advisory Committee that reviewed the agency's evaluation of risks associated with lead. He earned his PhD in decision sciences from Harvard University.

Adam M. Finkel is professor of environmental and occupational health at the University of Medicine and Dentistry of New Jersey School of Public Health and executive director of the Penn Program on Regulation at the University of Pennsylvania Law School. From 2004 to 2007, he was also a visiting professor at Princeton University's Woodrow Wilson School of Public and International Affairs. His research interests include quantitative risk assessment of health hazards in the workplace and general environment, regulatory design and policy, scientific-integrity issues, human susceptibility to carcinogenesis, and occupational and environmental regulation and enforcement. From 1995 to 2005, he was a senior executive at the U.S. Occupational Safety and Health Administration (OSHA), serving as OSHA's national director of regulatory programs and later as chief OSHA administrator in the six-state Rocky Mountain region, based in Denver, CO. He has developed methods to quantify and communicate uncertainties in risk and cost estimation and to explore the variation in environmental and medical risks that people face because of differences in susceptibility, exposure, and other factors. Dr. Finkel received his ScD in environmental health sciences from the Harvard School of Public Health.

Gary Ginsberg is a senior toxicologist in the Division of Environmental Epidemiology at the Connecticut Department of Public Health, an assistant clinical professor at the University of Connecticut School of Medicine, and an adjunct faculty member at the Yale University School of Medicine. Dr. Ginsberg is involved with the use of toxicology and risk-assessment principles to evaluate human exposures to chemicals in air, water, soil, food, and the workplace. He provides risk-assessment expertise to the department and other state agencies in standard-setting and site-remediation projects. Dr. Ginsberg is a member of the Federal Advisory Committee on Children's Health Protection, which reports to the administrator of the Environmental Protection Agency. He served on the National Research Council Committee on Human Biomonitoring for Environmental Toxicants. He received his PhD in toxicology from the University of Connecticut.

Bruce K. Hope is a senior environmental toxicologist in the Air Quality Division of the Oregon Department of Environmental Quality. Dr. Hope's expertise includes preparation and review of human, ecologic, and probabilistic risk assessments; exposure modeling; development of air-toxics benchmarks and risk-assessment strategies; and evaluation and communication of health and environmental risk associated with chemical releases. He has been an adjunct faculty member of the Oregon Health & Science University, where he taught courses in risk communication, toxicology, and risk assessment. Dr. Hope served on a number of Environmental Protection Agency (EPA) Science Advisory Board committees. Recently, he served as a panelist in the Workshop on Ecological Risk Assessment—An Evaluation of the State-of-the-Practice and on EPA's Regulatory Environmental Modeling Guidance Advisory Panel. He received his PhD in biology from the University of Southern California.

Jonathan I. Levy is an associate professor of environmental health and risk assessment in the Department of Environmental Health and the Department of Health Policy and Management at the Harvard School of Public Health and an affiliate of the Harvard Center for Risk Analysis. His research interests include quantitative risk assessment with a focus on air-pollution–related health risks in urban environments, development of quantitative measures of environmental equity suitable for risk assessment and benefit-cost analyses, and development and application of exposure models for multiple pollutants in urban low-income settings. Dr. Levy previously served on the National Research Council Committee on the Effects of Changes in New Source Review Programs for Stationary Sources of Air Pollutants. He received his ScD from the Harvard School of Public Health in environmental science and risk management.

Thomas E. McKone is senior staff scientist and deputy department head at the Lawrence Berkeley National Laboratory and an adjunct professor and researcher at the University of California, Berkeley School of Public Health. Dr. McKone's research interests include the use of multimedia compartment models in health-risk assessments, chemical transport and transformation in the environment, and measuring and modeling the biophysics of contaminant transport from the environment into the microenvironments with which humans have contact and across the human-environment exchange boundaries—skin, lungs, and gut. One of Dr. McKone's most recognized achievements was his development of the CalTOX risk-assessment framework for the California Department of Toxic Substances Control. He has been a member of several National Research Council committees, including the Committees on Environmental Decision Making: Principles and Criteria for Models, EPA's Exposure and Human Health Reassessment of TCDD and Related Compounds, Toxicants and Pathogens in Biosolids Applied to Land, and Toxicology. Dr. McKone was recently appointed by California Governor Arnold Schwarzenegger to the California Scientific Guidance Panel. He is a fellow of the Society for Risk Analysis, former president of the International Society of Exposure Analysis, and a member the Organizing Committee for the International Life-Cycle Initiative, a joint effort of the UN Environment Program and the Society for Environmental Toxicology and Chemistry. He earned his PhD in engineering from the University of California at Los Angeles.

Gregory M. Paoli is a co-founder and principal risk scientist at Risk Sciences International based in Ottawa, Canada. He has experience in the development and application of risk analysis methods in diverse risk domains including microbiologic, toxic, and nutritional hazards, climate-change adaptation, air quality, drinking water, engineering devices, risk-based sampling and inspection, and a number of comparative risk assessment applications. His consulting activities also include risk management and risk communication, primarily for public-sector clients. Mr. Paoli previously served on the National Research Council Committee on the Review of the USDA *E. coli* 0157:H7 Farm-to-Table Process Risk Assessment. He serves on numerous expert panels including expert consultations convened by the World Health Organization (JEMRA), advisory panels of Canada's National Roundtable on the Environment and the Economy, Health Canada's Expert Advisory Committee on Antimicrobial Resistance Risk Assessment and the Canadian Standards Association's Technical Committee on Risk Management. Mr. Paoli is a member of the editorial board of *Risk Analysis* and served as a councilor of the Society for Risk Analysis. Mr. Paoli earned a master of applied science degree in systems design engineering from the University of Waterloo.

Charles Poole is associate professor in the Department of Epidemiology at the University of North Carolina School of Public Health. Previously, he was with the Boston University School of Public Health. Dr. Poole's work focuses on the development and use of epidemiologic methods and principles, including problem definition, study design, data collection, statistical analysis, and interpretation and application of research results. His research experience includes studies in environmental and occupational epidemiology. Dr. Poole was an epidemiologist in the Environmental Protection Agency Office of Pesticides and Toxic Substances for 5 years and worked for a decade as an epidemiologic consultant. Dr. Poole was a member of the Institute of Medicine Committee on Gulf War and Health: Review of the Literature on Pesticides and Solvents and the National Research Council Committees on Estimating the Health-Risk-Reduction Benefits of Proposed Air Pollution Regulations, on Fluoride in Drinking Water, and on the Review the OMB Risk Assessment Bulletin. He received his ScD in epidemiology from the Harvard School of Public Health.

Joseph V. Rodricks is a founding principal of ENVIRON International Corporation. Dr. Rodricks has expertise in toxicology and risk analysis and in their uses in regulation. He was formerly deputy associate commissioner for health affairs and toxicologist for the Food and Drug Administration, and he is now a visiting professor at the Johns Hopkins University Bloomberg School of Public Health. Dr. Rodricks's experience includes chemical products and contaminants in foods, food ingredients, air and water pollution, hazardous wastes, the workplace, consumer products, and medical devices and pharmaceutical products. He has consulted for manufacturers, government agencies, and the World Health Organization. He has more than 150 publications on toxicology and risk analysis, and he has lectured nationally and internationally on these topics. He has been a diplomate of the American Board of Toxicology since 1982. Dr. Rodricks has served on numerous National Research Council and Institute of Medicine committees and currently serves on the Board on Environmental Studies and Toxicology. He earned his PhD in biochemistry from the University of Maryland.

Bailus Walker, Jr., (IOM) is professor of environmental and occupational medicine at Howard University College of Medicine. His research interests include lead toxicity, environmental carcinogenesis, and the social and economic dimensions of environmental-risk management strategies. He was the commissioner of public health for the Commonwealth of Massachusetts and, earlier, state director of public health for Michigan. In other regulatory and service work, Dr. Walker was director of the Health Standards Division of the U.S. Occupational Safety and Health Administration (OSHA). In academe, his assignments have included being a professor of environmental health and toxicology at the University at Albany, State University of New York at Albany, and dean of the Faculty of Public Health at the University of Oklahoma Health Sciences Center, Oklahoma City. Dr. Walker has also served as chairman of the Board of Scientific Counselors of the Agency for Toxic Substance and Disease Registry and is senior science adviser on environmental health to the National Library of Medicine. He is a past president of the American Public Health Association and a Distinguished Fellow of the Royal Society of Health (London, England) and the American College of Epidemiology. Dr. Walker is a member of the Institute of Medicine and served for two terms on the Board of Environmental Studies and Toxicology (BEST) of the National Research Council. In addition, he served on a number of other National Research Council committees, including being chair of the Committee on Toxicology and a member of the Committee on Estimating Mortality Risk Reduction Benefits from Decreasing Tropospheric Ozone Exposure. Dr. Walker received his PhD in occupational and environmental medicine from the University of Minnesota at Minneapolis.

Terry F. Yosie is president and CEO of the World Environment Center, a nonprofit, nonadvocacy organization whose mission is to advance sustainable development through the private sector in partnership with government, nongovernment organization, academic, and other stakeholders. From 2001 through 2005, Dr. Yosie served as the American Chemistry Council's vice president for the Responsible Care initiative, a performance program that includes environmental, health, and safety management; product stewardship; security; and other aspects of the business value chain. He has about 25 years of professional experience in managing and analyzing the use of scientific information in the setting of environmental standards. He was the first executive director of the Clean Air Scientific Advisory Committee, which is responsible for reviewing the scientific basis of national ambient air quality standards. He served as director of the Environmental Protection Agency (EPA) Science Advisory Board from 1981 to 1988 and instituted policies and procedures for enhancing the use of scientific information in regulatory decision-making. Dr. Yosie was vice president for health and environment at the American Petroleum Institute and executive vice president of Ruder Finn consultancy, where he was responsible for the firm's environmental-management practice. He has served on a number of National Research Council committees and boards, including the Committee to Review the Structure and Performance of the Health Effects Institute, the Committee on Research Priorities for Airborne Particulate Matter, and the Board on Environmental Studies and Toxicology. He is the author of about 60 publications on the use of scientific information in the development of public health and environmental policies. He earned his doctorate from the College of Humanities and Social Sciences at Carnegie Mellon University in 1981.

Lauren Zeise is chief of the reproductive and cancer hazard assessment branch of the California Environmental Protection Agency. Her current work focuses on cancer and reproductive hazard risk assessments, assessment methods, cumulative impact analysis, and the California Environmental Chemical Biomonitoring Program. She has served on advisory boards of the U.S. Environmental Protection Agency, the World Health Organization, the Office of Technology Assessment, and the National Institute of Environmental Health Sciences. She has also served on several Institute of Medicine and National Research Council committees, including the Committees on Risk Characterization, on Toxicity Testing and Assessment of Environmental Agents, on Comparative Toxicology of Naturally Occurring Carcinogens, on Copper in Drinking Water, and on Review of EPA's Research Grants Program. Dr. Zeise is a member of the National Research Council Board on Environmental Studies and Toxicology. She received her PhD from Harvard University.

Appendix B

Statement of Task of the Committee on Improving Risk Analysis Approaches Used by the Environmental Protection Agency

An NRC committee will develop scientific and technical recommendations for improving the risk analysis approaches used by the Environmental Protection Agency (EPA). Taking into consideration past evaluations and ongoing studies by the NRC and others, the committee will conduct a scientific and technical review of EPA's current risk analysis concepts and practices. The committee will consider analyses applied to contaminants in all environmental media (water, air, food, soil) and all routes of exposure (ingestion, inhalation, and dermal absorption). The committee will focus primarily on human health risk analysis and will comment on the broad implications of its findings and recommendations to ecological risk analysis. In making recommendations, the committee will indicate practical improvements that can be made in the near term (2-5 years) and improvements that would be made over a longer term (10-20 years). The committee will address topics such as the following:

- Increased role for probabilistic analysis in risk analysis, including the potential expanded role for expert elicitation.
- Scientific bases for and alternatives to default assumption choices made in areas of uncertainty.
- Quantitative characterization of uncertainty resulting from all steps in the risk analysis.
- Approaches for assessing cumulative risk resulting from multiple exposures to contaminant mixtures, involving multiple sources, pathways, routes.
- Variability in receptor populations, especially sensitive subpopulations and critical life stages.
- Biologically relevant modes of action for estimating dose-response relationships, and quantitative implications of different modes.
- Improvements in environmental transport and fate models, exposure models, physiologically based pharmacokinetic (PBPK) models, and dose-response models.

- How the concepts and practices of ecological risk analysis can help inform and improve the concepts and practices of human health risk analysis, and vice versa.
- Scientific basis for derivation of uncertainty factors.
- Use of value-of-information analyses and other techniques to identify priorities and approaches for research to obtain relevant data to increase the utility of risk analyses.

Appendix C

Timeline of Selected Environmental Protection Agency Risk-Assessment Activities

TABLE C-1 Timeline of Selected EPA Risk-Assessment Activities

Date and Title of Milestone	Comments[a]
EPA 1976 *Interim Procedures and Guidelines for Health Risk and Economic Impact Assessments of Suspected Carcinogens*	First agency "inference" guidelines on cancer risk. "How likely is the risk to occur, and if it does occur, what are the consequences? How likely is an agent to be a human carcinogen? How much cancer might be produced by the agent if it remains unregulated?"
NRC 1983 *Risk Assessment in the Federal Government: Managing the Process*	Seminal risk-assessment report that established the four organizing principles for government risk efforts: hazard identification, dose-response assessment, exposure assessment, and risk characterization. Report also recommended that uniform inference guidelines be developed and that regulatory agencies take steps to establish and maintain a clear distinction between risk-assessment and risk-management activities. *Definition of Risk Assessment:* characterization of potential adverse health effects of human exposure to environmental hazards.
EPA 1984 *Risk Assessment and Management: Framework for Decision Making*	EPA's response to NRC (1983), *Risk Assessment in the Federal Government: Managing the Process.* Discusses EPA's activities to address recommendations in the 1983 NRC report, including establishing the Risk Assessment Forum and efforts to develop six risk-assessment guidelines. Risk-management activities were expanded to include cost-effectiveness tools that could be used in risk management, the importance of strengthening communication in risk management, and risk-management principles, such as consistency of approach in making decisions. Prompted training program for EPA senior managers with emphasis on the distinction between risk-assessment and risk-management activities. *Definition of Risk Assessment:* In simplest sense, population risks posed by toxic pollutants are a function of two measurable factors: hazard and exposure. To cause a risk, a chemical has to be both toxic (present as intrinsic hazard) and present in the human environment at some substantial level (provide opportunity for human exposure). Risk assessment interprets evidence on the two points, judging whether an adverse effect will occur and (if appropriate) making the necessary calculations to estimate the extent of total effects.
1984 Risk Assessment Forum Charter	In 1984, the Risk Assessment Forum (RAF) is established in response to an NRC (1983) recommendation "to promote consensus on risk assessment issues." RAF convenes risk-assessment experts to study and report on risk-assessment issues. RAF has produced risk-assessment guidelines, technical panel reports on special risk-assessment issues, and peer-consultation and peer-review workshops (EPA 2002a).
OSTP 1985 *Chemical Carcinogens: Review of the Science and Its Associated Principles*	Report details 31 principles developed by interagency group for carcinogenicity evaluations in regulatory settings.
EPA 1986a *Memorandum: Establishment of the Risk Assessment Council*	The Risk Assessment Council is established in 1986 by Lee Thomas to "oversee virtually all aspects of the Agency's risk assessment process, to identify issues and problems with that process" (EPA 1986a), and to ensure that EPA programs use risk assessment in a consistent and scientifically credible fashion.

TABLE C-1 Continued

Date and Title of Milestone	Comments[a]
EPA 1986b *Guidelines for Carcinogen Risk Assessment*	The 1986 guidelines, developed to address an NRC (1983) recommendation to craft cancer inference guidelines, incorporate concepts and approaches established since the previous cancer guidelines were released in 1976.
EPA 1986c *Guidelines for Mutagenicity Risk Assessment*	The guidelines state that "a consistent approach to the evaluation of mutagenic risk from chemical substances arises from the authority conferred upon the Agency by a number of statutes to regulate potential mutagens" (EPA 1986c, p. 2). *Definition of Risk Assessment:* Risk assessment comprises hazard identification, dose-response assessment, exposure assessment, and risk characterization (NRC 1983). Hazard identification is qualitative risk assessment, dealing with the inherent toxicity of a chemical substance. A qualitative mutagenicity assessment answers the question of how likely an agent is to be a human mutagen. The three remaining components constitute quantitative risk assessment, which provides a numerical estimate of the public-health consequences of exposure to an agent. The quantitative mutagenicity risk assessment deals with the question of how much mutational damage is likely to be produced by exposure to a given agent under particular exposure scenarios.
EPA 1986d *Guidelines for Chemical Mixtures Risk Assessment*	Details agency approaches to assessing risks posed by complex chemical mixtures with supplementary update in EPA (2000a).
EPA 1987 *Unfinished Business: A Comparative Assessment of Environmental Problems*	Assesses agency resource allocations relative to magnitude of risks and protection gained. "Many new [environmental] problems are difficult to evaluate; many involve toxic chemicals that can cause cancer or birth defects at levels of exposure that are hard to detect; and many involve persistent contaminants that can move from one environment medium to another, causing further damage even after controls have been applied for one medium. The complexity and gravity of these issues make it particularly important that EPA apply its finite resources where they will have the greatest effect. Thus, the Administrator of EPA commissioned a special task force of senior career managers and technical experts to assist him and other policy makers in the task. The assignment was to compare the risks currently associated with major environmental problems" (EPA 1987, p. xiii).
EPA 1989 *Risk Assessment Guidance for Superfund (RAGS)*	Provides guidance on conducting site-specific risk assessments at Superfund sites. About four pages are devoted to planning and scoping. See EPA 1989 *Risk Assessment Guidance for Superfund, Vol. 1—Human Health Evaluation Manual*, Parts A-E; Baseline Assessment (EPA 1989), Community Involvement (EPA 1999); Preliminary Remediation Goals (EPA 1991a); Remedial Alternatives (EPA 1991b); Standardized Planning and Reporting, and Dermal Risk Assessment (EPA 2001).

Continued

TABLE C-1 Continued

Date and Title of Milestone	Comments[a]
NRC 1989 *Improving Risk Communication*	Risk communication is a two-way process involving participation of and information exchange between the scientist and the public. *Definition of Risk Assessment:* Generally refers to characterization of potential adverse effects of exposures to hazards. Characterization of potential adverse effects of exposures to hazards; includes estimates of risk and of uncertainties in measurements, analytic techniques, and interpretive models; quantitative risk assessment characterizes risk in numerical representations.
EPA SAB 1990 *Reducing Risk: Setting Priorities and Strategies for Environmental Protection*	Science Advisory Board peer review of 1987's *Unfinished Business*—"National policy affecting the environment must become more integrated and more focused on opportunities for environmental improvement than it has been in the past. . . . Integration in this case means that government agencies should assess the range of environmental problems of concern and then target protective efforts at the problems that seem to be the most serious. . . . The concept of environmental risk can help the nation develop environmental policies in a consistent and systematic way" (EPA SAB 1990, pp. 1-2).
1990 Amendments to the Clean Air Act	To expedite control of air toxics, Congress switches EPA's approach from a risk-assessment–oriented program to a technology-oriented regulatory approach with a mandate to study "residual risks" posed by 189 air toxics 8 y after technology controls are put into place.
EPA 1991c *Guidelines for Developmental Toxicity Risk Assessment*	Guidelines outline principles and methods to characterize risks posed by environmental exposures during human development. They address relationship between maternal and developmental toxicity, characterization of health-related database for developmental-toxicity risk assessment, use of reference dose or reference concentration for developmental toxicity, and use of benchmark dose. *Definition of Risk Assessment:* Process by which scientific judgments are made concerning the potential for toxicity to occur in humans.
EPA 1991d *Alpha2u-Globulin: Association with Chemically Induced Renal Toxicity and Neoplasia in the Male Rat*	EPA's Risk Assessment Forum is among first to describe animal tumors not found in humans; related volume on thyroid follicular-cell tumors is published in 1998.
EPA 1992a *Guidance on Risk Characterization for Risk Managers and Risk Assessors*	Agencywide guidance includes a statement of confidence about data and methods used to develop assessment; need to provide basis of greater consistency and comparability in risk assessments across agency programs; and role of professional scientific judgment in overall statement of risk.
EPA 1992b *Developing a Work Scope for Ecological Assessments*	Develops a framework for ecologic risk assessment. Describes process in detail and demonstrates how it could be applied to broad array of situations. Defines ecologic risk assessment as "a process that evaluates the likelihood that adverse ecological effects may occur or are occurring as a result of exposure, to one or more stressors" (EPA 1992b).
EPA 1992c *Guidelines for Exposure Assessment*	Guidelines, which pertain to both human and wildlife exposures to chemicals, provide general information on exposure assessment, including definitions and guidance on planning, conducting exposure-assessment studies, presenting results, and characterizing uncertainty. State that exposure estimates will be fully detailed in risk assessments, including assumptions, uncertainties, and rationale for each.

TABLE C-1 Continued

Date and Title of Milestone	Comments[a]
EPA 1992d *Dermal Exposure Assessment: Principles and Applications*	Summarizes current state of knowledge regarding dermal exposure to water, soil, and vapors; presents methods for estimating dermal absorption stemming from contact with these media; and elaborates on their associated uncertainties. Focuses on evaluating exposures from waste-disposal sites or contaminated soils.
EPA 1993 *Memorandum: Creation of a Science Policy Council*	Science Policy Council (SPC) is created in 1993 to replace RAC and is chaired by assistant administrator for Office of Research and Development (ORD). It is tasked with an expanded mission to "implement and ensure the success of selected initiatives recommended by external advisory bodies such as the National Research Council and the Science Advisory Board, as well as others such as the Congress, industry and environmental groups, and Agency staff." SPC has developed a number of guidance documents and policies for the agency.
NRC 1993a *Pesticides in the Diets of Infants and Children*	Concluded that children consume more air, water, and food on a body-weight basis than adults and engage in other behaviors that make them more susceptible to environmental exposures, including hand-to-mouth and object-to-mouth behaviors. The publication of this report is one of the factors that prompted the 1996 Food Quality Protection Act for pesticides.
NRC 1993b *Issues in Risk Assessment*	This report examines the scientific basis, inference assumptions, and regulatory uses o and research needs in risk assessment in two parts. First, use of maximum tolerated dose in animal bioassays for carcinogenicity addresses whether the maximum tolerated dose should continue to be used in carcinogenesis bioassays. Second, two-stage models of carcinogenesis, stems from efforts to identify improved means of cancer risk assessment that has resulted in the development of a mathematical dose-response model.
EPA 1994a *Guidance Manual for the IEUBK Model for Lead in Children*	Given that there is no reference dose for lead, the EPA risk reduction goal for contaminated sites is to limit the probability of a child's blood lead concentration exceeding 10 µg/dL to 5% or less after cleanup. Blood lead concentration can be correlated with exposure and adverse health effects. The Integrated Exposure Uptake Biokinetic Model for Lead in Children is used to predict blood lead concentration and the probability of a child's blood lead concentration exceeding 10 µg/dL, considering a multimedia exposure scenario and toxicokinetics.
NRC 1994 *Science and Judgment in Risk Assessment*	Report makes a variety of recommendations to EPA, many directed at the Office of Air and Radiation, including that EPA explicitly identify each use of a default option in risk assessments, the agency should conduct quantitative analyses of uncertainty, that risk managers be given characterizations of risk that are both qualitative and quantitative, and that EPA make uncertainties explicit and present them as accurately and fully as is feasible and needed for risk-management decision-making. *Definition of Risk Assessment:* Risk assessment entails evaluation of information on the hazardous properties of substances, on the extent of human exposure to them, and on the characterization of the resulting risk. Risk assessment is not a single, fixed method of analysis. Rather, it is a systematic approach to organizing and analyzing scientific knowledge and information on potentially hazardous activities or on substances that might pose risks under specified conditions. In brief, according to the Red Book, risk assessment can be divided into four steps: hazard identification, dose-response assessment, exposure assessment, and risk characterization.

Continued

TABLE C-1 Continued

Date and Title of Milestone	Comments[a]
EPA 1994b *Interim Methods for Development of Inhalation Reference Concentrations (RfCs)*	Provides guidance on how to model lung dosimetry across species for setting RfCs. The method includes consideration of respiratory anatomy, physiochemical properties of the agent, and portal-of-entry considerations, such as comparative pulmonary toxicity.
EPA 1994c *Report of the Agency Task Force on Environmental Regulatory Modeling: Guidance, Support Needs, Draft Criteria and Charter*	The report concludes that there is a need for training, additional technical support, and agency guidance on external peer review of environmental regulatory modeling, among others.
EPA 1995 *Memorandum: Policy for Risk Characterization at the U.S. Environmental Protection Agency*	Reaffirms the principles and guidance in the agency's 1992 policy (*Guidance on Risk Characterization for Risk Managers and Risk Assessors*). The policy statement and associated guidance were designed to "ensure that critical information from each stage of a risk assessment is used in forming conclusions about risk and that this information is communicated from risk assessors to risk managers (policy makers), from middle to upper management, and from the Agency to the public" (EPA 1995, p. 1). Policy and guidance discuss key aspects of risk characterization, including the need to bridge risk assessment and risk management, discuss confidence and uncertainties in data, and present several types of risk information. Emphasizes the need for an iterative approach to risk assessment and makes recommendations for promoting clarity, comparability, and consistency in risk assessment.
EPA 1996 *Guidelines for Reproductive Toxicity Risk Assessment*	Guidance provides principles and procedures to be used when conducting risk assessments for reproductive toxicity.
1996 Passage of Food Quality Protection Act (FQPA)	Modernizes pesticide risk assessment by requiring accelerated licensing reviews, consideration of aggregate pesticide exposure (drinking water, residential, lawn, and food uses), and sophisticated analysis and regulation of cumulative risk of chemicals that share a mode of toxic action. In addition, mandates developing screens for potential "endocrine disruptors." FQPA also requires EPA to invoke an additional safety factor of 2-10 to account for children's risks in regulating pesticides when data are lacking.
1996 Passage of Safe Drinking Water Act amendments	Requires explicit consideration of susceptible subpopulations in setting maximum contaminant levels for drinking-water pollutants in addition to consideration of technical feasibility and costs. SDWA mandates "endocrine disruptor" screens and tests.
NRC 1996 *Understanding Risk: Informing Decisions in a Democratic Society*	Recommends that risk characterization be a "decision-driven activity, directed toward informing choices and solving problems" (NRC 1996, p. 155). Also recommends a focus on problem formulation during the initial stages of risk-assessment planning.
EPA 1997a *Guiding Principles for Monte Carlo Analysis*	Documents EPA's position "that such probabilistic analysis techniques as Monte Carlo analysis, given adequate supporting data and credible assumptions, can be viable statistical tools for analyzing variability and uncertainty in risk assessments" (EPA 1997a, p. 1) and presents an initial set of principles to guide the agency in using probabilistic analysis tools.

TABLE C-1 Continued

Date and Title of Milestone	Comments[a]
EPA 1997b *Policy for Use of Probabilistic Analysis in Risk Assessment at the U.S. Environmental Protection Agency*	Includes guiding principles to support the use of various techniques for characterizing variability and uncertainty and defines eight conditions for acceptance. The conditions are required "for ensuring good scientific practice in quantifying uncertainty and variability" (EPA 1997b, p. 1).
PCCRARM 1997a *Framework for Environmental Health Risk Management—Volume 1*	The commission was tasked under Section 303 of the Clean Air Act Amendments of 1990 to investigate the policy implications and appropriate uses of risk assessment and risk management in regulatory programs.
PCCRARM 1997b *Risk Assessment and Risk Management in Regulatory Decision-Making—Volume 2*	The Commission on Risk Assessment and Risk Management helped to stimulate agency policies, legislation, and private-sector activities that improved risk assessment and risk management. Commission's recommendations are cited in EPA policy changes on probabilistic analysis, risk characterization, and cumulative risk. The Food Quality Protection Act and the Safe Drinking Water Act Amendments of 1996 reflect commission proposals.
	"To make an effective risk management decision, risk managers and other stakeholders need to know what potential harm a situation poses and how great is the likelihood that people or the environment will be harmed. Gathering and analyzing this information is referred to as *risk assessment*. The nature, extent, and focus of a risk assessment should be guided by the risk management goals" (PCCRARM 1997b, p. 19). "For this reason, the Commission recommends that a risk assessment characterize the scientific aspects of a risk and note its subjective, cultural, and comparative dimensions [see "How Should Risks Be Analyzed?" on page 24]. While this expands risk assessment beyond its traditional, more narrowly scientific scope, including these additional dimensions will help educate all stakeholders about key factors affecting the perception of risk" (p. 21).
EPA 1997c *Guidance on Cumulative Risk Assessment—Part 1, Planning and Scoping*	1997 memorandum from Science Policy Council states: "This guidance directs each office to take into account cumulative risk issues in scoping and planning major risk assessments and to consider a broader scope that integrates multiple sources, effects, pathways, stressors and populations for cumulative risk analyses in all cases for which relevant data are available" (EPA 1997d).
EPA 1997e *Exposure Factors Handbook*	The purposes of the handbook are to: "(1) summarize data on human behaviors and characteristics which affect exposure to environmental contaminants, and (2) recommend values to use for these factors" (EPA 1997e, p. 1).
Executive Order 13045 1997 *Protection of Children From Environmental Health Risks and Safety Risks*	Primary directive to federal agencies and departments to "make it a high priority to identify and assess environmental health risks and safety risks that may disproportionately affect children." States that those agencies should "ensure that policies, programs, activities, and standards address disproportionate risks to children that result from environmental health risks or safety risks" [Sec. 1-101(a)(b)]. Establishes Task Force on Environmental Health Risks and Safety Risks to Children.
EPA 1998a *Guidelines for Neurotoxicity Risk Assessment*	Guidelines provide principles and procedures for evaluating neurotoxic risks due to chemical exposures.

Continued

TABLE C-1 Continued

Date and Title of Milestone	Comments[a]
EPA 1998b *Guidelines for Ecological Risk Assessment*	Guidelines incorporate slight modifications to the process described in 1992 (*Developing a Work Scope for Ecological Assessments*). They emphasize the importance of problem formulation in the risk-assessment process as recommended in the 1996 NRC report *Understanding Risk*. They state: "During planning, risk managers and risk assessors are responsible for coming to agreement on the goals, scope, and timing of a risk assessment and the resources that are available and necessary to achieve the goals. Together they use information on the area's ecosystems, regulatory requirements, and publicly perceived environmental values to interpret the goals for use in the ecological risk assessment. . . . The characteristics of an ecological risk assessment are directly determined by agreements reached by risk managers and risk assessors during planning dialogues. These agreements are the products of planning. They include (1) clearly established and articulated management goals, (2) characterization of decisions to be made within the context of the management goals, and (3) agreement on the scope, complexity, and focus of the risk assessment, including the expected output and the technical and financial support available to complete it" (EPA 1998b, pp. 13-15). Guidelines state that many of the difficulties with risk assessment can be traced back to issues with problem formulation. Successful ecologic risk assessment is more likely if there is an up-front discussion of what is at risk, what the assessment end points are, how they are measured, and what constitutes unacceptable risk.
NSTC 1999 *Ecological Risk Assessment in the Federal Government*	Developed by interagency work group under auspices of Committee on Environment and Natural Resources to discuss major uses of ecologic risk assessment by federal agencies. The report discussed "examples of current ecological risk assessment areas (established uses), potential uses where components of ecological risk assessment are used, and related ecological assessments and other scientific evaluations that might benefit from the use of ecological risk assessment methodologies. Recommendations were made to improve the science, enhance information transfer, and improve risk management coordination" (NSTC 1999, p. 10-5).
EPA 2000b *Risk Characterization: Science Policy Council Handbook*	Handbook provides a "single, centralized body of risk characterization implementation guidance for Agency risk assessors and risk managers to help make the risk characterization process transparent and the risk characterization products clear, consistent and reasonable" (EPA 2000b, p. vii). It implements EPA's 1992a *Guidance on Risk Characterization for Risk Managers and Risk Assessors* and its 1995 *Policy for Risk Characterization*. The handbook emphasizes the need for planning in the risk assessment process and clearly displaying all relevant information and policy choices, and it reinforces general guidance on variability and uncertainty, including distinguishing between them.
EPA 2000c *Benchmark Dose Technical Guidance Document*	Provides guidance on the "application of the benchmark dose approach to determining the point of departure (POD) for linear or nonlinear extrapolation of health effects data. Guidance discusses computation of benchmark doses and benchmark concentrations (BMDs and BMCs) and their lower confidence limits, data requirements, dose-response analysis, and reporting requirements" (EPA 2000c, p.1). Guidance provides an alternative to reliance on no-observed-adverse-effect levels as a POD.

TABLE C-1 Continued

Date and Title of Milestone	Comments[a]
EPA SAB 2000 *Toward Integrated Environmental Decision-Making*	Effort by EPA's SAB. Attempt at integrating ecology, human health, and economic valuation to develop holistic assessments.
EC 2000 *First Report on the Harmonisation of Risk Assessment Procedures*	Report of the Scientific Steering Committee Working Group on Harmonisation of Risk Assessment Procedures in the Scientific Committees advising the European Commission in human and environmental health. *Definition of Risk Assessment:* Process of evaluation that includes identification of attendant uncertainties, of the likelihood and severity of adverse effects/ or events occurring in humans or the environment after exposure under defined conditions to a risk sources. A risk assessment comprises hazard identification, hazard characterization, exposure assessment, and risk characterization.
EPA 2002b *A Review of the Reference Dose and Reference Concentration Processes*	Provides comprehensive guidance on setting reference values and recommends different exposure metrics (subchronic and acute) for IRIS.
OMB 2002 *OMB Guidelines for Ensuring and Maximizing the Quality, Utility, and Integrity of Information Disseminated by Federal Agencies*	Establishes governmentwide standards for the quality of data used and disseminated by the federal government. EPA releases its own guidelines for information quality based on OMB's guidelines in same year (see below).
EPA 2002c *Guidelines for Ensuring and Maximizing the Quality, Objectivity, Utility, and Integrity of Information Disseminated by the Environmental Protection Agency*	Developed in response to OMB's information-quality guidelines. EPA's guidelines discuss EPA's procedures developed for "ensuring and maximizing the quality of information [EPA] disseminate[s]" and "administrative mechanisms for EPA pre-dissemination review of information products" (EPA 2002c, p. 3).
EPA 2002d *OSWER Draft Guidance for Evaluating the Vapor Intrusion to Indoor Air Pathway from Groundwater and Soils*	"Vapor intrusion is the migration of volatile chemicals from the subsurface into overlying buildings. Volatile chemicals in buried wastes and/or contaminated groundwater can emit vapors that may migrate through subsurface solids and into air spaces of overlying buildings" (EPA 2002d, p. 4). "In extreme cases, the vapors may accumulate in dwellings or occupied buildings to levels that may pose near-term safety hazards... [or] acute health effects" (p. 5).
EPA 2003a *A Summary of General Assessment Factors for Evaluating the Quality of Scientific and Technical Information*	Document was developed to "raise the awareness of the information-generating public about EPA's ongoing interest in ensuring and enhancing the quality of information available for Agency use. Further, it complements the *Guidelines for Ensuring and Maximizing the Quality, Objectivity, Utility, and Integrity of Information Disseminated by the Environmental Protection Agency* (EPA Information Quality Guidelines). This summary of Agency practice is also an additional resource for Agency staff as they evaluate the quality and relevance of information, regardless of source" (EPA 2003a, p. iv).

Continued

TABLE C-1 Continued

Date and Title of Milestone	Comments[a]
EPA 2003b *Framework for Cumulative Risk Assessment*	Framework was developed to provide a consistent approach to cumulative risk assessment and identifies basic elements of the process, including a flexible structure for conducting and evaluating cumulative risk assessments and providing definitions for key terms. It also describes the three main phases of cumulative risk assessment: planning, scoping, and problem formulation; analysis; and risk characterization. Discusses planning and scoping as one distinct activity and problem formulation as another.
EPA 2003c *Human Health Research Strategy*	Strategy presents a conceptual framework for human health research by ORD and includes two strategic research directions to be pursued over the next 5-10 y: (1) research to improve the scientific foundation of human health risk assessment, including harmonizing cancer and noncancer risk assessments, assessing aggregate and cumulative risk, and determining risk to susceptible human subpopulations; and (2) research to enable evaluation of public-health outcomes of risk-management decisions.
EPA 2004a *Boron and Compounds*	EPA's IRIS assessment for boron and compounds is the first for an oral reference dose that includes a nondefault value for interspecies extrapolation and the first IRIS assessment that divides the uncertainty factor for intraspecies uncertainty (UFH) into toxicokinetic and toxicodynamic components; the assessment also develops a nondefault value for intraspecies variability (DeWoskin et al. 2007).
EPA 2004b *An Examination of EPA Risk Assessment Principles and Practices*	EPA staff paper that includes recommendations as to how EPA could strengthen and improve its risk-assessment practices. *Definition of Risk Assessment:* Referring to the NRC Red Book, this document defines it as "a process in which information is analyzed to determine if an environmental hazard might cause harm to exposed persons and ecosystems" (EPA 2004b, p. 2).
EPA 2004c *Air Toxics Risk Assessment Reference Library*	Provides "descriptions of the major methods and technical tools that are commonly used to perform air toxics risk assessments. Specifically, the manual attempts to cover all the common basic technical approaches that are used to evaluate: how people in a particular place (e.g., a city or neighborhood) may be exposed; what chemicals they may be exposed to and at what levels; how toxic those chemicals are; and how likely it is that the exposures may result in adverse health outcomes. Topics include uncertainty and variability, basic toxicology and dose-response relationships, air toxics monitoring and modeling, emissions inventory development, multipathway risk assessment, and risk characterization" (EPA 2004c, Vol.1, Part 1, p. 1-5). It provides separate and extensive guidance on planning and scoping and on problem formulation and discusses them as distinct activities. States that "planning and scoping may be the most important step in the risk assessment process. Without adequate planning, most risk assessments will not succeed in providing the type of information that risk management needs to make a well-founded decision" (EPA2004c, Vol. 1, Part 2, p. 5-9).

TABLE C-1 Continued

Date and Title of Milestone	Comments[a]
EPA 2005a *Guidelines for Carcinogen Risk Assessment*	Revises cancer guidelines, inviting mechanistic data review and consideration of early-life exposures (mutagens trigger additional safety factors). Does not discuss planning and scoping or problem formulation. *Definition of Risk Assessment:* Page 1-3: Publications by the Office of Science and Technology (OSTP 1985) and the National Research Council (NRC 1983, 1994) provide information and general principles about risk assessment. Risk assessment uses available scientific information on the properties of an agent and its effects in biologic systems to provide an evaluation of the potential for harm as a consequence of environmental exposure. The 1983 and 1994 NRC documents organize risk-assessment information into hazard identification, dose-response assessment, exposure assessment, and risk characterization. This structure appears in these cancer guidelines, with additional emphasis on characterization of evidence and conclusions in each part of the assessment.
EPA 2005b *Human Health Risk Assessment Protocol for Hazardous Waste Combustion Facilities*	The protocol is an "approach for conducting multi-pathway, site-specific human health risk assessments on Resource Conservation and Recovery Act hazardous waste combustors" (EPA 2005b, p. 1-1). Does not discuss planning and scoping or problem formulation.
Expansion of IRIS program	Planned expansion of the Integrated Risk Information System (IRIS) program with toxicity-assessment reviews to include broader input of federal partners, OMB, and other parties. (See Risk Policy Report 2005a,b)
EPA 2005c *Aging and Toxic Response: Issues Relevant to Risk Assessment*	Identifies data gaps and research needs to assist ORD in characterizing risks to the aging population from exposure to environmental toxicants.
EPA 2006a *Child-Specific Exposure Factors Handbook*	Provides non-chemical-specific data on exposure factors for childhood age groups with respect to breast-milk ingestion, food ingestion, drinking-water ingestion, soil ingestion, hand-to-mouth and object-to-mouth activity, such dermal exposure factors as surface areas and soil adherence, inhalation rates, duration and frequency in different locations and various microenvironments, duration and frequency of consumer-product use, and body weight.
OMB 2006 *Proposed Risk Assessment Bulletin*	Was developed in an effort to "enhance the technical quality and objectivity of risk assessments prepared by federal agencies by establishing uniform, minimum standards" (OMB 2006, p. 3). Includes language related to conducting uncertainty analyses, seven standards for conducting general risk assessments, and nine special standards for influential risk assessments. *Definition of Risk Assessment:* Risk assessment refers to a document that assembles and synthesizes scientific information to determine whether a potential hazard exists and/or the extent of possible risk to human health, safety, or environment.

Continued

TABLE C-1 Continued

Date and Title of Milestone	Comments[a]
GAO 2006 *Human Health Risk Assessmentg*	GAO evaluated EPA's progress in human risk assessment since release of the 1994 NRC report *Science and Judgment*. Indicates that EPA has strengthened its risk-assessment process by, for example, increasing planning for assessments, using new methods, developing guidance documents, improving its ability to characterize variability, and initiating steps to address cumulative risk. However, improvements are needed, including in the planning process, training for staff, and transparency in documenting analytic choices.
2006 EPA Changes to development of risk ranges for estimates in IRIS database	Office of Research and Development sets priorities for development of risk ranges for estimates in IRIS chemical risk value database to reflect uncertainty (see Risk Policy Report 2006a,b).
2006 European Parliament passes REACH legislation (Registration, Evaluation and Authorisation of Chemicals)	Sweeping new chemical regulation (REACH) places burden of assessing safety on industry for high-production-volume chemicals.
NRC 2007 *Scientific Review of the Proposed Risk Assessment Bulletin from the Office of Management and Budget*	Reviews OMB 2006 and recommends that it be withdrawn. One criticism concerned OMB's definition of risk assessment as documents that synthesize science. Recommends reverting to NRC Red Book definition as a process involving hazard identification, dose-response assessment, exposure assessment, and risk characterization.
2006 EPA Immunotoxicity Guidelines, In development (personal communication, EPA's Mary Jane Selgrade 12/15/06)	First-time effort will address challenging subject of immune-system biology and toxicants.
EPA 2006b *Framework for Assessing Health Risks of Environmental Exposures to Children*	Emphasizes need to account for potential exposures to environmental agents during all stages of development and to consider relevant adverse health outcomes that may occur as a result of such exposures.
EPA SAB 2007 *Consultation on Enhancing Risk Assessment Practice and Updating EPA's Exposure Guidance*	The SAB recommends that the Agency "incrementally replace the current system of single-point uncertainty factors with a set of distributions, using probabilistic methods."

[a]Included are definitions of *risk assessment* cited in the documents to illustrate the various definitions discussed in Chapter 3.

REFERENCES

DeWoskin, R.S., J.C. Lipscomb, C. Thompson, W.A. Chiu, P. Schlosser, C. Smallwood, J. Swartout, L. Teuschler, and A. Marcus. 2007. Pharmacokinetic/physiologically based pharmacokinetic models in integrated risk information system assessments. Pp. 301-348 in Toxicokinetics and Risk Assessment, J.C. Lipscomb, and E.V. Ohanian, eds. New York: Informa Healthcare.

EC (European Commission). 2000. First Report on the Harmonisation of Risk Assessment Procedures [online]. Available: http://ec.europa.eu/food/fs/sc/ssc/out83_en.pdf [accessed June 3, 2007].

EPA (U.S. Environmental Protection Agency). 1976. Interim Procedures and Guidelines for Health Risk and Economic Impact Assessments of Suspected Carcinogens. U.S. Environmental Protection Agency, Washington, DC.

EPA (U.S. Environmental Protection Agency). 1984. Risk Assessment and Management: Framework for Decision Making. EPA 600/9-85-002. Office of the Administrator, U.S. Environmental Protection Agency, Washington, DC.

EPA (U.S. Environmental Protection Agency). 1986a. Establishment of the Risk Assessment Council. Memorandum to Assistant Administrators, Associate Administrators, Regional Administrators, and General Counsel, from Lee M. Thomas, Office of the Administrator, U.S. Environmental Protection Agency, Washington, DC. June 30, 1986 [online]. Available: http://www.epa.gov/OSA/spc/pdfs/creation.pdf [accessed Oct. 11, 2007].

EPA (U.S. Environmental Protection Agency). 1986b. Guidelines for Carcinogen Risk Assessment. EPA/630/R-00/004. Risk Assessment Forum, U.S. Environmental Protection Agency, Washington, DC [online]. Available: http://cfpub.epa.gov/ncea/cfm/recordisplay.cfm?deid=54933 [accessed June 3, 2007].

EPA (U.S. Environmental Protection Agency). 1986c. Guidelines for Mutagenicity Risk Assessment. EPA/630/R-98/003. Risk Assessment Forum, U.S. Environmental Protection Agency, Washington, DC [online]. Available: http://cfpub.epa.gov/ncea/raf/recordisplay.cfm?deid=23160 [accessed June 3, 2007].

EPA (U.S. Environmental Protection Agency). 1986d. Guidelines for Chemical Mixtures Risk Assessment. EPA/630/R-98/002. Risk Assessment Forum, U.S. Environmental Protection Agency, Washington, DC [online]. Available: http://cfpub.epa.gov/ncea/raf/recordisplay.cfm?deid=20533 [accessed June 3, 2007].

EPA (U.S. Environmental Protection Agency). 1987. Unfinished Business: A Comparative Assessment of Environmental Problems. EPA 230287025a. Office of Policy Analysis, Office of Policy Planning and Evaluation, U.S. Environmental Protection Agency, Washington, DC.

EPA (U.S. Environmental Protection Agency). 1989. Risk Assessment Guidance for Superfund: Volume I—Human Health Evaluation Manual (Part A). EPA/540/1-89/02. Office of Emergency and Remedial Response, U.S. Environmental Protection Agency, Washington, DC. December 1989 [online]. Available: http://www.epa.gov/oswer/riskassessment/ragsa/pdf/rags-vol1-pta_complete.pdf [accessed June 3, 2007].

EPA (U.S. Environmental Protection Agency). 1991a. Risk Assessment Guidance for Superfund: Volume I—Human Health Evaluation Manual (Part B, Development of Risk-Based Preliminary Remediation Goals). Interim. Publication 9285.7-01B. EPA/540/R-92/003. Office of Emergency and Remedial Response, U.S. Environmental Protection Agency, Washington, DC. December 1991 [online]. Available: http://www.epa.gov/oswer/riskassessment/ragsb/index.htm [accessed Oct 11, 2007].

EPA (U.S. Environmental Protection Agency). 1991b. Risk Assessment Guidance for Superfund: Volume I—Human Health Evaluation Manual (Part C, Risk Evaluation of Remedial Alternatives). Interim. Publication 9285.7-01C. Office of Emergency and Remedial Response, U.S. Environmental Protection Agency, Washington, DC. October 1991 [online]. Available: http://www.epa.gov/oswer/riskassessment/ragsc/ [accessed Oct. 11, 2007].

EPA (U.S. Environmental Protection Agency). 1991c. Guidelines for Developmental Toxicity Risk Assessment. EPA/600/FR-91/001. Risk Assessment Forum, U.S. Environmental Protection Agency, Washington, DC [online]. Available: http://www.epa.gov/NCEA/raf/pdfs/devtox.pdf [accessed June 3, 2007].

EPA (U.S. Environmental Protection Agency). 1991d. Alpha2u-Globulin: Association with Chemically Induced Renal Toxicity and Neoplasia in the Male Rat. EPA/625/3-91/019F. Risk Assessment Forum, U.S. Environmental Protection Agency, Washington, DC.

EPA (U.S. Environmental Protection Agency). 1992a. Guidance on Risk Characterization for Risk Managers and Risk Assessors. Memorandum to Assistant Administrators, and Regional Administrators, from F. Henry Habicht, Deputy Administrator, Office of the Administrator, Washington, DC. February 26, 1992 [online]. Available: http://www.epa.gov/oswer/riskassessment/pdf/habicht.pdf [accessed Oct. 10, 2007].

EPA (U.S. Environmental Protection Agency). 1992b. Developing a Work Scope for Ecological Assessments. Ecological Update. Pub. 9345.0-051. Intermittent Bulletin, Vol. 1(4). Office of Solid Waste and Emergency Response, U.S. Environmental Protection Agency, Washington, DC [online]. Available: http://www.epa.gov/oswer/riskassessment/ecoup/pdf/v1no4.pdf [accessed June 3, 2007].

EPA (U.S. Environmental Protection Agency). 1992c. Guidelines for Exposure Assessment. EPA/600/Z-92/001. Risk Assessment Forum, U.S. Environmental Protection Agency, Washington, DC. May 1992 [online]. Available: http://cfpub.epa.gov/ncea/cfm/recordisplay.cfm?deid=15263 [accessed Oct. 10, 2007].

EPA (U.S. Environmental Protection Agency). 1992d. Dermal Exposure Assessment: Principles and Applications. Interim Report. EPA/600/8-91/011B. Office of Health and Environmental Assessment, U.S. Environmental Protection Agency, Washington, DC [online]. Available: http://rais.ornl.gov/homepage/DERM_EXP.PDF [accessed Oct. 10, 2007].

EPA (U.S. Environmental Protection Agency). 1993. Creation of a Science Policy Council. Memorandum to Assistant Administrators, Associate Administrators, and Regional Administrators, from Carol M. Browner, Office of the Administrator, U.S. Environmental Protection Agency, Washington, DC. December 22, 1993 [online]. Available: http://www.epa.gov/OSA/spc/pdfs/memo1222.pdf [accessed June 3, 2007].

EPA (U.S. Environmental Protection Agency). 1994a. Guidance Manual for the IEUBK Model for Lead in Children. EPA OSWER 9285.7-15-1. NTIS PB93-963510. Office of Solid Waste and Emergency Response, U.S. Environmental Protection Agency, Washington, DC [online]. Available: http://www.epa.gov/superfund/programs/lead/products/toc.pdf [accessed June 3, 2007].
EPA (U.S. Environmental Protection Agency). 1994b. Interim Methods for Development of Inhalation Reference Concentrations (RfCs). EPA/600/8-90/066A. Environmental Criteria and Assessment Office, Office of Research and Development, U.S. Environmental Protection Agency, Research Triangle Park, NC.
EPA (U.S. Environmental Protection Agency). 1994c. Report of the Agency Task Force on Environmental Regulatory Modeling: Guidance, Support Needs, Draft Criteria and Charter. EPA 500-R-94-001. Office of Solid Waste and Emergency Response, U.S. Environmental Protection Agency, Washington, DC.
EPA (U.S. Environmental Protection Agency). 1995. Policy for Risk Characterization at the U.S. Environmental Protection Agency. Memorandum from Carol M. Browner, Office of the Administrator, U.S. Environmental Protection Agency, Washington, DC. March 21, 1995 [online]. Available: http://64.2.134.196/committees/aqph/rcpolicy.pdf [accessed Oct. 10, 2007].
EPA (U.S. Environmental Protection Agency). 1996. Guidelines for Reproductive Toxicity Risk Assessment. EPA/630/R-96/009. Risk Assessment Forum, U.S. Environmental Protection Agency, Washington, DC [online]. Available: http://www.epa.gov/ncea/raf/pdfs/repro51.pdf [accessed June 3, 2007].
EPA (U.S. Environmental Protection Agency). 1997a. Guiding Principles for Monte Carlo Analysis. EPA/630/R-97/001. Risk Assessment Forum, U.S. Environmental Protection Agency, Washington, DC [online]. Available: http://www.epa.gov/NCEA/pdfs/montcarl.pdf [accessed June 3, 2007].
EPA (U.S. Environmental Protection Agency). 1997b. Policy for Use of Probabilistic Analysis in Risk Assessment at the U.S. Environmental Protection Agency. U.S. Environmental Protection Agency, Washington, DC [online]. Available: http://www.epa.gov/osa/spc/pdfs/probpol.pdf [accessed June 3, 2007].
EPA (U.S. Environmental Protection Agency). 1997c. Guidance on Cumulative Risk Assessment. Part 1. Planning and Scoping. Science Policy Council, U.S. Environmental Protection Agency, Washington, DC. July 3, 1997 [online]. Available: http://www.epa.gov/osa/spc/pdfs/cumrisk2.pdf [accessed Oct. 10, 2007].
EPA (U.S. Environmental Protection Agency). 1997d. Cumulative Risk Assessment Guidance—Phase I. Memorandum to Assistant Administrators, General Counsel, Inspector General, Associate Administrators, Regional Administrators, and Staff Office Directors, from C.M. Browner, Administrator, and F. Hansen, Deputy Administrator, Office of Administrator, U.S. Environmental Protection Agency, Washington, DC. July 3, 1997 [online]. Available: http://www.epa.gov/swerosps/bf/html-doc/cumulrsk.htm [accessed Aug. 13, 2008].
EPA (U.S. Environmental Protection Agency). 1997e. Exposure Factors Handbook, Vols. 1-3. EPA/600/P-95/002F. Office of Research and Development, National Center for Environmental Assessment, U.S. Environmental Protection Agency, Washington, DC [online]. Available: http://www.epa.gov/ncea/efh/ [accessed June 3, 2007].
EPA (U.S. Environmental Protection Agency). 1998a. Guidelines for Neurotoxicity Risk Assessment. EPA/630/R-95/001F. Risk Assessment Forum, U.S. Environmental Protection Agency, Washington, DC [online]. Available: http://www.epa.gov/ncea/raf/pdfs/neurotox.pdf [accessed June 3, 2007].
EPA (U.S. Environmental Protection Agency). 1998b. Guidelines for Ecological Risk Assessment. EPA/630/R-95/002F. Risk Assessment Forum, U.S. Environmental Protection Agency, Washington, DC [online]. Available: http://cfpub.epa.gov/ncea/cfm/recordisplay.cfm?deid=12460 [accessed June 3, 2007].
EPA (U.S. Environmental Protection Agency). 1999. Risk Assessment Guidance for Superfund: Volume I—Human Health Evaluation Manual Supplement to Part A: Community Involvement in Superfund Risk Assessments. OSWER 9285.7-01E-P. EPA 540-R-98-042. Office of Emergency and Remedial Response, U.S. Environmental Protection Agency. March 1999 [online]. Available: http://www.epa.gov/oswer/riskassessment/ragsa/pdf/ci_ra.pdf [accessed Oct. 11, 2007].
EPA (U.S. Environmental Protection Agency). 2000a. Supplementary Guidance for Conducting Health Risk Assessment of Chemical Mixtures. EPA/630/R-00/002. Risk Assessment Forum, U.S. Environmental Protection Agency, Washington, DC. August 2000 [online]. Available: http://www.epa.gov/NCEA/raf/pdfs/chem_mix/chem_mix_08_2001.pdf [accessed Feb. 15, 2008].
EPA (U.S. Environmental Protection Agency). 2000b. Risk Characterization: Science Policy Council Handbook. EPA 100-B-00-002. Office of Science Policy, Office of Research and Development, U.S. Environmental Protection Agency, Washington, DC [online]. Available: http://www.epa.gov/OSA/spc/pdfs/rchandbk.pdf. [accessed June 3, 2007].
EPA (U.S. Environmental Protection Agency). 2000c. Benchmark Dose Technical Guidance Document. EPA/630/R-00/001. Risk Assessment Forum, U.S. Environmental Protection Agency, Washington, DC [online]. Available: http://www.epa.gov/ncea/pdfs/bmds/BMD-External_10_13_2000.pdf [accessed June 3, 2007].

EPA (U.S. Environmental Protection Agency). 2001. Risk Assessment Guidance for Superfund: Volume I—Human Health Evaluation Manual (Part D, Standardized Planning, Reporting, and Review of Superfund Risk Assessments). Final. Publication 9285.7-47. Office of Emergency and Remedial Response, U.S. Environmental Protection Agency, Washington, DC [online]. Available: http://www.epa.gov/oswer/riskassessment/ragsd/tara.htm [accessed Oct. 11, 2007].

EPA (U.S. Environmental Protection Agency). 2002a. U.S. Environmental Protection Agency Risk Assessment Forum Charter [online]. Available: http://cfpub.epa.gov/ncea/raf/raf-char.cfm [accessed June 3, 2007].

EPA (U.S. Environmental Protection Agency). 2002b. A Review of the Reference Dose and Reference Concentration Processes. EPA/630/P-02/002F. Risk Assessment Forum, U.S. Environmental Protection Agency, Washington, DC [online]. Available: http://www.epa.gov/IRIS/RFD_FINAL%5B1%5D.pdf [accessed Oct 10, 2007].

EPA (U.S. Environmental Protection Agency). 2002c. Guidelines for Ensuring and Maximizing the Quality, Objectivity, Utility, and Integrity of Information Disseminated by the Environmental Protection Agency. EPA/260R-02-008. Office of Environmental Information, U.S. Environmental Protection Agency, Washington, DC [online]. Available: http://www.epa.gov/QUALITY/informationguidelines/documents/EPA_InfoQualityGuidelines.pdf [accessed Oct. 10, 2007].

EPA (U.S. Environmental Protection Agency). 2002d. OSWER Draft Guidance for Evaluating the Vapor Intrusion to Indoor Air Pathway from Groundwater and Soils (Subsurface Vapor Intrusion Guidance). EPA530-D-02-004. Office of Solid Waste and Emergency Response, U.S. Environmental Protection Agency, Washington, DC [online]. Available: http://www.epa.gov/correctiveaction/eis/vapor/complete.pdf [accessed June 3, 2007].

EPA (U.S. Environmental Protection Agency). 2003a. A Summary of General Assessment Factors for Evaluating the Quality of Scientific and Technical Information. EPA 100/B-03/001. Science Policy Council, U.S. Environmental Protection Agency, Washington, DC. June 2003 [online]. Available: http://www.epa.gov/osa/spc/pdfs/assess2.pdf [accessed Oct. 12, 2007].

EPA (U.S. Environmental Protection Agency). 2003b. Framework for Cumulative Risk Assessment. EPA/630/P-02/001F. Risk Assessment Forum, U.S. Environmental Protection Agency, Washington, DC [online]. Available: http://cfpub.epa.gov/ncea/cfm/recordisplay.cfm?deid=54944 [accessed June 3, 2007].

EPA (U.S. Environmental Protection Agency). 2003c. Human Health Research Strategy. EPA/600/R-02/050. Office of Research and Development, U.S. Environmental Protection Agency, Washington, DC [online]. Available: http://www.epa.gov/nheerl/humanhealth/HHRS_final_web.pdf [accessed June 3, 2007].

EPA (U.S. Environmental Protection Agency). 2004a. Boron and Compounds (CASRN 7440-42-8). Integrated Risk Information System, U.S. Environmental Protection Agency [online]. Available: http://www.epa.gov/iris/subst/0410.htm [accessed June 3, 2007].

EPA (U.S. Environmental Protection Agency). 2004b. An Examination of EPA Risk Assessment Principles and Practices. EPA/100/B-04/001. Office of the Science Advisor, U.S. Environmental Protection Agency, Washington, DC [online]. Available: http://www.epa.gov/OSA/pdfs/ratf-final.pdf [accessed June 3, 2007].

EPA (U.S. Environmental Protection Agency). 2004c. Air Toxics Risk Assessment Reference Library. EPA-453-K-04-001. Office of Air Quality Planning and Standards, U.S. Environmental Protection Agency, Research Triangle Park, NC [online]. Available: http://www.epa.gov/ttn/fera/risk_atra_main.html [accessed Oct. 10, 2007].

EPA (U.S. Environmental Protection Agency). 2005a. Guidelines for Carcinogen Risk Assessment. EPA/630/P-03/001B. Risk Assessment Forum, U.S. Environmental Protection Agency, Washington, DC [online] Available: http://www.epa.gov/iris/cancer032505.pdf [accessed June 3, 2007].

EPA (U.S. Environmental Protection Agency). 2005b. Human Health Risk Assessment Protocol for Hazardous Waste Combustion Facilities. EPA530-R-05-006. Office of Solid Waste and Emergency Response, U.S. Environmental Protection Agency, Washington, DC [online]. Available: www.epa.gov/epaoswer/hazwaste/combust/risk.htm [accessed June 3, 2007].

EPA (U.S. Environmental Protection Agency). 2005c. Aging and Toxic Response: Issues Relevant to Risk Assessment. EPA/600/P-03/004A. National Center for Environmental Assessment, Office of Research and Development, U.S. Environmental Protection Agency, Washington, DC [online]. Available: http://cfpub.epa.gov/ncea/cfm/recordisplay.cfm?deid=156648 [accessed Oct. 11, 2007].

EPA (U.S. Environmental Protection Agency). 2006a. Child-Specific Exposure Factors Handbook (External Review Draft). EPA/600/R/06/096A. U.S. Environmental Protection Agency, Washington, DC [online]. Available: http://cfpub.epa.gov/ncea/cfm/recordisplay.cfm?deid=56747 [accessed Oct. 11, 2007].

EPA (U.S. Environmental Protection Agency). 2006b. Framework for Assessing Health Risks of Environmental Exposures to Children (External Review Draft). EPA/600/R-05/093A. National Center for Environmental Assessment, Office of Research and Development, U.S. Environmental Protection Agency, Washington, DC. March 2006 [online]. Available: http://cfpub.epa.gov/ncea/cfm/recordisplay.cfm?deid=150263 [accessed Oct. 11, 2007].

EPA SAB (U.S. Environmental Protection Agency Science Advisory Board). 1990. Reducing Risk: Setting Priorities and Strategies for Environmental Protection. SAB-EC-90-021. Science Advisory Board, U.S. Environmental Protection Agency, Washington, DC. September 1990 [online]. Available: http://yosemite.epa.gov/sab/sabproduct.nsf/28704D9C420FCBC1852573360053C692/$File/REDUCING+RISK++++++++++EC-90-021_90021_5-11-1995_204.pdf [accessed Aug. 13, 2008].

EPA SAB (U.S. Environmental Protection Agency Science Advisory Board). 2000. Toward Integrated Environmental Decision-Making. EPA-SAB-EC-00-011. Science Advisory Board, U.S. Environmental Protection, Washington, DC.

EPA SAB (U.S. Environmental Protection Agency Science Advisory Board). 2007. Consultation on Enhancing Risk Assessment Practice and Updating EPA's Exposure Guidance. EPA-SAB-07-003. Letter to Stephen L. Johnson, Administrator, from Rebecca T. Parkin, Chair, Integrated Human Exposure, Committee and Environmental Health, and Granger Morgan, Chair, Science Advisory Board, U.S. Environmental Protection Agency, Washington, DC. February 28, 2007 [online]. Available: http://yosemite.epa.gov/sab/SABPRODUCT.NSF/55E1B2C78C6085EB8525729C00573A3E/$File/sab-07-003.pdf [accessed Aug. 13, 2008].

GAO (U.S. Government Accountability Office). 2006. Human Health Risk Assessment. GAO-06-595. Washington, DC: U.S. Government Printing Office [online]. Available: http://www.gao.gov/new.items/d06595.pdf [accessed Oct. 11, 2007].

NRC (National Research Council). 1983. Risk Assessment in the Federal Government: Managing the Process. Washington, DC: National Academy Press.

NRC (National Research Council). 1989. Improving Risk Communication. Washington, DC: National Academy Press.

NRC (National Research Council). 1993a. Pesticides in the Diets of Infants and Children. Washington, DC: National Academy Press.

NRC (National Research Council). 1993b. Issues in Risk Assessment. Washington, DC: National Academy Press.

NRC (National Research Council). 1994. Science and Judgment in Risk Assessment. Washington, DC: National Academy Press.

NRC (National Research Council). 1996. Understanding Risk: Informing Decisions in a Democratic Society. Washington, DC: National Academy Press.

NRC (National Research Council). 2007. Scientific Review of the Proposed Risk Assessment Bulletin from the Office of Management and Budget. Washington, DC: The National Academies Press.

NSTC (National Science and Technology Council). 1999. Ecological Risk Assessment in the Federal Government. CENR/5-99/01. Committee on Environment and Natural Resources, National Science and Technology Council, Washington, DC. May 1999 [online]. Available: oaspub.epa.gov/eims/eimscomm.getfile?p_download_id=36384 [accessed June 3, 2007].

OMB (Office of Management and Budget). 2002. OMB Guidelines for Ensuring and Maximizing the Quality, Utility, and Integrity of Information Disseminated by Federal Agencies. Washington, DC: Office of Management and Budget, Executive Office of the President.

OMB (U.S. Office of Management and Budget). 2006. Proposed Risk Assessment Bulletin. Released January 9, 2006. Washington, DC: Office of Management and Budget, Executive Office of the President [online]. Available: http://www.whitehouse.gov/omb/inforeg/proposed_risk_assessment_bulletin_010906.pdf [accessed Oct. 11, 2007].

OSTP (Office of Science and Technology Policy). 1985. Chemical Carcinogens: Review of the Science and Its Associated Principles. Washington, DC: Office of Federal Register.

PCCRARM (Presidential/Congressional Commission on Risk Assessment and Risk Management). 1997a. Framework for Environmental Health Risk Management, Vol. 1. Washington, DC: U.S. Government Printing Office [online]. Available: http://www.riskworld.com/nreports/1997/risk-rpt/pdf/EPAJAN.PDF

PCCRARM (Presidential/Congressional Commission on Risk Assessment and Risk Management). 1997b. Risk Assessment and Risk Management in Regulatory Decision-Making, Vol. 2. Washington, DC: U.S. Government Printing Office [online]. Available: http://www.riskworld.com/Nreports/1997/risk-rpt/volume2/pdf/v2epa.PDF [accessed Oct. 11, 2007].

Risk Policy Report. 2005a. Senators Fear EPA Toxics Review Plan Cedes Science To Outside Groups. Inside EPA's Risk Policy Report 12(27):1, 6.

Risk Policy Report. 2005b. EPA Plan For Expanded Risk Reviews Draws Staff Criticism, DOD Backing. Inside EPA's Risk Policy Report 12(17):1, 8.

Risk Policy Report. 2006a. NAS Dioxin Study May Boost Science Chief's Bid for Uncertainty Analysis. Inside EPA's Risk Policy Report 13(29):1, 6.

Risk Policy Report. 2006b. EPA Science Chief Pushes Plan For Risk Ranges Amid Mixed Responses. Inside EPA's Risk Policy Report 13(5):1.

Appendix D

Environmental Protection Agency Response to Recommendations from Selected NRC Reports: Policy, Activity, and Practice

Table D-1 was developed as an information resource to illustrate the kinds of policies and activities that the Environmental Protection Agency (EPA) has undertaken in response to previous National Research Council recommendations (NRC 1983, 1994, 1996) for the list of bulleted topics presented below. *This is not a comprehensive review*. Rather, it presents representative ***recommendations*** from these key National Research Council reports, beginning with the so-called Red Book; related EPA ***policies*** as reflected in guidance documents and other materials; and related implementation ***activities***, along with an ***assessment*** of some of these guidance documents and implementation activities as summarized in a 2006 report from the Government Accountability Office (GAO).

Many of the individual National Research Council reports and EPA documents address the risk-assessment issues below repeatedly and with some variations in a single report. As a result, passages quoted or summarized in the table are highly selected "snapshots" and are not the only examples for the indicated topic in a given report. In addition, the "response" to recommendations in the table is considered somewhat loosely, as it simply considers whether EPA addressed the issue at some point in time. For a full picture on any topic of interest, the committee advises readers to begin with pages cited in the table and to look beyond those citations for related information. Note also that several National Research Council recommendations and EPA policy statements cover multiple topics (such as both "risk characterization" and "uncertainty" or both "models" and "defaults"). Several issues are therefore discussed under several topic headings.[1,2]

---

[1]Empty cells indicate only that the committee could not easily identify and isolate a representative quotation, not that related policies or implementation activities do not exist.

[2]As explained in Chapter 2, the report cited as "NRC 1994" (*Science and Judgment in Risk Assessment)* gave special attention to issues arising under the Clean Air Act Amendments of 1990, and many of the recommendations in that report focused on air issues. A recommendation directed mainly to the air program is designated by "(Directed to Air Program)." Similarly, a recommendation directed mainly to the IRIS program is designated by "(Directed to IRIS Program)."

- Aggregate and Cumulative Risk
- Default Assumptions and Options
- Distinguishing and Linking Risk Assessment and Risk Management
- Distinguishing Science and Science Policy
- Exposure Assessment (and Methods Validation)
- Health-Risk and Toxicity Assessment for Cancer and Other End Points
- Inference Guidelines
- Interagency and Outside Collaboration
- Iterative Approach to Risk Assessment
- Models and Model Validation
- Peer Review and Expert Panels
- Priority-Setting and Data-Needs Management
- Problem Formulation and Ecologic Risk Assessment
- Public Review and Comment; Public Participation
- Risk Characterization
- Risk Communication in Relation to Risk Management
- Uncertainty Analysis and Characterization
- Variability and Differential Susceptibility

TABLE D-1 Environmental Protection Agency Response to National Research Council Recommendations of 1983-2006: Policy, Activity, and Practice

Topic	NRC Report: Recommendation[a]	EPA Response: Stated Policy[b]	EPA Response: Implementation Activity[c]
Aggregate and Cumulative Risk	**NRC 1994 at 240:** "EPA should consider using appropriate statistical (e.g., Monte Carlo) procedures to **aggregate** cancer risks from exposure to multiple compounds."	**EPA 1997a Science Policy Council Memorandum:** "This guidance directs each office to take into account cumulative risk issues in scoping and planning major risk assessments and to consider a broader scope that integrates multiple sources, effects, pathways, stressors and populations for **cumulative risk** analyses in all cases for which relevant data are available." **EPA1997b Cumulative Risk Assessment Guidance:** "Agency managers need to place special emphasis on **cumulative risk** (that is, the potential risks presented by **multiple stressors in aggregate**). The specific elements of risk evaluated need to be determined as an explicit part of the Planning and Scoping (PS) stage of each risk assessment. . . . The Agency will support research to improve our understanding of **cumulative risks** and to develop methods to account for the multiple elements of risks that affect humans, animals, plants and their environment. In addition, the Science Policy Council will support workshops for risk assessors and managers to discuss implementation opportunities and problems, and solutions." **EPA 2000a Supplementary Guidance for Conducting Health Risk Assessment of Chemical Mixtures at xiv:** This guidance updates the 1986 agencywide guidance on chemical mixtures and "describes more	**EPA 2003a: Human Health Research Strategy at E-2:** "ORD's research program on **aggregate and cumulative risk** will address the fact that humans are exposed to mixtures of pollutants from multiple sources. Research will provide the scientific support for decisions concerning exposure to a pollutant by multiple routes of exposure or to multiple pollutants having a similar mode of action. ORD will also develop approaches to study how people and communities are affected following exposure to multiple pollutants that may interact with other environmental stressors." **Also:** The research strategy identified the following research objectives related to **cumulative risk**: "(1) Determine the best and most cost effective ways to measure human exposures in all relevant media, including pathway-specific measures of multimedia human exposures to environmental contaminants across a variety of relevant microenvironments and exposure durations and conditions; (2) Develop exposure models and methods suitable for EPA and the public to assess aggregate and cumulative risk, including mathematical and statistical relationships among sources of environmental contaminants, their environmental fate, and pathway specific concentrations; models linking dose and exposure from biomarker data; and approaches to assess population-based **cumulative risk**, including those involving exposure to stressors other than

continued

TABLE D-1 Continued

Topic	NRC Report: Recommendation[a]	EPA Response: Stated Policy[b]	EPA Response: Implementation Activity[c]
		detailed procedures for **chemical mixture** assessment using data on the mixture of concern, data on a toxicologically similar **mixture**, and data on the mixture component chemicals. [It] is organized according to the type of data available to the risk assessor, ranging from data rich to data poor situations. . . . An evaluation of the data may lead the user to decide that only a qualitative analysis should be performed. This generally occurs in cases where data quality is poor, inadequate quantitative data are available, data on a similar **mixture** cannot be classified as "sufficiently similar" to the **mixture** of concern, exposures cannot be characterized with confidence, or method-specific assumptions about the toxicologic action of the mixture or of its components cannot be met. When this occurs, the risk assessor can still perform a qualitative assessment that characterizes the potential human health impacts from exposure to that **mixture**."	pollutants; and (3) Provide the scientific basis to predict the interactive effects of pollutants in mixtures and the most appropriate approaches for combining effects and risks from pollutant mixtures." **GAO 2006 at 50:** "The extent to which program offices assess the effects of **cumulative** and **aggregate** exposures is related to the regulatory responsibilities of each office and by the availability of data. For example, the hazardous air pollutant office routinely analyzes a mix of chemicals from various emitting sources, such as petroleum refineries, to regulate hazardous air pollutants. Similarly, as mentioned above, the Office of Pesticide Programs is required to consider exposure to pesticides from various pathways, such as food, drinking water, and residential uses, and various routes, such as eating, breathing, and contact with skin." **Note:** The Toxic Substances Control Act does not require the Office of Pollution Prevention and Toxics to assess the risks of a new chemical that may occur through its interaction with other chemicals. The office also assesses the risks of existing chemicals but cannot conduct cumulative risk assessment for classes of chemical that share a common mode of action because no data exist.
		EPA 2003b Framework for Cumulative Risk Assessment at xvii: "a simple, flexible structure for conducting and evaluating **cumulative risk assessment** within the EPA. . . . The framework describes three main phases to a **cumulative risk** assessment: (1) planning, scoping, and problem formulation, (2) analysis, and (3) risk characterization...Research and development needs are also discussed, including understanding the timing of exposure and its relationship to effects;	**GAO 2006 at 49:** "The branch of the Office of Air Quality Planning and Standards that regulates hazardous air pollutants employs the **Multiple Pathways of Exposure model** to

understanding the composition and toxicity of **mixtures**; applying the risk factor approach; using biomarkers; considering hazards presented by nonchemical stressors; methods for combining different types of risk; and development of default values for **cumulative risk** assessments, among others."

EPA 2001 and 2002: General Principles for Performing Aggregate Exposure and Risk Assessments: This document "focus[es] on describing principles to guide the way in which **aggregate exposure** and risk assessment may be performed when more extensive distributional data and more sophisticated exposure assessment, methods and tools are available. . . . [The guidance] looks beyond the Interim Guidance to encompass the use of distributional data for all pathways of exposure when data are available. A distributional data analysis (as opposed to a point estimate approach) is preferred because this tool allows an aggregate exposure assessor to more fully evaluate exposure and resulting risk across the entire population, not just the exposure of a single, high-end individual." (EPA 2001, p. 4) The 2002 guidance (EPA 2002a, p. ii) "provides guidance to OPP scientists for evaluating and estimating the potential human risks associated with such multichemical and **multipathway** exposures to pesticides."

64 Fed. Reg. 38705[1999]: The **Integrated Urban Air Toxics Strategy** includes guidance on assessing **cumulative risks** on both the national and the urban-

assess and predict the movement and behavior of chemicals in the environment. [It] includes procedures to estimate human exposures and health risks that result from the transfer of pollutants from the air to soil and surface water bodies and the subsequent uptake of the pollutant by plants, animals, and humans. The model specifically **addresses exposures from breathing; consuming food, water, and soil; and contact with skin.**"

GAO 2006 at 49: EPA developed the Total Risk Integrated Methodology (TRIM) and created the TRIM Fate, Transport, and Ecological Exposure model that describes the movement of air pollutants emitted from any type of stationary source as well as their transformation over time in water, air and soil.

continued

TABLE D-1 Continued

Topic	NRC Report: Recommendation[a]	EPA Response: Stated Policy[b]	EPA Response: Implementation Activity[c]
		neighborhood scales. It provides "an overview of EPA's national effort to reduce air toxics, including stationary and mobile source standards, **cumulative risk** initiatives, assessment approaches, and education and outreach." The "national air toxics program includes activities under multiple Clean Air Act (Act) authorities to reduce air toxics emissions from all sources, including major industrial sources, smaller stationary sources, and mobile sources such as cars and trucks. By integrating activities under different parts of the Act, EPA can better address **cumulative public health risks** and adverse environmental impacts posed by exposures to multiple air toxics in areas where the emissions and risks are most significant." **EPA 2004a: Air Toxics Risk Assessment Reference Library at 14-1:** The guidance states that "**multipathway risk assessment** may be appropriate generally when air toxics that persist and which also may bioaccumulate and/or biomagnify are present in releases. These generally will focus on the persistent bioaccumulative hazardous air pollutant (PB-HAP) compounds (Exhibit 14-1), but specific risk assessments may need to consider additional chemicals that persist and which also may bioaccumulate and/or biomagnify. For these compounds, the risk assessment generally will need to consider exposure pathways other than inhalation—in particular, pathways that involve deposition of air toxics onto soil and plants and	

into water, subsequent uptake by biota, and potential human exposures via consumption of contaminated soils, surface waters, and foods. Substances that persist and bioaccumulate readily transfer between the air, water, and land. Some may travel great distances, and linger for long periods of time in the environment." The guidance provides information on planning, scoping, problem formulation, data analysis, and risk characterization.

EPA 2002a Guidance on cumulative risk assessment of pesticide chemicals that have a common mechanism of toxicity: Provides "guidance to OPP scientists for evaluating and estimating the potential human risks associated with such multichemical and multipathway exposures to pesticides. . ." (p. ii). "Cumulative risk assessments may play a significant role in the evaluation of risks posed by pesticides, and will enable OPP to make regulatory decisions that more fully protect public health and sensitive subpopulations, including infants and children. . . . The purpose of this guidance is to set forth the basic assumptions, principles, and analytical framework that are recommended for use by OPP risk assessors in conducting cumulative risk assessments. It is also intended to inform decision makers and the public of the principles and procedures generally followed in the conduct of cumulative risk assessments on pesticide chemicals" (p. 7).

continued

TABLE D-1 Continued

Topic	NRC Report: Recommendation[a]	EPA Response: Stated Policy[b]	EPA Response: Implementation Activity[c]
Default Assumptions and Options (see also Risk Characterization, Models, Uncertainty Analysis)	**NRC 1994 at 8:** "EPA should continue to regard the use of **default** options as a reasonable way to deal with uncertainty about underlying mechanisms in selecting methods and models for use in risk assessment."	**EPA 2005a Carcinogen Risk Assessment Guidelines, Appendix A (71 FR 17809-12):** The guideline "covers [five] major **default options** commonly employed when data are missing or sufficiently uncertain in a cancer risk assessment. . . . These options are predominantly inferences that can help use the data observed under empirical conditions in order to estimate events and outcomes under environmental conditions."	**GAO 2006 at 41:** "To a large degree, the use of **defaults** is intertwined with EPA's ability to get the data it needs. As was discussed previously, EPA has targeted research, both within EPA and through its grant programs, to understand variability and uncertainty in the data derived from studies of laboratory animals, and this research may further reduce EPA's need to rely on **default** options."
	NRC 1994 at 8: "EPA should explicitly identify each use of a **default** option in risk assessments." **NRC 1994 at 8:** "EPA should clearly state the scientific and policy basis for each **default** option."	**EPA 2004b at 51:** "EPA's current practice is to examine all relevant and available data first when performing a risk assessment. When the chemical- and/or site-specific data are unavailable (i.e., when there are data gaps) or insufficient to estimate parameters or resolve paradigms, EPA uses a **default** assumption in order to continue with the risk assessment. Under this practice EPA invokes **defaults** only after the data are determined to be not usable at that point in the assessment—this is a different approach from choosing **defaults** first and then using data to depart from them. The default assumptions are not chemical- or site-specific, but are relevant to the data gap in the risk assessment. They are based on peer reviewed studies and extrapolation to address specific data gaps. These **defaults** are based on published studies, empirical observations, extrapolation from related observations, and/or scientific theory." **EPA 1996 Proposed Carcinogen Risk Assessment Guidelines at 61 FR 18000:**	**GAO 2006 at 40:** "The majority of IRIS assessments completed since 1997 describe the **defaults** used in the analysis and any departures from those **defaults**. "Despite the increased focus on more transparency in the use of **defaults**, EPA acknowledges it could more consistently describe how the **default** was developed and explain why it is a reasonable assumption. In its staff paper, EPA acknowledges it needs to ensure that the **defaults** are supported by the best available data and should look for opportunities to increase certainty and confidence in the **defaults** and extrapolations used." **EPA 2004a:** The Office of Air Quality Planning and Standard's Air Toxics Risk Assessment Reference Library (EPA 2004a) discusses **defaults** that should be used when preparing risk assessments. This is discussed, for example, when conducting screening analyses: "For complete or potential exposure pathways identified in the exposure pathway evaluation, the screening analysis may involve

	Risk characterization includes "risk estimates and their attendant uncertainties, including key uses of **default assumptions** when data are missing or uncertain." **Also at 17966-17970ff:** Explaining the scientific and policy bases of five "major" **default** options. **Also at 17964ff:** "Pursuant to [the National Research Council recommendation related to criteria for departure from defaults] the following discussion presents . . . general policy guidance on using and **departing from defaults** in specific risk assessments."	comparing media concentrations at points of exposure to 'screening' values (based on protective **default** exposure assumptions) and estimating exposure doses based on study area-specific exposure conditions. The assessor then compares estimated doses with health-based guidelines to identify substances requiring further evaluation."
NRC 1994 at 8: "The agency should consider attempting to give greater formality to its criteria for a departure from **default** options, in order to give greater guidance to the public and to lessen the possibility of ad hoc, undocumented **departures from default** options that would undercut the scientific credibility of the agency's risk assessments. At the same time, the agency should be aware of the undesirability of having its guidelines evolve into inflexible rules."	**EPA 2005a Carcinogen Risk Assessment Guidelines at 71 FR 17770ff:** "Rather than viewing **default** options as starting points from which departures may be justified by new scientific information, *these cancer guidelines view a critical analysis of all of the available information that is relevant to assessing the carcinogenic risk as the starting point from which a default option may be invoked if needed to address uncertainty or the absence of critical information* [emphasis in original]." **Also Appendix A at 17809ff:** Discusses **default** options and alternative approaches. **EPA 2000b Risk Characterization Handbook at 21:** Directs risk assessors to "describe the uncertainties inherent in the risk assessment and the **default** positions used to address these uncertainties or gaps in the assessment."	

continued

TABLE D-1 Continued

Topic	NRC Report: Recommendation[a]	EPA Response: Stated Policy[b]	EPA Response: Implementation Activity[c]
		Also at 41: "Risk assessors should carefully consider all available data before deciding to rely on **default** assumptions. If **defaults** are used, the risk assessment should reference the Agency guidance that explains the **default** assumptions or values."	
	NRC 1994 at 186: "EPA sometimes attempts to 'harmonize' risk-assessment procedures between itself and other agencies, or among its own programs, by agreeing on a single common model assumption, even though the assumption chosen might have little more scientific plausibility than alternatives (e.g., replacing FDA's body-weight assumption and EPA's surface-area assumption with body weight to the 0.75 power). . . . Rather than 'harmonizing' risk assessments by picking one assumption over others when several assumptions are plausible and none of the assumptions is clearly preferable, EPA should maintain its own **default** assumption for regulatory decisions but indicate that any of the methods might be accurate and present the results as an uncertainty in the risk estimate or present multiple estimates and state the uncertainty in each. However, 'harmonization' does serve an important purpose in the context of uncertainty analysis—it will help, rather than hinder, risk assessment if agencies cooperate to choose and validate a common set of *uncertainty distributions.*"	**EPA 2005a Carcinogen Risk Assessment Guidelines at 70 FR 17808:** "Important features [of the risk characterization] include the constraints of available data and the state of knowledge, significant scientific issues, an significant science and science policy choices that were made when alternative interpretation of data exist [citations omitted]. Choices made about using data or **default** options in the assessment are explicitly discussed in the course of the analysis, and if a choice is a significant issue, it is highlighted in the summary."	**Note:** Although both the 1996 and 2005 guidelines refer to the scaling-factor issue (at 61 FR 17968 and 71 FR 17796, respectively), it is not clear whether EPA has addressed interagency harmonization to the extent recommended.

	NRC 1994 at 241: "EPA's guidelines should clearly state a **default** option of nonthreshold low-dose linearity for genetic effects on which adequate data (e.g., data on chromosomal aberrations or dominant or X-linked mutations) might exist. This **default** option allows a reasonable quantitative estimate of, for example, first-generation genetic risk due to environmental chemical exposure."	**EPA 2005a Carcinogen Risk Assessment Guidelines at 70 FR 17811:** "The linear approach is used when a view of the mode of action indicates a linear response, for example, when a conclusion is made that an agent directly causes alterations in DNA, a kind of interaction that not only theoretically requires one reaction but also is likely to be additive to ongoing, spontaneous gene mutation."	
Distinguishing Linking Risk Assessment and Risk Management (see also Problem Formulation)	**NRC 1983 at 7:** "Regulatory agencies should take steps to establish and maintain a clear conceptual distinction between assessment of risks and the consideration of **risk management** alternatives; that is, the scientific findings and policy judgments embodied in risk assessments should be explicitly distinguished from the political, economic, and technical considerations that influence the design and choice of regulatory strategies." **NRC 1983 at 49:** "Two kinds of policy can potentially affect risk assessment: that which is inherent in the assessment process itself and that which governs the selection of regulatory options. The latter, risk management policy, should not be allowed to control the former, risk-assessment policy."	**EPA 1984 at 3:** "Scientists assess a risk to find out what the problems are. The process of deciding what to do about the problems is '**risk management**.' . . . The *distinction* between the two activities has become an attractive means for understanding and improving upon the two fundamental processes involved in environmental decision-making." **EPA 1984 at 30:** "First, we want to obtain a better and more consistent information base for making decisions about the control of risk. Second, we want to use the various analytic methods associated with **risk management** whenever appropriate in developing environmental policy; we also want to place more emphasis on figuring out what we have achieved in terms of risk reduction through past efforts and on locating and efficiently managing the serious risks remaining. Third, we must communicate to the public what we are doing, why we are doing it in **risk management** terms, and how the **risk management** approach will improve the way that EPA carries out its mission." **EPA 1986 Guidelines for Carcinogen Risk Assessment at 51 FR 33993:** "Regulatory	Administrators William Ruckelshaus and Lee Thomas mandated and funded a series of training programs for (1) the entire SES corps and other senior management and (2) agency staff in all program and regional offices. The training materials were based in materials developed initially by Bernard Goldstein and Jack Moore. There was a heavy financial investment in the program, which ran for about 5 y (approximately 1987-1992), with remnants and updates continuing sporadically even today.

continued

TABLE D-1 Continued

Topic	NRC Report: Recommendation[a]	EPA Response: Stated Policy[b]	EPA Response: Implementation Activity[c]
		decision-making involves two components: **risk assessment and risk management.** . . . The risk assessments will be carried out independently from considerations of the consequences of regulatory action."	
	NRC 1994 at 267: "EPA should increase institutional and intellectual linkages between **risk assessment** and **risk management** so as to create better harmony between the science-policy components of **risk assessment** and the broader policy objectives of **risk management**. This must be done in a way that fully protects the accuracy, objectivity, and integrity of its **risk assessments**—but the committee does not see these two aims as incompatible."	**EPA 1995a Agency-wide Policy Memorandum:** "Risk characterization, the last step in risk assessment, is the starting point for **risk management** considerations and the foundation for regulatory decision-making, but it is only one of several important components in such decisions. As the last step in **risk assessment**, the risk characterization identifies and highlights the noteworthy risk conclusions and related uncertainties. Each of the environmental laws administered by EPA calls for consideration of other factors at various stages in the regulatory process. As authorized by different statutes, decisionmakers evaluate technical feasibility (e.g., treatability, detection limits), economic, social, political, and legal factors as part of the analysis of whether or not to regulate and, if so, to what extent. Thus, regulatory decisions are usually based on a combination of the technical analysis used to develop the **risk assessment** and information from other fields." **EPA 2002b Lessons Learned from Planning and Scoping Environmental Risk Assessments at vi:** Was developed "to provide early feedback to agency scientists and managers regarding our experiences with **planning and scoping** as the first step	**1985:** EPA established the Risk Assessment Forum, an interoffice standing committee of senior scientists. From 1985 to 2006, the forum published more than 50 peer-reviewed reports on science-policy issues for the information of risk assessors, risk managers, and the public. EPA established the Risk Management Council as a multioffice decision-making body to deal with the multioffice forum products. In 1993, the RMC became the Science Policy Council; over the years, it has taken on a broader range of activities. When the RMC was set up, only senior science managers and political appointees with a science background were members; the SPC is more broadly based, but it is an extension of the original RMC. **GAO 2006 at 29:** "The Office of Air and Radiation recognized the need for planning and developed planning guidance as part of its Air Toxics Risk Assessment Reference Library, issued in 2004. EPA acknowledged in its 2004 staff paper that it needs to continue to stress the importance of concerted and conscious planning with risk assessors and risk managers before **a risk assessment** is started. According to EPA, risk assessors need to outline early in the development of

in conducting environmental assessments since the 1997 'Guidance on Cumulative Risk Assessment—Part 1'. . . . This handbook is meant to reinforce the concept that formal planning and dialogue prior to the conduct of an environmental assessment can improve the final assessment product in terms of relevancy to an environmental decision and addressing the concerns of decision makers, scientists, economists and stakeholders (where applicable). This handbook is also meant to be a catalyst to encourage agency managers to adopt formal planning and scoping as part of EPA's culture, especially when conducting significant and/or unique environmental assessments."

EPA 1998, Ecological Risk Assessment Guidelines at 13: "The characteristics of an ecological risk assessment are directly determined by agreements reached **by risk managers and risk assessors** during **planning** dialogues. These agreements are the products of planning. They include (1) clearly established and articulated management goals, (2) characterization of decisions to be made within the context of the management goals, and (3) agreement on the scope, complexity, and focus of the **risk assessment**, including the expected output and the technical and financial support available to complete it."

EPA 2000b Risk Characterization Handbook at 28: "Planning and scoping provides the opportunity for the **risk manager(s), the risk assessor(s)**, and other members of the 'team' to define what is expected to be covered in the risk a **risk management** what will and will not be addressed and how they will develop the **risk management.**"

continued

TABLE D-1 Continued

Topic	NRC Report: Recommendation[a]	EPA Response: Stated Policy[b]	EPA Response: Implementation Activity[c]
		assessment and to explain the purposes for which the **risk assessment** information will be used. During the planning and scoping phase of the **risk assessment** process **risk assessors and risk managers** should engage in a dialog to identify: a) Motivating need for the risk assessment (regulatory requirements? public concern? scientific findings? other factors?); b) Management goals, issues, and policies needing to be addressed; c) Context of the risk; d) Scope and coverage of the effort; e) Current knowledge; f) What and where are the available data; g) An agreement about how to conduct the assessment . . .; h) Plans for how the results will be communicated to senior managers and to the public; and i) Information needs/data for other members of the 'team' to conduct their analyses (e.g., economic, social, or legal analyses)."	
Distinguishing Science and Science Policy	**NRC 1983 at 153:** "Before an agency decides whether a substance should or should not be regulated as a health hazard, a detailed and comprehensive written risk assessment should be prepared and made publicly accessible. This written assessment should clearly distinguish between the **scientific basis and the policy basis** for the agency's conclusions." **NRC 1994 at 27:** "**Science-policy** choices are distinct from the policy choices associated with ultimate decision-making . . . The science-policy choices that regulatory agencies make in carrying out risk assessment have considerable influence	**EPA 1996 Proposed Carcinogen Assessment Guidelines at 61 FR 17968:** "The default to include benign tumors observed in animal studies in the assessment of animal tumor incidence is they have the capacity to progress to the malignancies with which they are associated. This treats the benign and malignant tumors as representative of responses to the test agent, which is scientifically appropriate. This is a **science policy** decision that is somewhat more conservative of public health than not including benign tumors in the assessment." **Also at 17977:** These guidelines adopt the **science-policy** position that tumor findings	

	on the results. . . ." **Also:** "Interagency and public understanding would be served by the preparation and release of a report on the science-policy issues and decisions that affect EPA's **risk-assessment** and **risk-management** practices."	in animals indicate that an agent may produce such effects in humans. **EPA 1998 Ecological Risk Guidelines at 110:** The risk-assessment report should discuss "**science policy judgments** or default assumptions used to bridge information gaps and the basis for these assumptions." **EPA 2005a Carcinogen Risk Assessment Guidelines, at 70 FR 17774:** "The agency considered both the advantages and disadvantages to extending recommended, age dependent adjustment factors for carcinogenic potency to carcinogenic agents for which the mode of action remains unknown EPA decided to recommend these factors only for carcinogens acting through a mutagenic mode of action based on a combination of analysis of available data **and long-standing science policy positions** which govern the Agency's overall approach to carcinogen risk assessment." **Also at 17808:** "Important features [of the risk characterization] include the constraints of available data and the state of knowledge, significant scientific issues, and significant **science and science policy** choices that were made when alternative interpretations of data exist."
Exposure Assessment (and Methods Validation)	**NRC 1994 at 217:** "The committee endorses the EPA's use of bounding estimates, but only in screening assessments to determine whether further levels of analysis are necessary. For further levels of analysis, the committee supports EPA's development of distributions of **exposure**	**EPA 1992a[d] Guidelines for Exposure Assessment, sec. 5.3.4:** "A common approach to estimating exposure and dose is to do a preliminary evaluation, or screening step, during which bounding estimates are used, and then to proceed to refine the estimates for those pathways

continued

TABLE D-1 Continued

Topic	NRC Report: Recommendation[a]	EPA Response: Stated Policy[b]	EPA Response: Implementation Activity[c]
	values based on available measurements, modeling results, or both. These distributions can also be used to estimate the **exposure** of the maximally exposed person. For example, the most likely value of the **exposure** to the most exposed person is generally the 100[(N - 1)/N] percentile of the cumulative probability distribution characterizing interindividual variability in **exposure**, where N is the number of persons used to construct the **exposure** distribution. This is a particularly convenient estimator to use because it is independent of the shape of the **exposure** distribution. The committee recommends that EPA explicitly and consistently use an estimator such as 100[(N - 1)/N], because it, and not a vague estimate 'somewhere above the 90th percentile,' is responsive to the language in CAAA-90 calling for the calculation of risk to 'the individual most exposed to emissions.'" **(Directed to Air Program)**	that cannot be eliminated as of trivial importance. "The method used for bounding estimates is to postulate a set of values for the parameters in the exposure or dose equation that will result in an exposure or dose higher than any exposure or dose expected to occur in the actual population. The estimate of exposure or dose calculated by this method is clearly outside of (and higher than) the distribution of actual exposures or doses. If the value of this bounding estimate is not significant, the pathway can be eliminated from further refinement. "There are two important points about bounding estimates. First, the only thing the bounding estimate can establish is a level to eliminate pathways from further consideration. It cannot be used to make a determination that a pathway is significant (that can only be done after more information is obtained and a refinement of the estimate is made), and it certainly cannot be used for an estimate of actual exposure (since by definition it is clearly outside the actual distribution). Second, when an exposure scenario is presented in an assessment, it is likely that the amount of refinement of the data, information, and estimates will vary by pathway, some having been eliminated by bounding estimates, some eliminated after further refinement, and others fully developed and quantified. This is an efficient way to evaluate scenarios. In such cases, bounding estimates must not be considered to be	

	equally as sophisticated as an estimate of a fully developed pathway, and should not be described as such." **Note:** The 1992 Guidelines do not address the 1994 recommendation for a 100[(N - 1)/N] estimator.
NRC 1994 at 218: "EPA should use the mean of current life expectancy as the assumption for the duration of individual residence time in a high-**exposure** area, or a distribution of residence times which accounts for the likelihood that changing residences might not result in significantly lower **exposure**. Similarly, EPA should use a conservative estimate for the number of hours a day an individual is exposed, or develop a distribution of the number of hours per day an individual spends in different **exposure** situations. Such information can be gathered through neighborhood surveys, etc. in these high-**exposure** areas. Note that the distribution would correctly be used only for individual risk calculations, as total population risk is unaffected by the number of persons whose **exposures** sum to a given total value (if risk is linearly related to **exposure** rate)."	**EPA 2005a Guidelines for Carcinogen Risk Assessment at 71 FR 17801:** "Unless there is evidence to the contrary in a particular case, the cumulative dose received over a lifetime, expressed as average daily exposure prorated over a lifetime, is recommended as an appropriate measure of exposure to a carcinogen."
"EPA has not provided sufficient documentation in its **exposure**-assessment guidelines to ensure that its point-estimation techniques used to determine the 'high-end **exposure** estimate' (HEEE) when data are sparse reliably yield an estimate at the desired location within the overall distribution of **exposure** (which, according to these guidelines, lies above the 90th percentile but not beyond the confines of	**EPA 1992a[d] Sec. 4.3.1:** "The Exposure Factors Handbook (EPA 1989a [updated later]) contains a summary of published data on activity patterns along with citations. Note that the summary data and the mean values cited are for the data sets included in the Handbook, and may or may not be appropriate for any given assessment." **Note:** Choice of parameters within EFH is left to the discretion of the assessor depending on the assessment goals, and so on. **EPA 1992a[d] Sec. 5.3.5.1:** "Some of the alternative methods for determining a high-end estimate of [exposure and] dose are: (1) If sufficient data on the distribution of doses are available, take the value directly for the percentile(s) of interest within the high end. If possible, the actual percentile(s) should be stated, or the number of persons determined in the high end above the estimate, in order to give

continued

TABLE D-1 Continued

Topic	NRC Report: Recommendation[a]	EPA Response: Stated Policy[b]	EPA Response: Implementation Activity[c]
	the entire distribution)." **(Directed to Air Program)**	the risk manager an idea of where within the high end-range the estimate falls. (2) If data on the distribution of doses are not available, but data on the parameters used to calculate the dose are available, a simulation (such as an exposure model or Monte Carlo simulation) can sometimes be made of the distribution. In this case, the assessor may take the estimate from the simulated distribution. (3) If some information on the distribution of the variables making up the exposure or dose equation (e.g., concentration, exposure duration, intake or uptake rates) is available, the assessor may estimate a value which falls into the high end by meeting the defining criteria of 'high end': an estimate that will be within the distribution, but high enough so that less than 1 out of 10 in the distribution will be as high. The assessor often constructs such an estimate by using maximum or near-maximum values for one or more of the most sensitive variables, leaving others at their mean values The exact method used to calculate the estimate of high-end exposure or dose is not critical; it is very important that the exposure assessor explain why the estimate, in his or her opinion, falls into the appropriate range, not above or below it. (4) If almost no data are available, it will be difficult, if not impossible, to estimate exposures or doses in the high end. One method that has been used, especially in screening-level assessments, is to start with a bounding estimate and back off the limits used until	

	the combination of parameter values is, in the judgment of the assessor, clearly in the distribution of exposure or dose. Obviously, this method results in a large uncertainty. The availability of pertinent data will determine how easily and defensibly the high-end estimate can be developed by simply adjusting or backing off from the ultra conservative assumptions used in the bounding estimates. This estimate must still meet the defining criteria of 'high end,' and the assessor should be ready to explain why the estimate is thought to meet the defining criteria."
NRC 1994 at 218: "EPA should provide a clear method and rationale for determining *when* point estimators for the HEEE can or should be used instead of a full Monte Carlo (or similar) approach to choosing the desired percentile explicitly. The rationale should more clearly indicate how such estimators are to be generated, should offer more documentation that such point-estimation methods do yield reasonably consistent representations of the desired percentile, and should justify the choice of such a percentile if it differs from that which corresponds to the expected value of **exposure** to the 'person most exposed to emissions.'" (**Directed to air program**)	**Note:** See box immediately above for related guidance in the 1992 Exposure Assessment Guidelines.
NRC 1994 at 240: "Health-risk assessments should generally consider all possible routes by which people at risk might be exposed, and this should be done universally for compounds regulated by EPA under the Clean Air Act Amendments of 1990. The agency's risk-assessment	**Note:** See the *Framework for Cumulative Risk Assessment* (EPA 2003b) and the pesticides cumulative risk guidance (EPA 2002a) for related discussion.

continued

TABLE D-1 Continued

Topic	NRC Report: Recommendation[a]	EPA Response: Stated Policy[b]	EPA Response: Implementation Activity[c]
	guidance for Superfund-related regulatory compliance (EPA 1989b) can serve as a guide in this regard, but EPA should take advantage of new developments and approaches to the analysis of multimedia fate and transport data. This will facilitate systematic consideration of multiroute **exposures** in designing and measuring compliance with Clean Air Act requirements." (**Directed to air program**)		
	NRC 1994 at 140: "EPA should explicitly consider the inclusion of noninhalation pathways, except where there is prevailing evidence that noninhalation routes—such as deposition, bioaccumulation, and soil and water uptake—are negligible." (**Directed to air program**)	**Note:** Nothing in the 1992 exposure assessment guidelines prevents an assessor from considering these pathways. In Sec. 7.3, a reviewer of the assessment is asked, "*Has the pathways analysis been broad enough to avoid overlooking a significant pathway?* For example, in evaluating exposure to soil contaminated with PCBs, the exposure assessment should not be limited only to evaluating the dermal contact pathway. Other pathways, such as inhalation of dust and vapors or the ingestion of contaminated gamefish from an adjacent stream receiving surface runoff containing contaminated soil, should also be evaluated as they could contribute higher levels of exposure from the same source."	
Health Risk and Toxicity Assessment for Cancer and Other End Points	**NRC 1994 at 141:** "In the absence of human evidence for or against carcinogenicity, EPA should continue to depend on **laboratory-animal data** for estimating the carcinogenicity of chemicals. However, laboratory-animal tumor data	**EPA 2005a Guidelines for Carcinogen Risk Assessment at 70FR 17772 Sec. 1.3.3:** "Data from epidemiological studies are generally preferred for characterizing human cancer hazard and risk."	

should not be used as the exclusive evidence to classify chemicals as to their human carcinogenicity if the **mechanisms** operative in laboratory animals are unlikely to be operative in humans; EPA should develop criteria for determining when this is the case for validating this assumption and for **gathering additional data** when the finding is made that the species tested are irrelevant to humans." **NRC 1994 at 142:** "Pharmacokinetic and pharmacodynamic data and models should be validated, and quantitative extrapolation from animal bioassays to human should continue to be evaluated and used in risk assessment." **NRC 1994 at 9:** "EPA should continue to explore and, when scientifically appropriate, incorporate pharmacokinetic **models** of the link between exposure and biologically effective dose (i.e., dose reaching the target tissue)."	**Also:** "The cancer guidelines emphasize the importance of **weighing all of the evidence** in reaching conclusions about the carcinogenic potential of agents. . . . Evidence considered includes tumor findings, or lack thereof, in humans and laboratory animal; an agent's chemical and physical properties; its structure-activity relationships (SARS) as compared with other carcinogenic agents; and studies addressing potential carcinogenic processes and **mode(s) of action**, either in vivo or in vitro." **Also at 17771:** "The use of **mode of action** in the assessment of potential carcinogens is a main focus of these cancer guidelines." **Also at 17788-91ff:** "The interaction between the biology of the organism and the chemical properties of the agent determine whether there is an adverse effect Thus, **mode of action** analysis is based on physical, chemical and biological information that helps to explain key events in an agent's influence on tumor development. The entire range of information developed in the assessment is reviewed to arrive at a reasoned judgment."	
NRC 1994 at 141: "EPA should continue to use the results of studies in mice and rats to evaluate the possibility of chemical carcinogenicity in humans." **NRC 1994 at 141:** "EPA and NTP are encouraged to explore the use of alternative species to test the hypothesis that results obtained in mice and rats are relevant to human carcinogenesis, the use of younger	**EPA 1996 Proposed Carcinogen Risk Assessment Guidelines at 61 FR 17976:** "The default assumption is that positive effects in animal **cancer** studies indicate that the agent under study can have **carcinogenic** potential in humans."	**Note:** It is not clear whether EPA has worked with NTP or other entities on the question of testing species other than mice and rats.

continued

TABLE D-1 Continued

Topic	NRC Report: Recommendation[a]	EPA Response: Stated Policy[b]	EPA Response: Implementation Activity[c]
	animals when unique sensitivity might exist for specific chemicals, and the age-dependent effects of exposure."		
	NRC 1994 at 142: "EPA should provide comprehensive narrative statements regarding the hazards posed by carcinogens, to include qualitative descriptions of both: 1) the strength of evidence about the risks of a substance; and 2) the relevance to humans of the animal models and results and of the conditions of exposure (route, dose, timing, duration, etc.) under which carcinogenicity was observed to the conditions under which people are likely to be exposed environmentally. EPA should develop a simple classification scheme that incorporates both these elements. A similar scheme to that set forth in Table 7-1 (NRC 1994) is recommended. The agency should seek international agreement on a classification system."	**EPA 2005a Guidelines for Carcinogen Risk Assessment at 70FR17775:** "The cancer guidelines emphasize the importance of a clear and useful characterization narrative that summarizes the analyses of hazard, dose response, and exposure assessment. These characterizations summarize the assessments to explain the extent and weight of evidence, major points of interpretation and rationale for their selection, strengths and weaknesses of the evidence and the analysis, and discuss alternative conclusions and uncertainties that deserve serious consideration [citing EPA's *Risk Characterization Handbook*]. See section 5.4 of the guidelines for more complete details."	
	NRC 1994 at 10: "EPA should develop a two-part scheme for classifying evidence on carcinogenicity that would incorporate both a simple classification and a narrative evaluation. At a minimum, both parts should include the strength (quality) of the evidence, the relevance of the animal model and results to humans, and the relevance of the experimental exposures (route, dose, timing, and duration) to those likely to be encountered by humans."	**EPA 1996 Proposed Guidelines for Carcinogen Risk Assessment at 61 FR 17985:** "Hazard classification uses three categories of descriptors for human **carcinogenic** potential. . . . The descriptors are presented only in the context of a weight of evidence narrative. . . . Using them within a narrative preserves and presents the complexity that is an essential part of the hazard classification." **EPA 2005a Guidelines for Carcinogen Risk Assessment at 70 FR 17772 Sec. 1.3.3:** "In order to provide some measure of clarity and consistency in an otherwise	

	free-form, narrative characterization, standard descriptors are used as part of the hazard narrative to express the conclusion regarding the **weight of evidence** for **carcinogenic** hazard potential. There are five recommended standard hazard descriptors: 'Carcinogenic to Humans,' 'Likely to Be Carcinogenic to Humans,' 'Suggestive Evidence of Carcinogenic Potential,' 'Inadequate Information to Assess Carcinogenic Potential,' and 'Not Likely to Be Carcinogenic to Humans.' Each standard descriptor may be applicable to a wide variety of data sets and weights of evidence and is presented only in the context of a **weight of evidence** narrative. Furthermore, as described in Section 2.5 of these cancer guidelines, more than one conclusion may be reached for an agent."
NRC 1994 at 143: "EPA should continue to use potency estimates—i.e., unit cancer risk—to estimate an upper bound on the probability of developing cancer due to lifetime exposure to one unit of a carcinogen. However, uncertainty about the potency estimate should be described."	**Also at 70 FR 17811-12:** The linear default is thought generally to provide an upper-bound calculation of potential risk at low doses, for example, a 1/1,000,000 to 1/100,000 risk. **And at 17802:** Assessments should discuss the significant uncertainties encountered in the analysis, distinguishing, if possible, among model uncertainty, parameter uncertainty, and human variation.
NRC 1994 at 13: "In the analysis of animal bioassay data on the occurrence of multiple tumor types, the cancer potencies should be estimated for each relevant tumor type that is related to exposure, and the individual potencies should be summed for those tumors."	**EPA 1996 Proposed Guidelines for Carcinogen Risk Assessment at 126:** "In analyzing animal bioassay data on the occurrence of multiple tumor types, these guidelines outline a number of biological and other factors to consider. The objective is to use these factors to select response data (including nontumor data as

continued

TABLE D-1 Continued

Topic	NRC Report: Recommendation[a]	EPA Response: Stated Policy[b]	EPA Response: Implementation Activity[c]
		appropriate) that best represent the biology observed. As stated in section 3 of the guidelines, appropriate options include use of a single data set, combining data from different experiments, showing a range of results from more than one data set, showing results from analysis of more than one tumor response based on differing modes of action, representing total response in a single experiment by combining animals with tumors, or a combination of these options. The approach judged to best represent the data is presented with the rationale for the judgment, including the biological and statistical considerations involved. The EPA has considered the approach of summing tumor incidences and decided not to adopt it. While multiple tumors may be independent, in the sense of not arising from metastases of a single malignancy, it is not clear that they can be assumed to represent different effects of the agent on cancer processes. In this connection, it is not clear that summing incidences provides a better representation of the underlying mode(s) of action of the agent than combining animals with tumors or using another of the several options noted above. Summing incidences would result in a higher risk estimate, a step that appears unnecessary without more reason." **EPA 2005a Guidelines for Carcinogen Risk Assessment at 71 FR 17801:** "When multiple estimates can be developed, all datasets should be considered and a judgment made about how best to represent	

NRC 1994 at 142: "EPA should develop **biologically based quantitative methods** for assessing the incidence and likelihood of **noncancer effects** in human populations resulting from chemical exposure. These methods should incorporate information on mechanisms of action and differences in susceptibility among populations and individuals that could affect risk."	the human **cancer** risk. Some options for presenting results include: adding risk estimates derived from different tumor sites" (NRC, 1994). **EPA 2000c Benchmark Dose Technical Guidance Document at 1:** "The purpose of this document is to provide guidance for the Agency and the outside community on the application of the benchmark dose approach to determining the point of departure (POD) for linear or nonlinear extrapolation of health effects data. This guidance discusses computation of benchmark doses and benchmark concentrations (BMDs and BMCs) and their lower confidence limits, data requirements, dose-response analysis, and reporting requirements. The document provides guidance based on today's knowledge and understanding, and on experience gained in using this approach. The Agency is actively applying this methodology and evaluating the outcomes for the purpose of gaining experience in using it with a variety of endpoints."	
NRC 1994 at 265: [Regarding IRIS], "EPA should enhance and expand the references in the data files on each chemical and include information on risk-assessment weaknesses for each chemical and the research needed to remedy such weaknesses. In addition, EPA should expand its efforts to ensure that IRIS maintains a high level of data quality. The chemical-specific files in IRIS should include references and brief summaries of EPA health-assessment documents		**GAO 2006 at 38:** "Since 1994, EPA has changed the IRIS assessment process in several ways. For example, each **IRIS** file now contains a discussion of the key studies, as well as a description of the decisions and **default** assumptions used in the assessment. EPA has also expanded the review that **IRIS** assessments undergo. For example, internal peer reviewers, including EPA senior health scientists representing program offices and regions, review the **IRIS** summary and accompanying detailed technical information.

continued

TABLE D-1 Continued

Topic	NRC Report: Recommendation[a]	EPA Response: Stated Policy[b]	EPA Response: Implementation Activity[c]
	and other major risk assessments of the chemicals carried out by the agency, reviews of these risk assessments by the EPA Science Advisory Board, and the agency's responses to the SAB reviews. Important risk assessments carried out by other government agencies or private parties should also be referenced and summarized."		After this review, ORD releases the document for external peer review. EPA makes draft assessments available to the public at this time and, following peer review, the **IRIS** assessment discusses the key issues reviewers raised and EPA's response. In addition, EPA has added a tracking system that allows **IRIS** users to readily determine where an individual assessment is in its development." IRIS Track is a compilation of status reports on EPA's IRIS assessments currently in progress and can be accessed at http://cfpub.epa.gov/iristrac/index.cfm. **GAO 2006 at 38:** "In September 2003, EPA completed a congressionally requested review to assess the need to update information in **IRIS**, based on concerns that EPA and state regulators rely on potentially outdated scientific information. Input from EPA program and regional offices, the public, and other stakeholders indicated that EPA should, among other things, increase the number of new or updated assessments completed each year to 50. To date, EPA has fallen considerably short of this goal. According to a program official, EPA completed 8 **IRIS** assessments in 2005, plans to complete 16 in 2006, and has approximately 75 assessments under way." "In 2004, the IRIS program also initiated a review of available scientific literature for the 460 chemicals in the database that are not under active reassessment to determine whether a reassessment based on new literature could significantly change existing

toxicity information. For 63 percent of the chemicals reviewed, no major new health effects studies were found. Such literature reviews will be conducted annually and the findings noted in the **IRIS** database. Some program offices maintain databases to enhance their risk assessments."

"EPA officials said a number of factors, such as the complexity of the assessment process, resource limitations, and extensive peer review, had limited EPA's ability to complete more assessments in 2005. EPA has increased the number of staff working on **IRIS** assessments from 6 to 23 and may ultimately increase the number to 29. The review also indicated that EPA needs to assign staff to develop health assessments for **IRIS**, and provide funding for extramural research and contracts to develop **IRIS** files and subject them to external peer review."

Also at 39: "The Office of Air Quality Planning and Standards (OAQPS) maintains a database of dose-response values developed by various sources, including IRIS, ATSDR, and the California Environmental Protection Agency, as an aide for its risk assessors. OAQPS staff update this database as better data become available. As part of its National Air Toxics Assessment—an ongoing comprehensive evaluation of hazardous air pollutants in the United States—EPA assessed 32 air pollutants plus particulate matter in diesel exhaust in 1996. The national assessment is designed to identify air pollutants with the greatest potential to harm human health, and the results will help set priorities for collecting additional data.

continued

TABLE D-1 Continued

Topic	NRC Report: Recommendation[a]	EPA Response: Stated Policy[b]	EPA Response: Implementation Activity[c]
			As part of its assessment, EPA compiled a national emissions inventory of hazardous air pollutants from outdoor sources, estimated population exposures to the pollutants, and characterized the potential cancer and noncancer health risks from breathing the pollutants."
Inference Guidelines	**NRC 1983 at 162:** "Uniform **inference guidelines** should be developed for the use of federal regulatory agencies in the risk assessment process."	**EPA 1984 at 19:** "In light of the NAS recommendations for developing risk assessment guidelines and procedures, we reviewed many of the technical issues that constitute components of risk assessment. These issues are numerous, diverse, and cover a broad spectrum of problems. To deal with problems like these, the Agency plans to complete new (or revise existing) guidelines on the following topics: carcinogenicity, mutagenicity, reproductive effects, systemic effects, chemical mixtures, and exposure assessment. **NRC 1994 at 5:** "In 1986, EPA **issued risk-assessment guidelines** that were generally consistent with the Red Book recommendations. The guidelines deal with assessing risks of carcinogenicity, mutagenicity, developmental toxicity, and effects of chemical mixtures. They include **default** options, which are essentially policy judgments of how to accommodate uncertainties. They include various assumptions that are needed for assessing exposure and risk, such as scaling factors to be used for converting test responses in rodents to estimated responses in humans."	**Note:** The recommendation for uniform "**inference guidelines**" (p. 7 ff; recommendations 5-9, pp. 162-169) *for all federal agencies* never really took hold, but EPA has been issuing and updating its version of such guidelines as "risk assessment guidelines" for the last 20 years. **GAO 2006 at 52:** "At least two-thirds of risk assessors responding to our survey who reported using **guidelines or reference documents** indicated that these documents were moderately to very helpful in preparing risk assessments. In addition, between one-third and two-thirds of respondents who reported using policy documents said these documents were moderately to very helpful in preparing risk assessments. More specifically, many risk assessors said agencywide guidelines and reference documents provide a framework to assess risks to human health that help make risk assessments more consistent. For example, some risk assessors noted the usefulness of agency reviewed or approved procedures to support their assessments. In addition, some risk assessors said the guidelines and reference documents helped clarify issues, and several assessors said they were a good source for data

GAO 2006 at 53: "The Office of Pesticide Programs periodically issues 'hot sheets' that describe how to apply general guidance to pesticide product risk assessments. In addition, the Office of Air and Radiation created the *Air Toxics Risk Assessment Reference Library* that provides information on how to analyze the risks from hazardous air pollutants."

needed to conduct assessments. Risk assessors responding to our survey cited the *Guidelines for Carcinogen Risk Assessment* as the document most frequently used when preparing human health risk assessments. More specifically, several risk assessors noted that the carcinogen guidelines provide a useful framework for preparing risk assessments. Many risk assessors commented that agencywide **guidelines and reference documents** are helpful or provide useful examples. For example, a few risk assessors stated that the *Exposure Factors Handbook* helps provide consistency among EPA offices that conduct exposure assessments because it defines standard values for exposure, and the rationale behind those values. Another assessor said that the *Review of the Reference Dose and Reference Concentration Processes* provides comprehensive guidance on setting reference values and contains a case study that serves as a model for concise and well-written hazard identification. Although risk assessors responding to our survey reported that guidance documents are generally helpful, many expressed concerns about them. For example, some risk assessors consider the documents too general or too difficult to decipher. In addition, 82 percent of the risk assessors whose offices have office specific guidance said that the guidance is very or moderately helpful with regard to preparing risk assessments. According to many risk assessors, office-specific guidance provides information in a format relevant to each office's specific needs. Over 65 percent of risk assessors reported that EPA and program offices were moderately to very effective at disseminating guidance."

continued

TABLE D-1 Continued

Topic	NRC Report: Recommendation[a]	EPA Response: Stated Policy[b]	EPA Response: Implementation Activity[c]
	NRC 1983 at 163: "The **inference guidelines** should be comprehensive, detailed, and flexible. They should make explicit the distinctions between the science and policy aspects of risk assessment. Specifically, they should have the following characteristics: • They should describe all components of hazard identification, dose-response assessment, and risk characterization and should require assessors to show that they have considered all the necessary components in each step. • They should provide detailed guidance on how each component should be considered, but permit flexibility to depart from the general case if an assessor demonstrates that an exception is warranted on scientific grounds. • They should provide specific guidance on components of data evaluation that require the imposition of risk assessment policy decisions and should clearly distinguish those decisions from scientific decisions. • They should provide specific guidance on how an assessor is to present the results of the assessment and the attendant uncertainties."	**1986-2005:** EPA issued **inference guidelines** for carcinogenic, reproductive, developmental, mutagenic, neurotoxic, and ecologic effects; for exposure assessment; and for chemical mixtures. Four guidelines first issued in 1986—on cancer, developmental toxicity, exposure, and chemical mixtures—have been updated and reissued. See tables of contents in the guidelines listed as references in this table for scope and contents of each guideline.	
	NRC 1983 at 166: "The process for developing, adopting, applying, and revising the recommended **inference guidelines** for risk assessment should reflect their dual scientific and policy nature.		**Note:** Although the congressionally chartered board recommended in the report was not established, EPA undertook some of the activities recommended for the board.

- An expert board should be established to develop recommended guidelines for consideration and adoption by regulatory agencies. The board's recommended guidelines should define the scientific capabilities and limitations in assessing health risks, delineate subjects of uncertainty, and define the consequences of alternative policies for addressing the uncertainties.
- The expert board's report and recommendations should be submitted to the agencies responsible for regulating the hazards addressed by the guidelines for their evaluation and adoption. The agencies, perhaps with central coordination, should, when possible, choose a preferred option from among the options that are consistent with current scientific understanding. The procedures for adoption should afford an opportunity for members of the public to comment.
- The process followed by the government for adoption of **inference guidelines** should ensure that the resulting guidelines are uniform among all responsible agencies and are consistently adhered to in assessing the risks of individual hazards.
- The resulting uniform guidelines should govern the performance of risk assessments by all the agencies that adopt them until they are re-examined and revised; they should not prevent members of the public from disputing their soundness or applicability in particular cases. In short, the guidelines should have the status of established

EPA 1984 Risk Assessment and Management: Framework for Decision Making: "We have established a Risk Assessment Forum to provide an institutional locus for the resolution of significant risk assessment issues as they arise, and to insure that Agency consensus on such issues is incorporated into the appropriate risk assessment guidelines. The Forum will also provide Agency scientists with a regular time and place to discuss problems of risk assessments in production. Peer advice and comment of this type will help improve the quality of risk assessments, with associated savings in time and resources."

Risk Assessment Forum–sponsored risk-assessment guidelines and all forum reports are peer-reviewed by independent panels in open meetings announced in the *Federal Register*. See, for example, 70 FR 17766, describing the peer-review process for the cancer risk-assessment guidelines issued in 2005: "In 1996, the Agency published proposed revisions to EPA's 1986 cancer guidelines for public comment. Since the 1996 proposal, the document has undergone extensive public comment and scientific peer review, including three reviews by EPA's Science Advisory Board [supplemented by the EPA Children's Health Protection Advisory Committee]. Review procedures for each risk assessment guideline are summarized in references listed for this table."

GAO 2006 at 36: "In addition to enhancing its scientific leadership, EPA has also increased its reliance on research advisory groups since 1994. The Science Policy

continued

TABLE D-1 Continued

Topic	NRC Report: Recommendation[a]	EPA Response: Stated Policy[b]	EPA Response: Implementation Activity[c]
	agency procedures, rather than binding regulations. • The guidelines should be reviewed periodically with the advice and recommendations of the expert board. The process for revising the guidelines, like the process for adoption, should afford an opportunity for comment by all interested individuals and organizations."		Council and the Risk Assessment Forum play key roles in advancing the practice of risk assessment at EPA. The council reviews the adequacy of existing policies, establishes science policy as needed, and coordinates EPA efforts related to methods, modeling, risk assessment, and environmental technology. The Science Policy Council staff facilitate ad hoc work groups, encourage communication and consensus building within the agency, and participate in technical work-group activities and deliberations. "The Risk Assessment Forum is a standing committee of senior EPA scientists established to promote agencywide consensus on difficult and controversial risk assessment issues and to ensure that this consensus is incorporated into guidance. According to an agency official, the forum is designed as a venue where staff can meet and discuss common risk assessment issues across program offices. One of the forum's main contributions to risk assessment at EPA has been the issuance of a series of risk assessment guidelines. The forum is currently working on new guidelines, such as one related to adverse effects on the immune system. When more specificity is needed on an existing guideline, the forum issues companion pieces, known as 'purple books' because of the color of their cover, that provide additional or updated information."
	NRC 1983 at 169: "The Committee recommends that guidelines initially be developed, adopted, and applied for assessment of cancer risks. Consideration	**1986-2005:** EPA issued guidelines for carcinogenic, reproductive, developmental, mutagenic, neurotoxic, and ecologic effects and for exposure and chemical mixtures.	

of other types of health effects should follow. It may not yet be feasible to draw up as complete a set of inferences guidelines for some other health effects. For these, defining the extent of scientific knowledge and uncertainties and suggesting methods for dealing with uncertainties would constitute a useful first step."	See reference list in this table.
NRC 1983 at 170: "Agencies should develop guidelines for exposure assessment. Because of diverse problems in estimating different means of exposure (e.g., through food, drinking water and consumer products), separate guidelines may be needed for each."	**1986:** Exposure Guidelines were issued. Revised guidelines were developed in 1992.
NRC 1994 at 266: "EPA should recognize that the conduct of risk assessment does not require any specific methodologic approach and that it is best seen not as a number or even a document, but as a way to organize knowledge regarding potentially hazardous activities or substances and to facilitate the systematic analysis of the risks that those activities or substances might pose under specified conditions. The limitations of risk assessment thus broadly conceived will be clearly seen as resulting from limitations in our current state of scientific understanding. Therefore, risk-assessment guidelines should be just that—guidelines, not requirements. EPA should give specific long-term attention to ways to improve this process, including changes in guidelines."	**EPA 1986[d] Guidelines for Carcinogen Risk Assessment at 51 FR 33993:** These guidelines describe the general framework to be followed in developing an analysis of carcinogenic risk and salient principles to be used in evaluating the quality of data and in formulating judgments concerning the nature and magnitude of the cancer hazard. . . . It is the intent of these Guidelines to permit sufficient flexibility to accommodate new knowledge and new assessment methods as they emerge.

continued

TABLE D-1 Continued

Topic	NRC Report: Recommendation[a]	EPA Response: Stated Policy[b]	EPA Response: Implementation Activity[c]
Interagency and Outside Collaboration	**NRC 1983 at 160:** "When two or more agencies share interest in and jurisdiction over a health hazard that is a candidate for regulation by the in the near term, a **joint risk assessment** should be prepared under the auspices of the National Toxicology Program or another appropriate organization. **Joint risk assessments** should be prepared primarily by scientific personnel provided by the agencies and assisted as necessary by other government scientists."		**GAO 2006 at 35:** "In 2004, EPA and ATSDR entered into a formal agreement to **coordinate** their efforts to develop toxicological assessments for ATSDR's work at specific highly contaminated locations and for EPA's Integrated Risk Information System (IRIS) database. EPA, NIEHS, and ATSDR also jointly develop and annually review a list of approximately 275 hazardous substances commonly found at the nation's highly contaminated sites and for which ATSDR will prepare toxicological assessments." **GAO 2006 at 36:** "Each toxicological assessment contains almost everything that is known about the chemical, including its potential to harm human health or the environment. A key difference between these toxicological assessments and the ones in EPA's IRIS database is that ATSDR includes chronic cancer and noncancer effects, as well as acute effects, while IRIS generally includes only chronic cancer and noncancer effects." **GAO 2006 at 35:** "Since 1994, EPA has strengthened and formalized **collaboration** with a range of other federal researchers to better leverage its limited research dollars and foster the development of data to improve human health risk assessments. Specifically, EPA has developed relationships with agencies such as the National Institute for Environmental Health Sciences (NIEHS) and the Agency for Toxic Substances and Disease Registry (ATSDR). For example, in 1998, EPA established a cooperative agreement with NIEHS to develop a body of research on the relationship between exposures and children's

	health. This **collaboration** jointly funded Children's Environmental Health Research Centers at seven U.S. universities and one medical center to research children's asthma and other respiratory diseases, as well as ways to reduce farm children's exposure to pesticides. "In addition, EPA works closely with ATSDR to help fill research gaps and develop chemical-specific toxicological assessments used in risk assessments. "At each annual review, agency staff may add chemicals to the list and identify priority research to fill gaps in knowledge. Of these 275 chemicals, approximately 150 have been identified by EPA as high priority needs."
NRC 1994 at 138: "EPA should conduct more **collaborative** efforts with outside parties to improve the overall risk-assessment process, and each step within that process."	**GAO 2006 at 57:** "Despite the improvements to **collaboration** at EPA, some risk assessors pointed out two barriers that limit **collaboration**. Specifically, assessors noted that conflicting priorities or goals among EPA offices and poor communication between some offices hinder the effectiveness of **collaboration**. For example, although some chemicals are studied by more than one office within EPA, the approaches and timelines differ among offices because the laws and responsibilities for each program office can differ significantly. As a result, what may be a priority chemical in one program office may not be a priority in another, thereby hindering timely **collaboration**. Furthermore, a couple of risk assessors found **collaboration** challenging because they could not find the right person in another office to communicate with on a specific issue."

continued

TABLE D-1 Continued

Topic	NRC Report: Recommendation[a]	EPA Response: Stated Policy[b]	EPA Response: Implementation Activity[c]
			Also at 57: "Several risk assessors suggested ways to improve and increase **communication among program offices**, ORD, and non-EPA organizations. For example, some risk assessors suggested more interagency work groups or meetings as a way to address research needs and foster information exchange on the development of methods. A few risk assessors suggested that a central library of risk assessment information would facilitate **collaboration** and avoid duplicating work already done by others. Specifically, one risk assessor said EPA could provide centralized databases of work conducted by different agencies and organizations, such as chemical-specific toxicity data, specific exposure or other values, and points of contact at each office."
			GAO 2006 at 37: "The Office of Pollution Prevention and Toxics has two programs to work with industry to develop data on contaminants that can be used to better understand risks. The first is the High Production Volume (HPV) Challenge Program. This program was officially launched in late 1998 to ensure that a baseline set of data would be made available to the public on approximately 2,800 chemicals that are manufactured or imported in amounts greater than 1 million pounds per year. Diverse stakeholders, including the American Chemistry Council, Environmental Defense, and the American Petroleum Institute participate in the program. The HPV Challenge Program provides an opportunity for all stakeholders, including the public, to

			comment on the tests and data summaries from the chemical sponsors—companies and consortia that volunteered to make publicly available screening-level data that allow EPA, industry, and other stakeholders to more effectively gauge the potential hazards of HPV chemicals. All comments are publicly available on the World Wide Web. As of January 2006, EPA had commitments from industry sponsors to provide data for 2,247 of the chemicals. The second program, the Voluntary Children's Chemical Evaluation Program, is designed to provide data that will allow the public to better understand the potential health risks to children associated with certain chemical exposures. EPA asked companies that manufacture or import 23 chemicals that have been found in human tissues in various biological monitoring programs to voluntarily sponsor the evaluation of specific chemicals in a pilot program. Thirty-five companies and 10 consortia volunteered to sponsor 20 chemicals. This program was developed only after considering comments and concerns from stakeholders. Of the 23 chemicals chosen for this pilot, data gathering has been completed for 9 and is under way for another 11. The remaining 3 chemicals in the pilot program have no sponsors."
Iterative Approach to Risk Assessment	**NRC 1994 at 14:** "EPA should develop the ability to conduct **iterative risk assessments** that would allow improvements to be made in the estimates until (1) the risk is below the applicable decision-making level, (2) further improvements in the scientific knowledge would not significantly change the risk estimate, or (3) EPA, the emission	**EPA 1998 Ecological Risk Guidelines at 92:** "If risks are not sufficiently defined to support a management decision, risk managers may elect to **proceed with another iteration** of one or more phases of the risk assessment process. Reevaluating the conceptual model (and associated risk hypotheses) or conducting additional	**GAO 2006 at 30:** "Some program offices have also adopted an **iterative**—or tiered—approach to risk assessment. . . . If this analysis indicates that the risk may be relatively high, assessors pursue more intensive analysis to determine if the risk is realistic or an artifact of the lower tier's conservative assumptions. Despite this move

continued

TABLE D-1 Continued

Topic	NRC Report: Recommendation[a]	EPA Response: Stated Policy[b]	EPA Response: Implementation Activity[c]
	source, or the public determines that the stakes are not high enough to warrant further analysis. **Iterative risk assessments** would also identify needs for further research and thus provide incentives for regulated parties to undertake research without the need for costly, case-by-case evaluations of each individual chemical. Iteration can improve the scientific basis of risk-assessment decisions while responding to risk-management concerns about such matters as the level of protection and resource constraints."	studies may improve the risk estimate."	toward greater use of an **iterative approach,** EPA acknowledges it could be clearer about when it is taking such an approach. For example, EPA could be more transparent about when and why it makes a **risk management** decision based on a screening level assessment rather than a more detailed assessment."
	NRC 1994 at 14: "EPA should develop and use an **iterative** approach to risk assessment. This will lead to an improved understanding of the relationship between risk assessment and **risk management** and an appropriate blending of the two."	**EPA 2005a Guidelines for Carcinogen Risk Assessment at 70 FR 17808:** "Risk assessment is **an iterative process** that grows in depth and scope in stages from screening for priority making to preliminary estimation to fuller examination in support of complex regulatory decision making. Default options may be used at any stage, but they are predominant at screening stages. . . . There are close to 30 provisions within the major statutes that require decisions based on risk, hazard or exposure assessment. . . . Given this range in the scope and depth of analysis, not all risk characterizations can or should be equal in coverage or depth."	**GAO 2006 at 30:** "When a screening assessment identifies a potential for a nontrivial risk, EPA decides if pursuing that risk is appropriate based on its current priorities and available resources. If EPA decides to pursue the risk, a more detailed, refined risk assessment is performed. The degree of refinement is based on the type of decision, the available resources, and the needs of the risk manager. After refinement of the estimate, EPA reviews it to see if it will be sufficient to answer the questions posed. Refinements proceed **iteratively** until the assessment provides an adequate answer for the decision maker within the resources available Both the revised cancer guidelines and EPA's 1995 *Policy for Risk Characterization* support an **iterative approach** to risk assessment."
	NRC 1994 at 264: "Rather than a tiered risk-assessment process, EPA should develop the ability to conduct **iterative risk assessments,** allowing improvements in the process until the risk, assessed conservatively, is below the applicable decision-making level (e.g., 1×10^{-6}, etc.); until further improvements would not significantly change the risk estimate; or until EPA, the source, or the public determines that the stakes are not high enough to warrant further analysis."	**EPA 2005a Guidelines for Carcinogen Risk Assessment at 70 FR 17808:** "Risk assessment is **an iterative process** that grows in depth and scope in stages from screening for priority making to preliminary estimation to fuller examination in support of complex regulatory decision making. Default options may be used at any stage, but they are predominant at screening stages. . . . There are close to 30 provisions within the major statutes that require decisions based on risk, hazard or exposure assessment. . . . Given this range in the scope and depth of analysis, not all risk characterizations can or should be equal in coverage or depth."	

		64 Fed. Reg. 38705 [1999]: "In analyzing residual risk, we'll conduct risk assessments consistent with the Agency's human health and ecosystem risk assessment technical guidance and policies. We'll use a **tiered approach**, usually first conducting a screening level assessment for a source category, and move to a refined assessment only where the risks identified in the screening assessment appear unacceptable. Depending on the characteristics of the HAPs, these assessments will address single or multiple pathways of exposure as well as human and ecological endpoints."	
Models and Model Validation	**NRC 1994 at 137:** "EPA should establish the predictive accuracy and uncertainty of the methods and **models** and the quality of data used in risk assessment with the high priority given to those which support the **default** options. EPA and other organizations should also conduct research on alternative methods and **models** that might represent deviations from the **default** options to the extent that they can provide superior performance and thus more accurate risk assessments in a clear and convincing manner."	**GAO 2006 at 41:** EPA's Agency Task Force on Environmental Regulatory **Modeling** published a report that concluded that a need existed for, among other things, training and technical support and agency guidance on external peer review of environmental regulatory **modeling**. **EPA 1994a at 4, Model Validation for Predictive Exposure Assessments:** "Presents in outline the methods and procedural steps of model validation and defines the role of validation in the overall process of developing a model. . . . [The document] discusses the significant role of expert opinion and qualitative judgment in determining the validation status of a model. Finally, [it] sets out the forms of evidence that will be necessary in implementing a protocol for judging whether a model can be said to have been validated." **GAO 2006 at 42:** "EPA has also	A National Research Council committee has been convened to "assess evolving scientific and technical issues related to the selection and use of computational and statistical **models** in decision making processes at the Environmental Protection Agency (EPA). The committee will provide advice concerning the development of guidelines and a vision of the selection and use of **models** at the agency. . . . The objective of the committee will be to provide a report that will serve as a fundamental guide for the selection and use of **models** in the regulatory process at the EPA." The committee's report was released in June 2007. **GAO 2006 at 41:** "In 1997, ORD and program offices conducted an agencywide conference, called the **Models** 2000 Workshop, to facilitate adherence to existing guidance on **modeling**, to define and implement improvements in how the agency developed and used **models**, and to recommend an implementation plan for

continued

TABLE D-1 Continued

Topic	NRC Report: Recommendation[a]	EPA Response: Stated Policy[b]	EPA Response: Implementation Activity[c]
		incorporated efforts to improve **models** in its research strategies and implementation plans. For example, in its plan for research on hazardous air pollutants, EPA established a long-term goal to reduce uncertainties in risk assessments through methods, data, and **models** of acute and chronic exposures and exposures through multiple pathways at both the national and regional levels."	improving **modeling** within the agency. **GAO 2006 at 43:** "EPA is beginning to embrace such new risk assessment **methodologies** as probabilistic risk assessment and mode of action analysis. Probabilistic risk assessment characterizes the variability or uncertainty in risk estimates as the range or distribution of the number of times each possible outcome will occur. In probabilistic risk assessment, one or more variables in the risk equation, such as the exposure rate, is defined as a distribution rather than as a single number. A primary advantage of probabilistic risk assessment is that it provides a quantitative description of the degree of variability or uncertainty. . . . EPA currently uses a number of models that include probabilistic analyses and is developing **a new modeling framework**, known as the Multimedia Integrated Modeling System, that will further enhance the agency's ability to probabilistically model uncertainty." **GAO 2006 at 41:** "EPA followed up these activities in 2000 by creating the Committee on Regulatory Environmental **Modeling** (CREM) to promote consistency and consensus within the agency on **modeling** issues (including **modeling** guidance, development, and application) and to enhance internal and external communications on **modeling** activities. CREM supports and enhances the existing **modeling** activities in the program offices and provides EPA with tools to support environmental decision

making. CREM also provides the public and EPA staff with a central point of inquiry about EPA's use of **models**. In 2000, CREM launched agencywide activities designed to enhance the development, use, and selection of regulatory environmental **models** at EPA. One such activity—a workshop to facilitate discussion of good **modeling** practices—resulted in the development of **modeling** guidance.

"In 2003, CREM developed guidance and created a database—called the **Models** Knowledge Base—of the **models** most frequently used in EPA."

GAO 2006 at 42: "One of ORD's laboratories established an exposure **modeling** research branch and develops population exposure **models**, such as the Stochastic Human Exposure and Dose Simulation **model** for inhalation and exposures of general and sensitive subpopulations through multiple pathways. EPA has also begun to use geographic information systems (GIS) to present risk information spatially. For example, a GIS system is being developed that maps all of the drinking water intakes in the United States and their associated watersheds, so that the agency can better assess risks to drinking water supplies stemming from activities in the related watershed. For risk assessments of hazardous air pollutants, GIS can display and analyze data during planning, scoping, and problem formulation, during the exposure assessment, and during the characterization of risks. GIS can also help communicate information to risk managers and other stakeholders."

continued

TABLE D-1 Continued

Topic	NRC Report: Recommendation[a]	EPA Response: Stated Policy[b]	EPA Response: Implementation Activity[c]
	NRC 1994 at 142: "EPA should continue to use the linearized multistage **model** as a **default** option but should develop criteria for determining when information is sufficient to use an alternative extrapolation model."	**EPA 1996 Proposed Guidelines for Carcinogen Risk Assessment at 125:** "The EPA proposes not to use a computer **model** such as the linearized multistage **model** as a **default** for extrapolation below the observed range. The reason is that the basis for **default** extrapolation is a theoretical projection of the likely shape of the curve considering mode of action. For this purpose, a computer **model** looks more sophisticated than a straight line extrapolation, but is not. The extrapolation will be by straight line as explained in the explanation of major **defaults**. This was also recommended by workshop reviewers of a previous draft of these guidelines (EPA 1994b). In addition, a margin of exposure analysis is proposed to be used in cases in which the curve is thought to be nonlinear, based on mode of action. In both cases, the observed range of data will be **modeled** by curve fitting in the absence of supporting data for a biologically based or case-specific **model**."	
Peer Review and Expert Panels	**NRC 1983 at 156:** "An agency's risk assessment should be reviewed by an independent scientific advisory panel before any major regulatory action or decision not to regulate. Peer review may be performed by science panels already established or authorized under current law or, in their absence, by panels created for this purpose." **Note:** By law, EPA is required to peer-review some categories of risk assessment.	**EPA 1992b, 1994c Peer Review Policy Memorandum:** "Major scientifically and technically based work products related to agency decisions normally should be peer reviewed. Agency managers within headquarters, Regions, laboratories, and field components determine and are accountable for the decision whether to employ peer review in particular instances and, if so, its character, scope, and timing" (EPA 1994c, p. 2).	**GAO 2006 at 26:** "In addition to enhancing its scientific leadership, EPA has also increased its reliance on research advisory groups since 1994." In 2003, Paul Gilman, assistant administrator for research and development at EPA and EPA science adviser, stated that "of the more than 800 products listed in our database as either having undergone **peer review** in 2002 or needing peer review in the next few years,

See, for example, CAA Sec. 109, FIFRA, Sec.25(d), SDWA Sec. 1412(b), and others.

EPA 2000d Peer Review Handbook 2nd edition at viii: "The goal of the Peer Review Policy and this Handbook is to enhance the quality and credibility of Agency decisions by ensuring that the scientific and technical work products underlying these decisions receive appropriate levels of peer review by independent scientific and technical experts."

EPA 2002c Guidelines for Ensuring and Maximizing the Quality, Objectivity, Utility, and Integrity of Information Disseminated by the Environmental Protection Agency: Discusses procedures for "ensuring and maximizing the quality of information [EPA] disseminate[s]" and "administrative mechanisms for EPA pre-dissemination review of information products."

EPA 2003c A Summary of General Assessment Factors for Evaluating the Quality of Scientific and Technical Information at iv: Was "intended to raise the awareness of the information-generating public about EPA's ongoing interest in ensuring and enhancing the quality of information available for Agency use. Further, it complements the Guidelines for Ensuring and Maximizing the Quality, Objectivity, Utility, and Integrity of Information Disseminated by the Environmental Protection Agency (EPA Information Quality Guidelines). This summary of Agency practice is also an additional resource for Agency staff as they evaluate the quality and relevance of information, regardless of source."

approximately 450 were slated for external **peer review**; 67 for internal review; 225 for refereed journal review; and for the balance, the review mechanism has not yet been determined (which is typical when products are a few years from completion)" (Gilman 2003, p. 6). Looking "more closely at the work products: 859 work products were reviewed by OSP [Office of Science Policy] OSP in 2002. Of that total, 113 had **peer reviews** completed in the past year; 273 products were designated as needing **peer review** sometime in the future (usually within the next 1-3 years, depending on where the product is in its development); 362 were scientific articles, or compilations of several articles, to be submitted to refereed scientific journals; and 111 were products that were deemed, usually because of their repetitive or routine nature, not to be candidates for **peer review**. Dividing 111 'peer review not needed' products by the 859 sum, we see that nearly 90 percent of our scientific and technical work products receive internal or external peer review" (p. 7). "By consistent and rigorous monitoring of the use of peer review across the Agency, led by ORD's annual evaluation of offices' peer review plans, the value of scientific **peer review** in ensuring the quality of EPA's scientific and technical products is now widely understood and accepted across the Agency" (Gilman 2003, p. 9).

EPA 2000e EPA Quality Manual for Environmental Programs at 2-5: EPA has taken a number of activities to help improve and ensure the quality of data and information, beginning in 2000, with the

continued

TABLE D-1 Continued

Topic	NRC Report: Recommendation[a]	EPA Response: Stated Policy[b]	EPA Response: Implementation Activity[c]
			EPA Quality Manual for Environmental Programs (EPA 2000e). The manual discusses EPA's role in managing and coordinating the data-quality system, including developing a quality-management plan, and "planning, directing, and conducting assessments of the effectiveness of the quality system being applied to environmental data operations and reporting results to senior management."
	NRC 1994 at 8: "EPA should continue to use the Science Advisory Board and other expert bodies. In particular, the agency should continue to make the greatest possible use of peer review, workshops, and other devices to ensure broad peer and scientific participation to guarantee that its risk-assessment decisions will have access to the best science available through a process that allows full public discussion and peer participation by the scientific community."		Risk Assessment Forum–sponsored risk-assessment guidelines and forum reports are **peer-reviewed** by **independent panels** in open meetings announced in the *Federal Register*. See, for example 70 FR 17766, describing the peer-review process for the cancer risk-assessment guidelines issued in 2005: "In 1996, the Agency published proposed revisions to EPA's 1986 cancer guidelines for public comment. Since the 1996 proposal, the document has undergone extensive public comment and scientific **peer review**, including three reviews by EPA's Science Advisory Board" supplemented by the EPA Children's Health Protection Advisory Committee.
			GAO 2006 at 27: "The Board of Scientific Counselors (BOSC) provides objective and independent advice, information, and recommendations about ORD's research program to ORD's assistant administrator. BOSC is composed of scientists and engineers from academia, industry, and environmental organizations who are recognized as experts in their fields. In 1998, BOSC completed a **peer review** of ORD's laboratories and centers. BOSC completed a second review

			of the laboratories and centers in 2002 and 2003 that identified key accomplishments of the laboratories and centers, as well as areas for future improvement. In addition, after EPA's Office of the Science Advisor issued its 2004 staff paper, it asked BOSC to host a workshop for EPA staff and other interested stakeholders, such as industry, environmental groups, and researchers, to provide feedback to refine EPA's current practices and to suggest alternative approaches for specific aspects of risk assessment."
	NRC 1983 at 171: "The Committee recommends to Congress that a Board on Risk Assessment Methods be established to perform the following functions: • To assess critically the evolving scientific basis of risk assessment and to make explicit the underlying assumptions and policy ramifications of the different inference options in each component of the risk assessment process. • To draft and periodically to revise recommended **inference guidelines** for risk assessment for adoption and use by federal regulatory agencies. • To study agency experience with risk assessment and evaluate the usefulness of the guidelines. • To identify research needs in the risk assessment field and in relevant underlying disciplines."	**EPA 1984 Risk Assessment and Management: Framework for Decision Making at 22:** "We have established a Risk Assessment Forum to provide an institutional locus for resolution of significant risk assessment issues as they arise, and to insure that Agency consensus on such issues is incorporated into the appropriate risk assessment guidelines. The Forum will also provide Agency scientists with a regular time and place to discuss problems of risk assessments in production. Peer advice and comment of this type will help improve the quality of risk assessments, with associated savings in time and resources."	**Note:** Although Congress did not establish the recommended board, EPA undertook agency-specific activities, such as the Risk Assessment Forum and risk-assessment guidelines, which are analogous to recommended board functions.
Priority-Setting and Data-Needs Management	**NRC 1994 at 10:** "EPA should compile an inventory of the chemical, toxicological, clinical, and epidemiological literature on		**GAO 2006 at 39:** "Some program offices maintain databases to enhance their risk assessments. For example, the Office of Air

continued

TABLE D-1 Continued

Topic	NRC Report: Recommendation[a]	EPA Response: Stated Policy[b]	EPA Response: Implementation Activity[c]
	each of the 189 chemicals identified in the 1990 Amendments [to the Clean Air Act]." [**Directed to air program**]		Quality Planning and Standards (OAQPS) maintains a database of dose-response values developed by various sources, including IRIS, ATSDR, and the California Environmental Protection Agency, as an aide for its risk assessors. OAQPS staff update this database as better data become available. As part of its National Air Toxics Assessment—an ongoing comprehensive evaluation of hazardous air pollutants in the United States—EPA assessed 32 air pollutants plus particulate matter in diesel exhaust in 1996. The national assessment is designed to identify air pollutants with the greatest potential to harm human health, and the results will help set **priorities** for collecting additional data. As part of its assessment, EPA compiled a national emissions inventory of hazardous air pollutants from outdoor sources, estimated population exposures to the pollutants, and characterized the potential cancer and noncancer health risks from breathing the pollutants."
	NRC 1994 at 10: "EPA should screen the 189 chemicals for **priorities** for the assessment of health risks, identify the data gaps, and develop incentives to expedite generation of the needed data by other public agencies (such as the National Toxicology Program, the Agency for Toxic Substances and Disease Registry, and state agencies) and by other organizations (industry, academia, etc.)." (**Directed to air program**)		**GAO 2006 at 36:** "In addition, EPA has begun to establish collaborative relationships with scientific and industry-related researchers. For example, EPA has cooperative agreements with the International Life Sciences Institute's Risk Science Institute (ILSI-RSI), an organization that researches critical scientific issues in risk assessment, such as the development of risk assessment methodologies. These cooperative agreements were specifically designed to engage the scientific community and bring together scientists from different affiliations (including

	academia, other parts of government, and the private sector including industry) to address risk assessment issues. Under one agreement, ILSI-RSI is to research risk assessment approaches for cumulative and aggregate exposures. In addition, EPA has used research provided by CIIT Centers for Health Research, a chemical research laboratory funded by EPA, industry, and other federal agencies, to provide information for its formaldehyde IRIS assessment. Furthermore, EPA and industry jointly fund the Health Effects Institute (HEI)—an organization that researches the health effects of various air pollutants, including airborne particulate matter and ozone. HEI has provided data for risk assessments and convened panels of experts to review and issue reports related to risk assessment, recently on diesel exhaust."
NRC 1994 at 158: "EPA should expand its efforts to gather emission and exposure **data** to personal monitoring and site-specific monitoring." (**Directed to air program**)	**GAO 2006 at 39:** "ORD also maintains personal monitoring **data** on the chemicals in the air, foods and beverages, water, and dust in an individual's personal indoor and outdoor environments. For example, in its National Human Exposure Assessment Survey (NHEXAS) program, which was completed in 1998, ORD collected human exposure data from hundreds of subjects from several areas of the country. NHEXAS provided data on background levels of total exposure to environmental contaminants that can be used as a baseline in exposure and risk assessments to estimate whether specific populations are exposed to increased levels of environmental contaminants."

continued

TABLE D-1 Continued

Topic	NRC Report: Recommendation[a]	EPA Response: Stated Policy[b]	EPA Response: Implementation Activity[c]
Problem Formulation and Ecologic Risk Assessment	**NRC 1996 at 3:** "Risk characterization is the outcome of an analytic-deliberative process. Its success depends critically on systematic analysis that is appropriate to the problem, responds to the needs of the interested and affected parties, and treats uncertainties of importance to the decision problem in a comprehensible way. Success also depends on deliberations that formulate the **decision problem**, guide analysis to improve decision participants' understanding, seek the meaning of analytic findings and uncertainties, and improve the ability of interested and affected parties to participate effectively in the risk decision process. The process must have an appropriately diverse participation or representation of the spectrum of interested and affected parties, of decision makers, and of specialists in risk analysis, at each step."	**EPA 1997a Memorandum: Cumulative Risk Assessment Guidance—Phase I Planning and Scoping:** "Recommendations from the National Research Council's (NRC) 'Understanding Risk: Informing Decisions in a Democratic Society' and a report from the Commission on Risk Assessment and Risk Management suggest that a variety of experts, including economists and social scientists, and stakeholders must be involved throughout the environmental risk assessment and risk management process. This guidance also recommends involving experts and stakeholders in the **planning and scoping** of risk assessments. The Agency is engaged in several activities that involve working with stakeholders. Experience from these activities will provide the solid basis for engaging interested and affected parties in risk assessment and risk management issues." **EPA 1998 Guidelines for Ecological Risk Assessment at 13:** "The characteristics of an ecological risk assessment are directly determined by agreements reached by risk managers and risk assessors during **planning** dialogues. These agreements are the products of **planning**. They include (1) clearly established and articulated management goals, (2) characterization of decisions to be made within the context of the management goals, and (3) agreement on the scope, complexity, and focus of the risk assessment, including the expected output and the technical and financial support available to complete it."	

	EPA 1998 Guidelines for Ecological Risk Assessment at 3: EPA (1998) also states that during its SAB review "most reviewers felt there was general compatibility between the Proposed Guidelines and the NRC report, although some emphasized the need for continued interactions among risk assessors, risk managers, and interested parties (or stakeholders) throughout the ecological risk assessment process and asked that the Guidelines provide additional details concerning such interactions. To give greater emphasis to these interactions, the ecological risk assessment diagram was modified to include 'interested parties' in the planning box at the beginning of the process and 'communicating with interested parties' in the risk management box following the risk assessment. Some additional discussion concerning interactions among risk assessors, risk managers, and interested parties was added, particularly to section 2 (planning)."
NRC 1996 at 6: "The analytic-deliberative process leading to a risk characterization should include early and explicit attention to problem formulation; representation of the spectrum of interested and affected parties at this early stage is imperative. The analytic-deliberative process should be mutual and recursive. Analysis and deliberation are complementary and must be integrated throughout the process leading to risk characterization: deliberation frames analysis, analysis informs deliberation, and the process benefits from feedback between the two."	**EPA 1998 at 13:** States that "during **planning**, risk managers and risk assessors are responsible for coming to agreement on the goals, **scope**, and timing of a risk assessment and the resources that are available and necessary to achieve the goals. Together they use information on the area's ecosystems, regulatory requirements, and publicly perceived environmental values to interpret the goals for use in the ecological risk assessment." **EPA 2003b at 63:** Includes discussion of the risk characterization recommendations

continued

TABLE D-1 Continued

Topic	NRC Report: Recommendation[a]	EPA Response: Stated Policy[b]	EPA Response: Implementation Activity[c]
		from the National Research Council 1996 report, including a box summarizing some of the points made in the report. Also states that "risk characterization is most efficiently conducted with early and continued attention to the risk characterization step in the risk assessment process (NRC 1996; EPA 2000b)."	
Public Review and Comment; Public Participation	**NRC 1994 at 267:** "EPA should provide a process for **public review** and **comment** with a requirement that it respond, so that outside parties can be assured that the methods used in risk assessments are scientifically justifiable."	**Note:** Consistent with requirements of the federal Administrative Procedure Act, as well as environmental laws administered by EPA, public notice and an opportunity for comment are provided in relation to all EPA actions subject to those laws.	**GAO 2006 at 29:** "Program offices involve stakeholders in various ways. For example, the branch of the Office of Air Quality Planning and Standards (OAQPS) responsible for setting certain air quality standards for six principal pollutants solicits input from stakeholders in the planning phase of its periodic updates to the standards it sets. In addition, the **public** may officially **comment** on draft air quality standards once they are publicly released. The Office of Water pursues stakeholder and **public involvement** that includes working with the environmental community, industry, trade associations, risk assessor organizations, states, and bordering countries. In addition, the office's periodic reviews of water quality standards and other nonregulatory actions, such as health advisories, are all open processes that allow for **public input** on various stages of the analysis.
	NRC 1996 at 30: "Successful risk characterization depends on input from three kinds of actors: public officials . . . analytic experts . . . and the interested and affected parties to the decision. The **interested and affected parties** have a right to influence which questions should be the subject of analysis and can contribute both to developing information and to the deliberative pars of the process."	**EPA 1998 Ecological Risk Assessment Guidelines at 63FR 11-12:** "In some risk assessments, **interested parties** also take an active role in planning, particularly goal development. . . . Interested parties may communicate their concerns to risk managers about the environment, economics, cultural changes, or other values potentially at risk from environmental management activities. . . . In some cases, **interested parties** may provide important information to risk assessors. Local knowledge, particularly in rural communities, and traditional knowledge of native peoples can provide valuable insights about ecological characteristics of a place, past conditions, and current changes."	"For risk assessments involving the reregistration of pesticides, the Office of Pesticide Programs (OPP) established a process that provides several opportunities for **public participation**. Depending on the potential health risks posed by a pesticide
		EPA 1997a Memorandum: Cumulative Risk Assessment Guidance—Phase I Planning	

and Scoping: In accordance with recommendations from NRC 1996, the agency is engaged in several activities that involve working with stakeholders. Experience with these activities will provide a solid basis for engaging interested and affected parties in risk-assessment and risk-management issues.

EPA 1997a Guidance on Cumulative Risk Assessment at 1, 2: "Directs each office to take into account cumulative risk issues in scoping and planning major risk assessments and to consider a broader scope that integrates multiple sources, effects, pathways, stressors and populations for cumulative risk analyses in all cases for which relevant data are available. . . .Our goal is to ensure that **citizens and other stakeholders** have an opportunity to help define the way in which an environmental or public health problem is assessed, to understand how the available data are used in the risk assessment, and to see how the data affect decisions about risk management."

product, the **public** has anywhere from one to four separate opportunities to **comment**. For example, if risk assessors estimate that the product poses little risk to human health, the **public** will have one opportunity to **comment** before OPP decides whether to approve the pesticide product. For higher-risk products, the **public** will have as many as four opportunities to **comment**. The first opportunity to **comment** occurs after OPP has completed a preliminary risk assessment. This preliminary assessment contains all of the elements of a risk assessment and has undergone internal review, but is not yet finalized. Notice of the opportunity to **comment** is distributed to people who have elected to sign up for such notifications, as well as through a 'notice of availability' published in the *Federal Register*. The **public** can also **comment** on risk assessments prepared by the Office of Pesticide Programs through the office's Science Advisory Panel—which holds periodic **public** meetings on pesticide-related risk assessment issues, such as methods to assess skin sensitivity to exposure to pesticides or **models** used to estimate dietary exposures."

GAO 2006 at 38: "EPA also changed how it sets priorities for which chemicals need new or updated **IRIS** assessments. Annually, EPA asks its program offices, regions, and the public to identify contaminants for which it should develop or revise **IRIS** assessments. EPA publishes the list in the *Federal Register* and requests the public and scientific community to submit any relevant data on substances undergoing review. EPA is currently reviewing ways to increase

continued

TABLE D-1 Continued

Topic	NRC Report: Recommendation[a]	EPA Response: Stated Policy[b]	EPA Response: Implementation Activity[c]
			coordination with other governmental agencies that develop chemical assessments, outreach to stakeholders earlier in the development of **IRIS** assessments, and consultation with independent external reviewers."
Risk Characterization	**NRC 1983 at 20:** "Risk characterization is the process of estimating the incidence of a health effect under the various conditions of human exposure described in exposure assessment. It is performed by combining the exposure and dose-response assessments. The summary effects of the uncertainties in the preceding steps are described in this step."	**EPA 1984 at 14, Risk Assessment and Management: Framework for Decision Making:** "The final assessment should display all relevant information pertaining to the decision at hand, including such factors as the nature and weight of evidence for each step of the process, the estimated uncertainty of the component parts, the distribution of risk across various sectors of the population, the assumptions contained within the estimates, and so forth." **EPA 1992b Agency-wide Policy Memorandum:** "Well-balanced **risk characterization** presents information for other risk assessors, EPA decision-makers, and the public regarding the strengths and limitations of the assessment." (NRC 1994, Appendix B). **EPA 1995a Agency-wide Policy Memorandum:** "Each risk assessment prepared in support of decision-making at EPA should include a **risk characterization** . . . that is clear, transparent, reasonable and consistent with other risk characterizations of similar scope prepared across programs in the agency. . . . To ensure transparency, risk characterizations	

should include a statement of confidence in the assessment that identifies all major uncertainties along with comment on their influence on the assessment, consistent with [EPA 1995b, Guidance in the *Risk Characterization Handbook*.]" (Reprinted as Appendix A in *Risk Characterization Handbook*.)

EPA 2000b Risk Characterization Handbook at 39: "At EPA, various risk assessment guidelines have been written to ensure a scientifically defensible and consistent approach to risk assessment. When you write the risk characterization portion of your assessment, indicate whether or not you followed the guidelines and describe the key assumptions you made during your assessment and the impact they have on the assessment outcome. . . . In years past, different EPA offices sometimes had different policies about how to assess risk (e.g., different uncertainty factors or different levels of regulatory concern). While the development of the various risk assessment guidelines and the establishment of the Science Policy Council have helped to eliminate such discrepancies, possibilities for policy choices affecting risk assessment outcomes still exist in EPA (i.e., different laws and their implementing regulations may still dictate divergent policies). Also, there may be important differences between EPA's risk assessment policy choices and those of other agencies. To the extent you are aware of such information be sure to describe it in the **risk characterization** portion of your assessment and to let your manager know of the impact the alternative policy choices have on the outcome of your assessment."

continued

TABLE D-1 Continued

Topic	NRC Report: Recommendation[a]	EPA Response: Stated Policy[b]	EPA Response: Implementation Activity[c]
	NRC 1994 at 5: "Risk characterization combines the assessments of exposure and response under various exposure conditions to estimate the probability of specific harm to an exposed individual or population. To the extent feasible, this characterization should include the distribution of risk in the population. When the distribution of risk is known, it is possible to estimate the risk to individuals who are most exposed to the substance in question."	**EPA 1986 Guidelines for Carcinogen Risk Assessment 51 FR 33999:** "The section of **risk characterization** should summarize the hazard identification, dose- response assessment, exposure assessment, and the public health estimates. Major assumptions, scientific judgments, and, to the extent possible, estimates of the uncertainties embodied in the assessment are presented."	
	NRC 1994 at 10: "EPA should continue to use as one of its risk characterization metrics, upper-bound potency estimates of the probability of developing cancer due to lifetime exposure. Whenever possible, this metric should be supplemented with other descriptions of cancer potency that might more adequately reflect the uncertainty associated with the estimates."	**EPA 1996 Proposed Carcinogen Assessment Guidelines at 125:** "The result of using straight line extrapolation is thought to be an upper bound on low-dose potency to the human population in most cases, but as discussed in the major defaults section, it may not always be. Exploration and discussion of uncertainty of parameters in curve-fitting a model of the observed data or in using a biologically based or case-specific model is called for in the dose response assessment and **characterization** sections of these guidelines."	
		EPA 2005a Guidelines for Carcinogen Risk Assessment at 70 FR 17801: "Linear extrapolation should be used in two distinct circumstances: (1) When there are data to indicate that the dose-response curve has a linear component below the POD [point of departure] and (2) as a default for a tumor site where the mode of action is not established. . . . The slope of this line, known as the *slope factor*, is an **upper-bound estimate of risk per increment of dose** that can be used to estimate risk probabilities for different exposure levels."	

	NRC 1994 at 27: "Risk characterization should also include a full discussion of the uncertainties associated with the estimates of risk."	**EPA 2005a Guidelines for Carcinogen Risk Assessment at 70 FR 17808:** "The **risk characterization** presents an integrated and balanced picture of the analysis of the hazard, dose-response, and exposure. The risk analyst should provide summaries of the evidence and results and describe the quality of available data and the degree of confidence to be placed in the risk estimates. Important features include the constraints of available data and the state of knowledge, significant scientific issues, and significant science and science policy choices that were made when alternative interpretations of data exist (EPA 1995a, 2000b). Choices made about using data or default options in the assessment are explicitly discussed in the course of analysis, and if a choice is a significant issue, it is highlighted in the summary. In situations where there are alternative approaches for risk assessment that have significant biological support, the decisionmaker can be informed by the presentation of these alternatives along with their strengths and uncertainties."	
Risk Communication in Relation to Risk Management	**NRC 1994 at 15:** "When EPA reports estimates of risk to decision-makers and the public, it should **present not only point estimates** of risk, but also the sources and magnitudes of uncertainty associated with these estimates." **NRC 1994 at 13:** "**Risk managers should be given** characterizations of risk that are both qualitative and quantitative, i.e., both descriptive and mathematical."	**EPA 1996 Proposed Carcinogen Risk Assessment Guidelines at 126:** "In part as a response to these recommendations, the Administrator of EPA issued guidelines for risk characterization and required implementation plans from all programs in EPA (EPA 1995a). The Administrator's guidance is followed in these cancer guidelines. The assessments of hazard, dose response, and exposure **will all have accompanying technical characterizations** covering issues of strengths and	**GAO 2006 at 64:** "Experts also said EPA risk assessments should clearly describe the sufficiency of the data and the scientific basis for its choice of a **default** assumption, method, or model. Some experts pointed out that risk assessments should identify and clearly discuss any data that are not available for the analysis, including the form the data need to be in and the most appropriate study design or methodology to obtain the needed data. In addition, several experts said EPA needs to more explicitly communicate

continued

TABLE D-1 Continued

Topic	NRC Report: Recommendation[a]	EPA Response: Stated Policy[b]	EPA Response: Implementation Activity[c]
		limitations of data and current scientific understanding, identification of defaults utilized in the face of gaps in the former, discussions of controversial issues, and discussions of uncertainties in both their qualitative, and as practicable, their quantitative aspects." **EPA 1998 Ecological Risk Guidelines at 109-110:** "When risk characterization is complete, risk assessors should be able to estimate ecological risks, indicate the overall degree of confidence in the risk estimates, cite lines of evidence supporting the risk estimates, and interpret the adversity of ecological effects. Usually this information is **included in a risk characterization report** . . . . **EPA 2005a Guidelines for Carcinogen Risk Assessment at 71FR 17807:** "The risk characterization includes a **summary for the manager in nontechnical discussion** that minimizes the use of technical terms. It is an appraisal of the science that informs the risk manager. . . . It also serves the needs of other interested readers. The summary is an information resource for preparing risk communication information, but . . . is not itself the vehicle for communication with every audience."	which **default** assumptions were used in a risk assessment and why the **defaults** were chosen. For example, one expert said that even though a risk assessment may be perfect, if the public does not understand the rationale behind the agency's choices, the risk assessment might be seen as flawed. Furthermore, in individual risk assessments, the agency could more transparently identify which critical studies would help the agency avoid relying on **default** assumptions. Some experts also suggested that EPA use as case studies completed assessments for which the agency had sufficient data to use models and other analytic tools rather than **default** assumptions to more accurately assess risks. Finally, some experts said that EPA should more transparently consider alternate methods and models in each risk assessment. For example, EPA should be more transparent about the judgments it makes when it employs certain methods, such as the benchmark dose method, which identifies the dose that produces a small increase in the risk of an adverse effect."
Uncertainty and Variability Analysis and Characterization	**NRC 1994 at 185:** "EPA should make **uncertainties** explicit and present them as accurately and fully as is feasible and needed for **risk management** decision-making. To the greatest extent feasible,	**EPA 1995a Agency-Wide Memorandum at 5:** "Particularly critical to full characterization of risk is a frank and open discussion of the **uncertainty** in the overall assessment and in each of its components.	**GAO 2006 at 43:** "EPA's 1997 policy states that probabilistic techniques, such as Monte Carlo analysis, can be viable statistical tools to analyze **variability** in risk assessments, when they are based on adequate supporting

(see also **Risk Characterization, Defaults**)	EPA should present quantitative, as opposed to qualitative, representations of **uncertainty**. However, EPA should not necessarily quantify model **uncertainty** (via subjective weights or any other technique), but should try to quantify the parameter and other **uncertainty** that exists for each plausible choice of scientific model. In this way, EPA can give its **default** models the primacy they are due under its guidelines, while presenting useful, but distinct alternative estimates of risk and **uncertainty**. In the quantitative portions of their risk characterizations (which will serve as one important input to standard-setting and residual-risk decisions under the Act), EPA risk assessors should consider only the **uncertainty** conditional on the choice of the preferred models for dose-response relationships, exposure, uptake, etc." **NRC 1994 at 13:** "Quantitative **uncertainty** characterizations conducted by EPA should appropriately reflect the difference between **uncertainty** and interindividual variability." **NRC 1994 at 185:** "EPA should develop **uncertainty** analysis guidelines—both a general set and specific language added to its existing guidelines for each step in risk assessment (e.g., the exposure assessment guidance). The guidelines should consider in some depth all the types of **uncertainty** (model, parameter, etc.) in all the stages of risk assessment. The **uncertainty** guidelines should require that the **uncertainties** in models, data sets, and parameters and their relative contributions to total **uncertainty**	The **uncertainty discussion** is important for several reasons. 1. Information from different sources carries different kinds of uncertainty and knowledge of these differences is important when **uncertainties** are combined for characterizing risk. 2. The risk assessment process, with management input, involves decisions regarding the collection of additional data (versus living with **uncertainty**); in the **risk characterization**, a discussion of the uncertainties will help to identify where additional information could contribute significantly to reducing uncertainties in risk assessment. 3. A clear and explicit statement of the strengths and limitations of a risk assessment requires a clear and explicit statement of related **uncertainties**." **EPA 1996 Proposed Guidelines on Carcinogenic Risk Assessment at 126:** "In part as a response to these recommendations [that EPA consider the limits of scientific knowledge], the Administrator of EPA issued guidelines for risk characterization and required implementation plans from all programs in EPA (EPA 1995b). The Administrator's guidance is followed in these cancer guidelines. The assessments of hazard, dose response, and exposure will all have accompanying technical characterizations covering issues of strengths and limitations of data and current scientific understanding, identification of **defaults** utilized in the face of gaps in the former, discussions of controversial issues, and discussions of **uncertainties** in both their qualitative, and as practicable, their quantitative aspects."	data and credible assumptions. The guidance presents a general framework and broad set of principles to ensure the use of good scientific practices when conducting probabilistic analyses of **variability and uncertainty**. In addition, the guidelines present a new cancer characterization system consisting of five summary descriptors, to be used in conjunction with narrative, to describe the extent to which available data support the conclusion that a contaminant causes cancer in humans and to justify the summary descriptor selected."

continued

TABLE D-1 Continued

Topic	NRC Report: Recommendation[a]	EPA Response: Stated Policy[b]	EPA Response: Implementation Activity[c]
	in a risk assessment be reported in a written risk-assessment document."	**EPA 2000b Risk Characterization Handbook at A-3:** "Key scientific concepts, data and methods (e.g., use of animal or human data for extrapolating from high to low doses, use of pharmacokinetics data, exposure pathways, sampling methods, availability of chemical-specific information, quality of data) should be discussed. To ensure transparency, risk characterizations should include a **statement of confidence** in the assessment that identifies **all major uncertainties along with comment on their influence on the assessment**, consistent with the Guidance on Risk Characterization." (See "Risk Characterization" section above for other relevant policy statements in EPA risk-assessment guidelines and other sources.)	
	NRC 1994 at 12: "EPA should conduct formal **uncertainty** analyses, which can show where additional research might resolve major **uncertainties** and where it might not." **NRC 1994 at 12:** "EPA should consider in its risk assessments the limits of scientific knowledge, the remaining **uncertainties**, and the desire to identify errors of either overestimation or underestimation."	**EPA 1997c Guiding Principles for Monte Carlo Analysis at 1:** "Such probabilistic analysis techniques as Monte Carlo analysis, given adequate supporting data and credible assumptions, can be viable statistical tools for analyzing **variability and uncertainty** in risk assessments and presents an initial set of principles to guide the agency in using probabilistic analysis tools."	
	NRC 1994 at 12: "Despite the advantages of developing consistent risk assessments between agencies by using **common assumptions** (e.g., replacing surface area	**EPA 1996 Proposed Guidelines on Carcinogen Risk Assessment at 125:** "The rationale for adopting the oral scaling factor of body weight to the 0.75 power	EPA did not adopt this recommendation in the 1996 guidelines.

with body weight to the 0.75 power), EPA should indicate other methods, if any, that might be more accurate."	has been discussed above in the explanation of major **defaults**. The empirical basis is further explored in **Federal Register 57(109): 24152** [1992]. The more accurate approach is to use a toxicokinetic model when data become available or to modify the **default** when data are available as encouraged under these guidelines. As the EPA [57 Fed. Reg. 24152 [1992] discussion explores in depth, data on the differences among animals in response to toxic agents are basically consistent with using a power of 1.0, 0.75, or 0.66. The Federal agencies chose the power of 0.75 for the scientific reasons given in the previous discussion of major **defaults**; these were not addressed specifically in the NRC report. It was also considered appropriate, as a matter of policy, for the agencies to agree on one factor. Again, the **default** for inhalation exposure is a model that is constructed to become better as more agent-specific data become available."
NRC 1994 at 12: "When ranking risks, EPA should consider the **uncertainties** in each estimate, rather than ranking solely on the basis of point estimate value. Risk managers should not be given only a single number or range of numbers. Rather, they should be given risk characterizations that are as robust (i.e., complete and accurate) as can be feasibly developed."	**EPA 2004b at 16:** "Since uncertainty and variability are present in risk assessments, EPA usually incorporates a 'high-end' hazard and/or exposure level in order to ensure an adequate margin of safety for most of the potentially exposed, susceptible population, or ecosystem. EPA's high-end levels are around 90% and above—a reasonable approach that is consistent with the NRC discussion (NRC 1994). This policy choice is consistent with EPA's legislative mandates (e.g., adequate margin of safety). Even with a high-end value, there will be exposed people or environments at greater risk and at lower risk. In addition to the high-end values,

continued

TABLE D-1 Continued

Topic	NRC Report: Recommendation[a]	EPA Response: Stated Policy[b]	EPA Response: Implementation Activity[c]
		EPA programs typically estimate central tendency values for risk managers to evaluate. This provides a reasonable sense of the range of risk that usually lies on the actual distribution."	
	NRC 1994 at 242: "The distinction between **uncertainty** and individual variability ought to be maintained rigorously at the level of separate risk-assessment components (e.g., ambient concentration, uptake, and potency) as well as at the level of an integrated risk characterization."	**EPA 2000b Risk Characterization Handbook at 40:** "The risk assessor should strive to distinguish between variability and **uncertainty** to the extent possible (see 3.2.8 for a discussion of **uncertainty**). **Variability** arises from true heterogeneity in characteristics such as dose-response differences within a population, or differences in contaminant levels in the environment. The values of some variables used in an assessment change with time and space, or across the population whose exposure is being estimated. Assessments should address the resulting variability in doses received by members of the target population. Individual exposure, dose, and risk can vary widely in a large population. Central tendency and high end individual risk descriptors capture the variability in exposure, lifestyles, and other factors that lead to a distribution of risk across a population (e.g., see Guidelines for Exposure Assessment)."	
Variability and Differential Susceptibility	**NRC 1994 at 11:** "Federal agencies should sponsor molecular, epidemiological, and other types of research to examine the causes and extent of interindividual **variability** in **susceptibility** to cancer and the possible correlations between	**EPA 1997d Exposure Factors Handbook:** Risk assessors have used the Exposure Factors Handbook to account for **variation** in exposure. The purposes of the handbook are to "(1) summarize data on human behaviors and characteristics which affect exposure to environmental contaminants,	**GAO 2006 at 47:** "Another way EPA addresses **variability** is through research. One of ORD's four strategic research directions in its *Human Health Research Strategy* is designed to improve the understanding of why some people and groups are more **susceptible** and highly exposed than others.

susceptibility and such covariates as age, race, ethnicity, and sex."

and (2) recommend values to use for these factors" (p. 1-1). The document includes over 150 data tables with information on exposure scenarios. It also discusses variability and attempts to characterize the variability of each of the exposure factors "(1) as tables with various percentiles or ranges of values; (2) as analytical distributions with specified parameters; and/or (3) as a qualitative discussion" (p. 1-5). The handbook discusses how risk assessors can identify the types of **variability** and ways that **variability** can be analyzed.

According to this strategy, ORD's research on subpopulations will focus on three factors—life stage, genetic factors, and pre-existing diseases—that have been identified by a program office and the scientific community as having a high priority for risk assessment. In 2000, ORD released its *Strategy for Research on Environmental Risks to Children* to strengthen the scientific foundation of risk assessment and management decisions that affect children and guide EPA's research needs and priorities over the following 5 to 10 years. Approximately 75 percent of the funding for this strategy will be dedicated to research grants under the STAR program, such as those designed to evaluate children's exposure to pesticides."

GAO 2006 at 46: "To further its understanding of **variability** in exposure, EPA has undertaken a number of research projects. For example, one of ORD's laboratories conducted the National Human Activity Pattern Survey to provide detailed human exposure information for specific populations and allow EPA to better understand actual human exposure to pollutants in real-world situations. The survey results are stored in the Consolidated Human Activity Database to help risk assessors estimate the time that exposed people spend in various environments and their inhalation, ingestion, and dermal absorption rates while in those environments. This laboratory also conducts research to define, quantify, and reduce the **uncertainty** associated with the exposure and risk assessments, to develop improved methods to more accurately measure exposure and dose, and to develop technical information

continued

TABLE D-1 Continued

Topic	NRC Report: Recommendation[a]	EPA Response: Stated Policy[b]	EPA Response: Implementation Activity[c]
			and quantitative tools to predict the nature and magnitude of human exposures to environmental contaminants. A recent EPA study was designed to identify chemicals commonly used in homes or day care centers, and whether children in these environments encountered the chemicals in the course of their daily activities. The research sought to identify the major routes (i.e., breathing and ingestion) and sources (i.e., dust, food, air, soil, and water) through which children come into contact with chemicals. "**Variability** also exists with regard to **susceptibility** to adverse affects because of inherent differences among humans."
	NRC 1994 at 11: "EPA should adopt a **default** assumption for differences in **susceptibility** among humans in estimating individual risks."	**EPA 1996 Proposed Guidelines for Carcinogen Risk Assessment at 125:** "The issue of a **default** assumption for human differences in **susceptibility** has been addressed under the major **defaults** discussion in section 1.3 with respect to margin of exposure analysis. The EPA has considered but decided not to adopt a quantitative **default** factor for human differences in **susceptibility** when a linear extrapolation is used. In general, the EPA believes that the linear extrapolation is sufficiently conservative to protect public health. Linear approaches (both LMS and straight line extrapolation) from animal data are consistent with linear extrapolation on the same agents from human data (Goodman and Wilson 1991; Hoel and Portier 1994). If actual data on human **variability** in sensitivity are	

	available they will, of course, be used."	
	EPA 2005a Guidelines for Cancer Risk Assessment at 17802: "The dose-response estimate strives to derive separate estimates for susceptible populations and lifestages so that these risks can be explicitly characterized. For a susceptible population, higher risks can be expected from exposures anytime during life, but this applies to only a portion of the general population. . . . In contrast, for a susceptible lifestage, higher risks can be expected from exposures during only a portion of the lifetime, but everyone in the population may pass through those lifestages."	
	Also at 17811: "As a default for oral exposure, a human equivalent dose for adults is estimated from data on another species by an adjustment of animal applied oral dose by a scaling factor based on body weight to the ¾ power. The same factor is used for children because it is slightly more protective than using children's body weight (see sec. 3.1.3)."	
NRC 1994 at 11: "The distinction between **uncertainty** and individual **variability** should be maintained rigorously in each component of risk assessment."	**EPA 2000b Risk Characterization Handbook at 40:** "The risk assessor should strive to distinguish between **variability** and **uncertainty** to the extent possible." **EPA 2000b Risk Characterization Handbook at 40:** "The risk assessor should strive to distinguish between **variability** and **uncertainty** to the extent possible (see 3.2.8 for a discussion of **uncertainty**). **Variability** arises from true heterogeneity in	**GAO 2006 at 45:** "All program offices address exposure **variability** in their risk assessments, although they do so in different ways. For example, risk assessors in the Office of Air Quality Planning and Standards who set certain air quality standards for six principal pollutants said they consider individual activity patterns for sensitive populations like children or asthmatics in exposure modeling by including a distribution of breathing rates to reflect **variability**

continued

TABLE D-1 Continued

Topic	NRC Report: Recommendation[a]	EPA Response: Stated Policy[b]	EPA Response: Implementation Activity[c]
		characteristics such as dose-response differences within a population, or differences in contaminant levels in the environment. The values of some variables used in an assessment change with time and space, or across the population whose exposure is being estimated. Assessments should address the resulting **variability** in doses received by members of the target population. Individual exposure, dose, and risk can vary widely in a large population. Central tendency and high end individual risk descriptors capture the **variability** in exposure, lifestyles, and other factors that lead to a distribution of risk across a population (e.g., see Guidelines for Exposure Assessment)." **EPA 2003b Framework for Cumulative Risk Assessment at 65:** "NRC (1994) notes a clear difference between **uncertainty and variability** and recommends that the distinction between these two be maintained: A distinction between **uncertainty** (i.e., degree of potential error) and **interindividual variability** (i.e., population heterogeneity) is generally required if the resulting quantitative risk characterization is to be optimally useful for regulatory purposes, particularly insofar as risk characterizations are treated quantitatively. The distinction between **uncertainty and individual variability** ought to be maintained rigorously at the level of separate risk assessment components (e.g., ambient concentration, uptake, and potency) as well as at the level of an integrated risk characterization."	inherent in the population. Furthermore, by modeling to protect the most sensitive or at-risk groups, they are assured of protecting the rest of the population. **Variability** in exposure to the six principal pollutants is generally described qualitatively in scientific summaries for each pollutant. The Office of Water includes an analysis of risks to various subpopulations and a narrative discussion of the strengths and weaknesses of the studies it used to estimate exposure, but generally does not include a quantitative analysis. The Office of Pesticide Programs considers 24 different population subgroups in its exposure estimates, including differences in age, gender, ethnicity, and geographic dispersion. When data allow, the Office of Pesticide Programs develops a distribution of exposures and risks for its more refined risk assessments."

NRC 1994 at 220: "If there is reason to believe that risk of adverse biological effects per unit dose depends on **age**, EPA should present separate risk estimates for adults and children. When excess lifetime risk is the desired measure, EPA should compute an integrated lifetime risk, taking into account all relevant age-dependent variables. "EPA does not usually explore or consider interindividual **variability** in key biologic parameters when it uses or evaluates various physiologic or biologically based risk-assessment models (or else evaluates some data but does not report on this in its final public documents). In some other cases, EPA does gather or review data that bear on human **variability**, but tends to accept them at face value without ensuring that they are representative of the entire population. As a general rule, the larger the number of characteristics with an important effect on risk or the more variable those characteristics are, the larger the sample of the human population needed to establish confidently the mean and range of each of those characteristics."	**EPA 2005b Supplemental Guidance for Assessing Susceptibility from Early-Life Exposure to Carcinogens at 1:** "The National Research Council (NRC, 1994) recommended that 'EPA should assess risks to **infants and children** whenever it appears that their risks might be greater than those of adults.' This document focuses on cancer risks from early-life exposure compared with those from exposures occurring later in life. Evaluating childhood cancer and childhood exposures resulting in cancer later in life are related, but separable, issues." **EPA 2004b at 42:** "Consideration of the variability among humans is a critical aspect of risk assessment. It is the goal of EPA risk assessments to identify all potentially affected populations, including human populations (e.g., **gender, nutritional status, genetic predisposition**) and **life-stages** (e.g., childhood, pregnancy, old age) that may be more susceptible to toxic effects or are highly or disproportionately exposed." **Also at 43:** "When data are available to describe toxicological differences for a **susceptible population or life-stage**, then those data are summarized and analyzed, and the decisions based on this information are presented. It is preferable to have population- and chemical-specific data to describe a susceptibility to toxic effects."	**GAO 2006 at 46:** "Legislation can also require EPA to consider potentially susceptible populations and life stages. For example, the Safe Drinking Water Act Amendments mandate that EPA consider risks to groups within the general population that are at greater risk of adverse health effects, including children, the elderly, and people with serious illnesses. In addition, the Food Quality Protection Act contains special provisions for the consideration of risks to children from pesticides. In 1995, EPA's Science Policy Council called for EPA to consider the risks to infants and children consistently and explicitly as part of its risk assessments. In 1997, the White House issued an executive order that required EPA and other federal agencies to identify and assess environmental health and safety risks that may disproportionately affect children and to ensure that policies, programs, activities, and standards address such disproportionate risks."

[a]Example of recommendation from NRC 1983, 1994, or 1996.
[b]Example of EPA policy bearing on issues raised in the recommendation in the form of written guidelines, reports, or policy memoranda.
[c]Commentary, practice, or activities related to issues raised in the National Research Council recommendation and related EPA guidance.
[d]These guidelines were not specifically in response to the National Research Council report but reflect agency policy related to this topic.

REFERENCES

EPA (U.S. Environmental Protection Agency). 1984. Risk Assessment and Management: Framework for Decision Making. EPA 600/9-85-002. Office of the Administrator, U.S. Environmental Protection Agency, Washington, DC.

EPA (U.S. Environmental Protection Agency). 1986. Guidelines for Carcinogen Risk Assessment. EPA/630/R-00/004. Risk Assessment Forum, U.S. Environmental Protection Agency, Washington, DC [online]. Available: http://cfpub.epa.gov/ncea/cfm/recordisplay.cfm?deid=54933 [accessed June 3, 2007].

EPA (U.S. Environmental Protection Agency). 1989a. Exposure Factors Handbook. EPA/600/8-89/043. NTIS PB90-106774/AS. Office of Health and Environmental Assessment, Office of Research and Development, U.S. Environmental Protection Agency, Washington, DC.

EPA (U.S. Environmental Protection Agency). 1989b. Risk Assessment Guidance for Superfund: Volume I—Human Health Evaluation Manual (Part A). Interim Final. EPA-540/1-89/002. Office of Emergency and Remedial Response, U.S. Environmental Protection Agency, Washington, DC [online]. Available: http://www.epa.gov/oswer/riskassessment/ragsa/pdf/rags-vol1-pta_complete.pdf [accessed Oct. 16, 2007].

EPA (U.S. Environmental Protection Agency). 1992a. Guidelines for Exposure Assessment. EPA/600/Z-92/001. Risk Assessment Forum, U.S. Environmental Protection Agency, Washington, DC. May 1992 [online]. Available: http://cfpub.epa.gov/ncea/cfm/recordisplay.cfm?deid=15263 [accessed Oct. 10, 2007].

EPA (U.S. Environmental Protection Agency). 1992b. Guidance on Risk Characterization for Risk Managers and Risk Assessors. Memorandum to Assistant Administrators, and Regional Administrators, from F. Henry Habicht, Deputy Administrator, Office of the Administrator, Washington, DC. February 26, 1992 [online]. Available: http://www.epa.gov/oswer/riskassessment/pdf/habicht.pdf [accessed Oct. 10, 2007].

EPA (U.S. Environmental Protection Agency). 1994a. Model Validation for Predictive Exposure Assessments. U.S. Environmental Protection Agency, Washington, DC. July 4, 1994 [online]. Available: http://www.epa.gov/ord/crem/library/whitepaper_1994.pdf [accessed Oct.15, 2007].

EPA (U.S. Environmental Protection Agency). 1994b. Report on the Workshop on Cancer Risk Assessment Guidelines Issues. EPA/630/R-94/005a. Office of Research and Development, Risk Assessment Forum, Washington, DC.

EPA (U.S. Environmental Protection Agency). 1994c. Peer Review and Peer Involvement at the U.S. Environmental Protection Agency. Memorandum to Assistant Administrators, General Counsel, Inspector General, Associate Administrators, Regional Administrators, and Staff Office Directors, from Carol M. Browner, Administrator, U.S. Environmental Protection Agency. June 7, 1994 [online]. Available: http://www.epa.gov/osa/spc/pdfs/perevmem.pdf [accessed Oct. 16, 2007].

EPA (U.S. Environmental Protection Agency). 1995a. Policy for Risk Characterization at the U.S. Environmental Protection Agency. Memorandum from Carol M. Browner, Office of the Administrator, U.S. Environmental Protection Agency, Washington, DC. March 21, 1995 [online]. Available: http://64.2.134.196/committees/aqph/rcpolicy.pdf [accessed Oct. 10, 2007].

EPA (U.S. Environmental Protection Agency). 1995b. Guidance for Risk Characterization. Science Policy Council, U.S. Environmental Protection Agency. February 1995 [online]. Available: http://www.epa.gov/osa/spc/pdfs/rcguide.pdf [accessed Oct. 15, 2007].

EPA (U.S. Environmental Protection Agency). 1996. Proposed Guidelines for Carcinogen Risk Assessment. EPA/600/P-92/003C. Office of Research and Development, U.S. Environmental Protection Agency, Washington, DC. April 1996 [online]. Available: http://www.epa.gov/ncea/raf/pdfs/propcra_1996.pdf [accessed Oct. 15, 2007].

EPA (U.S. Environmental Protection Agency). 1997a. Cumulative Risk Assessment Guidance—Phase I Planning and Scoping. Memorandum to Assistant Administrators, General Counsel, Inspector General, Associate Administrators, Regional Administrators, and Staff Office Directors, from Carol M. Browner, Administrator, and Fred Hansen, Deputy Administrator, Office of Administrator, U.S. Environmental Protection Agency, Washington, DC. July 3, 1997 [online]. Available: http://www.epa.gov/swerosps/bf/html-doc/cumulrsk.htm [accessed Oct. 15, 2007].

EPA (U.S. Environmental Protection Agency). 1997b. Guidance on Cumulative Risk Assessment. Part 1. Planning and Scoping. Science Policy Council, U.S. Environmental Protection Agency, Washington, DC. July 3, 1997 [online]. Available: http://www.epa.gov/osa/spc/pdfs/cumrisk2.pdf [accessed Oct. 10, 2007].

EPA (U.S. Environmental Protection Agency). 1997c. Guiding Principles for Monte Carlo Analysis. EPA/630/R-97/001. Risk Assessment Forum, U.S. Environmental Protection Agency, Washington, DC [online]. Available: http://www.epa.gov/NCEA/pdfs/montcarl.pdf [accessed June 3, 2007].

EPA (U.S. Environmental Protection Agency). 1997d. Exposure Factors Handbook, Vol. 1. General Factors. EPA/600/P-95/002F. Office of Research and Development, National Center for Environmental Assessment, U.S. Environmental Protection Agency, Washington, DC [online]. Available: http://www.epa.gov/ncea/efh/ [accessed June 3, 2007].

EPA (U.S. Environmental Protection Agency). 1998. Guidelines for Ecological Risk Assessment. EPA/630/R-95/002F. Risk Assessment Forum, U.S. Environmental Protection Agency, Washington, DC [online]. Available: http://cfpub.epa.gov/ncea/cfm/recordisplay.cfm?deid=12460 [accessed June 3, 2007].

EPA (U.S. Environmental Protection Agency). 2000a. Supplementary Guidance for Conducting Health Risk Assessment of Chemical Mixtures. EPA/630/R-00/002. Risk Assessment Forum, U.S. Environmental Protection Agency, Washington, DC. August 2000 [online]. Available: http://www.epa.gov/ncea/raf/pdfs/chem_mix/chem_mix_08_2001.pdf [accessed Oct. 15, 2007].

EPA (U.S. Environmental Protection Agency). 2000b. Risk Characterization: Science Policy Council Handbook. EPA 100-B-00-002. Office of Science Policy, Office of Research and Development, U.S. Environmental Protection Agency, Washington, DC [online]. Available: http://www.epa.gov/OSA/spc/pdfs/rchandbk.pdf. [accessed June 3, 2007].

EPA (U.S. Environmental Protection Agency). 2000c. Benchmark Dose Technical Guidance Document. EPA/630/R-00/001. Risk Assessment Forum, U.S. Environmental Protection Agency, Washington, DC [online]. Available: http://www.epa.gov/ncea/pdfs/bmds/BMD-External_10_13_2000.pdf [accessed June 3, 2007].

EPA (U.S. Environmental Protection Agency). 2000d. Per Review Handbook., 2nd Ed. EPA 100-B-00-001. Science Policy Council, Office of Science Policy, Office of Research and Development, U.S. Environmental Protection Agency, Washington, DC. December 2000 [online]. Available: http://www.epa.gov/osa/spc/pdfs/prhandbk.pdf [accessed Oct. 16, 2007].

EPA (U.S. Environmental Protection Agency). 2000e. EPA Quality Manual for Environmental Programs. 5360A1. Office of Environmental Information, U.S. Environmental Protection Agency, Washington, DC. May 5, 2000 [online]. Available: http://www.epa.gov/quality/qs-docs/5360.pdf [accessed Oct. 15, 2007].

EPA (U.S. Environmental Protection Agency). 2001. General Principles for Performing Aggregate Exposure and Risk Assessments. Office of Pesticide Programs, U.S. Environmental Protection Agency, Washington, DC. November 28, 2001 [online]. Available: http://www.epa.gov/pesticides/trac/science/aggregate.pdf [accessed Oct. 15, 2001].

EPA (U.S. Environmental Protection Agency). 2002a. Guidance on Cumulative Risk Assessment of Pesticide Chemicals That Have a Common Mechanism of Toxicity. Office of Pesticide Programs, U.S. Environmental Protection Agency, Washington, DC. January 14, 2002 [online]. Available: http://www.epa.gov/pesticides/trac/science/cumulative_guidance.pdf [accessed Oct. 16, 2007].

EPA (U.S. Environmental Protection Agency). 2002b. Lessons Learned on Planning and Scoping for Environmental Risk Assessments. Prepared by the Planning and Scoping Workgroup of the Science Policy Council Steering Committee, U.S. Environmental Protection Agency, Washington, DC. January 2002 [online]. Available: http://www.epa.gov/OSA/spc/pdfs/handbook.pdf [accessed Oct. 16, 2007].

EPA (U.S. Environmental Protection Agency). 2002c. Guidelines for Ensuring and Maximizing the Quality, Objectivity, Utility and Integrity of Information Disseminated by the Environmental Protection Agency. EPA/260R-02-008. Office of Environmental Information, U.S. Environmental Protection Agency, Washington, DC [online]. Available: http://www.epa.gov/QUALITY/informationguidelines/documents/EPA_InfoQualityGuidelines.pdf [accessed Oct. 10, 2007].

EPA (U.S. Environmental Protection Agency). 2003a. Human Health Research Strategy. EPA/600/R-02/050. Office of Research and Development, U.S. Environmental Protection Agency, Washington, DC [online]. Available: http://www.epa.gov/nheerl/humanhealth/HHRS_final_web.pdf [accessed June 3, 2007].

EPA (U.S. Environmental Protection Agency). 2003b. Framework for Cumulative Risk Assessment. EPA/630/P-02/001F. Risk Assessment Forum, U.S. Environmental Protection Agency, Washington, DC [online]. Available: http://cfpub.epa.gov/ncea/cfm/recordisplay.cfm?deid=54944 [accessed June 3, 2007].

EPA (U.S. Environmental Protection Agency). 2003c. A Summary of General Assessment Factors for Evaluating the Quality of Scientific and Technical Information. EPA 100/B-03/001. Science Policy Council, U.S. Environmental Protection Agency, Washington, DC. June 2003 [online]. Available: http://www.epa.gov/osa/spc/pdfs/assess2.pdf [accessed Oct. 12, 2007].

EPA (U.S. Environmental Protection Agency). 2004a. Air Toxics Risk Assessment Reference Library, Vol. 1-Technical Resource Manual, Part III: Human Health Risk Assessment: Multipathway. EPA-453-K-04-001. Office of Air Quality Planning and Standards, U.S. Environmental Protection Agency, Research Triangle Park, NC [online]. Available: http://www.epa.gov/ttn/fera/risk_atra_vol1.html#part_iii [accessed Oct. 12, 2007].

EPA (U.S. Environmental Protection Agency). 2004b. An Examination of EPA Risk Assessment Principles and Practices. EPA/100/B-04/001. Office of the Science Advisor, U.S. Environmental Protection Agency, Washington, DC [online]. Available: http://www.epa.gov/OSA/pdfs/ratf-final.pdf [accessed June 3, 2007].

EPA (U.S. Environmental Protection Agency). 2005a. Guidelines for Carcinogen Risk Assessment (Final). EPA/630/P-03/001F. Risk Assessment Forum, U.S. Environmental Protection Agency, Washington, DC. March 2005 [online]. Available: http://cfpub.epa.gov/ncea/cfm/recordisplay.cfm?deid=116283 [accessed Oct. 16, 2007].

EPA (U.S. Environmental Protection Agency). 2005b. Supplemental Guidance for Assessing Susceptibility from Early-Life Exposure to Carcinogens. EPA/630/R-03/003F. Risk Assessment Forum, U.S. Environmental Protection Agency, Washington, DC [online]. Available: http://www.epa.gov/iris/children032505.pdf [accessed Oct. 19, 2007].

GAO (U.S. Government Accountability Office). 2006. Human Health Risk Assessment. GAO-06-595. Washington, DC: U.S. Government Printing Office [online]. Available: http://www.gao.gov/new.items/d06595.pdf [accessed Oct. 11, 2007].

Gilman, P. 2003. Statement of Paul Gilman, Assistant Administrator for Research and Development and EPA Science Advisor, U.S. Environmental Protection Agency, before the Committee on Transportation and Infrastructure, Subcommittee on Water Resources and the Environment, U.S. House of Representatives, March 5, 2003 [online]. Available: http://www.epa.gov/ocir/hearings/testimony/108_2003_2004/2003_0305_pg.pdf [accessed Feb. 9, 2007].

Goodman, G., and R. Wilson. 1991. Predicting the carcinogenicity of chemicals in humans from rodent bioassay data. Environ. Health Perspect. 94:195-218.

Hoel, D.G., and C.J. Portier. 1994. Nonlinearity of dose-response functions for carcinogenicity. Environ. Health Perspect. 102(Suppl 1):109-113.

NRC (National Research Council). 1983. Risk Assessment in the Federal Government: Managing the Process. Washington, DC: National Academy Press.

NRC (National Research Council). 1994. Science and Judgment in Risk Assessment. Washington, DC: National Academy Press.

NRC (National Research Council). 1996. Understanding Risk: Informing Decisions in a Democratic Society. Washington, DC: National Academy Press.

Appendix E

Environmental Protection Agency Program and Region Responses to Questions from the Committee

In January 2007 the NRC committee sent EPA a list of questions (see below) to gather additional information on their risk assessment practices. EPA responses were provided by the Office of Air and Radiation (OAR); Office of Prevention, Pesticides, and Toxic Substances (OPPTS), Region 2; and the Office of Solid Waste and Emergency Response (OSWER); and the Office of Water (OW). The EPA responses do not represent the views of the committee on these issues.

QUESTIONS FOR EPA FROM THE NRC COMMITTEE

Give an example of a risk assessment from your office that you would consider an example of "best practice," and an example of a risk assessment that you think could have been improved (and if so, how).

What improvement in EPA risk assessment practices would you find particularly helpful in the short term (2-5 years) and in the longer term (10-20 years)? If these improvements were to be implemented, how do you foresee the changes impacting your office?

Please describe the risk assessment paradigm(s) used by your office. Do these paradigms adequately address environmental problems faced by the country? If not, how might current paradigms be modified or new paradigms identified to address these problems?

Describe problems that arise when using risk assessment to support regulatory decision making. Do you encounter similar problems when using risk assessment in non-regulatory decisions? Please provide specific examples to illustrate your points.

How would you recommend improving the presentation of EPA risk assessments for decision-making?

How have you addressed and communicated uncertainty in risk assessments?

Please discuss the adequacy of default assumption choices, and efforts to use alternatives to these default assumptions.

Please describe the ways in which children and potentially unique or vulnerable populations are specifically considered in your office's risk assessments. Please provide examples.

AGENCY RESPONSES TO QUESTIONS

OFFICE OF AIR AND RADIATION (OAR)

Current Practice

- Statutory basis/current approach and paradigms for risk assessment (specific to each program office)
  - Examples and best practices
  - Gaps and problems
- Uncertainty analysis
  - Examples
  - Communication of risk and uncertainty
- Sensitive and vulnerable subpopulations (e.g., children, elderly, tribes, endangered species)
  - Examples of physical attributes and unique exposures that impact risk
  - Problems and challenges
- Challenges for risk assessment in a regulatory process
  - Examples
  - Problems and challenges

General Comment

The 2004 Agency document "An Examination of EPA Risk Assessment Principles and Practices" (EPA 2004a) provides a good resource for understanding the Agency as well as OAR's approach to risk assessment. Consistent to the focus of the NAS committee charge this response does not address ecological risk assessment. Protection of ecosystems from adverse impacts from of air pollution is an important mission of our Office and we could provide additional information in this area if requested.

There are two programs within OAR that best illustrate the use of risk assessment in our Office. First, are assessment activities that support the development of national ambient air quality standards (NAAQS) for the 6 "criteria" air pollutants, and, second, those conducted in consideration of emissions controls for hazardous air pollutants (HAPs or air toxics).

National Ambient Air Quality Standards (NAAQS)

The "criteria" air pollutants are the six pollutants—ozone, particulate matter, carbon monoxide, nitrogen dioxide, sulfur dioxide and lead—the presence of which in the ambient air results from numerous or diverse sources, and for which there are established public health concerns at historic ambient levels. These pollutants have been extensively studied over time and health-based National Ambient Air Quality Standards (NAAQS) have been developed for each. Human exposure and/or health risk assessments and ecological risk assessments are performed during the periodic reviews of these standards.

The process under which exposure and/or risk assessments are performed for the criteria pollutants is largely driven by statutory language and legislative history and involves substantial external peer and public review. Each NAAQS review includes a full review of the underlying scientific database which supports the quantitative exposure and/or risk assessments (for an example, see the Air Quality Criteria for Ozone and Other Photochemical Oxidants [EPA 2008a]). The health-effects databases for criteria pollutants are generally very rich and include: epidemiological studies of normal exposures to the ambient mix of air pollutants, controlled-human exposure studies, and animal studies (short- and long-term exposures). Risk assessments for criteria air pollutants also benefit from extensive exposure related information including monitoring data and well developed exposure models.

Hazard characterization involves a weight-of-evidence approach, using all relevant information and considering the nature and severity of effects, patterns of human exposure, nature and size of sensitive populations, the kind and degree of uncertainties, and the consistency or coherence across all types of available evidence. "Dose"-response evaluations are based on the nature of available evidence from human studies, generally with no discernable thresholds (effects observed at current ambient concentrations). For example, for PM, ambient concentration-response functions are employed, for ozone, exposure-response and concentration-response relationships are used and for CO and lead, internal dose-metrics are used. When ambient concentration-response functions are used, simulations of "just meeting" alternative standards are used to examine levels of risk. When exposure or internal dose-response metrics are used, exposure modeling is relied upon that includes air quality monitoring/modeling and simulations of "just meeting" alternative standards, pollutant concentrations within relevant microenvironments (home, yard, car, office), amount of time in different microenvironments and level of exertion (time-activity and breathing rate data), population demographics (census data, commuting patterns), probabilistic assessment (including uncertainty and variability), and sensitivity analyses. This modeling provides the ability to identify, and characterize exposure distributions for sensitive and/or at risk groups.

Risk characterization for criteria pollutants includes both qualitative and quantitative approaches. There is an integration of evidence on acute and chronic health effects (strengths, weaknesses, uncertainties). Expert judgments are made on adversity of effects (severity, duration, frequency). There are qualitative and quantitative assessments of population exposures of concern and/or risks to public health. The risk characterizations are primarily based on available evidence from human studies and "real-world" air quality and exposure analyses; no need for traditional "uncertainty" or "safety" factors.

Risk assessments and characterizations for criteria pollutants, while considering the general population, include focus on the susceptible and/or the more highly exposed subpopulations (e.g., asthmatics and children are groups focused on in the current ozone NAAQS review). However, exposures and risks do not focus on maximum exposed individuals or maximum individual risk given the legislative history indicating that standards are to protect most of the sensitive population group but not the most sensitive individual.

Uncertainty in criteria pollutant risk assessments is routinely addressed using probabilistic assessment (including uncertainty and variability) and sensitivity analyses. For an example of the type of exposure and risk assessments conducted for the NAAQS reviews see the final OAQPS Staff Paper for Ozone (EPA 2008b) and the human exposure, health risk assessment, and exposure, risk and impacts assessment for vegetation technical support documents (EPA 2008c).

Risk assessments for criteria pollutants generally include quantitative sensitivity analyses of exposure and health risk estimates as mentioned above, and also include qualitative discussion of contributing uncertainties.

Key Issues and Challenges

Key issues and challenges in carrying out quantitative risk assessments for criteria pollutants have included: (1) how to appropriately reflect and characterize model uncertainty, especially with respect to the shape and location of concentration-response relationships for which epidemiological studies are often failing to discern population thresholds, even at ambient levels approaching background levels; and (2) how to appropriately address and consider multi-pollutant health effect models and to disentangle the likely interaction among air pollutants, many of which are correlated and come from common sources (e.g., combustion of fossil fuels) in causing various health effects.

In the area of exposure analysis, these challenges include how to use the human activity data base which consists of over 20,000 individual daily diaries to construct human activity sequences over months or an entire year. There is very little longitudinal data, so it is difficult to know if we are appropriately taking into account the repeated activities that individuals engage in. There also are few exposure field studies that include representative population sampling that would allow evaluation of the regulatory exposure models used by EPA in its NAAQS assessments. In addition, there are challenges in determining how "just meeting" hourly or daily standards will affect the overall distribution of pollutant concentrations across all hours and days. For non-threshold pollutants, the choice of method used in simulating attainment can have potentially large impacts on the estimated risks.

Hazardous Air Pollutants

The hazardous air pollutants (HAPs or "air toxics") are 187 substances listed in CAA (e.g., benzene, methylene chloride, cadmium compounds, etc.) which have been associated with, or for which data suggest, the potential for serious adverse health and/or environmental effects, and for which there are specific source-based statutory requirements. Although several HAPs have substantial health and/or ecological effects data bases, most others have very limited data, much of it based solely on knowledge of health effects on exposed animals rather than humans. HAPs are regulated through source-oriented technology and risk-based emissions standards.

HAP risk assessments are performed for consideration of risk-based emissions standards (residual risk standards) for source categories for which technology-based controls have already been applied (a good example of which may be found in the docket supporting the proposed residual risk rule for the source category called "Halogenated Solvent Cleaners" (look in ICF International 2006). Rather than focusing on the risks associated with exposure to an individual chemical, these risk assessments commonly examine cumulative risks associated with exposures resulting from the combination of pollutants emitted by a particular type of industry. By statutory language and regulatory history, these risk assessments include both a maximum individual risk (i.e., presuming an individual were exposed to the maximum level of a pollutant for a lifetime), as well as a characterization of a representative population risk.

HAP risk assessments may also be performed for other programmatic purposes. For example, national-scale assessments have been performed based on the 1996 and 1999 emissions inventories as part of the National Air Toxics Assessment (NATA) activities (EPA

2002a, 2003a). As another example, risk assessments may be performed to support decisions on petitions to list or delist individual HAPs or source categories from Clean Air Act regulatory consideration.

The scope of HAP risk assessments varies with the characteristics of the pollutants and sources being assessed. Inhalation and, as appropriate, other routes of exposure are assessed, and both chronic and acute time scales are considered. Ecological risks are also considered for residual risk decision-making. Routinely, a tiered approach is employed for efficiency, with lower tiers using simpler, more conservative tools and assumptions to identify important sources and pollutants, and higher tiers using more refined tools and site-specific data to determine where emission controls may be appropriate. Lower-tier risk assessments generally support decisions not to regulate or assist decisions to focus resources on a small number of stressors and sources for next iteration. They alone generally do not support decisions to mandate additional control of emissions. Such decisions, which can have significant economic implications, usually require more refined assessment.

Hazard and dose-response assessments for HAPs generally rely on the most current existing assessments that have undergone scientific peer review and public review. The dose-response metrics used are acute or chronic reference concentrations (RfCs), and cancer inhalation unit risk (IUR) estimates. The sources for these values include U.S. EPA (e.g., IRIS), U.S. ATSDR, California EPA, etc. The common qualities across the sources employed are: development under a defined scientific process, use of independent external peer review, and a reflection of the state of knowledge at the time of the assessment.

Risk assessments for HAPs routinely include, as a first step, derivation of risk estimates for conservative exposure scenarios (e.g., continuous lifetime exposure). Where this first step suggests risks in a range of potential concern, more refined assessments which utilize more of the available data are performed. The most refined assessments attempt to provide a probabilistic distribution of risk (including uncertainty and variability) and sensitivity analyses. The use of probabilistic assessments is currently limited to certain exposure assessment variables (i.e., those describing daily activity and long-term migration behaviors), and does not typically include variables describing emission rates, release conditions, meteorology, fate and transport, or dose-response.

Consideration of the most exposed receptors (individuals) is accomplished by estimating chronic exposures at the Census block level and acute exposures at the offsite location with the highest 1-hour concentration. OAR in its HAPs assessment is a user of Hazard/Dose response information (e.g., such as that produced under the IRIS program). Thus, consideration of sensitive subpopulations is considered in so far as it is explicitly built into the dose-response metrics that EPA uses to estimate risk (i.e., where data supporting such distinctions are available). Unit risk estimates typically incorporate protective low-dose extrapolation assumptions and are based on statistical upper confidence limits. Reference concentrations employ uncertainty factors that account for differences among species, within human populations, and database deficiencies (e.g., failure to identify no-effect doses and absence of chronic studies). These uncertainty factors are intended to ensure that the reference concentration represents an exposure that is likely to be without appreciable risk of adverse effects in human populations, including sensitive sub-populations.

Risk assessments for HAPs may include quantitative sensitivity analyses of exposure as mentioned above, and also include qualitative discussion of contributing uncertainties. However, the dose response information provided in IRIS (or other sources of dose response information) typically does not have information suitable quantitative analysis of either uncertainty or variability.

Key Issues and Challenges

Key issues and challenges in carrying out risk assessments for hazardous air pollutants include both lack of data and how to appropriately reflect and characterize uncertainty and variability in assessments.

As described above, risk assessments for the HAP program decisions routinely address multiple pollutant exposure and risk for multiple similar sources. Limitations associated with current assessments may contribute to uncertainties in resultant risk estimates. Examples of these are listed below as areas where improvements in risk assessment methods, tools or inputs might lead to reduced uncertainty in risk estimates.

- As described above, the single greatest challenge in risk analysis for most hazardous air pollutants is the need to rely primarily on animal or limited human data for the development of hazard and dose response assessments. The interpretation and implications of such data for potential risk is typically one of the greatest sources of uncertainty in such assessments.
- One of the significant sources of uncertainty to risk assessments is the source characterization, including emissions estimates. This is particularly true for source categories that have large numbers of sources and where "representative" data may not exist. For modeling purposes, source data should include site-specific release parameter/characterization information as well as better source emission estimates. For example, such parameters include map coordinates, release heights and temperatures, emissions data measured or estimated (and approved) directly by the facilities, annual and maximum hourly emission rates, and quantitative estimates of the uncertainties associated with each.
- We are limited in methods to consider the effects on source-specific exposure of longer-term population mobility. While such data on migration behavior on a local scale are available, they have not been developed into tools or analyses that are readily applicable to our risk assessment methods.
- Atmospheric deposition data, which would contribute to improved/enhanced assessment of non-inhalation exposures and risk, are limited.
- Methods for estimating and presenting uncertainty in a manner easily understood by decision makers are limited.
- Use of the Agency's traditional exposure-response assessments (e.g., cancer unit risk factors and RfCs) contribute to our limitations with regard to incorporating quantitative uncertainty and variability of response into risk estimates.
- Limitations with regard to spatial coverage of air toxics monitoring networks affect performance evaluation capabilities for local-scale air modeling used in HAP risk assessments.
- Our ability to evaluate mixtures and potential interactions (other than that provided under EPA's current mixtures guidance) is limited.
- Because of the number of hazardous air pollutants emitted from the many sources considered and the time required for updating the hazard and dose-response assessments, the development of those updated assessments can not kept up with the need to make regulatory decisions. Thus, OAR is often confronted with making such decisions with out the benefit of final IRIS assessments.

Future Directions: Addressing Gap, Limitations, and Needs

Both the Criteria and Hazardous air pollutant program operate under the risk assessment paradigm developed by the NRC in its 1983 "Red Book" report. The overall approach to risk assessment in the Hazardous Air pollutant program has also been guided by the 1994 NRC report, "Science and Judgment," which, for example, outlined a tiered approach to the assessment of risk from toxics air emissions from affected sources. We believe the basic paradigm for risk assessment remains sound.

In developing recommendations for improvements, we ask that the Committee consider that the agency must operate within mandated timeframes and growing resource constraints. Thus, any guidance on prioritization of recommendations or on those circumstances where potentially more resource intensive approaches are suggested, would be useful.

The "key issues and challenges" discussions in Part I of this submission (for both the NAAQS process and hazardous air pollutants) provide useful insight into areas where the Committee might focus in looking at future directions and needs. In addition to those points we would add the following few comments:

> The issue of needed data and tools for improving NAAQS assessments are to some extent addressed in the NAAQS review process. Of particular note is the role played by our external scientific review group, the Clean Air Scientific Advisory Committee (CASAC), that explicitly identifies policy-relevant research needs to improve our capabilities for the next cycle of review. This has led to a continuous improvement in our assessment capabilities.

Within the NAAQS program the application of additional methods for uncertainty analysis (e.g. expert elicitation) has particular promise in this program. However, the Agency is still in an early stage of considering how best to incorporate such approaches into its assessments, where appropriate, and how to consider such assessments relative to data driven assessments. Whatever approaches are adopted to characterize uncertainties, it is important to communicate how much weight to accord across the distribution of exposure and/or risk estimates, and not simply provide lower and upper uncertainty bounds.

OFFICE OF PREVENTION, PESTICIDES AND TOXIC SUBSTANCES (OPPTS)

Current Practice: Risk Assessment at the EPA

Statutory Basis/Current Approach and Paradigms for Risk Assessment (Specific ro Each Program Office)

A response to this question can be found at our websites (EPA 2008d,e) along with current practices and recommendations to improve risk assessment (EPA 2002b, 2007a, 2008f).

Very briefly, as an example, the passage of the 1996 Food Quality Protection Act requires that EPA consider, among other things, the best available data and information on the following: aggregate exposure to the pesticide (including exposure from food, water, and residential pesticide uses to a single pesticide), cumulative effects from other pesticides sharing a common mechanism of toxicity (including exposure from food, water, and residential pesticide uses to a multiple pesticides), whether there is an increased susceptibility from exposure to the pesticide to infants and children, and whether the pesticide produces an effect in humans similar to an effect produced by a naturally occurring estrogen, or other endocrine effects.

Like other EPA offices, OPPTS relies on the basic 4 component NAS paradigm from the

Red Book/Science and Judgment) (NRC 1983, 1994) in assessing aggregate and cumulative risks (hazard, dose response, exposure assessment and risk characterization). OPPTS follows EPA approaches for risk assessment described in Agency risk assessment guidelines. In order to reduce the application of default assumptions and default uncertainty/extrapolation factors, in the areas of animal to human extrapolation and high to low dose extrapolation, OPPTS has used physiologically based pharmacokinetic (PBPK) models, data-derived uncertainty factors, and mode of action data, and human biomonitoring data in their risk assessments. OPPTS has been a leader in developing and implementing newer and sophisticated approaches and tools such as probabilistic methods for assessing exposures in food, water, and from residential pathways. Key examples of the implementation of all of these approaches include the Organophosphate Pesticide (OP) and N-methyl carbamate cumulative risk assessments (EPA 2002c, 2007b), PFOA draft risk assessment (EPA 2005a), and draft lead risk assessment (EPA 2007c).

It should be noted that not all assessments need to be of the same depth and scope. We use an iterative and tiered process that considers exposure and sensitivity analyses to balance resources against the need to refine the assessment and reduce uncertainty where appropriate.

Uncertainty Analysis

OPPTS uses sensitivity analyses in the exposure component of risk assessments, particularly in those assessments that inform or support potentially consequential actions (e.g., pesticides and major industrial compounds). As noted below, OPPTS is working closely with ORD to develop more advanced methods of quantitative uncertainty analysis (e.g., 2-dimensional Monte Carlo). For example, OPPTS and ORD are planning to discuss science issues surrounding the implementation of 2-dimensional Monte Carlo into ORD's SHEDs model (Stochastic Human Exposure and Dose Simulation Model) with the FIFRA Science Advisory Panel in 2007. Current methods for the hazard component provide some quantitative measure of experimental data variability. For example, in the cumulative risk assessments for the OP and N-methyl carbamate pesticides, OPPTS quantified upper and lower confidence bounds on potency estimates for each chemical. For those risk assessments that utilize PBPK models, uncertainty/sensitivity analysis of the input parameters can be performed. Currently, however, uncertainty due to missing toxicological data is qualitatively described and established methods for quantifying that uncertainty are lacking.

Sensitive and Vulnerable Subpopulations (e.g., Children, Elderly, Tribes, Endangered Species)

A response to this question can be extracted from NCEA's Framework for Children's Health Risk Assessment (EPA 2006) and the RAF document on the RfD/RfC methodology (EPA 2002b) which OPPTS uses as guidance. For pesticides, it should be noted however, that the FQPA includes the statutory requirement of an additional 10X safety factor to protect infants and children. This 10X factor can only be reduced or removed if it is determined that the hazard and exposure analyses are protective of infants and children. OPP's guidance for implementing the FQPA factor can also be found via the web (EPA 2002d).

OPP also assesses the potential effect of pesticides to non-target species, including federally listed threatened and endangered species (listed species) and habitat deemed critical to their survival. The assessment is conducted consist with scientific methodology described in EPA's Overview Document (EPA 2004b) and endorsed by the U.S. Fish and Wildlife Service and National Marine Fisheries Service (FWS/NMFS 2004). This assessment results in an "effects determination" for a species—a determination of whether a particular pesticide's use has "no effect," is "not likely to adversely affect," or is "likely to adversely affect" the listed species on a geographically specific basis. Consistent with Departments of Interior and Commerce regulations governing federal agency responsibilities relative to listed species, EPA consults with the U.S. Fish and Wildlife Service and National Marine Fisheries Service (the Services), as appropriate, for any determination other than "no effect." Consultation and resulting input from the Services, informs OPPs decision on whether changes to the pesticide's registration are necessary to ensure protection of federally listed threatened or endangered species and their critical habitat.

Challenges for Risk Assessment in a Regulatory Process

There are many challenges for risk assessment in a regulatory process. One key issue is the training of staff to implement new tools (e.g., MOA analyses) and prepare risk characterizations that provide transparent weight of evidence analyses. Another one is accounting for missing toxicological data via quantitative uncertainty analyses and to move the evaluation of toxicological effects into probabilistic and multi- endpoint analyses. Lastly, an important overall direction for OPPTS is to improve and refine how we integrate all available and relevant toxicology, human studies/epidemiology, biomonitoring, and exposure information into a paradigm that balances resources with the needs of the risk assessment (i.e., sustainable).

Future Directions: Addressing Gap, Limitations, and Needs

Issues to Be Addressed: Needed Improvements and Recommendations

Short-term: 2-5 years

OPPTS is working closely with ORD to develop more advanced methods of quantitative uncertainty analysis (e.g., 2-dimensional Monte Carlo) and incorporating these into exposure models. As knowledge expands, these methods will need further refinement and improvements. There is a need to continue to promote the development of PBPKmodels and other approaches which allow for the replacement of default assumptions uncertainty/extrapolation and to develop methods to quantify uncertainty and variability for the hazard/effects component of risk assessment.

Long-term: 10-20 years

Replacement or reduction of animal testing and moving toward an "integrated" risk paradigm by improving QSAR approaches, developing methods for interpreting and incorporating "omics" data, *in silico*, etc approaches into risk analyses.

Address Media-Specific Needs for Risk Assessment, For Example:

Do Current Paradigms Adequately Address Environmental Problems Faced by the Country?

See above response to short and long term needs. OPPTS continues to develop and use alternatives to defaults by incorporating PBPK modeling and data derived uncertainty factors, mode of action data, probabilistic exposure modeling, and biomonitoring data. For example, As an alternative to the RfD, OPPTS also uses characterization of risk for specific age groups and evaluates exposures across different durations of exposure (e.g., single day to lifetime).

REGION 2 AND THE OFFICE OF SOLID WASTE AND EMERGENCY RESPONSE

Introduction

This report is primarily based on Chapter 5 of EPA's Office of the Science Advisor's Staff Paper titled: "Risk Assessment Principles and Practices" (EPA 2007a). The Chapter provides information regarding current practices for site and chemical specific risk assessments in EPA's Office of Solid Waste and Emergency Response (OSWER). As described on the OSWER homepage (EPA 2008g):

> OSWER provides policy, guidance and direction for the Agency's solid waste and emergency response programs. We develop guidelines for the land disposal of hazardous waste and underground storage tanks. We provide technical assistance to all levels of government to establish safe practices in waste management. We administer the Brownfields program which supports state and local governments in redeveloping and reusing potentially contaminated sites. We also manage the Superfund program to respond to abandoned and active hazardous waste sites and accidental oil and chemical releases as well as encourage innovative technologies to address contaminated soil and groundwater.

This chapter provides a perspective on site-specific risk assessments conducted within the Superfund program.

Current Practice

Statutory Basis/Current Approach and Paradigms for Risk Assessment (Specific to Each Program Office)

The Superfund Program

To understand the Superfund program and its application in OSWER and the Regions it is important to first take a look at the legislation that governs this regulatory program. The Comprehensive Environmental Response Compensation and Liability Act (CERCLA) was enacted in 1980 and is commonly referred to as the Superfund program. The Act was amended in 1986 under the Superfund Amendments and Reauthorization Act of 1986. These laws require that action selected to remedy hazardous waste sites be protective of human health and the environment. The National Oil and Hazardous Substances Pollution Contingency Plan, or NCP, establishes the overall approach for determining appropriate remedial action at Superfund sites across the country and mandates that a risk assessment is performed to characterize current and potential threats to human health and the environment (40 CFR § 300.430 (d)(4)[2004]). The preamble to the NCP (55 Fed. Reg. 8709[1990]) provides more detail on the general goals and approach for Superfund risk assessments.

The Superfund process involves a number of steps as shown in Figure E-1 from site discovery, listing on the National Priorities List (NPL), Remedial Investigation and Feasibility Study (RI/FS), Record of Decision (ROD) to final NPL deletion. Within the Superfund program, the range of activities at sites includes Removal Actions where actions are necessary in a short timeframe and longer remedial investigations of complex sites. This discus-

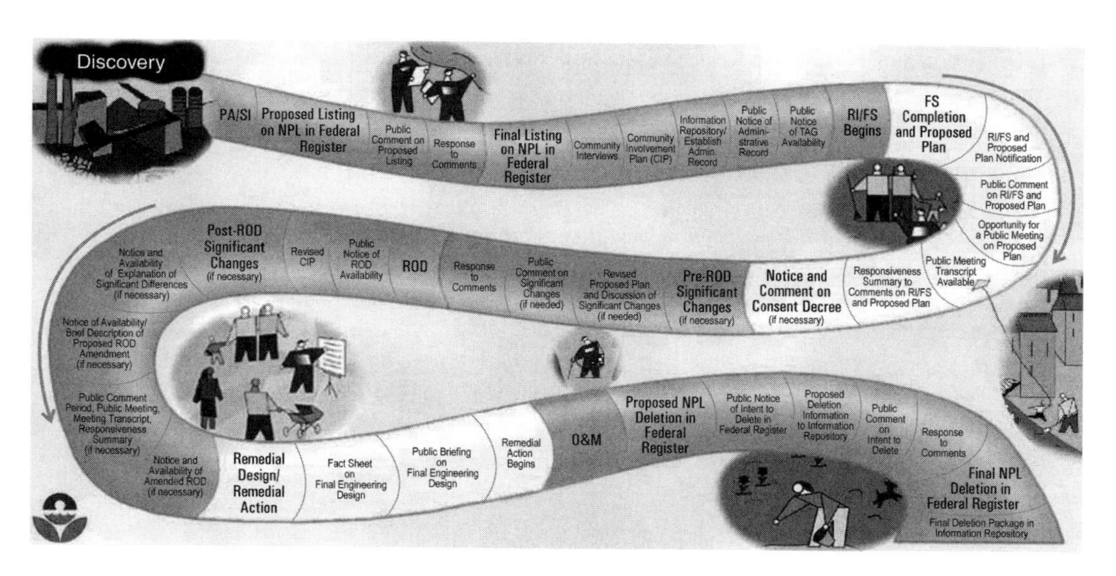

FIGURE E-1 Community involvement activities at NPL sites. Source: EPA 2001a.

sion will concentrate primarily on the latter type of investigation, i.e., sites that are on the NPL. Currently, across the country, there are 1,557 current and deleted sites on the NPL. The NPL is the list of national priorities among the known releases or threatened releases of hazardous substances, pollutants, or contaminants throughout the United States and its territories. The NPL is intended primarily to guide the EPA in determining which sites warrant further investigation. Further details regarding the Superfund program are available on the Superfund homepage (EPA 2008h).

At each site risk assessments are developed to assess both human health and ecological risks during the RI/FS. The risk information is used to determine whether remedial action is needed at the site. All decisions at Superfund sites must meet the nine criteria provided in Table E-1. The Threshold Criteria that must be met at all sites are protection of public health and the environment and meeting the Applicable or Relevant and Appropriate Requirements (ARARs) or statutory requirements. Risk assessment plays a critical role in determining that these criteria are met.

Risk Assessment in the Superfund Program

The Superfund program uses risk assessment to determine whether remedial action is necessary at a specific site and to determine the levels of remedial action where actions are required. The program protects human health and the environment from current and potential future threats posed by uncontrolled hazardous substances releases. Decisions at Superfund sites involve consideration of cancer risks, non-cancer health hazards, and site-specific information associated with both current and future land use conditions. Consideration of future land use and future risks is included in the risk assessment because CERCLA mandates that remedies are protective in the long-term.

The human health and ecological risk assessments developed at sites follow peer-reviewed guidelines, policies and guidance specific to the OSWER program as well as those for the Agency. The OSWER documents regarding risk assessment are available online (EPA

TABLE E-1 Nine Evaluation Criteria for Superfund Remedial Alternatives

THRESHOLD CRITERIA

Overall protection of human health and the environment determines whether an alternative eliminates, reduces, or controls threats to public health and the environment through institutional controls, engineering controls, or treatment.

Compliance with ARARs evaluates whether the alternative meets federal and state environmental statutes, regulations, and other requirements that pertain to the site, or whether a waiver is justified.

PRIMARY BALANCING CRITERIA

Long-term effectiveness and permanence considers the ability of an alternative to maintain protection of human health and the environment over time.

Reduction of toxicity, mobility, or volume of contaminants through treatment evaluates an alternative's use of treatment to reduce the harmful effects of principal contaminants, their ability to move in the environment, and the amount of contamination present.

Short-term effectiveness considers the length of time needed to implement an alternative and the risks the alternative poses to workers, residents, and the environment during implementation.

Implementability considers the technical and administrative feasibility of implementing the alternative, including factors such as the relative availability of goods and services.

Cost includes estimated capital and annual operation and maintenance costs, as well as present worth cost. Present worth cost is the total cost of an alternative over time in terms of today's dollar value. Cost estimates are expected to be accurate within a range of +50% to -30%.

MODIFYING CRITERIA

State acceptance considers whether the state agrees with the EPA's analyses and recommendations, as described in the RI/FS and Proposed Plan.

Community acceptance considers whether the local community agrees with EPA's analyses and preferred alternative. Comments received on the Proposed Plan are an important indicator of community acceptance.

2008i). The guidance provides an overall approach to developing risk assessments at a wide variety of sites across the country. The site specific risk assessments include assessment of contamination in multiple media (air, surface and groundwater, soil, fish, etc.) that occurs during the Remedial Investigation phase where the nature and extent of contamination are determined. Typically, site-specific risk assessments evaluate exposures to multiple chemicals through multiple routes of exposure (i.e., ingestion, inhalation, dermal contact, etc.). Receptors evaluated at sites include young children, adolescents, and adults depending and the current and future landuse.

Within the Superfund program we follow the basic risk assessment paradigm developed in the 1983 Framework document, i.e. the four steps of hazard identification, dose response assessment, exposure analysis, and risk characterization. Over the years, this paradigm has been expanded to include Problem Formulation, communication with risk managers, and early and continuous community involvement. On a site-specific basis evaluations regarding exposures and the availability of site-specific information (i.e., site-specific chemical sampling, activity patterns, creel surveys, etc.) are evaluated for inclusion in the risk assessment. For toxicity values, Superfund primarily relies on EPA's National Center for Environmental Assessment (NCEA) and the Superfund Technical Support Center assessments.

A typical Superfund site does not exist. Sites range from small contaminated parcels where groundwater and soil are impacted to large contaminated river systems or lakes that cover hundreds of miles. In general, most sites include multiple media, multiple chemicals, and multiple exposure pathways that are evaluated to determine the risks to the Reasonably Maximally Exposed individual or RME individual. The RME individual is defined as someone who is exposed to the highest exposure that is reasonably expected to occur at a Superfund site. As described in the National Contingency Plan, the regulation under which the Superfund program acts, the RME will

> result in an overall exposure estimate that is conservative but within a realistic range of exposures. Under this policy, EPA defines "reasonable maximum" such that only potential exposures that are likely to occur will be included in the assessment of exposures. The Superfund program has always designed its remedies to be protective of all individuals and environmental receptors that may be exposed to a site; consequently, EPA believes it is important to include all reasonably expected exposures in its risk assessments....

Uncertainty Analysis, Default Assumptions, Use of Alternatives, Probabilistic Risk Assessment and Communication of Risk, and Evaluation of Alternative Remediation Strategies and Superfund Process Post Remedial Investigation

Uncertainty Analysis. Within the Superfund program uncertainty in the risk assessments is addressed by discussing risks to the Reasonably Maximally Exposed Individual and the Central Tendency or average exposed individual. As described above, decisions are based on the RME individual. The presentation of the risks to the RME and CTE individual provides a bounding estimate of risks. In addition, site-specific risk assessment provide a qualitative discussion of uncertainties such as data limitations, where toxicity data is missing, where risk is potentially overestimated based on the data i.e., a screening level assessment, and discuss the impacts of these risk estimates. Risks are typically compared to the risk range identified in the National Contingency Plan or NCP, the Superfund regulation.

Default Assumptions

Risk assessments incorporate both *default* assumptions and *site-specific* information. The supplemental guidance document, "Standard Default Exposure Factors" (OSWER Directive 9285.6-03, March 25, 1991), presents the Superfund program's default exposure factors for calculating RME exposure estimates (EPA 1991a). This guidance was developed in response to requests that EPA make Superfund risk assessments more transparent and their assumptions more consistent. However, the guidance clearly states that the defaults should be used where "there is a lack of site-specific data or consensus on which parameter to choose, given a range of possibilities." These default exposure assumptions are supplemented with data from the Exposure Factors Handbook (EPA 1997a), and Child Specific Exposure Factors Handbook (EPA 2002e) where EPA compiled and analyzed scientific literature on exposure to develop ranges of exposure variables for risk assessments.

Table E-2 (EPA 2004a, Table 5-1) presents examples of default exposure values and the percentile of the population the values represent, as well as the peer reviewed studies supporting these assumptions. The RME approach uses default values designed to estimate the exposure of a high-end individual in the 90th percentile of exposure or above (EPA 1992). Consistent with this guidance, relevant default assumptions for various activity levels and age groups are used for drinking water consumption rates, soil ingestion rates, residence times, body weight, and inhalation rates. The table illustrates the range of percentiles—some defaults included the 50th percentile (e.g., body weight), 80th, 90th, and 95th percentiles.

Although the Superfund program routinely uses default assumptions to assess the risk to the RME individual at many sites, the characteristics of the surrounding population change from site to site. For example, the distributions of individual residence times will

TABLE E-2 Examples of Default Exposure Values With Percentiles

Exposure Pathway	Percentile	Source of Data
Drinking water consumption: 2 liters/day	90th	Approximately a 90th percentile value (EPA 2000).
Soil ingestion rate for children: 200 mg/day	65th	Analyses and distributions constructed by Stanek and Calabrese (1995a,b, 2000) places the 200 mg ingestion rate around the 65th percentile of average daily intakes throughout the year. The Stanek and Calabrese analyses suggests that ingestion rates for children in the top 10% (i.e., the high end) of the distribution would be greater than 1,000 mg/day.
Residence duration: 30 years	90th 80th 90th–95th	For home owners, farms, and rural populations; 30 years is greater than the 95th percentile residence time for renters and urban populations.
Body weight: 70 kg	50th	For males and females 18 to 75 years old (NCHS 1987)

Source: EPA 2004a, p. 100, Table 5-1.

vary depending on whether the site is located in a rural or an urban area. Individuals in rural communities are likely to have longer residence times than individuals in urban communities. Thus, a default value of 30 years may fall at the 80th percentile for farmers but above the 95th percentile for renters in an urban setting. The extent to which a single default value will impact the final exposure estimate depends on the values and variabilities of all the parameters used to estimate exposure. The goal is to estimate an individual exposure that actually occurs and is above the 90th percentile. In some cases, use of default assumptions may produce an estimate near the 90th percentile; in others, the estimate may be higher in the range.

In general, Superfund's default factors are designed to be reasonably protective of the majority of the exposed population. The assumptions used in Superfund's risk assessments are consistent with the 90th percentile or above and the Agency's exposure assessment guidelines (EPA 1992). Default exposure factors used to assess the RME are a mix of average and high-end estimates (see Table E-1). The use of these default exposure assumptions does not automatically result in an overestimation of exposures. The Principles and Practices Document (EPA 2004a) provides several other examples that may be of interest to the reader regarding exposure assumptions.

Probabilistic Risk Assessment Guidance

Development of the OSWER probabilistic risk assessment guidance illustrates the process used in the Superfund program to develop guidance to address uncertainty (EPA 2001b). In that case, Superfund identified the emerging science, developed an EPA workgroup to evaluate the available science and its application within the Superfund program, released the draft guidance document for public comment, and conducted an external peer review before the document was completed. The guidance document provides program-specific information regarding the conduct of probabilistic risk assessments and supplements the earlier policy on this issue (EPA 1997b). In addition, EPA has developed training courses on the application of this methodology within the Superfund program. To date, probabilistic risk assessment methods have been used or are being developed at several sites to evaluate exposures in relation to both cancer risks and non-cancer health hazards (TAM Consultants, Inc. 2000).

For example, at one regional site, a point estimate was presented along with the results from a probabilistic risk assessment to provide a comparison of results. As part of the community involvement, results from both assessments were shared and the results discussed regarding the relative impacts of varying exposure assumptions in a probabilistic assessment on the decision. The Region presented the data incorporating the point estimate and showing that when other exposure assumptions were used the risk remained above the risk range described for Superfund above. We found that it was important to work with the community before the final risk results from both the point estimate and probabilistic assessment were presented to highlight this tool and its application (i.e., what kind of data was used, why this technique was included, how the results of the deterministic and probabilistic risk assessment were comparable, and how this information is used in the decision-making process).

Evaluation of Alternative Remedial Strategies

Risk assessment is one of several tools used to inform risk management decisions. Risk managers weigh a number of factors, including uncertainties in exposure and risk estimates, when developing health and environmental protective decisions. EPA considers a variety of alternatives to protect human health and the environment at sites and evaluates them by considering the balancing criteria and modifying criteria presented in Table E-1 (i.e., long-term effectiveness, use of treatment, short-term effectiveness, implementability, and cost). EPA then proposes a protective, cost-effective remedy that is, compliant with the Applicable or Relevant and Appropriate Requirements (ARAR), which it may modify based on state and public comments (see also CERCLA § 121, 42 U.S.C. § 9621[1986] and 40 CFR § 300.430[e][9]). CERCLA establishes a preference for remedial actions in which treatment permanently and significantly reduces the volume, toxicity, or mobility of the hazardous substances, pollutants, and contaminants is a principal element [CERCLA § 121 (b)(1)]. This paragraph goes on to require a consideration of permanent solutions and alternative treatment technologies or resource recovery technologies in the remedy selection process. CERCLA also directs Superfund to consider long-term maintenance costs, potential for future remedial actions if the remedy should fail. CERCLA § 121(b)(1) also establishes as one of the fundamental remedy selection criteria that we select remedies that "utilize permanent solutions and alternatives to treatment technologies or resource recovery technologies to the maximum extent practicable." For evaluating and selecting remedies, the NCP at 40 CRF§ 300.430 (e) (9) (C) [long-term effectiveness and permanence] and (D) [reduction of toxicity, mobility, or volume through treatment] require consideration of "magnitude of residual risk...;" "adequacy and reliability of controls such as containment systems and institutional controls..;" "...the degree to which alternative employ recycling or treatment that reduces toxicity, mobility, or volume..;" "...the amount of hazardous material that will be destroyed, treated or recycled...;" "...the type and quantity of treatment residuals considering the persistence, toxicity, mobility, and propensity to bioaccumulate...;" "the degree to which treatment reduces the inherent hazards posed by principal threats at the site."

EPA initiatives are also looking at cross-program coordination in EPA's Land Revitalization Office, to return contaminated land to safe and beneficial uses (EPA 2007d).

Superfund Process Following Remedial Investigation

Following the completion of the Remedial Investigation (RI) during which the risk assessment is conducted, EPA develops a feasibility study (FS) that evaluates remedial alternatives for action at the site (EPA 1988). Among other objectives, the FS evaluates the risks in the absence of remedial actions or institutional controls. This provides a baseline for comparison with other remedial alternatives. The FS includes the development of Remedial Action Objectives, including Preliminary Remediation Goals (PRGs) that are developed based on the RME exposure assumptions used in the risk assessment. The PRGs provide concentration levels that are protective of the RME individual who is currently exposed or may be exposed in the future. EPA's guidance "The Role of the Baseline Risk Assessment" provides further information regarding risk management decisions on sites (EPA 1991b).

During the FS, remedial alternatives are developed to achieve the program goals through a variety of different methods, generally including containment and treatment alternatives. The alternatives reflect the scope and complexity of the site problem. The Superfund program evaluates these alternatives using nine criteria described by the NCP (see Table E-1). The criteria address protectiveness, effectiveness, implementability, and acceptability issues. The criteria were derived from remedy selection criteria provided by Congress in SARA 121. The detailed analysis consists of an assessment of the individual alternatives against each of the nine evaluation criteria and a comparative analysis focusing upon the relative performance of each alternative against those criteria. In addition to viable remedial alternatives, EPA evaluates a no-action remedial alternative at all sites. The no-action alternative provides a baseline for comparison of the various alternatives that are appropriate for a specific site. All of this information is provided in a Proposed Plan, which is released with the RI/FS for public review and comment.

EPA provides opportunities for community involvement and public review of this information. A public meeting is held to discuss the proposed remedial alternatives and to obtain comments. Public comments are addressed at the meeting and in the Response to Comments that is developed as part of the Record of Decision (ROD). The ROD identifies remedial actions that have been selected for the site.

Following the ROD, EPA begins the remedial design process and the implementation of construction. Depending on the nature of the remedial actions and the amount of time required to complete the construction, EPA may conduct 5-year reviews to determine the protectiveness of the remedy (EPA 2001c). Throughout this process, information is shared with the community regarding the progress of the remedial actions.

Sensitive and Vulnerable Subpopulations (e.g., Children, Elderly, Tribes, Endangered Species)

Children

A common question asked of EPA is why Superfund risk assessments evaluate "dirt eating kids": Why should Superfund sites be cleaned up to levels such that children can safely "eat" the soil there? Actually, EPA does not typically assume that children are eating the dirt; rather, EPA assumes that they are exposed to contaminants through the course of normal activities of play on the ground, exposure to dust in the home, and incidentally through mouthing behavior (EPA 1996, 2005b).

It is commonly observed that young children suck their thumbs or put toys and other objects in their mouths. This behavior occurs especially among children from 1 to 3 years old (Charney et al. 1980; Behrman and Vaughan 1983). This "hand-to-mouth" exposure is well documented in the scientific literature for children under 6, and is especially prevalent among children 1½ to 3 years old, a critical period for brain development. This time period is of special concern regarding potential exposure, since children may be at special risk of exposure to specific chemicals, e.g., lead (CDC 1991). Superfund experience has taught us that children do incur exposures to contaminated soil, as is evident at lead-contaminated sites in which elevated blood levels occur in children residing at those sites (EPA 1996, 2005b).

Scientists agree that because of this behavior, children may incidentally or accidentally take in soil and dust (Calabrese et al. 1989; Davis et al. 1990; van Wijnen et al. 1990). Where children are likely to be exposed to contaminated soils (in residential areas, for example), it is appropriate for EPA to evaluate potential risks and set cleanup levels that will protect children for this widely recognized pathway of exposure, especially during this sensitive developmental period in the child's lifetime.

The basis of EPA's default soil ingestion rate is generally a point of contention. EPA has developed soil ingestion rates that are used as "default exposure assumptions" for adults and children. For young children (6 years or younger), the Superfund program default value is 200 milligrams of soil and dust ingested per day (EPA 1991a, 1996). EPA's risk estimates address the "incidental" ingestion that might occur when a child puts a hand or toy in his or her mouth, or eats food that has touched a dusty surface. Although this default assumption is often presented as an overly conservative value, the amount (200 milligrams per day) represents a small amount of soil ingested. It is less than 1/100 of an ounce (or one-fifth of the contents of a single-serving packet of sugar) a day. This peer reviewed value is applied in estimates of RME exposures (EPA 1989a, 1991a, 1997a).

In Superfund risk assessments, this soil ingestion rate for young children is combined with site-specific assumptions about exposure frequency (days per year) to estimate an average intake over the 6-year exposure period. Exposure frequency varies depending on site-specific current and future land uses. Soil ingestion studies report daily averages; the amount of soil ingested cannot be prorated on an hourly basis. Also, soil ingestion is episodic in nature and dependent upon a child's activity patterns, so prorating by time is not always appropriate. This is a common misapplication of soil ingestion rates in risk assessment.

Some children deliberately eat soil and other non-food items (a behavior known as pica). Pica behavior has been identified in children at rates of up to 5,000 milligrams per day (Calabrese et al. 1991; ATSDR 1996, 2001). The Agency for Toxic Substances and Disease Registry uses this pica ingestion rate when calculating Environmental Media Evaluation Guides, which are used to select contaminants of concern at hazardous waste sites (ATSDR 1996). EPA itself does not routinely address this form of exposure unless site-specific information is available. The default soil ingestion rate of 200 milligrams per day applied in Superfund risk assessments is intended to ensure reasonable protection of children in cases where they are likely to become exposed to contaminated soils and dust associated with a Superfund site.

At sites, depending on land use consideration may also be given to evaluating risks to adolescent trespasser. The adolescent trespasser is typically older than the young child described above (i.e., 10 to 18 years) and has shorter exposure frequency and duration than the young child resident.

Sensitive Populations

Assessment of fish consumption patterns is an area where young children and sensitive subpopulations may be exposed to contaminants. In some cases site-specific surveys have been conducted to evaluate the consumption patterns for specific populations that the published surveys do not capture. These surveys found considerably higher consumption rates among these populations than if the standard default assumptions from the 1997 Exposure Factors Handbook were used (EPA 1997a). For example, a 3½-year site-specific creel survey (Toy et al. 1996) included information on whether or not adults harvested fish and shellfish from Puget Sound. The survey included 190 adults and 69 children between the ages of 0 and 6. The study found that tribal seafood consumption rates were considerably higher than Exposure Factors Handbook values. Among the Squaxin, the average consumption rate was 72.8 grams per day and the 90th percentile ingestion rate was 201.6 grams per day. Among the Tulalips, the average consumption rate was 72.7 grams per day and the 90th percentile was 192.3 grams per day. Other site-specific consumption surveys found similar differences in consumption rates (Chiang 1998; EPA 2001d; Sechena et al. 2003).

In cases where EPA has conducted individual surveys to identify fish consumption rates, EPA has found it important to include the community in the process (EPA 1999a). EPA and other agencies (both private and governmental) have spent considerable resources and time to plan and implement these studies. The surveys (Chiang 1998; EPA 2001d; Sechena et al. 2003) were all conducted using one-on-one interviews, as opposed to creel or mail surveys. The people conducting the interviews were always specially trained members of the ethnic group or community being surveyed.

Challenges for Risk Assessment in the Regulatory Process

The challenges faced in developing risk assessments include:

Communication of Complex Scientific Concepts

This was an issue identified by Bill Farland when he was with the Agency. Within the Superfund program there is extensive communication with the community regarding the remedial investigation, risk assessment, remedial actions, and Superfund process. One of the challenges that is faced at all sites is the explanation of complex scientific concepts such as hydrodynamic modeling, groundwater issues, changes in the understanding of the toxicity of chemicals, and application of ranges of toxicity values.

Training of Risk Assessors/Risk Managers in New Scientific Advancements

With the advances in areas such as genomics, other "omics," nanotechnology, understanding of mutagenic modes of action, and all of the emerging areas of science there are new challenges in training staff in these emerging areas, especially risk managers who are often more accustomed to addressing engineering concepts and questions. The challenge is how to provide adequate background information in these areas and bring both risk assessors and managers up to speed with consideration of the current time and resource constraints. The use of the Hazardous Waste Clean-Up Information (CLU-IN) Web Site provides information about innovative treatment and site characterization technologies to the hazardous waste remediation community; web based seminars, annual meetings, conference calls etc. have proven effective and are continuing to be used. Another part of this challenge is knowing what to do with the information that is developed. For example, using genomics to determine that some member of a population at a site may be particularly susceptible does not indicate a regulatory response to that information is appropriate or necessary. In some cases, there may not be the regulatory authority to act or to do the population sampling necessary to determine biomarkers. Typically, the Agency for Toxic Substances and Disease Registry (ATSDR) is responsible for taking clinical samples.

Lack of Toxicity Data

At sites, there are typically a number of chemicals that can not be assessed quantitatively in the assessment based on a lack of peer-reviewed toxicity values. Typically these chemicals are addressed qualitatively in the risk assessment. Development of peer-reviewed toxicity data to include in the quantification of cancer risks and non-cancer health hazards obviously is quite important in the development of risk assessments.

Future Directions: Addressing Gaps, Limitations, and Needs

Issues to be Addressed in the Short Term (2-5 years) and the Long Term (10-20 years)

Overarching challenges for EPA including OSWER are to address the need to reach regulatory conclusions in a timely and cost effective manner with limited data and limited resources for analyses. In addition, EPA needs to develop transparent, clear, consistent, and reasonable presentations and procedures to support and explain its analyses. Briefly noted here are a few key areas.

Planning and Scoping

Over the last several years, as noted in the EPA Staff Paper, EPA has increasingly emphasized the importance of identifying as early as possible in our processes, through dialogue between risk assessors and risk managers, the scope and level of effort that is appropriate for a planned assessment. And that this may need to be done repeatedly. It seems likely that greater reliance on these interactions and efforts will play an increasingly important role as assessments continue to grow in complexity, and in the amount of review and scrutiny that they may receive.

Toxicity Data

In the Superfund program, we rely on NCEA including the Integrated Risk Information System (IRIS) and the Superfund Technical Support Section as the source for toxicity values. Typically, regions do not develop site-specific toxicity values. OSWER has defined a hierarchy for using other toxicity values when these are not available (EPA 2003b). In brief, such sources should be the most current, with a basis that is transparent and publicly available, and that has been peer reviewed. Sources for these toxicity values include California toxicity values, ATSDR minimal risk levels, and others. In the absence of toxicity values we rely on a qualitative discussion of the uncertainties in the risk assessment.

The current developments in the areas of Informatics, gene arrays and related areas hold the possibility of improving our understanding of Quantitative Structure Activity Relationships (QSAR) and so to reduce uncertainty, to help bound potential toxicity values and to reduce the need to conduct toxicity tests to support those values.

In addition, as noted above, this is another area where early identification of data gaps and needs would allow for the possibility of data generation to support the assessment.

Short-Term Exposures

Toxicity values and analyses are needed for short-term and mid-term exposures. These toxicity values are important in Removal Actions at sites.

Mixtures

Typically at Superfund sites we evaluate exposures to multiple chemicals through multiple pathways. EPA program offices and regional risk assessors have a great need for both assessment information and risk assessment methods to evaluate human health and ecological risks from exposure to chemical mixtures.

Exposure Assumptions

Superfund recognizes the most accurate way to characterize potential site-specific exposures to populations around Superfund sites would be to conduct a detailed census of each site considering both current and future land uses. Theoretically, this should involve interviewing all potentially exposed individuals regarding their lifestyles, daily patterns, water usage, consumption of local fish and game and procedures, working locations and exposure conditions while collecting environmental samples. Although site-specific data are collected on environmental media (e.g., soil, groundwater air, etc.) as appropriate during the Remedial Investigation, such collection has significant limitations. The three almost insurmountable difficulties are time, expense and intrusion on privacy. In the absence of site-specific information, Superfund relies on the Standard Default Exposure Factors and the Exposure Factors Handbook as sources for exposure information for use at sites. The Exposure Factors Handbook and its updates have been very important sources of information on exposures to a variety of populations (i.e., children, anglers, and others) through multiple media. The recent addition of the Child-specific Handbook has also been helpful in understanding risks to sensitive populations such as children. Because we assess future potential risks, we often want information that can not be directly measured such as potential changes in behavior following remediation of an area.

Probabilistic Risk Assessment

Superfund has developed peer-reviewed specific guidance for conducting site-specific probabilistic risk assessment. At all sites, both the RME and CTE (or average exposures) are evaluated to provide a range of risks and inform the risk management decision. The RME, however, under the NCP is the basis for the decision. In some cases, site-specific assessments have used the tiered approach in the guidance beginning with a deterministic risk assessment and then progressing to a more refined technique such as the one dimensional and two dimensional analysis. At the present time, site-specific probabilistic risk assessments have been conducted at several sites to examine exposure assessments.

Superfund is currently working on the Risk Assessment Forum project to look at the application and use of probabilistic risk assessment in decision making. The project is also looking at ways to better communicate the application of these techniques to risk managers to help identify areas where this technique is more applicable.

Improving Communication

Consistent with EPA Superfund goals of improving the transparency of the process, the methods for summarizing risk information are found in the RAGS Part D (EPA 2001e). Superfund continues to update guidance documents to improve the transparency of risk information.

With the advancements in science described above, there are new challenges associated with the summarization and presentation of data. With advances in Geographic Information Systems it is possible to demonstrate areas within and exceeding specific risk ranges. Current ongoing activities to digitize data locations with samples will facilitate the process of process of providing this data for further analysis..

EPA guidance and educational materials help illustrate the ways that citizens can be involved in the risk assessment process (EPA 1999a,b). For example: Community-specific information on fishing preferences helped to identify exposure areas for sampling and fish species consumed by people who fish in a contaminated bay. Information from farmers on pesticide applications helped EPA determine why certain contaminants were present in an aquifer. Discussions with farmers about certain harvesting practices helped EPA refine exposure models and assumptions at another site (EPA 1999b).

EPA uses a range of communication tools to include the community in the Superfund process. These include newsletters, fact sheets, site-specific home pages, public meetings, public availability sessions, and 1-800- numbers to contact EPA staff. EPA strives to communicate information about the RI, the results of the risk assessment, proposed actions at the site, and the proposed and final decisions for remedial actions. The Record of Decision (ROD) includes a responsiveness summary that addresses comments including those from the community. During the period of the remedial action, communication with the community continues, including updates during the 5-year review process.

OFFICE OF WATER (OW)

Current Practice

Statutory Basis/Current Approach and Paradigms for Risk Assessment (Specific to Each Program Office)

Office of Water (OW) follows the 1983 paradigm for human health risk assessments for chemicals and radiation, as explicated in the published U.S. EPA Risk Assessment Guidelines and other Agency guidance.

OW also does assessment of human health risk from microbial disease, from consuming drinking water, using water for recreation, and consuming aquatic organisms, and from contact with waste water. The paradigm for microbial risk assessment involving host/parasite interactions is still evolving. There is an EPA Risk Assessment panel that is developing Guidelines based on a proposed framework and collaboration with other Agencies. And important component of the microbial disease assessment is risk/risk tradeoff, such as was considered in the development of linked drinking water regulations for limitation of microbes and disinfection by-products. Lastly, OW engages in ecological risk assessment, following the paradigm published in the *Guidelines for Ecological Risk Assessment* (Figure E-2) (EPA 1998).

The Risk Assessment "Staff Paper" (EPA 2004a) compiles many of the general and specific risk assessment practices used by OW.

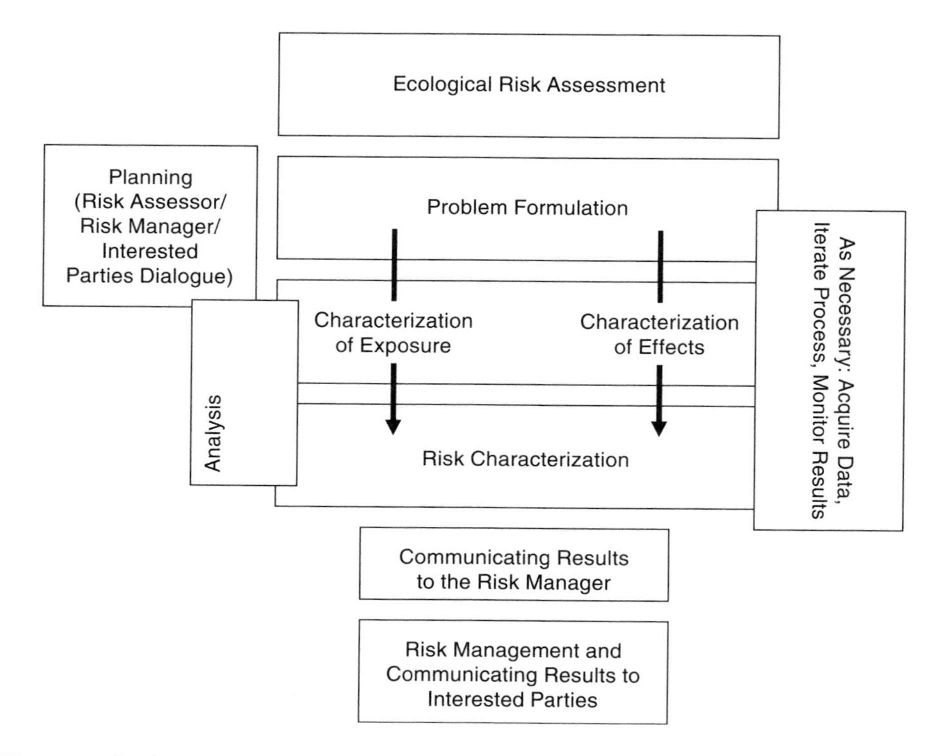

FIGURE E-2 The framework for ecological risk assessment (Modified from EPA 1998).

Office of Water operates under several pieces of enabling legislation. We have obligations under the following:

- Safe Drinking Water Act (Amended 1996)
- Clean Water Act
- Food Quality Protection Act (1996) (FQPA)
- Beaches Environmental and Coastal Health Act (BEACH Act) (2000)
- Coastal Zone Management Act
- Endangered Species Act

FQPA amended the Federal Insecticide, Fungicide and Rodenticide Act (FIFRA) in 1996; this was specifically to highlight risks to children from pesticides. As pesticides are found in drinking water source waters, OW adopts the risk assessments done under FQPA by the Office of Pesticides Programs, at least as far as hazard identification and dose response; exposure assessment will differ given the purview of the legislation under which the risk assessment is conducted.

The BEACH act is a 2000 amendment to the Clean Water Act (CWA). These changes set new requirements for recreational criteria and standards for coastal areas and the Great Lakes.

The Endangered Species Act requires that EPA engage in consultation with the U.S. Fish and Wildlife Service on any actions which may affect endangered plant or animal species.

The major pieces of enabling legislation for water programs are the CWA and the Safe

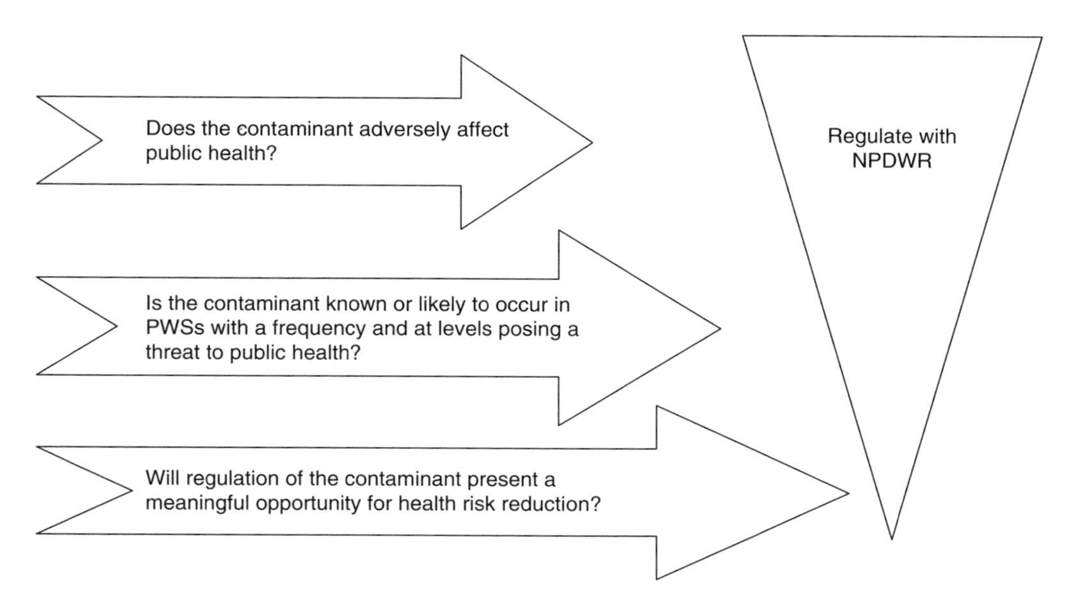

FIGURE E-3 Conditions for regulation under SDWA 1996.

Drinking Water Act (SDWA) as amended in 1996. SDWA deals with all uses of water from the tap, but only tap water (albeit from source to last public connection). Under SDWA, EPA establishes a list of chemical and microbial contaminants for potential regulation. EPA is obliged to revise this list on a regular basis; furthermore, EPA must make regulatory decisions on five agents on the list every five years. The bases for regulation are illustrated in Figure E-3. In order to regulate a contaminant in drinking water, EPA must establish the following: the contaminant can adversely affect public health; the contaminant occurs or is likely to occur in public water systems at levels that can affect public health; and there is a meaningful opportunity for public health improvement as a result of the regulation.

In answering these questions OW conducts quantitative risk assessments to determine nonenforceable Maximum contaminant level goals (MCLGs). OW then sets enforceable Maximum contaminant levels (MCLs) as close as technically feasible to the MCLGs after taking costs into consideration.

SDWA also requires that EPA conduct a Health Risk Reduction and Cost Analysis (HRCCA) for each proposed rule. There are seven elements of the HRRCA

1. Quantifiable and non-quantifiable health risk reduction benefits;
2. Quantifiable and non-quantifiable health risk reduction benefits form reduction in co-occurring contaminants;
3. Quantifiable and non- quantifiable costs;
4. Incremental costs and benefits;
5. Effects of the contaminant on the general population as well as sensitive subpopulations including infants, children, pregnant women, the elderly, individuals with a history of serious illness or others that may be at greater risk;
6. Any increase in health effects as a result of compliance including co-occurring contaminants;

7. The quality and extent of information, the uncertainties in the analyses and factors with respect to the degree and nature of the risk.

After completion of the HRCCA, analysis of technical feasibility of contaminant control, and determining appropriate monitoring, OW may propose and promulgate a National Primary Drinking Water Rule (NPDWR). These rules must be reviewed every six years by OW to determine if there is sufficient reason (e.g. new data, new risk assessment methods) to revise the rule.

The CWA provides broad outlines for controlling discharges to ambient waters from point sources of pollution and diffuse sources of contamination (e.g. run-off from agricultural lands, mining sites, etc). CWA requires that States and authorized Tribes designate uses for waterbodies (such as drinking water source water, fishable/swimable waterbody). The States then are required to take specific actions to ensure that those uses are attained; such as setting standards, issuing permits, defining total maximum daily loads of a contaminant to a water body. Under CWA, OW publishes ambient water quality criteria (AWQC) for both human health and aquatic life. These are risk assessments that the States and Tribes may choose to adopt; EPA determines whether State or Tribal standards are scientifically justified.

In deriving national AWQC, OW follows EPA published methodologies including the *Methodology for Deriving Ambient Water Quality Criteria for the Protection of Human Health* (EPA 2000), and the *Guidelines for Deriving Numerical National Water Quality Criteria for the Protection of Aquatic Organisms and Their Uses* (EPA 1985). The latter document is being updated. The Human Health Methodology is being expanded through Technical Support Documents. A series of technical documents deals with bioaccumulation through aquatic food webs, as human health criteria specifically identify consumption of contaminated seafood as a pathway in exposure assessment. The Human Health Methodology also describes the concept of relative source contribution (RSC), a method for apportioning the "allowable risk" such as an RfD over all plausible routes of exposure. OW also applies the RSC in calculating MCLGs under SDWA. For example in the risk assessment for chloroform, inhalation of vapors and concentrations in foods were considered in developing the MCLG. Ultimately the EPA default process had to be used in the chloroform RSC, as there were insufficient data on which to base a specific value.

Other examples of best practices can be seen in the economic analyses in support of NPDWRs such as the 2005 Long Term 2 Enhanced Surface Water Treatment Rule (LT2) and the 2006 Groundwater Rule (GWR). Both of these rules were based on assessment of human risk from a variety of microbial contaminants including protozoa, bacteria and viruses.

Uncertainty Analysis

Regarding the presentation of alternative risk estimates SDWA says the following:

The Administrator shall, in a document made available to the public in support of a regulation promulgated under this section, specify, to the extent practicable:

1. Each population addressed by any estimate of public health effects;
2. The expected risk or central estimate of risk for the specific populations;
3. Each appropriate upper-bound or lower-bound estimate of risk ... (OW; SDWA § 300g-1 (b)(3)).

OW describes areas of uncertainty and variability in the risk assessment documents supporting our regulatory and other risk management decisions. Some of these analyses included quantitative estimates of uncertainty and variability; this is most commonly done for exposure data. Recent economic analyses done in support of SDWA include assessments of uncertainty in occurrence or exposure data (for example, LT2, the arsenic NPDWR, GWR). Discussion of uncertainty in dose response assessment was published in the context of these rules as well. In addition OW discussed uncertain the effectiveness of drinking water treatment (LT2) as well as uncertainty in the measurements or indicators used in risk-targeted regulatory strategies (LT2 and GWR). These analyses are peer-reviewed and subject to public comment before publication of the final economic analysis.

OW has published sensitivity analyses and presentations of alternative risk estimates; for example in the Regulatory Impact Analysis (RIA) supporting the Arsenic NPDWR. Note that the preamble to this rule also included an extensive discussion of uncertainty in the dose response data and modeling. OW has also used published uncertainty analyses; for example, the assessment of variability in pharmacokinetic parameters presented by NRC (2000) was incorporated into the reference dose for methylmercury used in the AWQC (EPA 2001f).

OW uses default procedures and assumptions as indicated in EPA documents including the 2005 Cancer Guidelines and Supplemental Guidance (EPA 2005c,d) and the Staff Paper (EPA 2004a). OW has also published analyses that permit the use of distributional approaches to exposure assessment; for example, analyses of Continuing Study of Food Intake by Individuals (CSFII) data on consumption of water from public water systems, in beverages and so on. This report also supports the use of 2l/day for adult exposure assessment as a reasonable default when distributional approaches are not warranted (EPA 2004c).

Sensitive and Vulnerable Subpopulations (e.g., Children, Elderly, Tribes, Endangered Species)

The SDWA Amendments mandate that EPA consider risks to groups within the general population that are identified as being at greater risk of adverse health effects; these include children, the elderly, and people with serious illness (Safe Drinking Water Act [1996]). To this end OW includes consideration of appropriate susceptible populations in the risk assessment documents supporting risk management. This is always described in the preamble to regulations (for, example Disinfection By-products Stage 1). For example specific consideration of immunocompromised persons was highlighted in the Long Term Enhanced Surface Water Treatment Rules.

OW specifically recommends that States and authorized Tribes use waterbody specific population and exposure data in their derivation of criteria and standards. OW recommends use of default exposure factors only in absence of any relevant data (EPA 2000). OW is conscious of Native American and other traditional lifestyles that may result in exposure parameters different from those considered to be the norm. The American Indian Environmental Office (AEIO/OW) and EPA Tribal Science Council are among the groups pursuing these issues.

Challenges for Risk Assessment in a Regulatory Process

Under the SDWA, costs vs. benefits of regulation are a factor in the choice to regulate or not as well as in the limits set by an MCL. An illustration of the methods and challenges of benefits assessment is the RIA for the arsenic NPDWR. It should be noted that identified but not quantified, and quantified but not monetized, benefits are difficult to characterize and compare with monetized benefits. Given that the standard non-linear low dose extrapolation procedure, calculation of an RfD, does not provide an estimate of risk, this is a major challenge. In the GWR economic analysis, OW made the case using a semi-quantitative approach that monetized benefits might be more than five-fold greater than those used, if bacterial disease could be better quantified.

Under the Clean Water Act, OW publishes AWQC for human health; these risk assessments do not consider the cost or technological feasibility of meeting these criteria. However, demonstration of quantifiable, monetized benefits has become increasingly important in the acceptance of any risk management choice. The problem of assessing benefits of an ecosystem remains a very serious one.

The major problem in conduct of OW risk assessments is insufficient resources. Chief among the resource lack is the lack of data. None of the enabling legislation for water programs provide a means to require that ecological or health effect data be generated. OW can establish requirements for monitoring of various kinds, depending on the law, but there is no way to acquire health effects data. There is further a requirement in SDWA that data serving as the basis for regulation be peer-reviewed and publicly available. OW risk assessments are most often limited by paucity of usable data on health effects and occurrence of contaminants in food and water.

Data to support microbial dose response assessment are lacking and are likely not to be forthcoming. New human challenge studies are extremely unlikely to be conducted, and even if available may not be usable by EPA given recent restrictions on use of human studies. Those studies that are complete may not be applicable to assessment of exposure in the general population for these reasons.

- The studies administered laboratory strains of microbes; that is healthy infectious organisms grown or concentrated from specific hosts. Environmental organisms are of more diverse origin and may be more or less potent than laboratory strains.
- Challenge studies are conducted in healthy volunteers, usually one gender, and only of a limited age range (typically 20-50).

Another challenge in assessing microbial pathogens is lack of data and models on secondary transmission. Dynamic disease transmission modeling is developing as a useful tool.

Time is also a limited resource. SDWA risk assessments must be done to deadlines for regulation proposal, promulgation and review. For both CWA and SDWA actions, there are often court-ordered deadlines to be met. OW may not delay these actions to await data generation or method development.

Under SDWA OW is concerned with contaminant mixtures in drinking water in response to requirements of the Safe Drinking Water Act Amendments of 1996, including mixtures of DBPs and of Contaminant Candidate List chemicals (e.g., organotins, pesticides, metals, pharmaceuticals). Information and methods are being developed to better evaluate the toxic mode of action, the risk posed by drinking water mixtures, exposure estimates for mixtures via multiple routes, and the relative effectiveness of advanced treatment technologies (EPA 2003c,d).

Whole-mixture studies are routinely used in ecological risk assessments. The Agency has developed subchronic toxicity tests for whole aqueous effluents and for contaminated ambient waters, sediments, and soils (EPA 1989b, 1991c, 1994a). Furthermore, the effects of mixtures in aquatic ecosystems are evaluated using bioassessment techniques that are equivalent to epidemiology, but more readily employed (Barbour et al. 1999). Similar bioassessment methods are sometimes used at Superfund sites (EPA 1994b). These empirical approaches to assessing ecological risks from mixtures are employed in National Pollutant

Discharge Elimination System permitting and the development of Total Maximum Daily Loads, and are often used in Superfund baseline ecological risk assessments.

Many uncertainty analyses account for parameter uncertainty, but ignore model uncertainty. When only one model can reasonably explain or be fit to the data, then there is need only to account for uncertainty in that specific model's parameter values. For example, a dose-response relationship might be known to be exponential, and data are used to estimate and characterize uncertainty about the exponential model's single parameter (r). If it is uncertain whether the model is exponential, beta-Poisson, or some other form, then the data are used to characterize uncertainty about the model as well as the models' parameter values. In OW's GWR and LT2 rules, model uncertainty was explored in sensitivity analyses; these showed that the choice of model did not significantly alter the results. Dealing with model uncertainty may be a significant challenge in future analyses under these conditions: (a) data do not clearly point to a single preferred model; or (b) the regulatory outcome or estimate is sensitive to model choice.

Future Directions: Addressing Gaps, Limitations, and Needs

The 1983 NRC paradigm for human health risk assessment for chemicals and radiation remains adequate. The 1998 paradigm for ecological risk assessment remains adequate. We look forward to a federal peer-reviewed, published microbial risk assessment paradigm.

Water programs need improved dose response methods, in particular for microbial disease causing agents.

While OW would like to see increased use of data from "omic" technologies, there is an enormous amount of work in that field to be done before such use will be either practical or will stand the test of the courts. Probably the first accepted use of "omics" in water programs will be in microbial source tracking and in rapid detection of contaminants (rather than in risk assessment).

Improved and accepted methods for quantifying ecological benefit, and human health benefits (beyond value of a statistical life), will be immediately useful.

Means to assess the utility and the lessons learned from various types of uncertainty analyses will be immediately useful, as will improved methods for communicating uncertainty to both decision makers and the (litigious) public.

The major limitations in applying any new risk assessment methods will be lack of data (particularly health and ecological effects data); and degree of acceptance of new methods by stakeholders.

REFERENCES

ATSDR (Agency for Toxic Substances and Disease Registry). 1996. ATSDR Public Health Assessment Guidance Manual. Agency for Toxic Substances and Disease Registry, Atlanta, GA.

ATSDR (Agency for Toxic Substances and Disease Registry). 2001. Summary Report for the ATSDR Soil-Pica Workshop, June 2000, Atlanta, GA. Prepared by Eastern Research Group, Lexington, MA. Contract No. 205-95-0901. Task Order No. 29. March 20, 2001 [online]. Available: http://www.atsdr.cdc.gov/NEWS/soilpica.html [accessed Jan. 30, 2008].

Barbour, M.T., J. Gerritsen, B.D. Snyder, and J.B. Stribling. 1999. Rapid Bioassessment Protocols for Use in Streams and Wadeable Rivers: Periphyton, Benthic Macroinvertebrates and Fish, 2nd Ed. EPA 841-B-99-002. Office of Water, U.S. Environmental Protection Agency, Washington, DC [online]. Available: http://www.epa.gov/owow/monitoring/rbp/ [accessed Jan. 31, 2008].

Behrman, L.E., and V.C. Vaughan, III. 1983. Nelson Textbook of Pediatrics, 12 Ed. Philadelphia, PA: W.B. Saunders.

Calabrese, E.J., R. Barnes, E.J. Stanek, III, H. Pastides, C.E. Gilbert, P. Veneman, X.R. Wang, A. Lasztity, and P.T. Kostecki. 1989. How much soil do young children ingest: An epidemiologic study. Regul. Toxicol. Pharmacol. 10(2):123-137.

Calabrese, E.J., E.J. Stanek, and C.E. Gilbert. 1991. Evidence of soil-pica behavior and quantification of soil ingestion. Hum. Exp. Toxicol. 10(4):245-249.

CDC (Centers for Disease Control and Prevention). 1991. Preventing Lead Poisoning in Young Children. U.S. Department of Health and Human Services, Public Health Service, Centers for Disease Control and Prevention, Atlanta, GA. October 1, 1991 [online]. Available: http://wonder.cdc.gov/wonder/prevguid/p0000029/p0000029.asp [accessed Jan. 30, 2008].

Charney, E., J. Sayre, and M. Coulter. 1980. Increased lead absorption in inner city children: Where does the lead come from? Pediatrics 65(2):226-231.

Chiang, A. 1998. A Seafood Consumption Survey of the Laotian Community of West Contra Costa County, CA. Oakland, CA: Asian Pacific Environmental Network.

Davis, S., P. Waller, R. Buschbom, J. Ballou, and P. White. 1990. Quantitative estimates of soil ingestion in normal children between the ages of 2 and 7 years: Population-based estimates using aluminum, silicon, and titanium as soil tracer elements. Arch. Environ. Health 45(2):112-122.

EPA (U.S. Environmental Protection Agency). 1985. Guidelines for Deriving Numerical National Water Quality Criteria for the Protection of Aquatic Organisms and Their Uses. EPA 822/R-85-100. U.S. Environmental Protection Agency, Office of Research and Development, Environmental Research Laboratories, Duluth, MN, Narragansett, RI, and Corvallis, OR [online]. Available: http://www.epa.gov/waterscience/criteria/85guidelines.pdf [accessed Jan. 30, 2008].

EPA (U.S. Environmental Protection Agency). 1988. Guidance for Conducting Remedial Investigations and Feasibility Studies under CERCLA. Interim Final. OSWER Directive 9355.3-01. EPA/540/G-89/004. Office of Emergency and Remedial Response, U.S. Environmental Protection Agency, Washington, DC. October 1988 [online]. Available: http://rais.ornl.gov/homepage/GUIDANCE.PDF [accessed Jan. 30, 2008].

EPA (U.S. Environmental Protection Agency). 1989a. Risk Assessment Guidance for Superfund, Vol. 1. Human Health Evaluation Manual (Part A). EPA/540/1-89/02. Office of Emergency and Remedial Response, U.S. Environmental Protection Agency, Washington, DC. December 1989 [online]. Available: http://www.epa.gov/oswer/riskassessment/ragsa/pdf/rags-vol1-pta_complete.pdf [accessed Jan. 30, 2008].

EPA (U.S. Environmental Protection Agency). 1989b. Short-Term Methods for Estimating the Chronic Toxicity of Effluents and Receiving Waters to Freshwater Organisms, 2nd Ed. EPA 600/4-89/001. Environmental Monitoring Systems Laboratory, U.S. Environmental Protection Agency, Cincinnati, OH.

EPA (U.S. Environmental Protection Agency). 1991a. Risk Assessment Guidance for Superfund, Vol. I: Human Health Evaluation Manual, Supplemental Guidance, "Standard Default Exposure Factors." Interim Final. OSWER Directive 9285.6-03. PB91-921314. Office of Emergency and Remedial Response, U.S. Environmental Protection Agency, Washington, DC. March 25, 1991 [online]. Available: http://www.epa.gov/oswer/riskassessment/pdf/OSWERdirective9285.6-03.pdf [accessed Jan. 30, 2008].

EPA (U.S. Environmental Protection Agency). 1991b. Role of the Baseline Risk Assessment in Superfund Remedy Selection Decisions. OSWER Directive 9355.0-30. Memorandum to Directors: Waste Management Division, Regions I, IV, V, VII, VIII; Emergency and Remedial Response Division, Region II; Hazardous Waste Management Division, Regions III, VI, IX; and Hazardous Waste Division, Region X, from Don R. Clay, Assistant Administrator, Office of Solid Waste and Emergency Response, U.S. Environmental Protection Agency, Washington, DC. April 22, 1991 [online]. Available: http://www.epa.gov/oswer/riskassessment/pdf/baseline.pdf [accessed Jan. 30, 2008].

EPA (U.S. Environmental Protection Agency). 1991c. Methods for Measuring the Acute Toxicity of Effluents and Receiving Waters to Freshwater and Marine Organisms, 4th Ed. EPA-600/4-90/027. Office of Research and Development, U.S. Environmental Protection Agency, Washington, DC. September 1991.

EPA (U.S. Environmental Protection Agency). 1992. Guidelines for Exposure Assessment. EPA/600/Z-92/001. Risk Assessment Forum, U.S. Environmental Protection Agency, Washington, DC. May 1992 [online]. Available: http://cfpub.epa.gov/ncea/cfm/recordisplay.cfm?deid=15263 [accessed Oct. 10, 2007].

EPA (U.S. Environmental Protection Agency). 1994a. ECO Update: Catalog of Standard Toxicity Tests for Ecological Risk Assessment. EPA 540-F-94-013. Pub. 9345.0-051. Office of Solid Waste and Emergency Response, Washington, DC. Intermittent Bulletin 2(2) [online]. Available: http://www.epa.gov/swerrims/riskassessment/ecoup/pdf/v2no2.pdf [accessed Jan. 30, 2008].

EPA (U.S. Environmental Protection Agency). 1994b. ECO Update: Field Studies for Ecological Risk Assessment. EPA 540-F-94-014. Pub. 9345.0-051. Office of Solid Waste and Emergency Response, Washington, DC. Intermittent Bulletin 2(3) [online]. Available: http://www.epa.gov/swerrims/riskassessment/ecoup/pdf/v2no3.pdf [accessed Jan. 30, 2008].

EPA (U.S. Environmental Protection Agency). 1996. Soil Screening Guidance: User's Guide, 2nd Ed. OSWER Pub. 9355.4-23. EPA540/R-96/018. Office of Solid Waste and Emergency Response, Washington, DC. July 1996 [online]. Available: http://www.epa.gov/superfund/health/conmedia/soil/pdfs/ssg496.pdf [accessed Jan. 30, 2008].

EPA (U.S. Environmental Protection Agency). 1997a. Exposure Factors Handbook, Vols. 1-3. EPA/600/P-95/002F. Office of Research and Development, National Center for Environmental Assessment, U.S. Environmental Protection Agency, Washington, DC [online]. Available: http://www.epa.gov/ncea/efh/ [accessed June 3, 2007].

EPA (U.S. Environmental Protection Agency). 1997b. Policy for Use of Probabilistic Analysis in Risk Assessment at the U.S. Environmental Protection Agency. U.S. Environmental Protection Agency, Washington, DC. May 15, 1997 [online]. Available: http://www.epa.gov/osa/spc/pdfs/probpol.pdf [accessed June 3, 2007].

EPA (U.S. Environmental Protection Agency). 1998. Guidelines for Ecological Risk Assessment. EPA/630/R-95/002F. Risk Assessment Forum, U.S. Environmental Protection Agency, Washington, DC [online]. Available: http://cfpub.epa.gov/ncea/cfm/recordisplay.cfm?deid=12460 [accessed June 3, 2007].

EPA (U.S. Environmental Protection Agency). 1999a. Risk Assessment Guidance for Superfund: Vol. I—Human Health Evaluation Manual (Supplement to Part A): Community Involvement in Superfund RiskAassessments. EPA 540-R-98-042. Office of Solid Waste and Emergency Response, U.S. Environmental Protection Agency, Washington, DC. March 1999 [online]. Available: http://www.epa.gov/oswer/riskassessment/ragsa/pdf/ci_ra.pdf [accessed Jan. 31, 2008].

EPA (U.S. Environmental Protection Agency). 1999b. Superfund Risk Assessment and How You Can Help [videotape]. EPA-540-V-99-002. Office of Solid Waste and Emergency Response, U.S. Environmental Protection Agency, Washington, DC. September.

EPA (U.S. Environmental Protection Agency). 2000. Methodology for Deriving Ambient Water Quality Criteria for the Protection of Human Health. EPA-822-B-00-004. Office of Water, Office of Science and Technology, Washington, DC. October 2000 [online]. Available: http://www.epa.gov/waterscience/criteria/humanhealth/method/complete.pdf [accessed Jan. 31, 2008].

EPA (U.S. Environmental Protection Agency). 2001a. Community Involvement Activities Diagram. Superfund, U.S. Environmental Protection Agency. January 2001 [online]. Available: http://www.epa.gov/superfund/community/pdfs/pipeline.pdf [accessed Feb. 1, 2008].

EPA (U.S. Environmental Protection Agency). 2001b. Risk Assessment Guidance for Superfund (RAGS), Volume III, Part A: Process for Conducting Probabilistic Risk Assessment. EPA 540-R-02-002. Office of Emergency and Remedial Response, U.S. Environmental Protection Agency, Washington, DC [online]. Available: http://www.epa.gov/oswer/riskassessment/rags3a/ [accessed Oct 10, 2007].

EPA (U.S. Environmental Protection Agency). 2001c. Comprehensive Five-Year Review Guidance. EPA 540-R-01-007. Office of Emergency and Remedial Response, U.S. Environmental Protection Agency, Washington, DC. June 2001 [online]. Available: http://www.epa.gov/superfund/accomp/5year/guidance.pdf [accessed Jan. 31, 2008].

EPA (U.S. Environmental Protection Agency). 2001d. Record of Decision: Alcoa (Point Comfort)/Lavaca Bay Site Point Comfort, TX. CERCLIS #TXD008123168. Superfund Division, Region 6, U.S. Environmental Protection Agency. December 2001 [online]. Available: http://www.epa.gov/region6/6sf/pdffiles/alcoa_lavaca_final_rod.pdf [accessed Jan. 31, 2008].

EPA (U.S. Environmental Protection Agency). 2001e. Risk Assessment Guidance for Superfund: Vol. I—Human Health Evaluation Manual (Part D, Standardized Planning, Reporting, and Review of Superfund Risk Assessments). Final. Publication 9285.7-47. Office of Emergency and Remedial Response, U.S. Environmental Protection Agency, Washington, DC [online]. Available: http://www.epa.gov/oswer/riskassessment/ragsd/tara.htm [accessed Oct. 11, 2007].

EPA (U.S. Environmental Protection Agency). 2001f. Water Quality Criterion for the Protection of Human Health: Methylmercury. Final. EPA-823-R-01-001. Office of Water, Office of Science and Technology, U.S. Environmental Protection Agency, Washington, DC. January 2001 [online]. Available: http://www.epa.gov/waterscience/criteria/methylmercury/merctitl.pdf [accessed Jan. 31, 2008].

EPA (U.S. Environmental Protection Agency). 2002a. Technology Transfer Network: 1996 National-Scale Air Toxics Assessment. U.S. Environmental Protection Agency [online]. Available: http://www.epa.gov/ttn/atw/nata/ [accessed Aug. 19, 2008].

EPA (U.S. Environmental Protection Agency). 2002b. A Review of the Reference Dose and Reference Concentration Processes. EPA/630/P-02/002F. Risk Assessment Forum, U.S. Environmental Protection Agency, Washington, DC. December 2002 [online]. Available: http://cfpub.epa.gov/ncea/cfm/recordisplay.cfm?deid=55365 [accessed Jan. 4, 2008].

EPA (U.S. Environmental Protection Agency). 2002c. Organophosphate Pesticides: Revised Cumulative Risk Assessment. Office of Pesticide Programs, U.S. Environmental Protection Agency. June 10, 2002 [online]. Available: http://www.epa.gov/pesticides/cumulative/rra-op/ [accessed Feb. 4, 2008].

EPA (U.S. Environmental Protection Agency). 2002d. Determination of the Appropriate FQPA Safety Factor(s) in Tolerance Assessment. Office of Pesticide Programs, U.S. Environmental Protection Agency, Washington, DC. February 28, 2002 [online]. Available: http://www.epa.gov/oppfead1/trac/science/determ.pdf [accessed Jan. 25, 2008].

EPA (U.S. Environmental Protection Agency). 2002e. Child-Specific Exposure Factors Handbook. Interim Report. EPA-600-P-00-002B. National Center for Environmental Assessment, Office of Research and Development, U.S. Environmental Protection Agency, Washington, DC. September 2002 [online]. Available: http://cfpub.epa.gov/ncea/cfm/recordisplay.cfm?deid=5514 [accessed Feb. 1, 2008].

EPA (U.S. Environmental Protection Agency). 2003a. Technology Transfer Network 1999 National-Scale Air Toxics Assessment: 1999 Assessment Result [online]. Available: http://www.epa.gov/ttn/atw/nata1999/nsata99.html [accessed Aug. 19, 2008].

EPA (U.S. Environmental Protection Agency). 2003b. Human Health Toxicity Values in Superfund Risk Assessments. OSWER Directive 9285.7-53. Memorandum to Superfund National Policy Managers, Regions 1-10, from Michael B. Cook, Director /s/ Office of Superfund Remediation and Technology Innovation, Office of Solid Waste and Emergency Response, U.S. Environmental Protection Agency, Washington, DC. December 5, 2003 [online]. Available: http://www.epa.gov/oswer/riskassessment/pdf/hhmemo.pdf [accessed Feb. 1, 2008].

EPA (U.S. Environmental Protection Agency). 2003c. Developing Relative Potency Factors for Pesticide Mixtures: Biostatistical Analyses of Joint Dose-Response. EPA/600/R-03/052. National Center for Environmental Assessment, Office of Research and Development, U.S. Environmental Protection Agency, Cincinnati, OH. September 2003.

EPA (U.S. Environmental Protection Agency). 2003d. The Feasibility of Performing Cumulative Risk Assessments for Mixtures of Disinfection By-Products in Drinking Water. EPA/600/R-03/051. National Center for Environmental Assessment, Office of Research and Development, U.S. Environmental Protection Agency, Cincinnati, OH. June 2003 [online]. Available: http://cfpub.epa.gov/ncea/cfm/recordisplay.cfm?deid=56834 [accessed Jan. 31, 2008].

EPA (U.S. Environmental Protection Agency). 2004a. An Examination of EPA Risk Assessment Principles and Practices. EPA/100/B-04/001. Office of the Science Advisor, U.S. Environmental Protection Agency, Washington, DC [online]. Available: http://www.epa.gov/OSA/pdfs/ratf-final.pdf [accessed June 3, 2007].

EPA (U.S. Environmental Protection Agency). 2004b. Overview of the Ecological risk Assessment Process in the Office of Pesticide Programs: Endangered and Threatened Species Effects Determinations. Office of Prevention, Pesticides and Toxic Substances, Office of Pesticides Programs, U.S. Environmental Protection Agency, Washington, DC. September 23, 2004 [online]. Available: http://www.epa.gov/oppfead1/endanger/consultation/ecorisk-overview.pdf [accessed Feb. 1, 2008].

EPA (U.S. Environmental Protection Agency). 2004c. Estimated Per Capita Water Ingestion and Body Weight in the United States—An Update. EPA-822-R-00-001. Office of Water, Office of Science and Technology, U.S. Environmental Protection Agency, Washington, DC. October 2004 [online]. Available: http://www.epa.gov/waterscience/criteria/drinking/percapita/2004.pdf [accessed Jan. 31, 2008].

EPA (U.S. Environmental Protection Agency). 2005a. Draft Risk Assessment of the Potential Human Health Effects Associated with Exposure to Perfluorooctanoic Acid and Its Salts. Office of Pollution Prevention and Toxics, U.S. Environmental Protection Agency. January 4, 2005 [online]. Available: http://www.epa.gov/oppt/pfoa/pubs/pfoarisk.pdf [accessed Feb. 4, 2008].

EPA (U.S. Environmental Protection Agency). 2005b. Integrated Exposure Uptake Biokinetic Model for Lead in Children (IEUBKwin v1.0 build 264). Software and Users' Manuals, U.S. Environmental Protection Agency, Washington, DC. December 2005 [online]. Available: http://www.epa.gov/superfund/lead/products.htm [accessed Jan. 31, 2008].

EPA (U.S. Environmental Protection Agency). 2005c. Guidelines for Carcinogen Risk Assessment. EPA/630/P-03/001F. Risk Assessment Forum, U.S. Environmental Protection Agency, Washington, DC. March 2005 [online]. Available: http://cfpub.epa.gov/ncea/cfm/recordisplay.cfm?deid=116283 [accessed Feb. 7, 2007].

EPA (U.S. Environmental Protection Agency). 2005d. Supplemental Guidance for Assessing Susceptibility for Early-Life Exposures to Carcinogens. EPA/630/R-03/003F. Risk Assessment Forum, U.S. Environmental Protection Agency, Washington, DC. March 2005 [online]. Available: http://cfpub.epa.gov/ncea/cfm/recordisplay.cfm?deid=160003 [accessed Jan. 4, 2008].

EPA (U.S. Environmental Protection Agency). 2006. A Framework for Assessing Health Risks of Environmental Exposures to Children. EPA/600/R-05/093F. National Center for Environmental Assessment, Office of Research and Development, U.S. Environmental Protection Agency, Washington, DC. September 2006 [online]. Available: http://cfpub.epa.gov/ncea/cfm/recordisplay.cfm?deid=158363 [accessed Feb. 1, 2008].

EPA (U.S. Environmental Protection Agency). 2007a. Risk Assessment Practice. Office of Science Advisor, U.S. Environmental Protection Agency [online]. Available: http://www.epa.gov/osa/ratf.htm [accessed Aug. 19, 2008].

EPA (U.S. Environmental Protection Agency). 2007b. Revised N-Methyl Carbamate Cumulative Risk Assessment. Office of Pesticide Programs, U.S. Environmental Protection Agency. September 24, 2007 [online]. Available: http://www.epa.gov/oppsrrd1/REDs/nmc_revised_cra.pdf [accessed Feb. 4, 2008].

EPA (U.S. Environmental Protection Agency). 2007c. Lead Human Exposure and Health Risk Assessment for Selected Case Studies (Draft Report). EPA-452/D-07-001. Office of Air Quality Planning and Standards, U.S. Environmental Protection Agency, Research Triangle Park, NC. July 2007 [online]. Available: http://yosemite.epa.gov/opa/admpress.nsf/68b5f2d54f3eefd28525701500517fbf/14ec9929489233f785257329006645c0!OpenDocument [accessed Feb. 4, 2008].

EPA (U.S. Environmental Protection Agency). 2007d. Cleaning Up Our Land, Water and Air. Office of Solid Waste and Emergency Response, U.S. Environmental Protection Agency [online]. Available: http://www.epa.gov/oswer/cleanup/index.html [accessed Feb. 1, 2008].

EPA (U.S. Environmental Protection Agency). 2008a. Ozone (O_3) Standards Documents from Review Completed in 2008 Criteria Documents. Technology Transfer Network National Ambient Air Quality Standards, U.S. Environmental Protection Agency [online]. Available: http://www.epa.gov/ttn/naaqs/standards/ozone/s_o3_cr_cd.html [accessed Aug. 19, 2008].

EPA (U.S. Environmental Protection Agency). 2008b. Ozone (O_3) Standards Documents from Review Completed in 2008–Staff Papers. Technology Transfer Network National Ambient Air Quality Standards, U.S. Environmental Protection Agency [online]. Available: http://www.epa.gov/ttn/naaqs/standards/ozone/s_o3_cr_sp.html [accessed Aug. 19, 2008].

EPA (U.S. Environmental Protection Agency). 2008c. Ozone (O_3) Standards Documents from Review Completed in 2008–Technical Documents. Technology Transfer Network National Ambient Air Quality Standards, U.S. Environmental Protection Agency [online]. Available: http://www.epa.gov/ttn/naaqs/standards/ozone/s_o3_cr_td.html [accessed Aug. 19, 2008].

EPA (U.S. Environmental Protection Agency). 2008d. Office of Pollution, Prevention and Toxics, U.S. Environmental Protection Agency [online]. Available: http://www.epa.gov/oppt/ [accessed Feb. 1, 2008].

EPA (U.S. Environmental Protection Agency). 2008e. Office of Pesticides, U.S. Environmental Protection Agency [online]. Available: http://www.epa.gov/pesticides/ [accessed Feb. 1, 2008].

EPA (U.S. Environmental Protection Agency). 2008f. Cancer Guidances and Supplemental Guidance Implementation. Science Policy Council, Office of Science Advisor, U.S. Environmental Protection Agency [online]. Available: http://www.epa.gov/osa/spc/cancer.htm [accessed Aug. 21, 2008].

EPA (U.S. Environmental Protection Agency). 2008g. About EPA's Office of Solid Waste and Emergency Response (OSWER). U.S. Environmental Protection Agency [online]. Available: http://www.epa.gov/swerrims/welcome.htm [accessed Aug. 21, 2008].

EPA (U.S. Environmental Protection Agency). 2008h. Superfund. U.S. Environmental Protection Agency [online]. Available: http://www.epa.gov/superfund/index.htm [accessed Feb. 1, 2008].

EPA (U.S. Environmental Protection Agency). 2008i. Superfund Risk Assessment. Office of Solid Waste and Emergency Response, U.S. Environmental Protection Agency [online]. Available: http://www.epa.gov/oswer/riskassessment/risk_superfund.htm [accessed Feb. 1, 2008].

FWS/NMFS (U.S. Fish and Wildlife Service and National Marine Fisheries Service). 2004. Letter to Susan B. Hazen, Principal Deputy Assistant Administrator, Office of Prevention, Pesticides and Toxic Substances, U.S. Environmental Protection Agency, Washington, DC, from Steve Williams, Director, U.S. Fish and Wildlife Service and William Hogarth, Assistant Administrator, National Marine Fisheries Service. January 26, 2004 [online]. Available: http://www.fws.gov/endangered/pdfs/consultations/Pestevaluation.pdf [accessed Feb. 1, 2008].

ICF International. 2006. Risk Assessment for the Halogenated Solvent Cleaning Source Category. Prepared for Office of Air Quality Planning and Standards, U.S. Environmental Protection Agency, Research Triangle Park, NC, by ICF International, Research Triangle Park, NC. EPA Contract Number 68-D-01-052. August 4, 2006 [online]. Available: http://www.regulations.gov/search/index.jsp (EPA Docket Document ID: EPA-HQ-OAR-2002-0009-0022) [accessed Feb. 1, 2008].

NCHS (National Center for Health Statistics). 1987. Anthropometric Reference Data and Prevalence of Overweight, United States, 1976-1980. Data from the National Health and Nutrition Examination Survey. Series 11, No. 238. DHHS Publication No. (PHS) 87-1688. U.S. Department of Health and Human Services, Public Health Service, National Center for Health Statistics, Hyattsville, MD (as cited in EPA 2004a).

NRC (National Research Council). 1983. Risk Assessment in the Federal Government: Managing the Process. Washington, DC: National Academy Press.
NRC (National Research Council). 1994. Science and Judgment in Risk Assessment. Washington, DC: National Academy Press.
NRC (National Research Council). 2000. Toxicological Effects of Methylmercury. Washington, DC: National Academy Press.
Sechena, R., S. Liao, R. Lorenzana, C. Nakano, N. Polissar, and R. Fenske. 2003. Asian American and Pacific Islander seafood consumption-A community-based study in King County, Washington. J. Expo. Anal. Environ. Epidemiol. 13(4):256-266.
Stanek, E.J. III, and E.J. Calabrese. 1995a. Daily estimates of soil ingestion in children. Environ. Health Perspect. 103(3):276-285.
Stanek, E.J., III, and E.J. Calabrese. 1995b. Soil ingestion estimates for use in site evaluations based on the best tracer method. Hum. Ecol. Risk Assess. 1(2):133-156.
Stanek, E.J., III, and E.J. Calabrese. 2000. Daily soil ingestion estimates for children at a Superfund site. Risk Anal. 20(5):627-635.
TAM Consultants, Inc. 2000. Phase 2 Report: Further Site Characterization and Analysis, Vol. 2F- Revised Human Health Risk Assessment Hudson River PCBs Reassessment RI/FS. Prepared for U.S. Environmental Protection Agency, Region 2, New York, NY, and U.S. Army Corps of Engineers, Kansas City District. November 2000 [online]. Available: http://www.epa.gov/hudson/revisedhhra-text.pdf [accessed Jan. 31, 2008].
Toy, K.A., N.L. Polissar, S. Liao, and G.D. Mittelstaedt. 1996. A Fish Consumption Survey of the Tulalip and Squaxin Island Tribes of the Puget Sound Region. Tulalip Tribes, National Resources Department, Marysville, WA [online]. Available: http://www.deq.state.or.us/WQ/standards/docs/toxics/tulalipsquaxin1996.pdf [accessed Jan. 30, 2008].
van Wijnen, J.H., P. Clausing, and B. Brunekreef. 1990. Estimated soil ingestion by children. Environ. Res. 51(2):147-162.

Appendix F

Case Studies of the Framework for Risk-Based Decision-Making

In Chapter 8, we proposed a framework for risk-based decision-making in which an initial problem formulation and scoping phase is used to develop the analytic scope necessary to compare intervention options, risks and costs under existing conditions and with proposed interventions are assessed, and risk-management options are analyzed to inform decisions. We provide here three brief examples to demonstrate how the approach in Figure 8-1 might lead to a process and an outcome different from those of a conventional application of risk assessment. The examples are not meant to capture specific and current regulatory decisions in all their technical detail (and are perhaps caricatures of current decision-making paradigms) but are meant simply to illustrate some types of problems and how the framework would, in principle, address them. Similarly, while these examples would in principle involve multiple state and federal agencies under a variety of regulatory structures, they are meant to be more abstract examples of how the approach in Figure 8-1 would address risk management decisions.

A CASE STUDY OF ELECTRICITY GENERATION

Suppose that a new peaking power plant has been proposed to be sited in a low-income neighborhood that already contains other power-generating capacity or sources of similar pollutants. A conventional application of risk-assessment methods in this context might lead the proponent of the power plant to conduct analyses to determine whether the facility would contribute to exceedances of predefined risk thresholds—for example, greater than a 10^{-6} risk from air toxics for the maximally exposed person, a violation of ambient air quality standards for criteria pollutants. Issues related to alternative sites would typically be addressed in a separate part of the analysis, with argument of why the selected site is preferable, and no formal evaluations of alternative technologies and their implications for costs or benefits would be considered. Environmental-justice issues would typically be discussed but with no functional connection to the risk assessment or decision.

The questions addressed by risk assessment applied in that fashion attempt to determine whether there will be a "significant" problem if the plant is built with the proposed orientation. That sets up an adversarial relationship between the plant proponent and the local community in which the community is attempting to understand the intricacies of the risk assessment (which may have shown no "significant" increases in health risks) and is often operating under the assumption that the analysis has been manipulated in ways that the community does not understand or has not appropriately taken account of exposure and susceptibility conditions in the community. Whether the power plant is ultimately sited or not and whether the risk assessment represents best practice or not, this approach does not make optimal use of the insights that risk assessment can provide in that it focuses on only one alternative other than the status quo and provides limited information to stakeholders.

An alternative orientation following Figure 8-1 would still use risk-assessment methods but as part of Phase I would instead ask about the best approach to fulfill a given societal need that would minimize net impacts (including health impacts, costs, and other dimensions). With this orientation, the regulatory body that would be permitting the proposed facility would first determine the societal objective of the facility, which could be to decrease the projected gap between electricity supply and demand in the region during periods of high electricity use. That objective could be met in numerous ways, including energy-efficiency efforts by the utility's suppliers or customers, increased use of existing power plants, different storage technologies to meet peak power needs, or new power plants using different technologies (that is, alternative fuels and control technologies) in different locations. A do-nothing strategy and its implications would also be evaluated. Risk assessment can play a key role in distinguishing among the various options considered in combination with other methods and information.

In phase I, the set of possible interventions would be determined collectively by all stakeholders with the end points that could inform decision-making (for example, effects on electricity cost per kilowatt-hour, population risk, distribution of risk among defined subpopulations, life-cycle impacts, and probability of blackouts and brownouts). Stakeholders may mutually decide that some end points are unimportant or that some should get greater weight than others, and this will inform the choice of methods.

A comprehensive consideration of options at the outset would ensure that all relevant stakeholders were present, avoiding NIMBY outcomes in which an alternative site is chosen in a community that has not been involved in the process. The risk assessments and economic, technical, and other analyses would be oriented around the proposed interventions and would allow for explicit consideration of the tradeoffs among different desirable attributes of the decision and upfront transparency about the solution set, methods, and criteria for decision-making. For example, a clear presentation of the probability of blackouts under the do-nothing strategy and with alternative new facilities would help to demonstrate the importance of new capacity.

One possible criticism of this approach is that stakeholder participation and evaluation of multiple competing options require substantial effort and could lead to delays in decision-making. However, the current paradigm often leads to intractable debates about minute details of the risk assessment (Did the proponent use the right dispersion model? Were emissions estimated appropriately? Where would the maximally exposed person live?) without consideration of whether a choice among options would be influenced by these details. An upfront investment of time and effort in developing options and scoping the problem should reduce debate and antagonism considerably in the long term, should reduce analytic effort by focusing it on the end points that would help to discriminate among options, and should allow more coordinated planning of multiple projects with the same general aims. It could also be argued that explicit presentation of the tradeoffs among cost, risk, blackout probability, and equity would make decisions impossible because stakeholders would weigh these components differently, and there are no obvious bright-line distinctions. However, the current decision paradigm considers some of the factors implicitly while ignoring others without any explicit attempt to set priorities, so it is hard to argue that better understanding of the implications of decisions would not be beneficial. A final critique could be that stakeholders are ultimately concerned with the decision rather than the method. If this approach resulted in a conclusion that building the power plant in the low-income community were the optimal solution, residents of the community would be unhappy; if this approach resulted in a decision not to build a new facility, the proponents of the power plant would be unhappy (even if the process and analysis were transparent and agreed on). That may be impossible to avoid, but upfront consideration of scoping and decision criteria will at least reassure stakeholders that the criteria were not determined post hoc, and the rationale for the decision will be clearly presented.

A CASE STUDY OF DECISION SUPPORT FOR DRINKING-WATER SYSTEMS

Decision-makers and stakeholders seeking safe drinking water carry out their work in the face of a daunting array of microbial, chemical, climatic, operational, security and financial hazards. The capacity of risk assessment to support the societal goal of the provision of safe drinking water is an example of the critical need to reorient current risk-assessment practices away from the support of a series of disconnected single-hazard standard-setting processes and toward the provision of analytic support to facilitate the integration of complex health, ecologic, engineering, and economic elements of decision-making involved in providing safe drinking water.

Risk-assessment activities that are directed toward the safety of drinking water primarily support standard-setting exercises. The setting of such standards does not represent the types of more concrete system-design risk-management decisions that have direct physical, biologic, and chemical impacts on the safety of drinking water, representing *distal* decisions with ambiguous connections to risk reduction rather than *proximal* decisions with clear causal connections to risk reduction.

It is now generally understood that drinking water is best protected by an integrated risk-management approach in which multiple barriers are applied to protect against exposure to the hazards. The intervention options for drinking-water risk management include a complex set of decisions that affect system components that include sewage treatment, source-water selection and protection, multiple stages of water treatment, investments in operator training and information-management systems, changes in laboratory and monitoring practices, protection of the water in the distribution system, household water-use practices, and the capacity for effective emergency response that needs to be engaged when other barriers fail. It is inevitably a complex design problem to reduce risk from multiple sources that are subject to numerous competing constraints. The constraints include the fact that reducing some risks can increase others (the now classic problem of toxicity from disinfection byproducts that are produced in some processes aimed at reducing microbial risks or in choosing among sources of raw water that have varied microbial and chemical risk profiles). Other constraints include financial resources available in the short term and long term, the political and economic implications of issuing boil-water advisories, and the need to provide adequate protection to highly susceptible sub-populations (for example, in the case of persons with HIV/AIDS and the risk of cryptosporidiosis).

The societal goal is ultimately not to set standards themselves but rather to minimize the net risk associated with the provision of drinking water given the aforementioned risks and constraints. To that end, a series of decisions are made by the owners and operators of drinking-water systems. Some are discrete events, such as major investments in watershed protection, water-treatment technology, or construction of pipelines from distant water sources; some are continuous processes, such as treatment adjustments based on monitoring or customer complaints related to aesthetic properties of water.

It is obvious that those decisions would ideally be made in the presence of the most complete understanding of their implications that can reasonably be provided. The decisions are complex, and the selected actions will inevitably balance competing public goals. In this context, the present committee's goal for the conduct of risk assessment is the assembly and provision of information that describes (quantitatively and qualitatively) the implications of a set of intervention options, the characterization of the implications in the form of risk measures, and the characterization of the net risk that would be predicted in connection with the decision-maker's choice of a particular change in the water-management system. In the recommended framework in Figure 8-1, the Environmental Protection Agency (EPA), subject to the continuing reality of standard-setting processes required by statute, would orient risk-assessment activities toward providing risk-informed decision-support tools to the more *proximal* risk managers and stakeholders. With the help of this reoriented form of risk assessment, locally accountable decision-makers and stakeholders would be empowered by EPA's decision-support tools to make risk-informed decisions in designing and operating drinking-water systems.

A CASE STUDY OF METHYLENE CHLORIDE IN TWO SECTORS

The third example is based loosely on the regulatory response during the 1990s to the problems posed by methylene chloride ($MeCl_2$), a ubiquitous solvent that is a neurotoxin and a rodent carcinogen and that exacerbates carboxyhemoglobin formation. The example considers some of the likely costs and benefits of various interventions to reduce $MeCl_2$ risks in the workplace and in the general environment; its main point is to show that the outcome would depend heavily on how the regulatory agency chose to formulate the problem and potential intervention options. It also emphasizes that a too-narrow formulation of the problem, without consideration of intervention options at the outset, could exacerbate or fail to identify risk-risk tradeoffs.

A conventional application of risk-assessment methods might attempt to determine the allowable $MeCl_2$ concentration in ambient air to meet a defined risk threshold. In this case, the risk assessment supports a *distal* decision to set a risk-specific concentration. However, nothing would prevent facilities from complying with the standard by transferring the $MeCl_2$ risk to other chemicals or populations. They could substitute an unregulated (but potentially more toxic) solvent or simply change the production conditions so that less $MeCl_2$ is emitted from stack and fugitive emission points but more is released into the workplace. Other tradeoffs are also possible; for example, the allegation has been made in the aircraft sector that one compliance strategy (reduction in the frequency of stripping and repainting) can lead to an increased safety risk if it compromises the airworthiness of the craft.

An alternative strategy could involve finding the best available technology to control $MeCl_2$ emissions. In this case, the exercise is reduced to arranging the existing control techniques in order of efficiency and choosing either the "best available technology" (the single most efficient) or some "good enough available technology," as is done in the Maximum Achievable Control Technology (MACT) program under the Clean Air Act, which seeks to mandate the technology that corresponds to the average of the best-performing 12% of all current sources. As with any purely technology-based decision, the absolute risk reduction achieved may be insufficient to be acceptable, or it might be too stringent in that its costs outweigh its benefits. In spite of the simplicity of the approach, it is unlikely to yield the optimal solution, and firms could still respond to the technology mandate by adverse substitution, risk-shifting, plant closure, or some other action.

If the committee's framework for risk-based decision-making (Figure 8-1) were used instead, the initial problem-formulation step could determine that the goal is to minimize the total impacts of the production and use of the products that currently consume $MeCl_2$ (such as assembled foam and repainted aircraft). Risk assessments (and economic and other analyses) would be used to compare the residual risks and economic costs of control of each of a set of possible interventions. If the analytic question is asked about the process or function rather than about the substance, the set of interventions can be more expansive, and risk-risk tradeoffs can be minimized (or at least confronted explicitly).

Hypothetically, both EPA and the Occupational Safety and Health Administration might agree that for foam assembly, local ventilation plus carbon adsorption is the optimal solution for controlling $MeCl_2$ or any similar solvent that might be substituted for it. Similarly, for aircraft repainting, the optimal solution might involve requiring (or encouraging) the use of nontoxic abrasive material rather than a volatile solvent to remove the old paint layer.

The framework in Figure 8-1 could also allow the agencies to think more expansively and to seek global rather than local optima. Setting aside questions of agency scope, if the societal function were redefined as providing air travel rather than providing frequently repainted aircraft, intervention options might emerge for discussion that included changing the incentives to repaint so often, and this might broaden the analysis to include the impacts of jet-fuel use (fuel savings resulting from the coating, rather than painting, of planes). Even broader discussions of incentives for reducing the need for air travel might ensue; it is only the makeup of the involved participants and their preferences, subject to time and other logistical constraints, that dictates the scope of the interventions contemplated in this paradigm.